Brain, Mind and Medicine:

Essays in Eighteenth-Century Neuroscience

Brain, Mind and Medicine:

Essays in Eighteenth-Century Neuroscience

Edited by

Harry Whitaker
Northern Michigan University, Marquette, MI, USA

C.U.M. Smith
Aston University, Birmingham, UK

Stanley Finger
Washington University, St. Louis, MO, USA

Harry Whitaker
Department of Psychology
Northern Michigan University
Marquette, MI 49855-5334
USA

C.U.M. Smith
Vision Sciences
Aston University
Aston Triangle
Birmingham B4 7ET
UK

Stanley Finger
Department of Psychology
Washington University
Psychology Building
St. Louis, MO 63130
USA

Library of Congress Control Number: 200792479

ISBN-13: 978-0-387-70966-6
e-ISBN-13: 979-0-387-70967-3

Printed on acid-free paper.

9 8 7 6 5 4 3 2 1

springer.com

'There are three principal means of acquiring knowledge....observation of nature, reflection, and experimentation. Observation collects facts; reflection combines them; experimentation verifies the result of that combination'

−Denis Diderot (1753)

'For is it not reasonable to think that the electrical machine would add greatly to the efficacy of the materia medica?'

−Morris (1753)

'People will not look forward to posterity, who never look backward to their ancestors'

−Edmund Burke (1790)

Contents

Section F: Cultural Consequences

Contributors list

Paola Bertucci
CIS, Dipartimento di Filosofia, Università di Bologna
Via Zamboni 38, 40126 Bologna, Italy

Hugh W. Buckingham
Department of Communication Sciences and Disorders, Louisiana State University, Baton Rouge, LA 70803

George R. Cybulski
Department of Neurosurgery, Northwestern University and Division of Neurosurgery, Cook County Stroger Hospital, Chicago, IL

James G. Donat
P.O. Box 268529, Chicago, IL 60626-8529

Diana Faber
School of Psychology, Eleanor Rathbone Building, Bedford Street South, Liverpool L69 7ZA, UK

Stanley Finger
Department of Psychology, Campus Box 1125, Washington University, Saint Louis, MO 63130-4899

Miriam Focaccia
Department of Philosophy, University of Bologna, Via Zamboni 38-40126 Bologna, Italy

Brian J. Ford
Gonville and Caius College, Cambridge University, Cambridge CB2 1TA, UK

Eugenio Frixione
Sección de Metodología y Teoría de la Ciencia, Centro de Investigación y de Estudios Avanzados IPN, Apartado Postal 14-740, México, D.F. 07000, Mexico

Christopher Gardner-Thorpe
Consultant Neurologist, The Coach House, 1a College Road, Exeter EX1 1TE, UK

Robert B. Glassman
Department of Psychology, Lake Forest College, 555 N Sheridan Road, Lake Forest, IL 60045

James T. Goodrich
Division of Pediatric Neurological Surgery, Montefiore Children's Hospital, and Albert Einstein College of Medicine, New York, NY

Timo Kaitaro
Department of Philosophy, University of Helsinki, P. O. Box 9, SF-0014 Finland

Peter J. Koehler
Department of Neurology, Atrium Medical Centre, P.O. Box 4446, 6401 CX Heerlen, The Netherlands

Lawrence Kruger
Department of Neurobiology, David Geffen School of Medicine, University of California, Los Angeles, Los Angeles CA 90095

Douglas J. Lanska
500 E Veterans Street, Tomah, WI 54600
Staff Neurologist, Veterans Affairs Medical Center, Tomah, WI
Professor of Neurology, University of Wisconsin, Madison, WI

Joseph T. Lanska
500 E Veterans Street, Tomah, WI 54600

Marjorie Perlman Lorch
School of Languages, Linguistics and Culture, Birkbeck College, University of London, 43 Gordon Square, London WCIH OPD England

Ulf Norrsell
Physiology Section, Institute of Neuroscience and Physiology, Sahlgren Academy, Göteborg University, P.O. Box 432, SE40530 Göteborg, Sweden

Marco Piccolino
Dipartimento di Biologia, Università di Ferrara, Via Borsari 46, 44100 Ferrara

Jonathan Reinarz
Centre for the History of Medicine, University of Birmingham, Birmingham B15 3TT, UK

Julius Rocca
Torsvikssvangen 14, 181, Lidingo, Sweden

George Rousseau
Modern History Research Unit (MHRU), Block 11-2, Radcliffe Infirmary, University of Oxford, Oxford OX2 6HE, UK

Raffaella Simili
Department of Philosophy, University of Bologna, Via Zamboni 38-40126 Bologna, Italy

C.U.M. Smith
Vision Sciences, Aston University, Birmingham B4 7ET, UK

James L. Stone
Departments of Neurosurgery and Neurology, University of Illinois at Chicago, 912 S. Wood St., 4th Floor, Chicago, IL 60612

Catherine E. Storey
Clinical Associate Professor, Northern Clinical School, University of Sydney
Department of Neurology, Royal North Shore Hospital, St Leonards, NSW, Australia

Larry W. Swanson
Department of Biological Sciences, University of Southern California, Los Angeles, CA 90089

Hannah Sypher Locke
Department of Psychology, Campus Box 1125, Washington University, Saint Louis, MO 63130-4899

Yves Turgeon
Chef – programme Vieillissement en santé, Restigouche Health Authority, 189, chemin Lily Lake Road, C.P./P.O. Box 910, Campbellton, New Brunswick, Canada E3N 3H3

Nicholas J. Wade
Department of Psychology, University of Dundee, Dundee DD1 4HN, Scotland

Harry A. Whitaker
Department of Psychology, Northern Michigan University, Marquette, MI 49855

Section A
Introduction

Introduction

Harry Whitaker, C.U.M. Smith, and Stanley Finger

The idea for a volume on eighteenth-century studies of brain and behavior originated during a joint International Society for the History of the Neurosciences (ISHN) and Theoretical and Experimental Neuropsychology/Neuropsychologie Expérimentale et Théorique (TENNET) symposium held in Montreal in June 2004. We believe that these essays provide unique contemporary insights into the science and medicine of the nervous system, hence "neuroscience," during the "long" eighteenth century – a century too often given short shrift in textbooks as well as in historical reviews of the nervous system.

The long eighteenth century, which in thematic ways is often perceived as stretching from the 1660s into the opening decades of the 1800s, was an age of transition in the neurosciences. It saw the classic and time-honored ideas of neurophysiology – animal spirits moving in hollow nerve conduits to and from the ventricles of the brain – being gradually replaced by ideas more in accord with anatomical reality. It also saw an enormous increase in interest in the nervous system as the source of many of the ills of both body and mind, along with new therapies. It even saw, at least in the upper strata of polite society, a new and at times even "neurotic" concern for the health and proper functioning of the nervous system. The chapters in this book tell these fascinating stories, and more.

The volume is divided into six sections. After this brief introductory section and chronological table, the second section deals with the background against which work on the nervous system took place. After an overview of the development of ideas about brain and mind during the "long cen-

tury," Brian Ford discusses the most revealing of eighteenth-century instruments – so far as anatomy is concerned – the microscope. Next, Jonathan Reinarz reviews the way in which medical education developed in association with the voluntary hospitals movement. Finally, Christopher Gardner-Thorpe uses the life and work of James Parkinson as a lens through which to examine medicine and its milieu during the last quarter of the eighteenth century.

The next section comprises six chapters that discuss eighteenth-century investigations of the anatomy and physiology of the nervous system. The section starts with an account of the neuroscience of one of the most important eighteenth-century anatomists, and anatomical teachers, John Hunter. James Stone, James Goodrich, and George Cybulski review Hunter's many contributions to neurology and neuroscience and by including many quotations from his own work allow him, for the most part, to speak for himself. Next, Larry Kruger and Larry Swanson discuss the important but rather little known work of Pourfour du Petit on the functional anatomy of the brain and nervous system. After this, Julius Rocca describes the work of William Cullen and Robert Whytt who helped establish the preeminence of Edinburgh's medical school and were among that epoch's most influential investigators of the nervous system. The important controversy between Haller and Robert Whytt on the nature of muscular contraction forms the subject of the next chapter, contributed by Eugenio Frixione. The final two chapters examine the origin and growth of what we now call electrophysiology. Marco Piccolino provides a detailed review of eighteenth-century research into electric fish, whilst Rafaella Simili and Miriam Focaccia provide a similarly scholarly account of Luigi Galvani's career and discoveries.

The fourth section of the book is devoted to long-standing complex issues of brain and behavior. Nick Wade opens with a fascinating chapter on the somewhat obscure late eighteenth-century Scottish physician and anatomist, William Porterfield, whose investigations ranged from the anatomy and physiology of the visual system to the nature of the sensations generated by phantom limbs. Robert Glassman and Hugh Buckingham review the influential mid-century physiological psychology of David Hartley, and Harry Whitaker and Yves Turgeon examine the work of Charles Bonnet on memory and hallucinations. In the next chapter of this section, Ulf Norsell shows that Swedenborg was not merely a religious thinker and mystic, but also a very perceptive student of the nervous system who collected evidence for, and correctly envisioned, cortical localization of function well before this "nineteenth-century" doctrine came into vogue.

The fifth and longest section deals with medical theory and practice arising from eighteenth-century neuroscience. First, Peter Koehler provides a scholarly account of the neuroscience and attendant medicine of Herman Boerhaave and his most famous pupil, Albrecht von Haller. Catherine Storey shows how the concept of apoplexy (stroke) altered during the eighteenth century. The next three chapters review the origin and application of one of the eighteenth century's greatest enthusiasms – electricity – for treating medical conditions. Stanley Finger discusses the work of perhaps the greatest "electrician" of them all – Benjamin Franklin – who personally assessed the use of electric shocks in treating various common disorders, including stroke-induced paralysis and the seizures associated with hysteria. In the next chapter, Hannah Locke and Stanley Finger discuss what people learned about medical electricity from *Gentleman's Magazine*, the first widely disseminated periodical in Great Britain. Paola Bertucci takes up this theme in the next chapter where she examines in detail the many attempts to use electricity in medicine in the latter part of the eighteenth century. John Wesley, in addition to his better known religious convictions and desire to save souls, was also involved in medical electricity as an inexpensive way to heal sick bodies, and James Donat examines this and his long-standing interest in nervous disorders in the next chapter. Next, Douglas and Joseph Lanska discuss Mesmer's related, but ultimately fraudulent, pseudoscience of animal magnetism and show how it was ultimately refuted by the careful use of scientific methodology. Finally, Diana Faber takes up the topic of hysteria and shows how it was transformed from an organic to more of a psychological disease during the eighteenth century.

The final section considers the cultural consequences of this developing interest in the nervous system. Timo Kaitaro examines the consequences for the mind–brain argument in French thought of the period, and Marjorie Lorch brings out the way that Jonathan Swift used eighteenth-century understandings of the brain and nervous system in some of his satirical writings. Finally, George Rousseau examines the work of a well-known twentieth-century psychologist, Jerome Kagan, and shows how one of his central concepts, that of temperament, was first defined in the eighteenth century.

The essays in this book provide a wide perspective on the development of neuroscience and its application in medicine during the "long" eighteenth century. They also show the impact of this interest on the general culture of the time not only on philosophical ideas but also on literature and social life. The developments in neuroscience during the "long" eighteenth century form the basis upon which the great advances of the nineteenth and then the twentieth centuries were made. We hope these essays will be of interest not only to practicing neuroscientists and neurologists, but also to others in the many disciplines which study eighteenth-century life and thought.

Chronology

C.U.M. Smith

Vision Sciences, Aston University, Birmingham B4 7ET, UK

In this chronological table of the 'long' eighteenth century I have sought to place scientific publications in the context of their cultural milieu. I have purposefully omitted birth and death dates of the great figures of the eighteenth century in preference for the dates when their most significant publications and/or other contributions appeared. I have also, rightly or wrongly, sought to keep things as simple as possible by omitting the coronation dates of Kings and Queens in favour of the publication dates of novels and plays, the first performances of music or display of painting and sculpture. My hope is to have caught most of the important events mentioned in the chapters of this book. Finally, in the science column I have attempted to differentiate between primarily neuroscientific works and those pertaining to the other sciences. This is sometimes a matter of opinion and I hope that I shall not have to endure too much censure if in some cases my opinion does not coincide with that of the reader.

Science (Neuroscience in bold)	Cultural context
1660 Foundation of Royal Society	1660 Restoration of British Monarchy
1660 Mariotte discovers eye's blind spot	
1660 Boyle: *New experiments physico-mechanical touching the spring of the air*	
1661 Malpighi: *De pulmonibus observationes anatomicae*	
1661 Boyle: *Skeptical Chymist*	
1662/4 Descartes: *L'Homme*	1662 Spinoza: *De Ethica*
	1663 Ottomans defeated at Vienna
1664 Willis: *Cerebri Anatome*	
1664 Swammerdam: frog nerve-muscle preparation	
1665 Hooke: *Micrographia*	1665 Great plague in London
1665 Malpighi: *De cerebro*	
1666 Foundation of Académie Royale des Sciences	1666 Molière: *Le Misanthrope*
	1666 Great fire of London
1667 Steno: *Elementorum myologiae specimen*	1667 Milton: *Paradise Lost*
	1668 Dryden named Poet Laureate
	1670 Pascal: *Pensées*
	1670 Louis XIV founds Les Invalides
1671/76 Perrault: *Histoire Naturelle*	

5

Science (Neuroscience in bold)	Cultural context
1672 Glisson: *Tractatus de Natura Substantiae energetica*	
1672 Willis: *De Anima Brutorum*	
1674 Van Leeuwenhoek's microscopical sections of optic nerve	
1676 Sydenham: *Observationes Medicae*	
1677 Glisson: *Tractatus de ventriculo et intestinis*	
1678 Lorenzini: *Observationi intorno alle topedini*	
1679 Bonet: *Sepulchretum sive Anatomia Practica*	
1680/81 Borelli: *De Motu Animalium*	
1682 Newton describes partial decussation of optic nerves	
1683 Van Leeuwenhoek sees bacteria	
1686 Ray: *Historia plantarum*	1685 Edict of Nantes revoked
1687 Newton: *Principia Mathematica*	1688 'Glorious Revolution' in England
	1690 Locke: *An Essay concerning human understanding*
1691 Ray: *The Wisdom of God manifested in the Works of Creation*	
	1692 Salem witch trials
1699 Tyson: *Orang-Outang, sive Homo sylvestris*	
1701 Grew: *Cosmologia Sacra: or, A discourse of the Universe as it is the Creature and Kingdom of God*	
1702 Baglivi: *Specimen quatuor librorum de fibra motrice et morbosa*	
1704 Newton: *Opticks*	1704 Swift: *Tale of a Tub*
1707 Stahl: *Theoria medica vera*	1707 Act of Union between England and Scotland
1708 Boerhaave: *Institutiones medicae*	
1709 Berkeley: *New Theory of Vision*	
1710 Petit: *Lettres d'un Medecin*	1710 Berkeley: *Principles of Human Knowledge*
1714 Boerhaave commences clinical teaching	1714 Leibniz: *Monadology*
	1718 Watteau: *Gilles*
	1719 Defoe: *Robinson Crusoe*
	1722 Defoe: *Moll Flanders*
	1725 Vico: *The New Science*
1726 Establishment of Faculty of Medicine at Edinburgh	1726 Swift: *Gulliver's Travels*
1727 Petit discovers functions of cervical sympathetics	
	1729 Bach: *St. Matthew Passion*
1731/5 Gray: *Electrical experiments*	
1732 Boerhaave: *Elementa Chemiae*	1732/4 Pope: *Essay on Man*
1733 Cheyne: *The English Malady*	
	1734 Voltaire: *Lettres Pilosophiques*

Science (Neuroscience in bold)

1735 Linnaeus: *Systema Naturae* (1st edn)

1737 Porterfield: *An essay concerning the motions of our eyes*

1737 Kinneir: *A new essay on the nerves . . .*

1737/8 Swammerdam (published by Boerhaave): *Bybel der Natuure (Biblia Naturae)*

1739 Bayne: *A new essay on the nerves*

1740/41 Swedenborg: *Oeconomia Regni Animalis*

1744 Trembley: *Mémoires*

1745 Whytt: *An Enquiry . . .*

1745/6 Invention of Leyden jar

1745 Kratzenstein's medical applications of electricity

1746 Nollet: *Essai sur l'Electricité des Corps*

1746 Monro (primus): *The Anatomy of Human Bones and Nerves*

1747 Haller: *Primae Lineae Physiologiae*

1747 La Mettrie: *L'Homme Machine*

1748 Needham: *Observations* (spontaneous generation of life)

1749 Buffon: *Histoire Naturelle* (vol.1)

1749 Hartley: *Observations on Man*

1751 Maupertuis: *Système de la Nature*

1751 Whytt: *An Essay on the Vital and other Involuntary Motions of Animals*

1751 Franklin: *Experiments and Observations on Electricity*

1753 Beccario: *Dell'elettricismo artificiale et naturale*

1754 Condillac: *Traité de la Lumière*

1754/5 Bonnet: *Essai de psychologie*

1755 Haller: *Dissertation on the Sensible and Irritable Parts of Animals*

1755 Whytt: *Observations on Sensibility and Irritability*

1755 Whytt: *Physiological Essays*

1756 Lovett publishes on medical electricity

1757 Adanson: *Histoire Naturelle du Senegal*

1757/66 Haller: *Elementa Physiologiae Corporis Humani*

1758 Linnaeus: *Systema Naturae* (10th edn)

1759 Porterfield: *A Treatise on the Eye*

Cultural context

1735 Hogarth: *The Rake's Progress*

1738/9 D. Scarlatti: *Keyboard sonatas ("exercises")*

1739/40 Hume: *Treatise of Human Nature*

1742 Handel: *Messiah*

1747 Richardson: *Clarissa*

1748 Gainsborough: *Mr and Mrs Andrews*

1750 Rousseau: *Discours sur les sciences et les arts*

1751 d'Alembert: *Discours préliminaire*

1751 Diderot: *Encyclopédie*, vol.1

1755 Lisbon earthquake

1755 Johnson: *Dictionary of the English Language*

1755 Rousseau: *Discourse on Inequality*

1756-1763 Seven Years' War

1759 Voltaire: *Candide*

Science (Neuroscience in bold)	Cultural context
1760 Wesley: *Desideratum, or electricity made plain and useful*	1760 Sterne: *Tristram Shandy*
1760 Bonnet: *Essai analytique sur les facultés de l'âme*	
1760 Bonnet describes his eponymous syndrome	
1761 Morgagni: *De sedibus*	
	1762 Gluck: *Orfeo et Euridice*
	1762 Rousseau: *Du Contract Social*
	1764 Reid: *Inquiry into the Human Mind*
	1764 Voltaire: *Dictionnaire Philosophique*
1765 Whytt: *Observations . . . on Nervous, Hypochondriac, or Hysteric Disorders*	1765 Lunar Society founded
1766 Mesmer: *De Planetarum Influxu*	
1767 Priestley: *History and Present State of Electricity*	1767 Fragonard: *The Swing*
1768 Sauvage: *Nosologia Methodica*	
1768 William Hunter founds the Great Windmill Street anatomy school	1768 Cook's first voyage
1769 Bancroft: *Natural History of Guiana*	1769 Arkwright's Spinning Jenny
1769 Bonnet *La Palingénésie Philosophique*	1769 James Watt invents steam engine
1770 Holbach: *Système de la Nature*	1770 Cook at Botany Bay
1770 Cullen: *Lectures on the Institutions of Medicine*	
1772 Walsh: *Experiments on the Torpedo or Electric Ray*	
1772 Priestley: *The History and Present State of Discoveries relating to Vision, Light and Colours*	
1773 Walsh: *On the electric property of the torpedo*	
1773 John Hunter: *Anatomical observations on the Torpedo*	1773 First cast–iron bridge at Ironbridge, Shropshire
	1774 Goethe: *Die Leiden des jungen Werther*
1775 Lavater: *Physiognomische Fragmente*	
1775 Mesmer demonstrates 'animal magnetism'	
1775 Hunter: *An account of Gymnotus Electricus*	
1776 Cavendish: *An account of some attempts to imitate the effects of the Torpedo by electricity*	1776 American Declaration of Independence
1776 Musgrave: *Speculations and Conjectures on the Qualities of the Nerves*	1776 Herder: *Sturm und Drang*
	1776 Smith: *Wealth of Nations*
	1776 Gibbon: *Decline and Fall of the Roman Empire*
1777 Cullen: *First lines in the practice of Physic*	1777 Goya: *El quitasol*
1778 Mesmer develops group treatment by animal magnetism	1778 Banks elected President of Royal Society
1779 Mesmer: *Mémoire sur la découverte du magnetisme animal*	1779 Lessing: *Nathan der Weise*
1779 Prochaska: *De structura nervorum*	
1780 d'Eslon: *Observations sur le magnétisme animal*	1780 Gordon riots in London

Science (Neuroscience in bold)	Cultural context
1780 Spallanzani: *Disssertazione di fisica animale e vegetabile*	
1781 Fontana's first microscopical observation of a nerve fibre	1781 Kant: *Kritik der reinen Vernunft*
1781 Fontana: *Traité sur le Venin de la Vipère*	1781 Schiller: *Die Rauber*
	1781 Houdon: *Bust of Voltaire*
1783 Monro Secundus: *Observations on the Structure and Functions of the Nervous System*	1783 Blake: *Poetical Sketches*
	1783 First human ascent in hot-air balloon
1784 Franklin commission on mesmerism	
1784 Vicq d'Azyr: *Recherches sur la structure du cerveau*	
1785 Hutton: *Theory of the Earth*	
1786 John Hunter: *Observations on certain parts of the animal economy*	1786 Kant: *Metaphysische Anfangsgründe der Naturwissenschaft*
	1786 Mozart: *Le Nozze di Figaro*
1787 Abernethy establishes a surgical curriculum at St Bartholomew's Hospital	
1789 Lavoisier: *Traité Elémentaire de Chimie*	1789 Revolution in France
1789 Pinel: *Nosographie Philosophique*	
1790 Goethe: *Versuch die Metamophose der Pflanzen zu erklären*	1790 Burke: *Reflections on the Revolution in France*
	1791 Paine: *The Rights of Man*
	1791 Mozart: *Die Zauberflöte*
1791 Galvani: *De riribas electricitatis in motu musculari*	1792 Stewart: *Elements of the Philosophy of the Human Mind*
1792 Volta: *Memoria sull'elettricità animale*	
1792 Wells: *An Essay upon Single Vision*	
1792 Kirkland: *A Commentary on Apoplectic and Paralytic Affections*	
1793 Young: *Observations on Vision*	1793 David: *Marat Assassinated*
1794-6 Darwin: *Zoonomia*	1794 Lavoisier guillotined
1793 Richard Fowler: Experiments and Observations relative to the influence lately discovered by M. Galvani and commonly called animal electricity	
1794-1801 Soemmerring: *De corporis humani fabrica*	
1794 John Hunter: *A Treatise on the blood, inflammation, and gun-shot wounds*	
	1797 Schelling: *Ideen zur Philosophie der Natur*
	1798 Malthus: *Essay on Population*
	1798 Wordsworth/Coleridge: *Lyrical Ballads*
1798 Jenner: *Inquiry into variolae vaccinae*	1798 Haydn: *Die Schöpfung*
	1799 Rosetta stone
	1799 Napoleon overthrows the Directory
1800 Volta: *On the electricity excited by the mere contact . . .*	1800 Mozart: *Requiem*
	1800 Schelling: *Transcendental Philosophy*
1801 Bichat: *Anatomie Générale Appliquée à la Physiologie et à la Médecine*	

Science (Neuroscience in bold)	*Cultural context*
1802 Paley: *Natural Theology*	
1802 Young: *On the theory of lights and colours*	1802 Coleridge: *Dejection: An Ode*
	1802 Constable: *Dedham Vale*
1802-22 Treviranus: *Biologie*	
1803 Darwin: *Temple of Nature*	
1804 Aldini: *Essai théorique et expérimentale sur le galvanisme*	1804-8 Beethoven: *5th Symphony*
1804 Wilkinson: *Elements of Galvanism*	1804 Jacquard invents eponymous loom
	1804 Trevithick invents first steam locomotive
	1805 Battle of Trafalgar
1807 Trotter: *A view of the nervous temperament* ...	1807 Hegel: *Phänomenologie des Geistes*
	1807 Abolition of Slave Trade Act passed in British Parliament
1808 Dalton: *New System of Chemical Philosophy*	1808 Goethe: *Faust part 1*
	1808 Goya: *El Tris Mayo*
1809 Rolando: *Saggio sopra la vera struttura del cervello dell'uomo e degl'animali*	1809 Davy invents first electric light
1809 Lamarck: *Philosophie Zoologique*	
1809-11 Oken: *Lehrbuch der Naturphilosophie*	
1810 Goethe: *Zur Farbenlehre*	
	1811 Luddite revolts in England
1810-19 Gall and Spurzheim: *Anatomie et physiologie du système nerveux. ... par la configuration de leur têtes* (first two volumes only with Spurzheim)	
1811 Bell: *Idea for a New Anatomy of the Brain*	
	1814 Austen: *Mansfield Park*
	1814 Stevenson's steam locomotive
	1815 Battle of Waterloo; Congress of Vienna
1817 Parkinson: *Essay on the Shaking Palsy*	1817 Coleridge: *Biographia Literaria*
	1818 Shelley: *Frankenstein*
	1819 Schubert: *Trout quintet*
1820 Cooke: *A Treatise of Nervous Diseases*	

Section B
Background

Introduction

This section consists of four chapters which sketch in some of the background against which the development of Neuroscience in the 'long eighteenth century' took place. The chronological table in the preceding section shows some of the major events in the wider world during the 'long century'. It has often been called 'the Age of Enlightenment' and the table shows how societies in the Western World emerged from the fundamentalisms of the seventeenth and earlier centuries – the Civil War in England during the 1640s, the Revocation of the Edict of Nantes in 1685, the Salem witch trials in the American colonies in the 1690s – to a more secular, not to say rational, settlement. Sterne's *Tristram Shandy*, Diderot's *Encyclopédie*, Gainsborough's *Mr and Mrs Andrews*, Fragonard's *The Swing*, Scarlatti's *Keyboard Sonatas*, all breathe a different air from that breathed by the fanaticisms of the earlier century. It did not last. In 1793 Charlotte Corday came from Normandy to assassinate Marat in his bath, in 1794 both Robespierre and Lavoisier were guillotined, in 1799 Napoleon came to power. The clear rational air of the eighteenth century thickened. In England, taste for the gothic climaxed in Mary Shelley's *Frankenstein*, in Spain Goya painted *El Tris Mayo*, in Germany, Goethe initiated the *Sturm und Drang* movement and Hegel published *Phänomenologie des Geistes*.

It is against this background that investigations of the brain and nervous system were carried out. This section does not, however, attempt to place eighteenth-century physicians, anatomists and physiologists in their wider cultural context but to focus more closely on their professional interests and technical abilities. The first chapter, contributed by C.U.M. Smith, sets the scene by reviewing the development of ideas concerning mind and brain during the 'long' century. He argues that the century, so far as physiological psychology is concerned, was a century of transition. The old ideas of spirit-filled cerebral ventricles and hollow nerves inherited from the medievals were discredited by anatomical and especially microscopical research. Yet until the very end of the century, when investigations of 'animal electricity' began to take hold, it was difficult to see with what the old ideas could be replaced. This dissonance between what was shown by the anatomist's scalpel and the microscopist's lens and what the general public believed persisted for a century and more. Nevertheless, both physicians and polite society became increasingly interested in the nervous system and nervous disease. The affect of this interest on the sophisticated classes can be seen in the best-selling novels of the time and in the development of fashionable spa society. Right at the end of the period evolutionary ideas began to challenge the established order and Smith's chapter ends with an account of Erasmus Darwin's evolutionary psychophysiology.

In the next chapter Brian Ford reviews the development of microscopy during the eighteenth century. He starts with the first great pioneers – Robert Hooke (1635–1703) and Anthony van Leeuwenhoek (1632–1723). Ford notes that some of Leeuwenhoek's first specimens were preparations of bovine optic nerve (1674) and that he could find no evidence that they were tubular. Other microscopists examined nerve fibres during the late seventeenth century but preparative techniques were not sufficiently well developed to allow a definitive answer to the question whether the fibres were hollow (as had long been believed) or solid, and the controversy lingered on into the eighteenth century. Ford next provides a valuable account of the design of eighteenth-century microscopes and their use by many of the luminaries of that century – Della Torre, Fontana, Monro (Primus and Secundus), Prochaska, etc. – and points out that although the resolution of the these microscopes was often surprisingly good, severe limitations were imposed by undeveloped preparative techniques. The chapter ends with a look forward into the nineteenth century when microtomes and the first achromatic compound microscopes became available, finely tooled from brass, and often collectors' items.

The topic of medical education and its association with the voluntary hospital movement in England is the subject of the next chapter by Jonathan Reinarz. Reinarz starts his account in the sixteenth century when British physicians received a university education but surgeons were taught through an apprenticeship scheme as befitted what seemed more like a craft. After a discussion of the famous Italian centres of medical education, especially Padua, the focus shifts to Leiden. Here graduates from Padua sought to replicate their own educational experience and initiated clinical teaching at the city's hospital. From Leiden many medical graduates migrated across Europe, especially to Scotland where William Cullen, at Edinburgh, lecturing in English rather than Latin, established a systematic medical curriculum. The scene then passes to London where the practice of clinical teaching at a number of the capital's hospitals, especially St Bartholomew's and Guy's, had already been established. Unlike the rather bureaucratic hospitals across the channel, London hospitals customarily allowed students much more freedom to develop their own programmes of study. With the exception of smallpox, they were allowed to see the full range of medical and surgical cases. The most famous London medical courses were given by physician William Hunter and his brother, surgeon John Hunter, at Great Windmill Street. The success of this school soon led to the founding of several other teaching establishments so that, by 1780, no less than 16 anatomical courses were advertised in the London newspapers. By the end of the century London had become a major world centre for medical education. Jonathan Reinarz traces these developments into the beginning of the nineteenth century when clinical teaching in a hospital environment had begun to spread to the English provinces.

In the final chapter in this section, Christopher Gardner-Thorpe examines the medical milieu in the last quarter of the eighteenth century through the lens provided by the life and work of James Parkinson (1755–1824). Although Parkinson is nowadays best known for his 1817 *Essay on the Shaking Palsy*, he had led a varied and eventful life before the *Essay* was published. Like others in the eighteenth century (one immediately thinks of Erasmus Darwin and Benjamin Franklin), his mind voyaged widely over the world being revealed by enlightenment thought and discovery. In addition to his medical interests he involved himself in politics, where he became a notable (though anonymous) pamphleteer, the church, where he was for many years a churchwarden, and geology and palaeontology where he was not only a founder member of the Geological Society but also published a highly regarded three volume work, *Organic Remains of a Former World*. Gardner-Thorpe shows how country physicians (like many country vicars) could sustain a wide-ranging interest in the world around them and yet make significant contributions to their own chosen profession. Parkinson was not only a founder member of London's Medical and Chirurgical Society but also much involved in domestic medicine (writing popular medical compendia for the general public), the regulation of 'Mad Houses', promoting good nutrition, studying gout and many other medical topics. Gardner-Thorpe's review provides a valuable insight into the medical and scientific world of the late eighteenth century. A world which, to paraphrase Thomas Wright's words in his 1750 discussion of astronomy, seemed to open up on all sides, revealing truths undreamt of in earlier centuries.

The Editors

1
Brain and Mind in the 'Long' Eighteenth Century

C.U.M. Smith

Introduction

How should the 'long' eighteenth century be defined? January 1, 1700 and December 31, 1799 are quite arbitrary dates. Why should they be chosen to segment our history rather than more significant periods of time, periods which have a coherent content, or are marked, perhaps, by the working out of a theme? Students of English literature sometimes take the long eighteenth century to extend from John Milton (*Paradise Lost*, 1667) to the passing of the first generation of Romantics (Keats (d. 1821), Shelley (d. 1822), Byron (d. 1824), Coleridge (d. 1834)). Students of British political history often take it to start with the accession of Charles II (the Restoration) in 1660 or, alternatively, the so-called *Glorious Revolution* of 1688 and to end with the great Reform Act of 1832. Others might choose different book ends. In the history of science and philosophy the *terminus a quo* is sometimes taken as the publication of Descartes' scientific philosophy or, in more Anglophone zones, the 1687 publication of Newton's *Principia* with its vision of a 'clockwork universe'. 'Nature and Nature's laws' as Alexander Pope enthused, 'lay hid in Night: God said, "Let Newton be!" and all was light!'.

But in the biological sciences, as the long eighteenth century wore on, the Newtonian illumination dimmed. After early enthusiasms, mechanistic interpretations of life-processes proved unfruitful. The models proposed by Descartes, Borelli and others began to seem absurdly simplistic. The reaction against 'clockwork' models took the form of Romantic biology and, especially in Germany,

drifted far from the clarity of Descartes and Newton. Kant published *Metaphysical Foundations of Natural Science* in 1786 and Goethe's *Metamorphosis of Plants* came out in 1790. In early nineteenth-century England, Samuel Taylor Coleridge, eschewing his fine poetical talent, sought to develop a science of life based on *Naturphilosophie* (Smith, 1999). The *terminus ad quem* of this long decline in mechanistic interpretations of the living process can, perhaps, be seen in the publication of Lawrence Oken's turgid *Lehrbuch der Naturphilosophie* in 1809–1811 (trans. Tulk: *Elements of Physiophilosophy* 1847). Thus, in the life sciences, we might adapt Squire's riposte to Pope's encomium and write: 'Then came the devil, howling "Ho! let Oken be!" and restored the status quo'.

Neuroscience could not be immune to these movements of thought. Perhaps the most obvious date to start is 1664 with the publication of Willis' *Cerebri Anatome*. The end point is less clear. Should it be with Gall and Spurzheim's phrenology (1810–1819) or Charles Bell's *Idea for a New Anatomy of the Brain* in 1811, or even later still, in the 1840s, with Emil du Bois Reymond's discovery of the action potential and the action current in nerve and muscle? What happened during this long period? It has been called the *Age of Enlightenment*. In France, Denis Diderot and Jean-le-Rond d'Alembert published the *Encyclopédie*, often taken to be the Enlightenment's master work; Diderot was sufficiently interested in physiology to write a treatise called *Eléments de Physiologie*, although this was never published. Roy Porter, perhaps the foremost of our recent historians of this period, saw it as an era when the old spiritual certainties, the old theologies, evaporated to be replaced by an uneasy materialism. Yet this materialism, as we shall see, was very different from the mechanistic materialism put forward by René Descartes and his followers in the seventeenth century. The thought world of Erasmus Darwin, at the end of the 'long' century, is very different from that of the Cartesians at the beginning! Nevertheless, the old ideas about the brain and its physiology refused to go without a tenacious rearguard action. Long before the eighteenth century, Vesalius, in the 1543 *Fabrica*, had strongly denied that the nerves were hollow: 'I have never seen a channel, even in the optic nerve' (p. 317), and in 1620, the Edinburgh medical student, John Moir, recorded in his lecture notes that 'nerves have no perceptible cavity internally, as the veins and arteries have' (French, 1975). Yet the notion of animal spirit travelling in nerve tubes was still current in popular culture 150 years later. For Tristram Shandy, in Laurence Sterne's novel of the 1760s, it was simply conventional wisdom. The long eighteenth century, for the historian of neuroscience, is an age of transition but not of revolution. The old framework of ideas, animal spirit, subtle fluids, spiritual substances and hollow nerves lived on, in spite of the evidence, because it was difficult to see until the very end of the period with what they could be sensibly replaced.

Setting the Scene

At the beginning of the Western tradition 2,500 years ago, Hippocrates expounded a view of the brain with which we would hardly be uncomfortable today. In his work on epilepsy, *On the Sacred Disease*, he located all the psychical functions (' – joys, delights, laughter, . . . sorrow, griefs, despondency, . . . the acquisition of wisdom and knowledge, . . . ethical understanding, . . . seeing and hearing, . . . bad dreams and delirium, . . .') in the brain.

At about the same time, on the other side of the Aegean, Plato's Pythagorean tract, the *Timaeus*, had a very different story to tell. A rational soul is confined within the skull whilst a mortal soul full of passion is confined in the torso and beneath that, separated by the diaphragm, a lower, concupiscent soul. The *Timaeus* is an explicit continuation of the *Republic* and its tripartite schematic mirrors the tripartite sociology of Plato's ideal state. Plato's famous pupil, Aristotle, far more a biologist than his master, also developed a tripartite psychophysiology. The three divisions of his animating principle – vegetable, animal and rational souls – clearly relate to his classification of the biological world, rather than the social stratification of an ideal state.

It is, however, only with the Alexandrians of the third century BC that we find the first physiological thought based on anatomical dissection. Both Herophilus and Erasistratus were aware of the cerebral ventricles and Erasistratus developed a physiology in which *pneuma zotikon* was transported by blood to the brain where it was transformed into *pneuma psychikon* to be distributed to the muscles by hollow nerves (see Smith, 1976). This idea reappears in Galen in the second century AD and variants (*pneuma psychikon* being translated as 'animal spirit') persisted for two millennia up until the Renaissance and beyond.

The origin of that other long-persisting notion, the 'cell' or 'ventricular' theory, where the 'rational' soul is divided into a number of parts and located in three cerebral ventricles or 'cells', is more obscure (1). There are hints both in Nemesius of Emesa (fl. fourth century AD) whose work, *On the Nature of Man*, synthesises ancient Hellenistic and Judaic thought, and in St. Augustine of Hippo at the beginning of the fifth century. It was only with the rebirth of anatomy in the sixteenth century that it was recognised that the medieval cell diagrams, valuable though they were as representations of psychology, bore little or no resemblance to the anatomy of the brain.

Leonardo's early sixteenth-century wax cast of the ox ventricles (unknown until the nineteenth century) is perhaps the first true representation. Leonardo writes, 'My works are the issue of pure and simple experience, who is the one true mistress'. Renaissance anatomists began to insist that function – in this case mental function – should be related to structure revealed by the scalpel. This brings us, rapidly, to the beginning of the modern era.

Descartes and Willis

Descartes' *L'Homme* was published just 2 years before the start of our period, in 1662, though it had been written long before. Descartes' anatomy was, of course, far inferior to that of Willis (Willis, 1664), although he had much first-hand experience of dissecting specimens from butchers' shops in Amsterdam. His psychophysiology is yet more speculative (see Smith, 1998). Both Descartes and Willis were, however, clear that the 'lower' souls of the ancients and medievals, the animal and vegetable souls, were wholly material.

Descartes, it will be remembered, argued that the rational soul swayed the pineal gland, thus directing animal spirit present in the ventricle into one or other nerve conduit, hence causing appropriate behavioural movements. Willis' theory is in many respects rather similar. Animal spirit, distilled from vital spirit in the blood by the grey matter of the brain, fills 'the medullar trunk' and passes from thence via the nerves to the body 'and so imparts to those bodies, in which the nervous fibres are interwoven, a motive or sensitive feeling of force' (Willis, 1681, p. 126). But in both cases an immaterial soul, unique to man, lurked somewhere beyond the scalpel (see Changeux, 1985, p. 11).

On the other hand neither Descartes nor Willis denied infra-human animals sensation. Descartes writes to the Marquess of Newcastle, 'As for the movements of our passions . . . it is . . . very clear that they do not depend on thought, because they often occur in spite of us. Consequently they can also occur in animals, even more violently than they do in human beings . . .' and in a letter to Henry More in 1649 he writes, 'I do not deny life to animals . . .; and I do not even deny sensation, insofar as it depends on a bodily organ' (Cottingham, Stoothoff, & Murdoch, 1985, p. 366). The fashionable notion that Descartes (though not some of his followers) regarded animals as unfeeling automata is plainly incorrect.

Willis, for his part, maintains that animals possess a 'corporeal soul . . . having extension and local parts' (Willis, 1672) which, although thoroughly material, nevertheless vivifies the body and is sensitive to the aches and pains and pleasures of life. It is fashionable to say, with Coleridge, that Descartes, in exorcising the lower spirits, transformed the body into an unfeeling machine. The truth, as ever, is more ambiguous. This ambiguity lived on to plague the eighteenth century.

Nerves are not Hollow Conduits

The ancient neurophysiology to which both Descartes and Willis subscribed – that the nerves are conduits linking the brain with the periphery along which the animal spirit travelled – was, even as their works were being published, on the point of being discredited. In the last decades of the seventeenth century both Jan Swammerdam and Giovanni Borelli provided experimental evidence against the idea that animal spirit travelled down nerve tubes like a wind to inflate the muscles. Swammerdam had shown by an ingenious and delicate experiment, as early as 1663, that frog muscles did not expand on contraction (2). However, although he demonstrated his experiment widely to academic audiences (Nordstrom, 1954), he tried to explain away its implications and his work was not placed in the public domain until Boerhaave published an edited version in the 1737/1738 *Bibjel der Nature (Biblia Naturae)* (see Cobb, 2002). Borelli also contested the old idea of nerves as hollow conduits. He believed the nerves were 'canals filled with a spongy material, like elder pith . . . moistened with a spirituous juice (succus nerveus) originating in the brain . . . saturated to turgescence'. Instead of a flow of 'spirit', he argued that a 'commotion', 'concussion' or 'undulation' was all that was transmitted (Borelli, 1680/1681; see also Glynn, 1999). Willis, although satisfied that nerves contained no cavity visible to the naked eye or simple microscope, nonetheless believed that they were like 'Indian canes' through which animal spirit could percolate (Willis, 1681).

The work of the physiologists was supplemented by that of microscopists (see Ford, this volume). At the end of the seventeenth century Antony van Leeuwenhoek used his microscope to examine sections of bovine optic nerve and concluded that no cavity could be perceived, although he later appears to have had second thoughts. During the next century improved microscopical techniques suggested ever more strongly that nerves contained no cavity (3). The old neurophysiology thus became highly questionable. Yet the old ideas held on tenaciously. In *The English Malady*, George Cheyne devotes many pages to a discussion of animal spirits, what they might be and how they might act (Cheyne, 1733, pp. 75–89). Ford (this volume) refers to an illustration depicting hollow nerves in a publication dated as late as 1842! The old neuro-physiology also lingered in popular culture. Tristram Shandy in 1760 is well versed in them. 'You have all, I dare say, heard of the animal spirit' he writes, 'Nine parts in ten of a man's sense and nonsense . . . depend on its motions and activity . . .' (Sterne, 1760, p. 1), and Jonathan Swift, in the 1704 *Mechanical Operation of the Spirit*, writes that 'it is the Opinion of Choice Virtuosi, that the Brain is only a Crowd of little Animals, but with Teeth and Claws extremely sharp, and therefore, cling together in the Contexture we behold, like the Picture of Hobbes's Leviathan, or like Bees in per-pendicular swarm upon a Tree, or like a Carrion corrupted into Vermin, still preserving the Shape and Figure of the Mother Animal'. Swift also famously defines 'punning' as 'the art of harmo-nious jingling with words, which passing in at the ears, excites a titillary motion in those parts; and this being conveyed by the animal spirit into the muscles of the face, raises the cockles of the heart'.

One important reason for the lingering of the old neurophysiology was the difficulty of knowing with what to replace it. The traditional understand-ing of the human being was at least a consistent system. Alexandre Koyré says the same of Aristotelian physics. He remarks that it '. . . forms an admirable and perfectly coherent theory which, to tell the truth, has only one flaw (besides that of being false) . . . that it is contradicted by the every-day practice of throwing' (Koyré, 1943). The same could be said of the ancient neurophysiology, sub-stituting 'the fact that nerves are not hollow tubes' for 'the everyday practice of throwing'. But replacing the usefully ambiguous psychophysical substance with straightforwardly physical substance, though seemingly inescapable, merely made the psy-chophysical problem more intractable.

There were a number of attempts to incorporate neurophysiology into the Aristotelian system's successor: Newtonian mechanism. Boerhaave pro-posed that nerve fluid consisted of the finest of all particles, far smaller than the large corpuscles mak-ing up other body fluids, and that these strung end to end communicated impetus along a nerve fibre much as a line of billiard balls in a tube. Because of Boerhaave's immense reputation, this idea helped prolong hydraulic neurophysiology well into the eighteenth century. David Hartley, taking a hint from Newton's *Optics*, proposed a rather

different idea (see Buckingham, this volume). He argued that the nervous system operated by way of vibrations and vibratiuncles (in many ways this reminds us of Borelli's 'undulations') (Hartley, 1749). Like most of his eighteenth-century con-temporaries he felt the pull of Newton's genius and wished to develop 'an experimental physics of the mind' (see Smith, 1987). However, although his associationist psychology was very influential, his neurophysiology did not find favour. Anatomists like Alexander Monro primus were quick to point out the anatomical infelicities of his theory: '. . . the nerves are unfit for vibrations because their extremities . . . are quite soft and pappy' (4). The beginnings of our modern understanding awaited the second part of the nineteenth century, first with the work of Emil du Bois Reymond and then with that of Helmholtz, Bernstein and others, and was not fully completed until the work of Hodgkin, Huxley and Katz in the mid-twentieth century. Indeed, it might even be said that complete under-standing awaited the Nobel Prize researches of Robert MacKinnon and colleagues at the beginning of the twenty-first century (MacKinnon, 2003; Smith, 2002b).

The Mind Escapes the Cells

The puzzle posed by transmission down 'solid' nerves was not the only puzzle facing eighteenth-century anatomists. Another, and equally acute, puzzle was that posed by the neural correlatives of mind. What and where were they? The new anatomy gave no hint. The ancient 'cell' or 'ven-tricular' psychology had long been recognised as having no anatomical basis. Descartes and many others insisted instead that the 'mind' had no phys-ical dimension and thus, as Henry More remarked, was strictly speaking, 'nowhere'. This seemed to point directly to atheism, a conclusion which did not escape the Holy Office. In 1663, it put Descartes' physio-philosophy on its Index Librorum Prohibitorum.

Interest in this problem was not confined to philosophers, theologians and anatomists. Laurence Sterne, for instance, has much to say about it in his novels. Tristram Shandy's opinionated father philosophises over this very question in intensely humorous passages, eventually insisting that the neural correlatives of mind were to be found in the fine material of the cerebellum. Accordingly he instructed his man-midwife, Dr Slop, to take great care to ensure that the back of Tristram's head was well protected at birth.

More seriously, Samuel Johnson, that epitome of clarity of thought and expression, writes in many places, not least in his *Dictionary*, that the union of psyche and soma is incomprehensible: 'Man is compounded of two very different ingredients, spirit and matter, but how two such unallied and disproportioned substances should act upon each other, no man's learning could yet tell him' (Johnson, 1755, quoting Collier). He was bitingly dismissive of those who would identify mind with brain: a 'quagmire', he remarked, whose 'clammy consistency' could have nothing to do with the 'motion of thought' (Porter, 2003, p. 169).

Johnson was burdened with many of the 'ills the flesh is heir to', half blind in one eye, half deaf in one ear, corpulent, subject to all sorts of tics and compulsions. Yet he was gifted with an astonishing memory and articulacy. It must have seemed obvious to him that mind and body had little to do with each other. Humans, for him, were, as they were for Plato, essentially embodied souls. He lived his life in ever-present fear of the hereafter, seldom far from thoughts of his end, more than once descending into black melancholia, striving always to act so as to be able, as he says, 'to render up my soul to God unclouded' (quoted in Porter, 2003, p. 188). He is said to have refused opiates at the end, wishing to pass over with a clear mind. His hero, Herman Boerhaave, thought much the same. Johnson points out in the biography he wrote for the *Gentleman's Magazine* (Johnson, 1739, p. 174) that 'he (Boerhaave) had never doubted of the spiritual and immaterial Nature of the Soul, but declared that he had lately had a kind of experimental Certainty of the distinction between Corporeal and Thinking Substances, which mere Reason and Philosophy cannot afford, and Opportunities of contemplating the wonderful and inexplicable Union of Soul and Body, which nothing but long Sickness can give. This he illustrated by a Description of the Effects which the Infirmities of his Body had upon his Faculties, which yet they did not so oppress or vanquish, but his Soul was always Master of itself, and always resigned to the Pleasure of its Maker.'

Animal Spirit Escapes the Nervous System

Not only did the 'mind' escape its traditional confinement in the cerebral cells, but the 'animal spirit' of the ancients also began to escape, this time from the nervous system itself. Towards the end of the seventeenth century, the first microscopes allowed Leeuwenhoek, Hooke (Willis' pupil), Swammerdam and others to discover the world of microbes and protista. Leibniz at the beginning of the eighteenth century, impressed by the work of Jan Swammerdam and Antony van Leeuwenhoek, saw continuity all the way from monad to man. He believed, furthermore, that there was no discontinuity between the plant and animal kingdoms. In a letter to Louis Bourget in 1715 he writes that 'Mr Swammerdam has supplied observations which show that insects are close to plants with respect to their organs and that there is a definite order of descent from animals to plants. But perhaps there are other beings between these two' (Loemaker, 1956, vol. 2, p. 1079) and in another place he writes that the existence of zoophytes is 'wholly in keeping with the order of nature' and 'the principle of continuity' (5).

Notions of a so-called 'great chain of being' had, of course, been around for centuries but, as Lovejoy remarks, they achieved great prominence in the eighteenth century (Lovejoy, 1930). Alexander Pope's lines are only the best known of a multitude of similar expressions:

Vast chain of being, which from God began,
Natures aethereal, human, angel, man,
Beast, bird, fish, insect! what no eye can see
No glass can reach! from Infinite to thee,
From thee to Nothing! . . .

Essay on Man, 237–241

Leibniz' prediction that zoophytes (links, as he supposed, between the animal and plant worlds) must exist was confirmed in 1739 when Abraham Trembley discovered the fresh water hydrozoa. Trembley went on to show that fresh water polyps could be subdivided indefinitely and that from each fragment a new polyp would regenerate (Trembley, 1744). This had an important implication, for it was taken to show that 'soul', the principle of life, was distributed throughout the body. Rieppel (1988) shows how strong an affect Leibniz' ideas had on the development of Bonnet's holistic thought. Bonnet concluded that it implied that body and soul could not be two distinct and separate substances but that animate beings constituted what he called an 'être mixte' (6). Julien Offray de La Mettrie also seized on this implication in his mid-century works, *l'Homme Machine* (1747) and *Traité de l'âme* (1751) (see Smith, 2002a). He concluded, like Bonnet, that the division of creation into two parts – body and soul – was absurd. Both, he writes, were created together, at the same instant, as if 'by a single brush stroke' (de La Mettrie, 1745, p. 2). To think otherwise was nothing more than a casuistry designed to throw dust into the eyes of the watching theologians (6). But this sort of panpsychism has, of course, tricky implications. Does all matter have this 'dual aspect'? Leibniz, at least, recognised this implication and was content to allow his fundamental units – the monads – to possess both attributes.

century this had to remain a mystery. Haller would have nothing to do with Stahl's mysterious animism and writes that muscle fibres are composed of nothing more than gluten and earth. The power of contraction could hardly be inherent in earth; *ergo* it must be a property of gluten. 'Hence', he concludes, 'the physical cause must depend upon the arrangement of the ultimate particles (of which gluten is composed), though the experiments we can make are too gross to investigate them'. It is interesting to note that Erasmus Darwin, later in the century, could not restrain his powerful speculative energy, and developed an interesting hypothesis to account for this contractile power (see below and Smith, 2005). But, as with the physical basis of the nerve impulse, a further two and half centuries were to elapse before experimental techniques had developed sufficient delicacy to provide an answer to Haller's question.

Irritability

Towards the middle of the eighteenth century another concept, that of 'irritability', began to make headway. The concept is, of course, of great antiquity. Francis Glisson had developed the notion and coined the term in the seventeenth century, but Albrecht von Haller made the idea very much his own (7). Indeed, in Tissot's 1755 preface to Haller's *Dissertation on the Sensible and Irritable Parts of Animals*, he apostrophises him as having made 'the great discovery of the present age' (von Haller, 1755, p. 3).

Haller writes of making a multitude of experiments designed to discover which parts of an animal are irritable. His stimuli included blowing, heat, spirit of wine, lapis infinalis, oil of vitriol, butter of antimony, touching, cutting, burning, etc. He concludes that 'it (irritability) does not depend on the nerves, but on the original fabrication of the parts which are susceptible of it' (p. 32). And, a little further on, he homes in on muscle fibres, '. . . there is nothing irritable in the animal', he writes, 'but the muscular fibre and the faculty of endeavouring to shorten itself when we touch it is proper to this fibre' (p. 37). This they do when quite isolated from the nervous system.

What is it about muscle fibres which give them this property? In the first half of the eighteenth

Sensibility

Sensibility is quite different. Haller writes that '. . . the sensible parts of the body are the nerves themselves, and those to which they are distributed in the greatest abundance; for by intercepting the communication between a part and its nerve, either by compression, by tying, or cutting, it is thereby deprived of sensation . . . Wherefore the nerves alone are sensible of themselves . . .' (von Haller, 1755, p. 31). Far from all, eighteenth-century physiologists agreed with Haller's distinction. William Cullen, in particular, believed that the fine endings of nerve fibres transformed into muscle fibres in the interior of muscles (Cullen, 1827; see Rocca, this volume). This somewhat bitter dispute seems to have been more about words and personalities than about the observations. Both Haller and Cullen agreed that the muscle fibres contained an inherent power of 'contractility', a *vis insita*, and this implied that the animate principle was diffused throughout the neuromuscular system, not confined to the brain. Similarly, both agreed that the webwork of nerves was endowed with 'feeling'. Robert Whytt, a colleague of Cullen's at Edinburgh, writes that 'we know certainly that the nerves are endowed with feeling' and the notable eighteenth-century English physician, George Cheyne, also educated at Edinburgh, agreed, writing that 'Feeling (physical sensibility) is nothing but the Impulse, Motion or Action of Bodies, gently or violently impressing the Extremities or Sides of the Nerves . . .' (Cheyne, 1733, p. 49).

This recognition that the nerves were not merely inanimate conduits for an animating principle originating in the brain influenced medical practice. Thomas Trotter writes in 1807, p. 17, that

the last century has been remarkable for the increase in a class of diseases little known in former times, and which had but lightly engaged the study of physicians prior to that period . . . Sydenham at the conclusion of the seventeenth century computed fevers to constitute two thirds of the diseases of mankind. But at the beginning of the nineteenth century we do not hesitate to affirm that nervous disorders have now taken the place of fevers and may be justly reckoned two thirds of the whole, with which civilised society is afflicted.

Trotter also notes that nervous disorders 'are to be found in abundance in large towns, or wherever luxurious habits have displaced simplicity' (p. 200). It has been remarked that these luxurious habits led to the growth of the fashionable 'spa society' so well captured in the novels of Jane Austen, Brinsley Sheridan and Tobias Smollet.

This new understanding of the nerves influenced late eighteenth-century English literature in other ways. In Samuel Richardson's best seller of the 1780s, Clarissa, the heroine dies because of her nervous sensibility (Stephanson, 1988). 'The origin of your disorder', the doctor tells Clarissa, is that 'you were born with weak Nerves . . . and then the Nerves have been wasted and relaxed by your sedentary life and thinking attentively'. Robert Whytt writes in 1765 that 'In some, the feelings, perceptions and passions are naturally dull, slow and difficult to be aroused . . . in others the opposite is the case on account of a greater delicacy and sensibility of the brain and nerves' (Whytt, 1765). George Cheyne agreed. 'Persons of slender and weak Nerves are generally of the first Class: the Activity, Mobility and Delicacy of their intellectual Organs make them so', he writes, and he goes on to say that nervous debility only attacks persons of this upper class, 'the brightest and most spiritual, and whose Genius is most keen and penetrating' (Cheyne, 1733, p. 105). The lower, plodding, labouring classes are spared these agonies. It was comforting for those living a cosseted life to be told by their medical advisors that their ailments were

not due to character or spiritual weaknesses but to real physical causes (see Porter, 2001).

Yet the old understanding of what it is to be a human being refused to go quietly. It may be that the age-old language of 'animal spirit' was becoming as metaphorical as the late-medieval cell theory, but the concept still pervaded popular culture. People felt differently about themselves in the eighteenth century than we do today in our computer-obsessed time. As John Sutton remarks, the language of spirit spills easily across the divide 'from fibres and pores to passions and feelings and conscious and unconscious motivations' (Sutton, 1998).

Electricity

Animal spirit refused to go quietly largely because there was, at the beginning of the long eighteenth century, no obvious successor and clearly *some* influence passed along the nerves, to and from the brain. However by the mid-eighteenth century, there was at last a contender which began to grow in popularity: electricity. The study of electricity had, of course, been set on its modern course by the publication of William Gilbert's *De Magnete* in 1600, but it only became a popular subject in the mid-eighteenth century. In the 1730s, Stephen Gray in England and Charles du Chisternay Dufay in France initiated what became an electrical 'craze'. A little later, both Gray and the Abbé Jean-Antoine Nollet devised a series of public shows in which they astounded audiences by electrifying boys and girls. Indeed this developed into a piece of theatre in both countries. According to Joseph Priestley the Abbé Nollet remarked that he would 'never forget the surprise when the first electrical spark was drawn from a human body' (Priestley, 1775, p. 47), and electrified young women provided more than the usual excitement to the young men who ventured to embrace them (see Bertucci, this volume).

But what was this mysterious new 'vertu'? In his great work on the *History and Present State of Electricity*, Priestley asks again and again whether it is identical to Sir Isaac Newton's aether (Priestley, 1775, especially pp. 448–450). Notions of 'subtle fluids' or aethers were, of course, common in the eighteenth century. Could the electrical fluid have medical applications? In the 1740s

Kratzenstein and others suggested that this might well be the case and soon many physicians and would-be physicians were trying their hand at 'the electrical cure' (see Bertucci, this volume). The study of electricity in the mid-eighteenth century hovered between science, quackery and entertainment. As Paola Bertucci remarks, it played to that taste for the marvellous and the inexplicable, which is so much a part of human nature and which pervaded the eighteenth-century learned world. No one yet understood this mysterious and powerful influence.

It was in this environment of intellectual uncertainty that Franz Mesmer (1734–1815) popularised the idea of animal magnetism (see Bloch, 1980). Mesmer had qualified in medicine in 1766 with a dissertation on the influence of the heavenly bodies on human health. Just as the planets were held in their courses by the mysterious force known as 'gravity', so he believed that human bodies were affected by another mysterious force-carrying aether, 'animal gravity'. This idea of an all-pervasive 'subtle fluid' recurred in his later work when, after being introduced to a new type of treatment using magnets by a Jesuit priest, Father Maximillian Hell, he replaced 'animal gravity' with 'animal magnetism' (see Lanska & Lanska, this volume).

In essence, he believed that good health depended on the free flow of the processes of life through the body's innumerable channels. He agreed with George Cheyne in regarding the human body as 'a Machine of an infinite Number and Variety of different Channels and Pipes, filled with various and different Liquors and Fluids, perpetually running, glideing (sic) or creeping forward, or returning backward, in a constant circle' (Cheyne, 1733, p. 4). When these channels were blocked, illness ensued. In 1774 Mesmer successfully treated a patient by getting her to swallow a solution containing iron and then attaching magnets to various parts of her body. Later he dispensed with iron and magnets and cures were alleged to be effected by direct control of the mysterious magnetic 'fluid'. The physician's task was to act as a conduit for this all-pervading magnetic aether and channel it out of the patient's body, rather like that other popular practice, 'blood-letting', so that a healthy equilibrium could be achieved. Mesmer believed that he was able to control the flows of the magnetic aether in a patient's body by staring fixedly into his or her

eyes and making certain passes with his hands until a 'magnetic crisis' was experienced, analogous to an electric shock, after which recovery would ensue. He also developed a device, the *baquet*, for concentrating the magnetic fluid which he regarded as analogous to the Leyden jar. Anton Mesmer epitomises the confusion which reigned during the latter part of the eighteenth century concerning the phenomena of magnetism and electricity. It was, as Priestley remarked, 'a field just opened . . . (where) there is great room to make new discoveries' (1775, preface x). Mesmer's belief that he had made one of these discoveries and was able to control the new 'fluid' proved as groundless as many of the other 'discoveries' of the time. He was unable to convince his fellow physicians in his native Vienna and, when he transferred to Paris, his practice was investigated by a commission set up by Louis XVI in 1784 which included Lavoisier, Guillotin and the American ambassador, Benjamin Franklin, and was shown to be without foundation (see Finger, this volume; for more detail on Mesmer see Lanska & Lanska, this volume).

Very different from the unfounded speculations of Mesmer were the sober researches of experimentalists interested in the electricity generated by electric fish such as *Gymnotus* and *Torpedo*. These investigations made considerable headway in the eighteenth century and in the next century led both to the science of neurophysiology and via the electric or voltaic 'pile' to the physics of electricity itself. Indeed when Volta, at the end of the century, constructed the first 'electric (voltaic) pile', he modelled it closely, on J. W. Nicholson's *artificial torpedo* which, in turn, had been modelled on Hunter's dissection of the *Torpedo*'s electric organ (Pancaldi, 1990). Indeed, in his letter to the Royal Society announcing his discovery he called it an 'artificial electric organ' and hoped to improve it by adjusting its structure more closely to that of the organ found in *Torpedo* (Volta, 1800). These researches are discussed in detail by Piccolino in this volume so it is unnecessary to expand on them here.

Finally, at the very end of the century, in the 1790s, Galvani and others published the results of their famous frog experiments (see Focaccia & Simili, this volume; Piccolino, 2006;). Richard Fowler, a pupil of Monro Secundus, repeated Galvani's experiments (as had Monro) and concluded that the effect was caused by the application of two dissimilar metals, in his case Zinc and Silver. The account which he published, *Experiments and Observations relative to the influence recently discovered by Galvani and commonly called Animal electricity*, details an immense number of experiments on frogs and shows how puzzling electricity and animal electricity were to acute eighteenth-century minds (Fowler, 1793). He writes, for instance, of his great surprise at finding that contractility persisted even when the heart had ceased beating; he could hardly credit his eyes and he writes that although it is so, he hardly expects to be believed. This shows how difficult it is to shake off old paradigms: in this case the idea of the brain filtering an animating principle from the blood and the entire psychophysiological theory of which this is a part. If the heart had ceased to beat, how could muscle still be induced to contract?

These developments lie outside the scope of this chapter and are dealt with in detail in Section B of this volume. But it did begin to seem that in this mysterious fluid, the time-honoured animal spirit might finally have found a physical basis (Galvani, 1791/1998, p. 42; see also Focaccia and Simili, this volume). It is clear, as Priestley remarks in his 1775 *History and Present State of Electricity* that the new discoveries sparked great general interest, not only in the medical profession but also throughout the general public, and all and sundry tried their hands at electrical experiments: '. . . no other part of the whole compass of philosophy', he writes, 'affords so fine a scene of ingenious speculation' (Priestley, 1775, p. 411).

Erasmus Darwin

Erasmus Darwin (1731–1802) was one of the most significant medical thinkers of the latter part of the eighteenth century. His great medical compendium, *Zoonomia*, running to some 300,000 words, was highly influential at the end of the century. It had four English editions and also editions and/or translations in America, France, Spain, Italy and Portugal, and was dedicated '. . . to all those who study the operations of the Mind as a Science, or who practice Medicine as a Profession'. He was also quite clear about the importance of theory in medicine. In the introductory chapter of *Zoonomia*, he writes that 'theory' is needed to connect together

the 'great number of unconnected facts' to make them easier to remember so that 'men of moderate ability' can practice. Darwin's 'take' on the neuropsychology of the end of the eighteenth century thus forms a suitable end point for this chapter (for more detail on Darwin's neurology see Gardner-Thorpe & Pearn, 2006).

Darwin studied medicine first in London and then in Edinburgh where he completed his degrees in 1756. He was thus one of the many to be influenced by both Robert Whytt and William Cullen. He was scornful of those who, like the Cartesians, saw 'the body as a hydraulic machine . . . forgetting *animation* (is) its essential characteristic' (Darwin, 1801, preface 1). Like Cullen and Haller he traced the principle of animation to fibres. 'The power of contraction' Darwin writes 'which exists in organised bodies, and distinguishes life from inanimation, appears to consist of an aethereal spirit which resides in the brain and nerves of living bodies, and is expended in the act of shortening . . .' (Darwin, 1803, 1, 245n). Unlike the attractive ether (gravitational) and the repulsive ether ('the matter of heat'), the spirit of animation (which Darwin also calls the *contractive aether*) requires at first the contact of a goad, or stimulus, which appears to draw it off from the contracting fibre, and to excite the sensorial power of irritation. This, Darwin writes, accounts for the otherwise odd quantitative incompatibility between stimulus and response: as when a horse is pricked by a spur. This expenditure of animal spirit also accounts for the fatigue of fibres after exertion and for the tremor shown by the muscular movements of the aged.

Is 'animal spirit' the same as Galvani's 'animal electricity'? Darwin is doubtful. To see why we have to turn to the reflections with which he begins *Zoonomia*. 'The whole of Nature' he writes 'may be supposed to consist of two essences or substances; one of which is termed *spirit*, the other *matter*. The former of these possesses the power to commence or produce motion, and the latter to receive and communicate it (vol. 1, p. 1). He compares the 'spirit of animation' with other 'spirits', such as electricity, magnetism, gravitation, heat. 'If two particles of iron', he writes,

lie near each other without motion and afterwards approach each other; it is reasonable to conclude that something besides the iron particles is the cause of their approximation; and this something is called magnetism.

In the same manner, if the particles, which compose an animal muscle, do not touch each other in the relaxed state of the muscle, and are brought into contact during the contraction of the muscle; it is reasonable to conclude, that some other agent is the cause of this new approximation. For nothing can act where it is not; for to act includes to exist; and therefore the particles of the muscular fibres (which in a state of relaxation are supposed not to touch) cannot affect each other without the influence of some intermediate agent; this agent is called here the spirit of animality (Darwin, 1801, vol. 1, p. 64).

This spirit, he continues, may be compared to electricity and magnetism. But the action of both the latter 'aethers' vary inversely with distance and this does not apply to muscular contraction. Indeed the force of contraction is sometimes greater when the muscle is stretched and its constituent particles (presumably) are further apart from each other. 'On this account', he writes, 'I do not think the experiments conclusive which were lately published by Galvani, Volta and others to shew the similitude between the spirit of animation which contract muscular fibres and the electrical fluid'.

Darwin's system provides a fascinating insight into the state of neuropsychology at the turn of the eighteenth century. Its fundamental unit is not, as it is for us, the cell but the fibre. Cell theory was over half a century into the future and Cajal's neuron doctrine an even more distant dream. The human sensory as well as nervous and muscular system is woven of fibres. Darwin supports his belief that sense organs consist of contractile fibres with a lengthy account of the retina and the phenomenon of ocular spectra (we would say 'after-images') (Darwin, 1801, vol. 2, pp. 373–374; also Darwin, 1786) (8). He describes an investigation of the ox retina, placing it in an alkali solution and showing it to apparently consist of fibres. He proposes that, when a succession of objects are presented to the visual system (he instances 'trumpets, horns, lords, ladies, trains and canopies'), the retina changes its configuration and these motions (he writes) are our visual ideas 'and the voluntary repetitions of them, when the object is withdrawn, our memory of them' (Darwin, 1801, vol. 4, Section XL). Ideas are thus, according to Darwin, 'movements of the nerves of sense' (1801, vol. 2, p. 263).

The brain itself he likens to a gland, such as the pancreas, which he suggests (a little tentatively), separates the aethereal animating principle from the blood for 'the purposes of motion and sensation' (9). We are aware once more of the tenacity of the old schematic. The time-honoured neurophysiology, initiated by the Alexandrians in the third century BC and accepted by Willis in the seventeenth century AD, still glimmers behind the new thought of the eighteenth century AD. The comparison of brain to gland was, of course, quite common in the eighteenth century. Indeed it seemed to be supported by Malpighi's microscopical investigation of the cerebral cortex published as long ago as 1666 (Malpighi, 1666). Luigi Galvani writes in *de Viribus* that 'the electric fluid is produced by the activity of the cerebrum where it is extracted in all probability from the blood . . .', and as late as 1835 Baillarger was writing that 'the analogy between the structure of the cerebral surface and the appearance of a galvanic apparatus' suggests that it secretes 'an electric fluid' into the underlying white matter (Baillarger, 1840). However, as we have seen, Darwin does not believe that his aethereal fluid is identical with electricity but only that it is of a similar nature and more tenuous. The brain secretes this tenuous fluid into its medullary substance which, he says, is continuous with 'the innumerable ramifications of the nerves to the various muscles and organs of sense' (Darwin, 1801, p. 10).

Not only is Darwin's aethereal fluid, his spirit of animation, physically distinguishable from Galvani's electric fluid, but it also retains the psychological dimension of the ancient animal spirit. 'A certain quantity of contraction' says Darwin (1801, vol. 1, p. 37) in his Vth law of animal causation 'if it be perceived at all, produces pleasure; a greater or lesser quantity of contraction, if it be perceived at all, produces pain; these constitute sensation'. For Darwin, as noted above, not only behavioural movement but also sensory reception was mediated by fibre contractions. Thus he concludes that both sensory and motor pleasure and pain are, at root, due to the contractions and relaxations of fibres. From this foundation he is able to proceed to his VIth law of animal causation: 'A certain quantity of sensation produces desire or aversion: these constitute volition'.

It is not difficult to see how Darwin develops a fibre-based associationist psychophysiology from these beginnings. An organism learns to seek pleasurable and avoid aversive stimuli. Fibre contractions responsible for pleasurable sensations will become associated with each other. Darwin gives many examples. He describes the development of associated muscular activities during babyhood. He goes on to anticipate Ivan Pavlov by well over a century in accounts of nutritional conditioned reflexes. He writes of '. . . those unconquerable antipathies . . . which some people have at the sight of peculiar kinds of food, of which in their infancy they had eaten to excess or by constraint' and notes that the sight of palatable food 'excites into action (the) salivary glands; as is seen in the slavering of hungry dogs' (p. 63). But not only are the fibres responsible for motor activity subject to the laws of association, but also those responsible for ideas: 'All the fibrous motions, whether muscular or sensual which are frequently brought into action together, either combined in tribes, or in successive trains, become connected by habit, that when one of them is reproduced the others have a tendency to succeed or accompany it . . . In like manner' he continues 'many of our ideas are originally excited in tribes (and) associated by habit . . . form complex ideas . . . as this book, or that orange'. Thus, he says, 'the taste of a pine-apple, though we eat it blindfold, recalls the colour and shape of it; and we can scarcely think of solidity without figure'. In the *Temple of Nature* he uses heroic couplets to give his associationist psychology poetic expression:

Last in thick swarms ASSOCIATIONS spring
Thoughts join to thoughts, to motions motions cling
Whence in long trains of catenation flow
Imagined joy, and voluntary woe.

Darwin's association psychology is, of course, very similar to that which John Locke and David Hartley taught earlier in the century, though without Hartley's explicit connection to neurophysiological mechanism. This cannot be surprising as Joseph Priestley, who edited Hartley's *Observations* and wrote a laudatory preface, was a prominent member of the Lunar Society of which Darwin was a founder and very active member. Indeed Darwin explicitly uses one of Hartley's ideas to develop the theory for which he and his grandson Charles are best known: organic evolution.

'The ingenious Dr Hartley in his work on man' writes Darwin in Section XXXIX of *Zoonomia*, 'is of the opinion that our immortal part acquires during this life certain habits of action or of sentiment,

which become for ever indissoluble, continuing after death in a future state of existence; and adds, that if these habits are of the malevolent kind, they must render the possessor miserable even in Heaven'. The world, for Hartley, is thus a moral training ground for the hereafter. In exactly the same way, thinks Darwin, the parental generation forms a training ground for the forthcoming filial generation and passes on its acquired characteristics. 'I would apply this ingenious idea' writes Darwin 'to the generation or production of the embryon, or new animal, which partakes so much of the form and propensities of the parent'. In appropriate conditions:

Where milder skies protect the nascent brood
And Earth's warm bosom yields salubrious food
Each new Descendent with superior powers
Of sense and motion speeds the transient hours;
Braves each season, tenants every clime
And Nature rises on the wings of Time.

This optimistic view of progressive evolution based on the inheritance of acquired characteristics was, of course, destined to be replaced by the darker vision of his grandson, Charles, where organic evolution was based on 'random variation and selective retention' and the notion of progress was reduced from an inevitability to a hope. But that is the next century, and already I have strayed too far from the subject of neuropsychology.

Conclusion

Thus, returning to our subject, we can see, in conclusion, that the long eighteenth century was an age of transition. To adapt Matthew Arnold's image, it witnessed 'the long withdrawing roar' of the old psychophysiology. In the exact middle of the eighteenth century, Thomas Wright published an account of the astronomy of his time saying that '. . . now, thanks to science, the scene begins to open to us on all sides, and truths scarce to have been dreamt of before persons of observation had proved them possible, invade us with a subject too deep for understanding . . .' (Wright, 1750, letter VII). The same might be said of our subject. The old psychophysiology was everywhere on the retreat but at the end of the century it was still not clear with what it could be replaced. Animal spirit, which had seemed discredited by experiments and observations a century earlier, refused to take the hint and depart. It was still active at the end of the century in Darwin's concept of an aethereal 'spirit of animality'. Nevertheless, widespread popular and scientific interest in electrical phenomena began to suggest that perhaps it might have a purely physical explanation. The ancient and useful psychophysical substance, 'with teeth and claws extremely sharp', began to evaporate into mere metaphor. But what was this mysterious electrical 'fluid'? No one knew and some, like Anton Mesmer, were given to wild speculation. Nearly 50 more years were to elapse before painstaking work in microscopy and electrophysiology allowed an adequate interpretation of animal electricity to be adduced (du Bois Reymond, 1848/1849), and nearly another 50 before Cajal enunciated the neuron doctrine.

E. A. Burtt in his classical account of the origins of modern science, *The Metaphysical Foundations of Modern Physical Science*, writes that post-seventeenth-century science 'reads Man completely out of Nature' (Burtt, 1931). We can see that the development of psychophysiology in the long eighteenth century is all of a piece with this great current of thought. The growing recognition that the brain and the rest of the nervous system are no more and no less than a remarkable chemistry, evolved, as Erasmus Darwin saw, over aeons of evolutionary time, is just as remarkable a transformation in our understanding of the world as any that Thomas Wright or E. A. Burtt chronicled in their ground-breaking books. For, the demise of the old psychophysiology throws the psychophysical problem, the relation of mind to brain, into sharp relief. Chalmers (1996) calls it the 'hard problem' and, in the late-twentieth and early twenty-first centuries, it has launched a thousand papers, books and conferences.

Notes

1. For a history of ventricular psychology see Glynn, 1999, Green, 2003, and Tascioglu and Tascioglu, 2005.
2. Experiments showing that muscles did not contract by 'ballooning' were also carried out by Francis Glisson in 1677 (see Kaitaro, this volume) but Swammerdam's experiments were far more sophisticated.
3. The notion that nerve fibres were in some sense hollow, in fact, lingered on well into the middle of the nineteenth century. Microscopes and preparative techniques were just not good enough to eliminate the possibility that nerve trunks did not contain minute channels. It was only with the work of Robert Remak in the 1840s that it became clear that axons contained no cavity along which animal spirit could travel and only with the work of Matteucci and du Bois Reymond in the 1840s and 1860s that the electrical nature of nerve messages was established. It must, of course, be recognised that we now know that the axoplasm within nerve fibres is the site of flows of particles in both directions and at different velocities.
4. Monro, A. (1781). In A. Monro (Secundus) (Ed.), *The works of Alexander Monro* (pp. 322–324). Edinburgh: Charles Elliot. Monro in fact gives three pertinent objections to the Hartley's theory, the third being that nerves just do not have the tension necessary to carry vibrations. Boerhaave also believed that Hartley's vibrationism could not be squared with the neuroanatomical facts. Hartley had, however, been more subtle than his detractors supposed. At the beginning of *Observations on Man* he insists that 'the nerves themselves should vibrate like musical strings' is highly absurd and that the vibrations and vibratiuncles of his theory only applied to intra-neuronal particles (Hartley, 1749, p. 12; see also Glassman and Buckingham, this volume).
5. Leibniz writes '. . . there is nothing monstrous in the existence of zoophytes, or plant animals . . . on the contrary it is wholly in keeping with the order of nature that they should exist. And so great is the force of the principle of continuity . . . that not only should I not be surprised to hear that such things had been discovered . . . but, in fact, I am convinced that there must be such creatures and that natural history will some day perhaps become acquainted with them . . .' (quoted Rieppel, 1988).
6. This view of the soul as being 'smeared' throughout the anatomy, although attractive to Georg Ernst Stahl and his followers, did not, of course, find universal or easy acceptance in many parts of the medico-scientific world. Albrecht von Haller, for instance, threw his great authority behind the contrary view that the soul is confined to the brain (see Koehler, this volume).
7. Haller defined what he means by irritability and sensibility at the beginning of his 1755: *Dissertation*: I call that part of the human body irritable which becomes shorter on being touched . . . I call that a sensible part of the human body, which upon being touched transmits the impression of it to the soul; and in brutes, in whom the existence of the soul is not so clear, I call those parts sensible, the irritation

of which occasions evident signs of pain and disquiet in the animal (pp. 7–9).

8. He asks us to concentrate on a red spot for a length of time and then to close the eyes and shade them with the hand. We will become aware of a green after-image. We would now explain this in terms of depletion of neurotransmitters in the synapses of specific pathways leading to the LGN. Darwin has an analogous explanation (unsupported by experimental evidence, of course) in terms of fibre fatigue. '. . . it appears', he writes, 'that a part of the retina which had been fatigued by contraction in one direction relieves itself by exerting the antagonist fibres, and producing a contraction in the opposite direction . . . (as) when we are tired with the long action of our arms in one direction, as in holding a bridle on a journey, we occasionally throw them into an opposite position to relieve the fatigued muscles'. He concludes that 'the muscular actions of the retina may constitute . . . (our visual ideas), and the voluntary repetitions of them, when the object is withdrawn, our memory of them' (Darwin, 1801, vol. 4, Section XL). He argues that a similar account may be constructed for the other senses.

9. It is likely that Darwin received this idea from Alexander Monro Primus, his anatomy teacher at Edinburgh in the 1750s. 'Let us now suppose it probable' writes Monro 'that the encephalon and spinal marrow secern a liquor from the blood which is sent into all the nerves, and that by means of this liquor the nerves perform all the offices commonly assigned to them' (Monro, 1781, p. 322).

References

Augustine (401). *The literal meaning of genesis* (J. H. Taylor, Trans.), 1983. *Ancient Christian Writers* (Vol. 2). Mahway, NJ: Paulist Press (Chapter 18).

Baillarger, J. G. F. (1840). Recherches sur la structure de la couche corticale des circonvolutions du cerveau. *Mémoires Académie Royale Médicin 8*: 149–183.

Bloch, G. (1980). *Mesmerism, a translation of the original scientific and medical writings of F. A. Mesmer*. Los Angeles: Kaufmann.

du Bois Reymond, E. (1848/1849). *Untersuchungen über theirische Elektricität* (2 vols.). Berlin: Reimer.

Borelli, G. (1680/1681). *De Motu Animalium* (2 vols.). Roma: A. Bernabo.

Burtt, E. A. (1931). *The metaphysical foundations of modern physical science*. London: Routledge and Kegan Paul.

Chalmers, D. J. (1996). *The conscious mind: In search of a fundamental theory*. Oxford: Oxford University Press.

Changeux, P. (1985). *Neuronal man: The biology of mind*. New York: Oxford University Press.

Cheyne, G. (1733). *The English malady*. London: G. Strahan.

Cobb, M. (2002). Exorcizing the animal spirit: Jan Swammerdam on nerve function. *Nature Reviews Neuroscience, 3*, 395–400.

Cottingham, J., Stoothoff, R., & Murdoch, D. (1985). *The philosophical writings of Descartes* (2 vols.). Cambridge: Cambridge University Press.

Cullen, W. (1827). In J. Thomson (Ed.), *The works of William Cullen* (Vol. 1). Edinburgh: Blackwood.

Darwin, R. W. (1786). New experiments on the ocular spectra of light. *Philosophical Transactions of the Royal Society, 76*, 313–348.

Darwin, E. (1801). *Zoonomia* (3rd ed. corrected). London: J. Johnson.

Darwin, E. (1803). *Temple of Nature*. London: J. Johnson.

Fowler, R. (1793). *Experiments and observations relative to the influence lately discovered by M. Galvani and commonly called animal electricity*. Edinburgh: Duncan.

French, R. K. (1975). *Anatomical education in a Scottish university, 1620: An annotated translation of the lecture notes of John Moir*. Aberdeen: Equipress.

Galvani, L. (1791). In A. Forni (Ed.), *De Viribus Electricitatis in Motu Musculari* (Facsimile ed.). Bologna: Ex typographia Instituti Scientiarum, 1998.

Gardner-Thorpe, C. & Pearn, J. (2006). Erasmus Darwin (1731–1802): Neurologist. *Neurology, 66*, 1913–1916.

Glynn, I. (1999). Two millennia of Animal Spirit. *Nature, 402*, 353.

Green, C. D. (2003). Where did the cerebral localisation of the mental faculties come from? *Journal of the History of the Behavioral Sciences, 39*, 131–142.

von Haller, A. (1755). *A dissertation on the sensible and irritable parts of animals*. London: Nourse.

Hartley, D. (1749). *Observations on man, his frame, his duty, and his expectations*. London: Thomas Tegg.

Hippocrates. *On the sacred disease in works by Hippocrates* (Francis Adams, Trans.). Adelaide: e-Books.

Johnson, S. (1739). Life of Dr Herman Boerhaave, late professor of physic in the University of Leyden in Holland. *Gentleman's Magazine, 9*, 37–176.

Johnson, S. (1755). *A dictionary of the English language*. London: Longman, Hitch, Hawes, Millan, R & D Dodsley.

Koyré, A. (1943). Galileo and Plato. *Journal of the History of Ideas, 4*, 400–428.

de La Mettrie, J. O. (1745). *Histoire Naturelle de l'Ame*. La Haye: Jean Neaulme.

Loemaker, L. E. (Trans. and Ed.). (1956). *Gottfried Wilhelm Leibniz: Philosophical papers and letters*. Chicago: University of Chicago Press.

Lovejoy, A. O. (1930). *The Great Chain of Being*. New York: Harper.

MacKinnon, R. (2003). Nobel Prize Lecture at http://www.nobelprize.org/chemistry/laureates/2003/mackinnon-lecture.html

Malpighi, M. (1666). *De cerebri cortice*. Bologna.

Monro, A. (1781). In A. Monro (Secundus) (Ed.), *The works of Alexander Monro*. Edinburgh: Charles Elliot.

Nemesius of Emesa. In W. Telfer (Ed.), (1955). *On the nature of man. Cyric of Jerusalem and Nemesius of Emesa*. Philadelphia: Westminster Press.

Nordstrom, J. (1954). Swammerdamiana: Excerpts from the travel journal of Olaus Borrichius and two letters from Swammerdam to Thevenot. *Lynchnos, 16*, 21–65.

Pancaldi, G. (1990). Electricity and life: Volta's path to the battery. *Historical Studies in the Physical Sciences, 21*, 123–159.

Piccolino, M. (2006). Luigi Galvani's path to animal electricity. *Comptes Rendus Biologies, 329*, 303–318.

Porter, R. (2001). Nervousness, eighteenth and nineteenth century style: From luxury to labour. In M. Gijswijt-Hofstra, & R. Porter (Eds.), *Cio Medica/The Wellcome series in the history of medicine* (Vol. 63). London: Rodopi.

Porter, R. (2003). *Flesh in the age of reason*. London: Penguin Books.

Priestley, J. (1775). *The history and present state of electricity with original experiments* (4th ed.). London: Bathurst.

Rieppel, O. (1988). The reception of Leibniz's philosophy in the writings of Charles Bonnet (1720–1793). *Journal of the History of Biology, 21*, 119–145.

Smith, C. U. M. (1976). *The problem of life: An essay in the origins of biological thought*. London: Macmillan.

Smith, C. U. M. (1987). David Hartley's Newtonian neuropsychology. *Journal of the History of the Behavioral Sciences, 23*, 87–101.

Smith, C. U. M. (1998). Descartes' pineal neuropsychology. *Brain and Cognition, 36*, 57–72.

Smith, C. U. M. (1999). Coleridge's "Theory of Life". *Journal of the History of Biology, 32*, 31–50.

Smith, C. U. M. (2002a). Jean Offray de La Mettrie: 1709–1751. *Journal of the History of the Neurosciences, 11*, 110–124.

Smith, C. U. M. (2002b). *Elements of molecular neurobiology* (3rd ed.). Chichester: Wiley.

Smith, C. U. M. (2005). All from fibres: Erasmus Darwin's evolutionary psychobiology. In C. U. M. Smith, & R. Arnott (Eds.), *The genius of Erasmus Darwin* (pp. 133–143). Aldershot: Ashgate.

Stephanson, R. (1988). Richardson's "Nerves": Physiology of sensibility in Clarissa. *Journal of the History of Ideas, 49*, 267–285.

Sterne, L. (1760). *Tristram Shandy*. London: R. & J. Dodsley.

Sutton, J. (1998). *Philosophy and memory traces: Descartes to connectionism*. Cambridge: Cambridge University Press.

Tascioglu, A. O. & Tascioglu, A. B. (2005). Ventricular anatomy: Illustrations and concepts from antiquity to Renaissance. *Neuroanatomy, 4*, 57–83.

Trembley, A. (1744). *Memoires pour servir à l'histoire d'un genre de polypes d'eau douce, a bras en formes de cornes*. Leiden: Verbeek.

Trotter, T. (1807). *A view of the nervous temperament*. London: Longman, Hurst, Rees and Orme.

Vesalius, A. (1543). *De Humani corporis Fabrica*. Impression Anastatique (1964), Bruxelles: Culture et Civilisation.

Volta, A. (1800). On the electricity excited by mere contact of conducting substances of different kinds. *Philosophical Transactions of the Royal Society, 90*, 403–431.

Whytt, R. (1765). *Observations on the nature, causes nervous, hypochondriac or hysteric, to which are prefixed some remarks on the sympathy of the nerves*. Edinburgh: Balfour.

Willis, T. (1664). *Cerebri anatome*. London: Martyn and Allestry; (1681). *Anatomy of the Brain*. London: Classics of Medicine Library.

Willis, T. (1672). *De Anima Brutorum* (S. Pordage, Trans.). London: Thomas Dring; (1684). *Dr Willis' practice of physic*. London: Thomas Dring.

Wright, T. (1750). *An original theory or new hypothesis of the universe*. Reprinted with an introduction by M. A. Hoskin. New York: Elsevier.

2
Enlightening Neuroscience: Microscopes and Microscopy in the Eighteenth Century

Brian J. Ford

Origins of Microscopical Neurology

Little has ever been written on the history of microscopical neurology. The topic is ordinarily ignored – indeed the terms 'microscope' (and microscopy), 'neuron' (or neurone), 'cell' and 'histology' are missing altogether from the index to the overview of the history of neurology by Riese (1959).

Microscopy was born in the years prior to the eighteenth century and nerve specimens were among the first to be examined. The late sixteenth century saw the first descriptions of a recognisable microscope and questions of priority persist, since the study of magnification and of refraction – which preceded the practical application of lenses in scientific instruments – was already a matter of some antiquity (Disney, Hill, & Watson Baker, 1928). The first microscope to be pictured was a compound instrument in 1631, and during the first few years these microscopes were utilised in the quest to unravel the structure of familiar objects – the sting of a nettle or a bee, the wings of a butterfly or bird. We must bear in mind that these were truly macroscopic, rather than microscopic, investigations. Observers were exploring everyday specimens, searching for details the eye could almost discern. Only when the high-power microscope emerged could investigators progress to the most far-reaching development in natural science – the recognition that there were forms of life, and marvellous structures, the existence of which nobody had previously recognised.

The first great pioneer of the microscopic – perhaps better *macroscopic* – world was Robert Hooke (1635–1703) who was appointed to be curator of experiments at the Royal Society of London in 1662. On 25 March 1663, Hooke was enjoined to begin a series of demonstrations with a view to publication, and on 1 April he was instructed to bring at least one microscopical observation before each meeting of the fellowship (Gunter, 1961). Hooke obtained a compound microscope magnifying some 40× from Christopher Cock, a London instrument manufacturer, and his studies with this instrument laid the groundwork for modern science. Hooke's pictures of flies and fungi, of seeds and spiders, needles, gnats and nettles, served to set natural philosophy afire. His large folio book *Micrographia,* published by the Society in 1665, gave readers a vivid insight into what he had seen (Hooke, 1665).

That much is well known to historians of science, but a crucial section of the Preface to his great work has been overlooked. In this key passage he described how to manufacture a microscope of much higher magnification. On page 22 of the (un-numbered), pages of the Preface appears a recipe for a microscope capable of magnifying hundreds, rather than tens, of times. This kind of instrument gives clear views of much smaller cells – bacteria, spermatozoa, erythrocytes – and could be made without specialist equipment. Curiously, Hooke never published an illustration of this microscope. But his description was seized upon by a convert to the cause who went on to make legion discoveries with this unrefined type of microscope, and who soon began a study of bovine optical nerve (Ford, 1991).

The enthusiastic newcomer was Antony van Leeuwenhoek (1632–1723), the draper of Delft, Netherlands. He became acquainted with Hooke's book on a visit to London about 1668, when the second edition of *Micrographia* had been published and the book was enjoying extraordinary popularity, and began with studies of whiteness in bodies like chalk (which Leeuwenhoek had encountered on his voyage up the Thames). By 1673, Reinier de Graff (1641–1706) was writing to the Royal Society about this 'most ingenious person' and his remarkable microscopes. They were diminutive instruments, little more than postage-stamped-sized rectangles of metal (typically brass or silver) between a perforation in which was held a small ground lens, little larger than the head of a dressmaker's pin (Fig. 1). Specimens were held on a tapered metal holder projecting from a small stage, itself about a centimetre long, and screws mounted on the plate allowed the user to adjust the position and the focus of the specimen. Solid specimens – insects, flowers, leaves – were held with wax on the end of the pin. Liquid materials, including aquatic microorganisms in pond-water, were confined with a flat capillary tube that was itself glued to the stage pin.

This design was explicitly set out in Hooke's *Micrographia* and the results that Leeuwenhoek obtained with it are remarkable. His earliest reports, sent to the Royal Society in 1673, were of specimens referred to by Hooke. As a rule, Leeuwenhoek's early accounts were sent in refutation of what 'a certain learned gentleman' had recently published. During the following year Leeuwenhoek continued his innovative investigations and on 1 June 1674 he sent to London his first selection of prepared microscopical specimens. Three of them – cork, elder pith and the white of a feather – were in direct response to observations Hooke had published in *Micrographia*.

There was one further specimen: a small packet containing slices of dried optic nerve. This was not stimulated by anything Hooke had described his *Micrographia*; these were examples of Leeuwenhoek's independent investigations. These are the first specimens from Leeuwenhoek's original research and they also served to launch the microscope as a tool of neurological investigation. Clifford Dobell, whose well-researched biography of Leeuwenhoek remains one of the most detailed

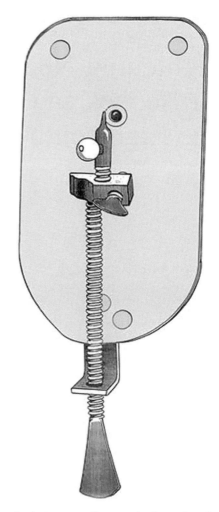

FIGURE 1. Antony van Leeuwenhoek produced microscopes, sometimes with plano-convex or aspheric biconvex lenses. The body plate and attachments were made by Leeuwenhoek at his home. These microscopes were still used for high-power microscopy by Home and others into the early nineteenth century

such works in the history of science, noted in 1932 that the specimen packets "have remained intact to the present day" but did not investigate what they might contain (Dobell, 1932). The presence of the optic nerve specimens in the Leeuwenhoek papers was noted by F. J. Cole, who in 1937 published a pioneering paper on Leeuwenhoek's zoological researches, but most writers did not refer to them (Cole, 1937). For example, a lengthy celebratory publication on Leeuwenhoek's researches was published in *Natura* in 1932, to commemorate the tercentenary of his birth, and – although this aimed

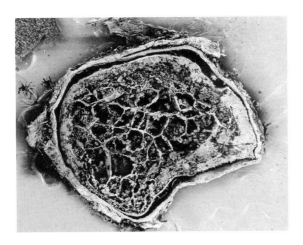

FIGURE 2. In 1674 Leeuwenhoek sent to London the first microscopical sections of nerve tissue. They were discovered by the author among the Leeuwenhoek papers of the Royal Society in 1981, and this electron micrograph was taken at Cardiff with a JSM 840A scanning electron microscope at 2kV. Field width = 4 mm (Ford, 1982)

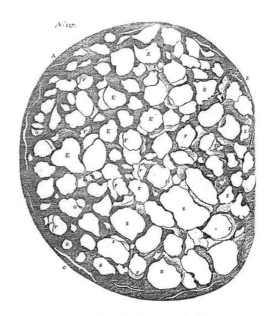

FIGURE 3. Leeuwenhoek's (unnamed) limner prepared pencil drawings of optic nerve and sent them to London. The studies were copied by engravers to produce this full-page image for publication in *Philosophical Transactions of the Royal Society*

to present a full chronology of all his researches – there was no mention of the surviving specimen packets. The definitive Dutch collection of the Leeuwenhoek correspondence (1932–present) mistakes some of his specimen packets for 'drawn rectangles' and misses others altogether. Although these extraordinarily well-prepared specimens represent the roots of modern bioscience, they remained lost to contemporary science. By the time I submitted them to both optical and scanning electron microscopical examination in 1981, they had lain essentially undisturbed for 308 years (Ford, 1981) (Figs. 2 and 3).

Early Microscopical Investigations

Microscopical investigation of the nervous system began with Leeuwenhoek's studies in 1674, when he made his first preparations of bovine optic nerve. His letter dated 7 September 1674 describes how he was encouraged to observe nerve specimens: 'I communicated my observations [of optic nerve] to Dr of anatomy Schravesande and he mentioned that since ancient times there has been some dissention among the learned about the optic nerve and that some anatomists affirmed [it] to be hollow; and that they themselves had seen the hollowness, through which they would have the animal

spirits that convey the visible species, represented in the eye, pass into the brain. I therefore concluded that such a cavity might be seen by me. . . . I solicitously viewed three optic nerves of cows, but could find no hollowness in them' (Anon, 1932).

This is a crucial moment in the history of neuroscience. The notion of a hollow nerve, analogous to a vessel transporting a fluid, had existed since ancient times. Galen (131–201 AD) viewed the nervous system as the distributive counterpart of the blood circulation, transporting vital spirits from the lung and heart around the body. René Descartes (1596–1650) published an account in which he wrote of the nerves conducting animal spirits between brain and musculature. In one edition of his book (Descartes, 1662) his descriptions are accompanied by an engraving by Florentio Schuyl; this figure clearly shows a cored, perhaps hollow, structure. The diagram was not by Descartes, and was omitted from the subsequent French translation of the book edited by Claude Clerselier (1664) and henceforth.

There had long been a tacit assumption that nerves were hollow until the writing of Andreas Vesalius (1514–1564). He described the optic nerve – in terms similar to those of Leeuwenhoek – as a solid structure. Yet it was not until the microscope was brought to bear on the topic that the true histological appearance of the optic nerve could finally be determined. On 4 December 1674, Leeuwenhoek wrote to London with an attempted resolution of the earlier theories. 'I took eight different optic nerves which did shrink up . . . upon which, a little pit comes to appear about the middle of the nerve, and it is this pit, in all probability, that Galen mistook for a cavity', he wrote. Leeuwenhoek made his transverse slices from dried specimen of bovine nerve, since the fresh material was too soft to be sectioned with a razor.

The drawings that he prepared display the anatomy of the optic nerve remarkably well. When cutting botanical material (including cork from *Quercus suber* and pith of the common elder *Sambucus nigra*) Leeuwenhoek used a rising, sawing motion with the razor edge. When the plant material began to become friable and break up, he cut slightly deeper. In this way the plant sections contained thicker supportive regions interspersed with thinner zones in which histological observations could be made. With the optic nerve, however, Leeuwenhoek writes that he used a single cut (not a 'sawing motion'), and the resulting specimens were thicker than his plant sections. He sensibly calls the nerve preparations 'slices' rather than 'sections' since they were >200 µm in thickness.

The nerve fibres comprising the fasciculi are missing from these preparations leaving a lattice of openings that accord well with Leeuwenhoek's description of a 'leathern sieve'. This appearance is due to the survival of the perineurium. The nerve sheath or epineurium is unique in the optic nerve because it derives from the pia, arachnoid and dura of the brain and thus has a three-layered structure. The separation of the layered epineurium is well portrayed in the studies of optic nerve that Leeuwenhoek sent to London and are testimony to his acute and accurate observation. It is noteworthy that Leeuwenhoek himself was no draughtsman; he employed a limner to make drawings on his behalf. We are reminded of this in a letter he sent on 25 December 1674 to his correspondent Mr C. Huijgens van Zuijlichem, in which he wrote: 'I enclose a copy of the optic nerve . . . as I saw though my own microscope, drawn to my order'.

A version of this study was published in *Philosophical Transactions* (Leeuwenhoek, 1675). Leeuwenhoek's description of the structure of optic nerve is set out in his letter of 4 December 1674:

"I have put before my microscope a piece of such a dried Optic Nerve of a Cow, and how it appeared, and you will see by the picture hereby transmitted unto you. ABCD is the circumference of the Optic Nerve, which did not dry round ways, but somewhat oblong on the side CD. E, and all the places that are left white and clear, are cavities in the dried Nerve which I imagine to have been filaments, and out of which, for the greatest part, the soft globules have been exhaled. F are particles or globules which are in the little holes of the filaments in many places, and such as have not been exhaled".

This is an interesting passage, notably for its insistence that the nerve fibres are 'filaments'. The term comes up again in 1677 when Leeuwenhoek wrote of nerves as comprising: 'diverse, very small threads or vessels lying by each other'. He speculated whether these 'conveyed the animal spirits throughout the spinal marrow.' We should *en passant* note that axoplasmic material can exude from the sectioned extremity of an axon, which might be held to support the notion of the nerve fibre as a hollow vessel that conducted a viscous fluid (Young, 1934). However, there are no records that microscopists of the era examined specimen material of this sort.

Leeuwenhoek's observations were not the only essays into nerve structure at this time. Comments on the microscopy of nerves were published by the Italian natural philosopher Giovanni Alfonso Borelli (1609–1679). Borreli (1681) reported that nerves were tubes filled with a moist and spongy substance. The first microscopist to move towards a true science of histology was also an Italian: Marcello Malpighi (1628–1694). He served as professor at Bologna, Pisa and Messina, and made some observations of the brain under the microscope (McHenry, 1969). Several years after Leeuwenhoek's pioneering observations, Malpighi injected blood-vessels to increase contrast, and concluded that the grey matter was made up of cellular follicles and the white matter comprised fine excretory ducts (Malpighi, 1686).

Notions of a hollow nerve, which Leeuwenhoek had satisfactorily dismissed through the use of the microscope, lingered on in the decades that followed. Three years after Malpighi's book appeared,

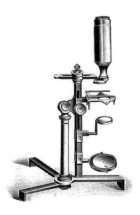

FIGURE 5. W. and S. Jones' 'Most improved' microscope characterises the brass and glass instrument of the late eighteenth century. It was produced in 1790 and influenced designs that were prevalent in the early nineteenth century

FIGURE 4. This typical compound microscope of the seventeenth century was described by Henry Baker (1743 plate III). Originally designed by Marshall and adapted by Culpeper, this model was designed by Edward Scarlett jr., Master of the Spectacles Company 1745–46

a book was posthumously published by students and colleagues of Theodoor Craanen of Leiden. It included a spurious engraving of hollow nerve fibres bound together with bands like bundles of bamboo (Craanen, 1689). It has been argued that this illustration was included to reinforce Craanen's strictly Cartesian view of nature in which nerve were believed to function as tubes that conducted animal spirits from the brain. Craanen was inclined to allow such preconceptions greatly to influence his interpretation of reality, claiming that "the subtlety of nature surpasses our powers of thought" (Ruestow, 1996) (Figs. 4 and 5).

Eighteenth-Century Microscopy

After the burgeoning interest in microscopy manifest during the latter half of the seventeenth century, a gathering of momentum might logically be assumed. But it was not to be. Microscopy made surprisingly little progress during this century, and neurological microscopy lay largely in the doldrums. Nerve cells are hard to observe in the freshly harvested state, and attention was instead captured by organisms like *Hydra* and by the intricacies of plant life. Nerve fibres were visible, though the impression gleaned through the microscopes of the period was largely misleading. The refractile myelin sheath was frequently mistaken for a hollow tube, and without a coherent approach to fixation and staining it was impractical to find ways to visualise components of the nervous system.

The notion of hollow nerve fibres was revived for many years. It reappeared throughout the eighteenth century and is typified by a figure published in 1761 by Martin Frobenius Ledermüller (1719–1769) that showed supposedly tubular nerves. The error continued on into the century that followed. Ledermüller's figure was reproduced in the decades following publication, and was still circulating in the middle nineteenth century, the most recent example of republication of this figure that I have traced being by F. A. Longet (1842) (Figs. 6 and 7).

Considerable interest in microscopical revelation was shown by many eighteenth century natural philosophers, though the microscopy of the nervous system was not greatly advanced during this century. Natural philosophers used their microscopes as gadgets, rather than as objects of special importance, and rarely described which instruments they employed for their work. The single lensed or simple microscope utilised by Leeuwenhoek remained a favourite instrument, though the design changed.

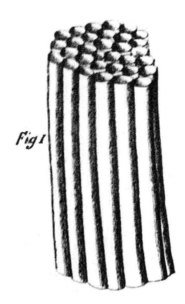

FIGURE 7. Tubular nerve elements were portrayed less ambiguously by Ledermüller (1761, Fig. 1). In this portrayal of nerve fibre bundles, a hollow appearance is clearly conveyed. This interpretation was erroneously published until the 1840s

FIGURE 6. Nerve fibres envisioned as tubes, published by Descartes (1662, Fig. vi). The appearance is close to that of myelinated fibres; the tube-like interpretation may lie in the artefact of refraction during observation

First, the single lens was constrained in a hand-held barrel design which was fitted with a lateral handle. Designed by an instrument manufacturer, it became known as the Wilson screw-barrel microscope. Before long an advanced design began to appear, in which the microscope was mounted on a brass stand. This made it easier to use; the stand usefully occupied the distance between the observer's eye when seated, and the height of the table on which the device was standing. Such microscopes were widely used by natural philosophers, and the development of the simple microscope (which permitted good magnifications at adequate resolution) proceeded in parallel with that of the compound instrument. Because of the simplicity of design, the specification, description and *modus operandi* of the simple microscope was rarely mentioned by their many users.

In Britain, the compound microscopes of the early eighteenth century had changed little in design from the type designed by Christopher Cock and used by Hooke in the 1660s. The Marshal microscope typifies the trend; its lenses produced an image of only moderate quality and it was the brass pillar and intricately tooled leather coating of the body tube that made it an object of desire. These were desk-top adornments as much as serious scientific instruments and they suffered from design difficulties – thus, it was almost impossible to observe wet preparations in a watch-glass. To this extent, the design of the instrument imposed limits on the level of inquiry that the investigator might pursue (Ford, 1985).

No microscopes of the eighteenth century were as good as those made by Leeuwenhoek, and his diminutive instruments continued to be in demand for high-power microscopical research. In a letter written on 20 August 1738, Lord Jersey commended the Leeuwenhoek microscopes for use in research on the structure of spermatozoa. Another enthusiast was Henry Baker (1698–1744), an enthusiastic amateur who wrote on the microscopy of the eighteenth century, often with more verve than veracity. Almost all of his writings were in the form of a simple rehash of what others had done before. His pictures were re-engraved from higher-quality originals that had previously been published. In one of his books (Baker, 1743) we find his account of Leeuwenhoek's work on the nervous system:

"Mr LEEUWENHOEK endeavoured to discover, by his *Microscope*, the *Structure* of the *Nerves*, in he Spinal Marrow of an Ox; and saw, with great Delight, that minute hollow Vessels, of an inconceivable Fineness, invested with their proper Membranes, and running out in Length parallel to one another, make up their Composition . . . He observed farther, that the Vessels in the Brain of a Sparrow are not smaller than those of an Ox; and argues from thence, that there is really no other Difference between the brain of a large Animal and that of a small one. . . ."

He published wise comments on the innervation of the insect eye (1743, p. 227). 'It is also reasonable to believe', Baker stated, 'that every *Lens* [of the compound eye] has a distinct Branch of the Optic Nerves administering to it: and yet, that Objects are not multiplied, or appear otherwise than single, any more than they do to us, who see not an Object double though we have two eyes'. The use of the microscope to unravel the problem of innervation is itself noteworthy, as is this prescient description of how compound eyes must surely function.

He also shows simple microscopes, for they remained the prime instruments of microscopy at higher powers, and described the Leeuwenhoek microscopes which were still being used by the investigators. Many workers were making the smallest lenses as spheroidal globules of glass, whereas Leeuwenhoek had pursued lens-grinding and had made fine biconvex lenses. Even as the mid-eighteenth century approached, Baker (1743, p. 7) was referring to the Leeuwenhoek microscopes with some sense of admiration, for they remained the best available:

"Several Writers represent the Glasses of Mr. Leeuwenhoek made use of in his Microscopes to be little Globules or Spheres of Glass, which Mistake most probably arises from their undertaking to describe what they had never seen; for at the Time I am writing this, the Cabinet of Microscopes left by that famous Man, at his Death, to the *Royal Society* as a legacy is standing upon my table; and I can assure the World, that every one of the twenty-six Microscopes contained therein is a double convex Lens, and not a Sphere or Globule."

Leeuwenhoek's supremacy continued even into the following century. Sir Everard Home (1756–1832), who worked on (and freely plagiarised) the papers of the anatomist John Hunter (1728–1793) took the Leeuwenhoek microscopes back to his apartment and used them for high-power microscopy in the 1820s. These delicate, silver instruments were probably destroyed in a disastrous fire at Home's apartment in the Chelsea Hospital, when he decided to burn the potentially incriminating Hunter manuscripts and set fire to his home in the process. No other microscopes have so long served a role in the history of science.

Microscope Design

Baker provides good evidence of the microscopes that were in vogue for this kind of research. His book (1743, p. 16) illustrates a 'double reflecting microscope' – i.e., one with an objective and a field lens or eyepiece, which could be fitted with a plano-concave substage mirror. The model he depicted was produced by Culpeper and Scarlet, being 'an alteration and improvement' of the design by Marshal.

The design was produced in many models by Edmund Culpeper from about 1730 onwards. Culpeper is believed to have served time as an instrument maker under Walter Hayes, who according to Clay and Court (1932) was well established as a maker of mathematical instruments from 1650 until his death in 1686. It seems that Culpeper then continued Hayes' business. In the eighteenth century, Culpeper began to produce a microscope of which the major design feature was the abandonment of a single pillar to support the body tube, in favour of a tripod design. This facilitates the alignment of objective, specimen and mirror and was altogether a more sturdy microscope. Yet, although it can be seen as an improvement upon the Marshal design, it was not a new concept. Microscopes from the seventeenth century were thus constructed; indeed the earliest known drawing of a microscope (dating from around 1590) depicts this same design principle. In time Culpeper microscopes became widely available. They were fitted with lenses ground from soda glass and had body tubes that were decorated with an ornate covering, typically shagreen. The design was widely adopted by manufacturers across Europe, and they became popular possessions of the wealthy.

Microscopes were widely utilised by the Italian natural philosophers. Dell Torre (1776) produced the first studies of peripheral nerves, which were delineated as rows of threads. Felice Gaspar Fontana (1730–1803) carried out extensive investigations of the stimulation of the nervous system using pressure, electricity and venom, and supplemented this work with microscopical observations of the nerves. He was able to observe the neuraxon and the myelin sheath many years before the findings of Remak and Schwann.

Alexander Monro *primus* turned his attention to the microscopical structure of nerves in 1732:

"The nervous Fibrils, which when examined with the best Microscope, appear only like so many small and distinct Threads lying parallel, without any Appearance of being Tubes: But in their Interstices and Membranes innumerable Branches of Vessels may be observed; the open Orifices of which, when seen in a Nerve cut transversely, are in hazard of making us believe, that we have discovered Cavities of the nervous Canals."

McHenry (1969) reports that Monro *secundus* (1783) described nerve fibres as 'one nine-thousandth of an inch' in diameter (approximately 3 μm) and of twisted, solid appearance. Wade (2004) reports that William Porterfield came up with similar dimensions in the same year: Porterfield proposed that the nerves were about '1 part in 7,200th inch' (approximately 3.5 μm). Wade reminds us that nerves were variously reported to consist of 'bundles of fibrils, filaments, capillaments, threads, or villi (as they were variously called)'. Only the microscope would eventually lead us to a proper understanding of the fine structure of the nervous system.

These pioneering investigations of the microscopy of the nervous system were largely prosecuted with simple microscopes that utilised a single magnifying lens. The development of simple microscopes led to a variation on the Marshal/Culpeper theme, for instruments with a single lens did not lend themselves so readily to a tripod construction. Tripod microscopes were easier to align, and typical simple microscope designs retained a lateral pillar to support the lens mount, stage and substage

mirror. Early examples were produced by John Cuff (1708–1772) and the lesser known James Ayscough (*c.* 1718–1759) whose design was an early example of a microscope that could be used as a simple or compound instrument. Makers of these microscopes included the two George Adams (father and son, who published in 1746 and 1787), Francis Watkins and Benjamin Martin. Jean Baptiste Amici produced a range of microscopes in Italy, and attempted to solve the problem of achromatism. He believed the task was impossible, and went on to produce reflecting microscopes that were based on the proposals by Sir Isaac Newton. In France, Dellebar's 'Microscope Universal' was based on the design popularised by Adams and Martin, being introduced to France by Lalande in 1762 (Montucla, 1802). Clay and Court (1932) report that Samuel Gottlieb Hoffman produced a similar microscope in Germany in 1772.

Binocular microscopes, which we take for granted in the modern laboratory, were little used during this period. The first was described by Chérubin d'Orléans about 1680, and consisted of two unrefined compound microscopes fitted side by side in a rectangular case. True binocularity was not obtainable, for two reasons. First, the lenses were individually made by hand and could not be accurately matched. Secondly, because such instruments produce an inverted image – movements of the specimen are reversed when observed – the image is inverted, and so is the binocular appearance. Thus, depressions appear proud and hills are seen as holes. This topic was considered by Charles Wheatstone, who published the first design for a successful binocular compound microscope in 1852. He coined the term 'pseudoscopic' for the effect.

The observation of unstained, fresh nerve fibres would have been undertaken by innumerable microscopical investigators using instruments of this kind. And not only nerve fibres: the neuron is widely believed to have been first identified by Purkyně, whose work certainly led to the first published figures depicting the neurons that now bear his name. There is, however, a widely overlooked report of a microscopical observation reported by Prochaska (1779). He pressed specimens of neural material between glass slides, in order to break up the cohesion and see what sub-units he might discern. He concluded that the nerve tissues were made up of innumerable globules. We cannot now be certain whether he was observing cellular particles or ground nerve fibres, though it is feasible that Procjaska was the first to observe discrete neurons.

Imagined Constraint

It is important to examine the real, and imagined, limitations that such microscopes imposed. Constraints on resolution limits and the problem of chromatism did not impose the barriers to progress that it is popular to propose. Compound lens arrays magnify aberrations more than they do specimens, and single lenses impose relatively few perturbations of image quality. I have shown that a Leeuwenhoek microscope can reveal microfibrils as fine as 0.7 μm – a result that is within a factor of four from the theoretical limits to optical resolution. A microscope of these specifications could clearly have revealed most of the epoch-making discoveries that were later to be made with optical microscopes. Chromatic aberration is not the problem we think it is, either; its main effect is to impart spurious colours to minute structural details within a specimen. The popular idea of rainbow-hued blurred images in which little detail can be discerned is unjustifiable, for images from diminutive biconvex lenses can be of surprising clarity and vividness. For example, a micrograph I have taken of an unstained blood smear, using the single-lensed Leeuwenhoek microscope at the Museum of the History of Science at Utrecht University, shows not only the erythrocytes with remarkable clarity, but even reveals the lobed nucleus within a polymorphonuclear granulocyte. Image acuity was clearly no barrier to histological observation (Fig. 8).

The limits lay not with the microscope, but rather with the techniques of specimen preparation.

Once nerve cells could be satisfactorily stained, even a simple microscope of the early eighteenth century could have allowed the presence of the cells to have been demonstrated. When McHenry (1969) writes: 'No real notion of the microscopic structure of the brain could be obtained until the invention of the compound microscope and microtome, along with the development of methods of fixation and staining of nervous tissue' he is only partially correct. In truth, the simple microscope was optically up to the task; it was the microtech-

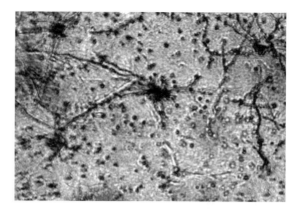

FIGURE 8. Cerebral section stained by the Golgi method and imaged by the author through a single-lensed microscope typical of those in use during the eighteenth century. Discrete neurons and their interconnections are clearly visible. Even primitive microscopes could allow the visualisation of neurons, though these cells remained unseen until staining procedures were available

nique (and staining in particular) that was lacking. Purkyně's successes were due to his readiness to adopt the latest, newly developed methods of fixation, staining and microtomy.

The Century Turns

It is in strange juxtaposition that I close this outline of eighteenth-century neurological microscopy with the work of Marie François Xavier Bichat (1771–1802) who is regarded as the 'father of histology'. Bichat read medicine and surgery at the University of Montpellier and at the Hôtel-Dieu, Paris. Even in his years as an undergraduate he conducted a prodigious amount of research into physiology, and went on to become a demonstrator, and then lecturer, at the Hôtel-Dieu. He died at the age of 31, just five years after graduating, yet amassed enough research to furnish three truly monumental works: *A Treatise on the Membranes* (1800), *Physiological Researches on Life and Death* (1800), and *General Anatomy* (1802). A web site summarising his work exists at http://www.bium.univ-paris5.fr/histmed/medica/bichat.htm and Bichat's reputation has long been secured in the history of medicine. The study of anatomy and physiology had been traditionally founded upon the bodily humors, or the study of organs. Bichat determined that pathology should better be

PREMIÈRE CLASSE. Fonctions relatives a l'Individu.

ORDRE PREMIER. *Fonctions de la Vie animale.*

Genre Ier.
Sensations.
- 1o. Des sensations générales, ou du tact. { extérieur. intérieur.
- 2o. Des sensations particulières. . . . . . { Vue. Ouïe. Odorat. Goût. Toucher.
- 3o. Du plaisir et de la douleur.

Genre IIe.
Fonctions cérébrales.
- 1o. relatives aux sensations. . . . . { De la perception. De l'imagination. De la mémoire.
- 2o. relatives à l'entendement. . . . . . . { De l'attention. Des idées. Du jugement. Du raisonnement, etc.
- 3o. relatives aux mouvemens. . . . . . . { De la volonté qui est déterminée par. . . { le jugement. les passions. } { De l'opposition de ces deux causes.
- — 4o. Connexion des fonctions cérébrales avec la vie . . . . . . { De la commotion. De l'apoplexie, etc.

Genre IIIe.
Locomotion.
- 1o. Des attitudes immobiles. . . . . { sur les pieds. . . . . . Station. sur les genoux. sur le bassin. sur la tête, etc., etc. — Prostration.
- 2o. Mouvemens . . . . {
  - des membres supérieurs. . . . . . . { Prépulsion. Répulsion. Diduction. Pression. Elévation, etc.
  - des membres inférieurs . . . . { Marche. Course. Saut.
  - du tronc. . . . . { Support, élévation des fardeaux.
  - de tout le corps. . . . . Natation.
  - — Du geste considéré comme sup- plément de la voix. . . . . { 1o. Gestes de la face. 2o. Gestes de la tête, en totalité. 3o. Gestes des membres supérieurs.

Genre IVe.
Voix.
- 1o. De la voix brute. . . . . Du mutisme.
- 2o. De la parole. . . . . { Du bégaiement. Du grassaiement, etc.
- 3o. Du chant. . . . . { juste. faux.
- 4o. De la déclamation.

Genre Ve.
Transmission. nerveuse.
- 1o. Transmission au cerveau des sensations. { générales. particulières.
- 2o. Transmission du mouvement. . . . . . { aux organes locomoteurs. aux organes vocaux.
- 3o. Mode de transmission.

De l'Intermittence des Fonctions de la Vie animale.

Sommeil.
- 1o. naturel . . . . { partiel . . . . . . { des sens. du cerveau . . . . . , Des sommeils sympathiques. des muscles. } général.
- 2o. contre nature.
- — 3o. Songes et somnambulisme.

FIGURE 9. Xavier Bichat (1800) published tables of tissue types and their properties which served to lay the ground-rules of the new science of histology. His work was made all the more remarkable by his persistent refusal to involve the microscope in his work

studied through tissue structure and his enthusiasms laid the groundwork for the modern science of histology. In his short career he dissected his way through 600 cadavers and identified 21 basic tissues (Fig. 9).

The paradox lies in the fact that Bichat did not use microscopes. He was unimpressed by them, believing them to distort reality, involve much subjective guesswork and to provide results that were not reliably repeatable. As Otis (2000) has commented, my earlier work has shown that many scientists who made key discoveries in pathology did not use microscopes at all. It is particularly curious that the father of a science which, in the modern world rests firmly on the expertise of the microscopist, himself turned away from microscopy.

In the decades following the eighteenth century, a swathe of microscopical discoveries was set to revolutionise neurology. Treviranus was to recognise the neurilemma of peripheral nerves using uncorrected microscopes in 1816; indeed derivatives of the Cuff microscope were used by many

investigators elsewhere in biology. Robert Brown carried out his work on the ubiquity of the cell nucleus, and on Brownian motion, using a similar simple microscope made by Bancks in London. We have already encountered Sir Everard Home using the original Leeuwenhoek microscopes for his micro-anatomical studies in the 1820s (*supra*). Indeed, Charles Darwin was still recommending a single-lensed, portable instrument as the microscope of choice as late as March 1848 (Ford 1985) (Figs. 10 and 11).

But from about 1830 achromatic microscopes were to become available. These were grander instruments, objects of desire, finely tooled from brass and easy to use. A burst of activity followed: Ehrenberg recorded ganglion cells in 1833, then Purkyně (1837) identified the neuron and in the following year His portrayed neurons with nuclei and dendrites. In due course Golgi developed in the kitchen at his home the staining system that bears his name and made the neuron visible to microscopists. It is important to note that it was

FIGURE 10. The Czech microscopist Johannes Evangelista Purkyně in his study at Prague during the 1850s. His microscope (right) which has much in common with the designs of W. and S. Jones (Fig. 5) has been identified for the author by Mr James Solliday of the Microscopical Society of Southern California, as being made by Plößl of Vienna

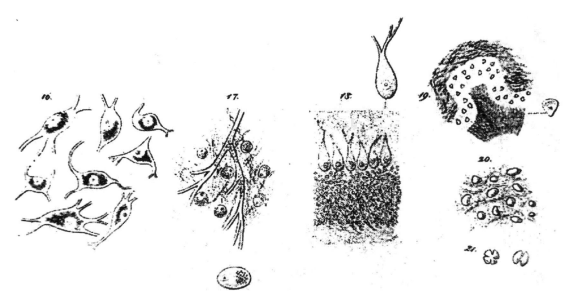

FIGURE 11. The table of nerve cells presented by Purkyně in his lecture to the Prague Congress of 1837 (Anon, 1962). The diagram numbered 18 depicts those now known as Purkyně cells. His talk gave a clear proposal of a cell theory (Körnchen), anticipating the work of Schwann. Purkyně also coined the term 'protoplasm'

primarily the newly developed staining techniques, rather than improvements in the microscope, that made neurological microscopy a reality. After a century in the doldrums, with these exciting new methods of specimen preparation and staining and with the achromatic microscope becoming generally available, microscopy of the nervous system was to embark upon its greatest era of expansion.

References

Adams, G. (elder) (1746). *Micrographia illustrata*. London: Adams.

Adams, G. (younger) (1787). *Essays on the microscope*. London: Adams.

Anon, (1932–present). *Antony van Leeuwenhoek, collected letters*. Amsterdam: Swets & Zeitlinger.

Anon (1962). *Jan Evangelista Purkyně*. Prague: State Medical Publishing House.

Baker, H. (1743). *The microscope made easy*. London: Dodsley, Cooper & Cuff.

Bichat, M. F. X. (1800). *Recherches physiologiques sur la vie et la mort*. Paris: Brosson, Gabon et Cie.

Bichat, M. F. X. (1800). *Treatise on the membranes in general, and of different membranes in particular*. Boston: Cummings and Hilliard, 1813. [Translated by John G. Coffin from the original, *Traité des membranes en général, et diverse membranes en particulier,* Paris].

Bichat, M. F. X. (1800). *Anatomie générale appliquée a la physiologie et a la médecine*, contenant les additions précédemment publiées par Béclard, et augmentée d'un grand nombre de notes nouvelles par F. Blandin. Paris: J. S. Chaud.

Borreli, G. A. (1681). *De Motu animalium*. Rome: Angelo Bernabo.

Clay, R. S., & Court, T. H. (1932). *The history of the microscope*. London: Charles Griffin.

Clerselier, C. (1664). *Descartes' L'Homme*. Paris: Angot.

Craanen, T. (1689). *Tractatus physico-medicus de homine*, etc., Leiden: Petrum vander Aa.

Dell Torre, P. D. G. M. (1776). *Nuove osservazione microscopiche*, Naples: con Licenza di Superiori.

Descartes, R. (1662). *De Homine figuris*. Leiden.

Disney, A. N., Hill, C. F., & Watson Baker, W. E. (1928). *Origin and development of the microscope*, London: Royal Microscopical Society.

Dobell, C. (1932). *Antony van Leeuwenhoek and his little animals*, London: John Bale, Sons & Danielsson.

Ford, B. J. (1981). The van Leeuwenhoek specimens. *Notes & Records of the Royal Society*, 36(1), 37–59.

Ford, B. J. (1982). Antony van Leeuwenhoek's sections of bovine optic nerve. *The Microscope*, 30:171–184.

Ford, B. J. (1985). *Single Lens – Story of the Simple Microscope*. London/New York: Heinemann/Harper & Row.

Ford, B. J. (1991). *The Leeuwenhoek legacy.* Bristol: Biopress & London: Farrand Press.

Gunter, R. T. (1961). Preface to reprint of *Micrographia, q.v.,* Dover Reprint, London: Constable.

Hooke, R. (1665). *Micrographia, or some physiological descriptions of minute bodies made by magnifying glasses,* etc., London: Martyn & Allestry.

Ledermüller, M. F. (1761). *Mikroskopische Gemüths- und Augen-Ergötzung, Bestehend, in ein hundert nach der Natur gezeichneten und mit Farben erleuchteten Kupfertafeln, sammt deren Erklärung,* Nürnberg: Selbstverlag.

Leeuwenhoek, L. van (1675). Study of bovine optic nerve published as Fig. 117, *Philosophical Transactions, 10.* See also the *Collected Letters* (supra).

Longet, F. A. (1842). *Anatomie et Physiologie du Système Nerveux de l'Homme et des Animaux Vertébrés,* Paris: Fortin, Masson.

McHenry, L. C. (1969). *Garrison's history of neurology,* Springfield, IL: Charles C Thomas.

Malpighi, M. (1686). *Omni Opera,* Londini: R. Scott.

Monro, A. (primus) (1732). *The anatomy of the humane bones. To which are added, an anatomical treatise of the nerves.* Edinburgh: William Monro.

Monro, A. (secondus) (1783). *Observations on the structure and functions of the nervous system.* Edinburgh: Creech.

Montucla, J-E. (1802). *Histoire des mathématiques,* Paris: Henri Agasse.

Otis, L. (2000). [disusses Brian J Ford work on achromatism and microscopy in] *Membranes: Metaphors of invasion in nineteenth-century literature* (p. 5). Baltimore, MA: John Hokpins University Press.

Prochaska, G. (1779). *De structura nervorum, tractatus anatomicus tabulis aëneis illustratis.* Vindobonoe: Groeffler.

Purkyně, J. (1837). Über die gangliösen Körperchen in Verschiedenen Teilen des Gehirns. *Berliner Versammlung deutscher Naturforscher:* 179.

Riese, W., 1959, *History of Neurology,* New York: MD Publications.

Ruestow, E. G. (1996). *The Microscope in the Dutch Republic.* Cambridge: University Press.

Wade, N. (2004). Visual science before the neuron. *Perception 33,* 869–889.

Wechsler, I. S. (1958). An Introduction to the history of neurology [pp. 701–735 in] *Textbook of clinical neurology.* Philadelphia/London: W. B. Saunders.

Young, J. Z. (1934). Structure of nerve fibres in *Sepia, Journal of Physiology,* 83: 27P–28P.

3
Corpus Curricula: Medical Education and the Voluntary Hospital Movement

Jonathan Reinarz

The centrality of hospitals to medical education is a relatively recent phenomenon in the history of medicine. Like many subjects in the history of medicine, connections can be traced to the eighteenth century, if not earlier. In order to understand significant changes in medical education, and especially in the field of anatomical instruction, one must look back even further, at least to the sixteenth century. The history of hospitals also has many turning points, including the fifteenth century, when such institutions began to proliferate, many more becoming principally dedicated to the sick. However, when these two subjects are considered jointly, the eighteenth century is not just significant, but central to the development of both institutions, especially in Western Europe. According to existing historiography, it was in this period that medical education and voluntary hospitals, at least in the United Kingdom, literally came together. The hospital was not only rapidly becoming the principal site for healing, but also one of learning about the sick and training prospective practitioners.

As historians have been quick to note, however, every history has its pre-history. For that reason, this chapter commences by considering some of the numerous false starts and birth pangs of hospital-based, or clinical, education in early modern Europe. It then considers eighteenth-century developments through the work of Hermann Boerhaave, among other less familiar staff at the University of Leiden medical school, who both embraced and popularised the clinical method of medical instruction, especially in the eighteenth century. Though not entirely an eighteenth-century figure,

Boerhaave's academic career is especially appropriate to the chronological parameters of this volume, having been appointed a lecturer in 1701.

From Leiden, many medical men took the lessons of Boerhaave and his colleagues to Paris, Vienna and across the channel and into the charitably funded, voluntary hospitals, the proliferation of which has repeatedly been identified as an eighteenth-century phenomenon, at least in England and Scotland (Porter, 1989, pp. 149–152). Rather than trace Leiden's influence on the development of medical education throughout Europe, the final section of this chapter will examine its impact on Britain. In particular it examines the way in which clinical training quickly developed in Edinburgh and, in successive decades, inspired London practitioners to attach schools to the hospitals to which they were affiliated. Though a new generation of provincial medical schools sought to retain local boys who might otherwise have travelled to hospital schools in Edinburgh and London, or even further afield, few instructors desired to change the way in which pupils were being educated. In less than a century, hospital-based instruction had become the tried and tested method of educating physicians.

Practical Pedagogy in Padua

Though traced to the ancient Greek and Roman physicians, bedside teaching was not incorporated into medical instruction at the earliest European universities. Instruction amongst the elite of the profession, as in other disciplines, including law and divinity, was largely theoretical in nature. For many practitioners, it was their textual learning of which they were most proud (Getz, 1995, p. 78). Although often limited to lecture notes, texts gave practitioners and students theory and helped them justify their medical practices. More importantly, it separated the learned practitioner from the empiric. Not surprisingly, the leading medical minds in the medieval period were translators, and medical education, for the university-educated physician at least, became text-based.

While physicians were trained at the early universities, surgeons and apothecaries, as in most manual, craft-oriented occupations, acquired their medical knowledge by way of apprenticeship. This involved the payment of a fee that bound a young man to a qualified practitioner, allowing him to learn his trade in a very practical and personal manner, actually living in the home of his master. On average, the period of training lasted between 5 and 7 years, during which time the practitioner conveyed well-established medical techniques, as well as acceptable professional behaviour. More importantly, those completing their period of training received an actual certificate from their master attesting to their skill (Lane, 1996). Though a pupil's education depended very much on the master to whom he was indentured, training also shifted with the culture and politics of the time, with monarchs and princes founding new universities and non-academic faculties that were empowered to licence local practitioners. For example, in England in 1518, Henry VIII granted a charter to the Royal College of Physicians, which controlled medical practice within 7 mile of London. In 1540, the barbers and surgeons were also incorporated into their own guild, or company, as is depicted in Holbein's famous portrait of the event. English apothecaries were granted a similar charter in 1617.

The advantages of this system are easy to see. Practitioners taught medicine because it allowed them to expand their practices, delegating many undesirable tasks to their pupils and benefit from their cheap labour. As this suggests, apprentices' experiences varied with each practice, some getting a very good training, others spending too long cleaning and manufacturing medicines and hardly any time with patients. This extreme variation remained a characteristic of medical education for many more centuries and no doubt contributed to the tremendous gap between upper and lower classes of surgeons in the seventeenth and eighteenth centuries (Lane, 1985, p. 76).

During the Renaissance the value of practical work began to be recognised by a number of university-based medical teachers, beginning with anatomists. Founded in 1212, the University of Padua was central to these changes in Europe. An important school from its first days, its reputation only increased with the appointments of Vesalius and Fallopius in the sixteenth century, the former having been the first to descend from the instructor's rostrum and take the knife into his own hands (O'Malley, 1970, p. 97). The influence of the school's anatomists very quickly extended to the rest of Europe. In England alone, 57 of the Fellows named on the Roll of the Royal College of Physicians in London had taken a degree in Padua (Guthrie, 1959, p. 108). The College's founder, Thomas Linacre, himself studied in Padua, taking his degree in 1492. Half a century later, John Caius, who would later also obtain a Cambridge MD, studied at the bedside of patients at the Hospital of San Francesco (Padua) under Giovanni Batista da Monte (1498–1551), known to many as Montanus (O'Malley, 1970, p. 93). Though Montanus's bedside methods would influence instruction outside Padua, many of his students, including Caius, appeared to value predominantly the anatomical instruction they received under his tutelage.

Following Montanus's death in 1561, clinical teaching declined somewhat while only dissection grew in popularity, especially among his former pupils. Inspired by anatomical work undertaken at Padua, Caius managed to obtain a royal grant in 1565, like that granted to the Company of Barber-Surgeons in 1540, permitting the Royal College of Physicians to dissect the bodies of two executed felons annually. Over the next two centuries, physicians began to work with their hands more regularly, while surgeons began to adopt aspects of the physician's theoretical form of training through lectures and texts. Though its place in the history of clinical medicine was secure, Padua's popularity declined with the Reformation, all non-Catholics having been excluded from Italian universities by Papal edict. Instruction at other schools by this time had become equally concerned with accurately examining and recording the disease phenomena at the patients' bedside.

Enter the Clinic

Inspired by Newton's *Principia* in 1687 and other works in the physical sciences, late-seventeenth-century surgeons attempted to discover the laws that governed the operation of disease and to classify these according to proven scientific principles. Not surprisingly, the literature and language of these practitioners became that of physics and mechanics, while the bedside increasingly became the appropriate place from which to observe the sick body. Nevertheless, another century was to pass before the Italian example of clinical teaching was replicated in other European countries.

Following Descartes in imagining the body as a machine, with laws that could be uncovered like those of the physical sciences by careful observation, a group of Paduan alumni in Northern Europe sought to replicate their masters' mode of instruction. Among these, the Dutch were pre-eminent. In his inaugural address (1636), Willem van der Straten (1593–1681) announced that medicine at the newly founded University of Utrecht would be taught at the bedside in the city's hospital. The following year, clinical lectures appeared in the syllabus at Leiden, where instructors feared a decline in student numbers, though only few taught with enthusiasm in the prescribed manner. These included Pieter Paaw (1564–1617) who, like William Harvey, studied under Fabricius. Paaw was Professor of Anatomy at Leiden when the school's first anatomical theatre was built in 1597. Another graduate of Padua was Johann van Heurne (1543–1601), who it is claimed was responsible for introducing the clinical method of teaching in Leiden. Interestingly, Padua did not actually offer clinical instruction during van Heurne's time there (Lindeboom, 1968, p. 284), so his influences clearly extended beyond his alma mater.

From their commencement, clinical lectures at Leiden were given in the old St. Caecilia 'Gast-en Proveniershuis', a former nunnery that had also served as a pesthouse and a madhouse before being transformed into a municipal hospital. From its records, it appears that two of the institution's six wards were regularly used for teaching, the increasing frequency of student visits having occasionally disturbed the working of the wards. In most cases, the earliest clinical courses lasted 5–10 years before being interrupted by death or retirement

of the tutor. A noticeable change came in 1660 with the appointment of Franz de la Boe, or Franciscus Sylvius, who made his rounds daily as opposed to twice weekly as had been common among his predecessors. The Socratic form of questioning which characterised Sylvius's teaching came to an end with his death in 1672, leading clinical instruction once again to become an intermittent part of medical education in Leiden.

In spite of earlier intermittent manifestations of the practice, clinical teaching only began to flourish in the eighteenth century, beginning with Boerhaave in 1714. Assisted by Herman Oosterdijk Schacht (1672–1744) from 1720, clinical teaching was carried out faithfully at Leiden 6 months of the year. Twice a week, on Wednesday and Saturday, the two town physicians would pick up the professor of medicine at his home and proceed to the hospital, where the students were already assembled (Lindeboom, 1968, pp. 287–288). Passing between the hospital's 12 teaching beds, the instructor used carefully chosen cases to display textual knowledge and underscored the importance of seeing real patients during medical studies. This is emphasised in Boerhaave's own writings, which, for example, advise students to consult textual works on the brain 3 or 4 days before actually undertaking a practical examination (Boerhaave, 1719, p. 205). At the conclusion of his ward demonstrations, the professor would join the matron and master of the hospital in the latter's apartment to determine patients' diets and drugs. Thus medical instruction was not just central to medical education in Leiden, it was also central to hospital treatment.

Students carefully recorded their instructors' comments at demonstrations. Surviving examples demonstrate that some patients were discussed repeatedly at different dates. They also reveal Boerhaave to have been true to the Hippocratic tradition, paying less attention to doctrines and systems than to signs and symptoms. In fact, it was Boerhaave who restored Hippocrates to students' reading lists at this time (Johnson, 1903, pp. 154–184). Drawing attention to the general appearance of patients, as well as nutrition, pulse and respiration, he regularly refined diagnostic methods, using a pocket lens and even a thermometer to determine the presence of fever. Though it has been said that Boerhaave relied on a limited therapeutic armamentarium (Snapper, 1956, p. 71), as a Hippocratic physician he treated patients carefully and as individuals, relying on drugs, diet, and basic physiotherapy. For example, faced with a patient showing cerebral symptoms and epileptic fits, he might advise the head to be sprinkled with water for half an hour, followed by a light massage (Lindeboom, 1968, p. 299).

Though Boerhaave's therapies were questioned a century later, his clinical sessions effectively completed the educations of his students; each student had learned the basic sciences before being introduced to physiology and pathology and then being conducted through the hospital's two wards where they learned to manage illness in a very practical manner. As Boerhaave's biographer has emphasised, 'nowhere else in Europe was such a medical curriculum available at such a level' (Lindeboom, 1968, p. 292). It is for this reason that so many foreign students came to learn in Leiden. As importantly, students were self-selecting, the university having opened its doors to all who came, regardless of race, nationality or religion. In turn, the example set by Leiden encouraged closer links between universities and hospitals, students being provided greater opportunities to observe patients and corpses. At the very least, university education had become more eclectic and the traditional divide between surgeons and physicians began to blur.

Leiden's Legacy

Over many generations, the small school in Leiden turned out what has been described as the elite of the medical profession. Boerhaave alone trained many prominent practitioners who went on to occupy key positions at medical schools and hospitals across Europe. These included Albrecht von Haller (1708–1777), who is famed for his work on the irritability of muscle tissue and brought similar fame to the new University of Goettingen; Gerhard van Swieten (1700–1772), who reconstructed the medical school of Vienna; and Alexander Munro, *primus* (1697–1767), a central, but by no means the only, figure in the foundation of the medical faculty at Edinburgh (Emerson, 2004: 183–218; Guthrie, 1959, p. 112). Haller himself rightly referred to Boerhaave as 'communis Europae praeceptor', the common teacher of Europe' (Lindeboom, 1968, p. 3). Though not listed separately in student registers,

Scottish students attending Leiden are estimated to have numbered 546, comprising a quarter of English-speaking medical students. During the time of Boerhaave (1701–1738), this proportion reached a third of students arriving from across the channel (Guthrie, 1959, p. 109).

Not surprisingly, Edinburgh very quickly became the centre of these new ideas in Britain, where medical teachers were attracted to the idea of the 'proto-clinic', as well as to the mechanical systems expounded by Boerhaave, soon after the establishment of its medical faculty in 1726. Associated with Alexander Munro in the faculty at Edinburgh were four other professors: John Rutherford (1695–1779), Andrew Plummer (1698–1756), Andrew St. Clair (1699–1760) and John Innes (1696–1733). All had been students at Leiden with Munro, except Rutherford who spent only a year there in 1718. As a result, the basic studies of botany, chemistry and anatomy in Edinburgh can be said to have been founded on knowledge obtained in Leiden (Guthrie, 1959, p. 115).

So too was the commitment of its instructors to clinical teaching. A formal course of clinical lectures had been commenced by Rutherford in 1748 in the hospital's amphitheatre to supplement limited bedside teaching on the wards of the Edinburgh Infirmary, founded nearly two decades earlier. More detailed and useful to students, Rutherford's lectures were judged a success and prompted the hospital's managers to open a special ten-bed teaching ward, which, as at Leiden, was kept open for half the year on the same two weekdays on which Boerhaave had made his rounds. By 1756, Rutherford was joined by three other professors, Alexander Munro, *secundus*, William Cullen and Robert Whytt. Each professor agreed to lecture twice a week for 5 weeks, the entire course being expanded to 6 months. The teaching ward was accordingly enlarged to 20 beds in two separate wards to accommodate an equal number of male and female patients (Risse, 1986, pp. 242–244).

Over the next decade, it was the physician and classifier William Cullen who would become the school's most influential clinical instructor. Following the death, retirement and departure of his colleagues, Cullen shared the teaching duties with John Gregory by the late 1760s. When the latter withdrew due to ill health early in 1772, Cullen carried the burden alone, taking on a gruelling teaching schedule. Besides being Edinburgh's most distinguished physician, Cullen carefully prepared his lectures from day to day, usually lecturing twice a week, though now on Tuesday and Friday evenings (Risse, 1986, p. 244). Dividing diseases into four classes (remnants of humoral theory perhaps), Cullen, like Boerhaave, treated each case as a distinct entity, while encouraging students to be guided by their own senses (Harrison, 2004, p. 55).

Interestingly, while hospital records appear to have had worse survival rates than patients in the eighteenth century, those from the Edinburgh Infirmary between 1770 and 1800 exist and give an indication of the types of cases students might have been presented with during their studies. Nervous and mental diseases, we find, comprised approximately 6% of all authorised hospital entries in these years (Risse, 1986, p. 124). This was roughly equivalent to the numbers of musculoskeletal and digestive disorders seen by staff during this period, but considerably less than genitourinary and infectious diseases, the most common cases, which comprised 20.6 and 15.6% of patients, respectively.

Among the nervous disorders admitted, the most common affliction was paralysis, or palsy (38 cases), which Cullen referred to 'as consisting of a loss of the power of voluntary motion', caused by an 'interrupted influx of the nervous power' and affecting certain parts of the body (Risse, 1986, pp. 153–154). Today, many of these cases might more commonly be described as the result of stroke, Parkinson's disease or advanced joint or arthritic diseases. Other ailments were cephalgia, or headache (24 cases), which, like hysteria (38), was deemed to afflict primarily women. Epilepsy (20), which Cullen placed in the category of neuroses, affected children and young adults, and regularly turned up in the teaching ward because of the therapeutic challenges posed by the associated fits (Risse, 1986, p. 154). Other cases, including chorea, vertigo, mania and melancholy, were seen in both the ordinary and teaching wards, though on only three or fewer occasions during the last three decades of the eighteenth century.

Unlike other instructors, Cullen did not deliver his clinical lectures in Latin, and is regarded as the first medical instructor in Great Britain to have lectured in the English tongue (Beekman, 1950a, p. 75). At the small clinic at the Royal Infirmary, sight soon took precedence over all other senses,

and another generation inspired by continental physicians attempted to correlate symptoms with anatomical evidence found in autopsy, rather than simply classify disease as Cullen had done. Though candidates for the university's MD degree had been regularly interrogated on practical medical matters since 1767, a decade later, in the 1780s, the official curriculum specifically included clinical lectures, with 50–55 students, or 20% of registrants, enrolling in the clinical course annually (Risse, 1986, p. 245).

London's Teaching Hospitals

London was not immune to these changes in medical education. Medical apprentices had long walked the wards of hospitals during their periods of indenture. In return for their voluntary services at hospitals, physicians and surgeons were permitted to take their own apprentices on ward rounds, as was first recorded at St. Bartholomew's in 1662 (Waddington, 2003, p. 18). As in Edinburgh, a number of influential London practitioners advocated the benefits of personal experiences in medical education. Nevertheless, medical staff at this and the other three hospitals in the metropolis – St. Thomas's, Bethlem and Christ's – appear to have been unusually conservative and committed to text-based learning, most early consultant posts having been occupied by Oxford-trained physicians. The most successful often resigned after short tenures to pursue private practice. Because of the preference for text-based tuition, library-study was prioritised over practical methods of instruction at Bart's in 1668 (Waddington, 2003, p. 24). Having failed a great proportion of the public during the recent plague outbreaks, practitioners, it has been argued, were more eager to elevate medicine and emphasise its scholarly nature. Additionally, ward rounds, when given, tended to be unstructured and remained fairly random in nature. Nevertheless, though unstructured, the flexible nature of hospital policy regarding students would permit enterprising medical staff to develop clinical training over the next generation.

By the first decades of the eighteenth century, hospital practitioners in London began to systematise training, offering lectures and establishing libraries and museums. As early as 1722, the staff at St. Bartholomew's had begun to push for the establishment of a dissection room (Waddington, 2003, p. 25). Already offering their students private instruction in anatomy, medical staff at the institution negotiated the allocation of two rooms for post-mortem examinations 4 years later. Shortly afterwards, Edward Nourse, an assistant surgeon at the institution, introduced lectures at the hospital. In an advertisement appearing in the *Evening Post* in 1734, Nourse claimed that, 'desirous to have no more lectures at my own house, I think it proper to advertise that I shall begin a course of Anatomy, Chirurgical Operations and Bandages on Monday, November 11, at St. Bartholomew's Hospital' (Waddington, 2003, p. 33). As this perhaps indicates, some surgeons made far greater use of hospitals for educational purposes than did most physicians (Bynum, 1985, pp. 118–119). Additionally, teaching still very much depended on personal initiative.

Though Nourse's course, like many delivered earlier on the continent, terminated shortly afterwards, a concerted effort was made by his apprentice, Percival Pott, to reintroduce teaching at the hospital in 1745. Two decades later, Pott transferred the lectures from his home to St. Bartholomew's, where he already offered bedside teaching to his apprentices. Similar instruction commenced at Guy's, one of the five new general hospitals built in London between 1720 and 1745, in 1770, and where many Edinburgh graduates filled staff positions (Ripman, 1951, pp. 59–60).

Although London would become better known as a 'lecturing empire' (Hays, 1983, pp. 91–119), the institutional structure of its medical schools clearly began to emerge in the middle decades of the eighteenth century. Even to those who might not have favoured such developments, including many hospital governors, the establishment of teaching faculties made economic sense, the pupils of medical practitioners having in most cases proved themselves an important source of free labour. Additionally, many practitioners quickly learned that teaching could significantly increase income, usually supplementing annual earnings by at least 10% (Gelfand, 1985, p. 148). Others emphasised that greater contact with the sick poor in these institutions would ultimately benefit medical knowledge (Bynum, 1985, p. 117).

In contrast to other English voluntary hospitals, the admissions policy at Bart's was comparatively liberal. Excluding only smallpox cases, the hospital's

wards permitted students to see the full range of medical and surgical cases; unfortunately, patient registers for this period do not exist and an exact breakdown of cases as exists for Edinburgh is not possible. Those who paid for the privilege, usually about £50, were permitted to accompany the physician or surgeon on an initial examination. Students also attended the daily ward rounds, which usually took place between 10 in the morning and 2 in the afternoon, after which time students were to vacate the wards. In addition to the basic structure of rounds, students were left to make the most of their opportunities to observe clinical cases. While much valuable information was acquired on the wards, clinical teaching had its share of problems. For example, surgical students rarely entered the medical wards unless they paid an additional fee to become a physician's pupil; neither were those cases that came before students always carefully presented, surgeons and physicians at all hospitals often having to make superficial examinations during their rushed rounds.

Though capable instructors clearly existed, enthusiasm for instruction was not always widely shared. With the most popular teachers attracting nearly a hundred students, only a few privileged students normally secured an unobscured view of patients and operations. In the third decade of the eighteenth century, when access to the teaching wards was extended, they became even more crowded (Waddington, 2003, pp. 49–51).

While many hospital governors in London, as at other hospitals, such as in Leiden at the beginning of the century, did not fully appreciate the benefits of hospital instruction, limitations imposed on fee-paying pupils were being removed. As wards grew more crowded, substantial sums began to be generated by some private instructors. Private instruction, whether in Paris or London, held a particular appeal to provincial and foreign students, who had money but little time to spend walking hospital wards. Unlike apprentices, students paid to acquire knowledge and skills from a master, not to undertake menial tasks in the home and practice of their instructor. Ambitious medical teachers had an open market for their services. Private teaching represented the triumph of the liberal economic values of free competition in medical education. In this respect, London hospitals served the needs of late-eighteenth-century clinical instruction better than their Paris counterparts because they better fit the prevailing private enterprise style of medical education (Bynum, 1985, p. 135). While French hospitals functioned as authoritarian and bureaucratic institutions, British hospitals relied on private initiative for their establishment and management (Gelfand, 1985, p. 137). Before century's end, four of the city's hospitals – St. Bartholomew's, the London, St. Thomas's and Guy's – housed recognised schools of medicine. Additionally, instruction remained flexible, permitting students to organise their own programme of studies, see medical patients and even enrol in private extramural courses.

Perhaps the most famous of these courses were those run by two Scots, William and John Hunter. William, a physician, and John, a surgeon–anatomist, taught their methods from an influential private anatomy school set up by William in Covent Garden, London in 1746. Apprenticed to William Cullen in 1736, William Hunter gained much of his medical knowledge at the bedside (Beekman, 1950a, p. 77). He then attended the University of Edinburgh where he learned the new methods of anatomy instruction, as well as the art of making anatomical preparations. Proceeding to London in 1740, he attended a further series of lectures with Frank Nicholls, the leading teacher of anatomy in the metropolis at the time (Beekman, 1950b, p. 182).

Critical of most courses he attended, Hunter productively redirected his frustrations and made the teaching of anatomy his career. In particular, he advocated that students should be provided the opportunity to investigate the anatomy of the human body personally, as he had experienced himself upon entering a course in anatomy in Paris under M. Antoine Ferrein (Beekman, 1950b, pp. 184, 192).

Returning to London in 1744, he commenced arrangements to establish his own school of anatomy in Covent Garden, a task made easier by the dissolution of the Company of Barber Surgeons in 1745, which had prohibited the dissection of human bodies outside the surgical guild's hall, not to mention the new College's hesitancy to found a practical school (Cope, 1959, p. 8). By September 1746, his first course, announced in the *London Evening Post*, earned him 70 guineas (Beekman, 1950b, p. 195). So highly was his anatomical instruction valued, his instructor at Edinburgh, Alexander Munro, sent his two sons to London so they might attend Hunter's lectures as a 'finishing touch to their education' (Beekman, 1950a, p. 84).

Soon many more schools opened and stimulated a trade in dead bodies, which inevitably attracted some scandal. Many were run by former pupils of the Hunters, including Joshua Brookes, who ran a school in Great Marborough Street, and John Sheldon, who opened a school in Great Queen Street, while another student, William Shippen Jr. established a lecture series in America, which imported and promoted the hands-on approach of the Hunters.

By 1780, 16 anatomical courses were advertised in the London papers; this rose to 57 in 1812 (Lawrence, 1988, p. 178). The promise of marketable skills, however, outweighed risks of social disapproval (Lawrence, 1995, p. 200). Attendance soared at Hunter's and other private schools and a new generation of students could stamp their regular training through their easy use of precise anatomical terms. Though medicine again appeared akin to a craft, alongside their new medical knowledge, surgeons in particular cultivated their social skills thereby making dissection acceptable.

As some historians, including Ruth Richardson have noted, all bodies of executed felons were offered to the Royal College of Physicians in 1752, but even this failed to keep up with demand, especially when pathological anatomy entered a new era with the work of Morgagni (1682–1771), who elucidated the relationship between the case history and morbid anatomy through post-mortem investigation. Frenchman Xavier Bichat (1771–1802), who laid the foundations for nineteenth-century pathological anatomy, took this one decisive step further with his doctrine of tissues, which he proposed were the analytical building blocks of anatomy, physiology and pathology (Richardson, 1993). Regularly conducting autopsies in his private Parisian anatomy classes, Bichat combined the scrutiny of dead bodies and specifically changes in bodily tissues in order to throw light on the causes of death (Foucault, 1993, p. 146). This new form of clinical training, soon reinforced by the introduction of new diagnostic tools, including the stethoscope, not to mention the low cost of a Paris education, attracted many foreign students to the French capital and gradually displaced the more individualistic, patient-centred medicine of the eighteenth century (Jewson, 1976, p. 238).

By the last decades of the eighteenth century, however, London's reputation as a centre for medical education had increased and had even begun to displace Paris as the capital of clinical education. According to one American student visiting London in the 1780s, the city had become 'the Metropolis of the whole world for practical medicine' (Gelfand, 1985, p. 138). Even the Paris surgeon Jacques Tenon (1724–1816), the leading European authority on hospitals, alluded to the advantages of British hospitals for medical education (Gelfand, 1985, p. 138).

Alongside such developments, private schools flourished, many becoming, as John Hunter would eventually prove, centres of research excellence with their vast anatomical collections. Having effectively assimilated the French clinical experience, many London medical instructors, like William and John Hunter, imported and adapted the Paris educational experience for consumption at home, making it less necessary for future generations to cross the channel. As a result, at exactly the same time that William Hunter's Great Windmill Street School had attained an international reputation, the Paris Academy of Surgery appeared on the wane (Gelfand, 1985, p. 143).

Meanwhile, the foundation and success of private anatomy schools in London only encouraged other practitioners to organise their teaching more effectively. Although a systematic form of instruction had been introduced at St. Bartholomew's by Percival Pott, it is the surgeon John Abernethy, a pupil of John Hunter, who is credited with having transformed the hospital into a school. Appointed to the post of assistant surgeon on the retirement of Pott in 1787, Abernethy not only wished to develop instruction at the hospital, but also had both an eccentric style and Parisian approach that appealed to the students. Originally delivered in his home, Abernethy's lectures were transferred to the hospital upon the completion of a 'Surgeon's Theatre' in 1795. With time and the cooperation of other practitioners, a comprehensive curriculum was eventually developed. So successful was instruction, that, by the second decade of the nineteenth century, surgical students alone numbered several hundreds. When a new anatomical theatre was opened in 1822, more than 400 people attended to hear Abernethy's inaugural lecture (Waddington, 2003, p. 39). While Hunter's pupils carried on his

teaching methods, the surgeon's greatest legacy was his museum, which was acquired by the College of Surgeons in 1799, thus providing its members with a world-class comparative anatomy resource. The following year, the renowned guild of surgeons acquired its 'royal' title.

By 1800, private anatomy schools continued to offer much competition to the hospital schools. Staff at hospitals, who held appointments in the Royal College of Surgeons, used their power to put an end to this competition. Rather than draw up a syllabus or requirements for medical candidates, the College refused to accept certificates of attendance on the private schools' popular summer anatomy courses (Cope, 1959, p. 43). It was the Apothecaries' Society that led the way in 1815 and encouraged a more organised form of teaching. In this way, it raised the status of the profession by creating a recognised qualification – the Licentiate of the Society of Apothecaries. Besides specifying that all apothecaries must hold the new degree, the Apothecaries Act (1815) also emphasised the central role played by hospitals in medical education, requiring students to fulfil 6 months' clinical work at a recognised institution. Two years earlier, besides courses on anatomical and surgical lectures, the Royal College of Surgeons had demanded a year's attendance on the wards of a London hospital before candidates were permitted to sit for their licence; this was reduced to 6 months after 1828.

In 1819, the College of Surgeons issued their first printed curriculum and for some years continued to develop their proposed curriculum on similar lines to those followed by the Society of Apothecaries (Cope, 1959, p. 135). Unlike the Apothecaries, however, in 1823, the surgeon's guild further hindered the growth of provincial schools by recognising only those in London, Dublin, Edinburgh, Glasgow and Aberdeen. Between 1821 and 1824, the Royal College of Surgeons passed a series of resolutions that recognised only those anatomy courses that were conducted in hospitals by a College examiner. When further limitations to hospital practice were introduced, both the College and the London teaching hospitals came under the attack of Thomas Wakley in a feud that unfolded in the pages of the *Lancet* (Cope, 1959, p. 44).

Despite the College of Surgeons making some concessions in 1826 and eventually removing their restrictions on provincial schools in 1829, the first decades of the nineteenth century saw the position of the London hospital schools, such as Bart's, greatly strengthened. Significantly, the Great Windmill Street School closed in these same years. A decade later, in 1838, when an attempt was made to standardise the surgical curriculum throughout the United Kingdom, among many other requirements, the Colleges of Surgeons of London, Edinburgh and Dublin agreed to a minimum of 3 years' study in a recognised school of surgery and 21 months' work in a hospital. Though it has been suggested that the decline of the private anatomy schools marked the end of an era in which private enterprise dominated advanced medical learning (Gelfand, 1985, p. 151), the private initiative which had led to the emergence of the London hospital schools was equally apparent in the development of provincial medical schools. As in the case of London, these too were quick to build affiliated teaching hospitals, and medical students would in future spend only proportionately greater periods of time on the wards of these institutions.

Concluding Remarks

Although alternatives to hospital-based training have been suggested in the past, as well as more recently, the combination of medical education and hospital instruction that unfolded throughout the eighteenth century continues to appear both inseparable and essential to the education of medical practitioners. While historians continue to uncover early evidence of bedside training in hospital environments, the sixteenth century appears to mark a distinctive change in the form of both medical education and the place of the hospital in this process. The seventeenth century witnessed the emergence of a new mechanistic model of the body which spurred both anatomical and physiological investigation. By the eighteenth century, medical education, formerly conducted on a one-to-one basis between master and pupil, was literally brought into the hospital and links between the two institutions grew only stronger throughout the century. As Risse has argued, as early as the 1770s, the role of hospitals in medical education seemed assured (Risse, 1986, p. 240). With the emergence of the provincial medical schools in the 1820s, the location of medical instruction in England may have changed yet again, but the hospital's pedagogical functions were cemented only more firmly.

Acknowledgements. The author is grateful to the volume's editors for their comments on earlier versions of this chapter. Equally, thanks are owing to Stephen Casper and Keir Waddington for their numerous helpful comments and suggested readings, as well as to Katie Ormerod for information concerning the archives of St. Bartholomew's Hospital.

References

Beekman, F. (1950a).William Hunter's early medical education, Part 1: His studies under William Cullen and at Edinburgh with Alexander Munro. *Journal of the History of Medicine and Allied Sciences, Winter*, 72–84.

Beekman, F. (1950b). William Hunter's early medical education, Part 2: He goes to London to complete his studies and remains there. *Journal of the History of Medicine and Allied Sciences, Spring*, 178–195.

Boerhaave, H. (1719). *A method of studying physick.* London: C. Rivington.

Bynum, B. (1985). Physicians, hospitals and career structures in eighteenth-century London. In W. F. Bynum, & R. Porter (Eds.), *William Hunter and the eighteenth-century medical world* (pp. 105–128). Cambridge: Cambridge University Press.

Cope, Z. (1959). *The Royal College of Surgeons of England: A history.* London: Anthony Blond.

Emerson, R. L. (2004). The founding of the Edinburgh Medical School. *Journal of the History of Medicine and Allied Sciences, 59*, 183–218.

Foucault, M. (1993). *The birth of the clinic.* London: Routledge.

Gelfand, T. (1985). 'Invite the philosopher, as well as the charitable': hospital teaching as private enterprise in Hunterian London. In W. F. Bynum, & R. Porter (Eds.), *William Hunter and the eighteenth-century medical world* (pp. 129–151). Cambridge: Cambridge University Press.

Getz, F. (1995). Medical education in later Medieval England. In V. Nutton, & R. Porter (Eds.), *The history of medical education in Britain* (pp. 76–93). Amsterdam: Rodopi.

Guthrie, D. (1959). The influence of the Leiden School upon Scottish medicine. *Medical History, 3*, 108–122.

Harrison, M. (2004). *Disease and the Modern World.* Cambridge: Polity Press.

Hays, J. N. (1983). The London Lecturing Empire, 1800–50. In I. Inkster, & J. Morrell (Eds.), *Metropolis and Province: Science in British culture, 1780–1850*. London: Hutchinson.

Jewson, N. D. (1976). The disappearance of the sick man from medical cosmology, 1770–1870. *Sociology, 10*, 225–244.

Johnson, S. (1903). *The Works of Samuel Johnson* (Vol. 14). New York: Pafraets.

Lane, J. (1985). The role of apprenticeship in eighteenth-century medical education in England. In W. F. Bynum, & R. Porter (Eds.), *William Hunter and the eighteenth-century medical world* (pp. 57–103). Cambridge: Cambridge University Press.

Lane, J. (1996). *Apprenticeship in England, 1600–1914*. London: UCL Press.

Lawrence, S. (1988). Entrepreneurs and private enterprise: The development of medical lecturing in London, 1775–1820. *Bulletin for the History of Medicine, 62*, 171–192.

Lawrence, S. (1995). Anatomy and address: Creating medical gentlemen in eighteenth-century London. In V. Nutton, & R. Porter (Eds.), *The history of medical education in Britain* (pp. 199–228). Amsterdam: Rodopi.

Lindeboom, G. A. (1968). *Hermann Boerhaave: The man and his work*. London: Methuen & Co.

Nutton, V., & Porter, R. (Eds.), (1995). *The history of medical education in Britain*. Amsterdam: Rodopi.

O'Malley, C. D. (1970). Medical education during the Renaissance. In C. D. O'Malley (Ed.), *The history of medical education* (pp. 89–102). Berkeley: University of California Press.

Porter, R. (1989). The gift relation: Philanthropy and provincial hospitals in eighteenth-century England. In L. Granshaw, & R. Porter (Eds.), *The hospital in history* (pp. 149–178). London: Routledge.

Richardson, R. (1993). *Death, dissection and the destitute*. London: Routledge.

Ripman, H. A. (1951). *Guy's hospital, 1725–1948*. London: Guy's.

Risse, G. (1986). *Hospital life in Enlightenment Scotland*. Cambridge: Cambridge University Press.

Snapper, I. (1956). *Meditations on medicine and medical education, past and present*. New York: Grune & Stratton.

Waddington, K. (2003). *Medical education at St Bartholomew's hospital*. Woodbridge: Suffolk, Boydell.

4

Some Thoughts on the Medical Milieu in the Last Quarter of the Eighteenth Century as Reflected in the Life and Activities of James Parkinson (1755–1824)

Christopher Gardner-Thorpe

Introduction

Many contributed to the advances in science and the arts during the latter part of the eighteenth century and prepared the way for thoughts and apparatus that would change the lifestyle of those who followed. Among those thinkers and practical persons was the apothecary James Parkinson (1755–1824). He was an accomplished doctor, well liked in his area of east London, and he contributed in several fields of medicine.

Parkinson is best known now for his *An Essay on the Shaking Palsy* (Parkinson, 1817) published in 1817. Essentially his contribution to neuroscience was just this one publication, one of the most famous in the whole of medicine. The description was clinical, based upon his observation of six persons, three being patients of his and the other three passers-by in the street. His classical description has stood the test of time – *involuntary tremulous motion, with lessened muscular power, in parts not in action and even when supported; with a propensity to bend the trunk forwards, and to pass from a walking to a running pace: the senses and intellects being uninjured.* This description of the shaking illness that was to be named after him by Jean-Martin Charcot (1825–1893), the famous Parisian neurologist, is his best-known medical contribution although he did publish on rabies, the prevention of head injury in children and other topics.

However, his medical work is far from his only claim to fame. Like others before and after him,

Parkinson contributed in many fields – in his case his interests lay in science, politics, the church and geology. This was not so unusual at the time and many enquiring minds roamed the fields of natural philosophy, expanding into various fields of science – physics, chemistry, biology, geology and more – as the decades passed. He was a polymath among other contemporaries.

Early Life

It is not known exactly where James Parkinson was born but it seems likely in his family home in the up-and-coming area of East London named Hoxton. After all, that was where his parents lived and home births were the norm. His father, John (1726–1784), a well-known surgeon, was prominent in the medical politics of the day for he was one of the two Anatomical Wardens of the Surgeons Company, an organisation that was powerful in its era (1745–1800) when the Worshipful Company of Barbers, one of the City Guilds, was becoming less influential and the Royal College of Surgeons was soon to become a major meeting ground for surgeons.

John lived with his wife, Mary, at 1 Hoxton Square when James was born on 11 April 1755. Of Mary, little is known. James was to gain two younger siblings, William (1761–1782) and Mary Sedgwick (1763–) and here the children grew up in a medical household.

Medical Man

James became apprenticed to his father, John. After perhaps a couple of years he joined other students at the London Hospital (now The Royal London Hospital) in the Mile End Road. Sir William Blizzard (1743–1874) and Dr. James Maddocks (–1786) had founded The London Medical College. The Minutes of the House Committee held on Tuesday, 20 February 1776 record: *The Committee admitted Mr. James Parkinson to be a dressing pupil for six months ensuing in this hospital, he being recommended by Mr. Grindale one of the surgeons of this Charity, the Chairman read to him the usual Charge whereunto he signed his Assent.*

These students also attended the private medical and surgical lectures of John Hunter (1728–1793), brother of William Hunter (1718–1783) physician and obstetrician whose collection forms the basis of the Hunterian Museum in Glasgow, at the Great Windmill Street School in London, later to become the Windmill Theatre in Soho's theatre land. John Hunter was also an enthusiast in geology and fossils, and probably whetted Parkinson's appetite.

Parkinson distinguished himself in many ways medically. A year after he was apprenticed at The London, on 28 October 1777, The Royal Humane Society awarded him a silver medal for the resuscitation of Brian Maxley. Maxley had hanged himself and, pulseless, appeared dead. The pulse returned after 30 minutes, respiration within 40 minutes and he regained consciousness in 90 minutes. The Society usually gave medals to those who refused fees: the scale was two guineas to the doctor for 2 hours of resuscitation, one guinea to the landlord of the public house where the patient was revived, 2/6d to the messenger and two guineas to the medical assistant when the patient recovered. On another occasion Parkinson treated a girl aged 14 months who was found face downwards in a tub of water such that her parents thought her dead but the child was rubbed with salt and brought to Parkinson who treated the child successfully.

Parkinson developed a keen sense of medical politics that is not surprising in view of his later political leanings and his dedication to his profession. Although it is not certain, it seems likely that *Observations on Dr. Hugh Smith's 'Philosophy of Physic'*, published in 1780, was his work (Parkinson, 1780).

In 1781 at the age of 26, Parkinson married Mary Dale who was well versed in medical life as her two uncles, grandfather and great uncle were medical men. We do not know where the newly married couple lived but they were to have six children though sadly, and common in those days, two were to die in infancy.

James' father, John, died in 1784, aged 58 years when James, aged 29, had already made a mark on the world. Shortly afterwards, James was awarded the diploma of the Company of Surgeons. Of James' mother we still know little for John's memorial stone mounted on the wall outside Sr. Leonard's Church in Shoreditch is badly worn and her own memorial, if such existed there, has not been identified.

After his apprenticeship and while in practice now on his own, in 1785 Parkinson attended the surgical lectures of John Hunter. John Hunter was Surgeon General to the Army and had one of the largest collections of fossils in the country; he observed that geological strata could be dated by the fossils they contain, although this important discovery is commonly attributed to Smith. In 1787 Parkinson became a member of the Medical Society of London, founded in 1773 and still a very active medical society today.

Lightning

Parkinson was interested in general science and natural phenomena and this is reflected in his paper on the effects of lightning published in 1789 (Parkinson, 1789). A house in Crabtree Row near Shoreditch Church was struck by lightning on 17 July 1787 and two men inside the house and another who was walking past were also struck. The house smelt of sulphur and the lead in the shop window melted. The inmates appeared dead. A striking finding was the discoloration of the extremities, especially the lower limbs, which were black but they changed to white and then to pink after rubbing. On another occasion Parkinson treated a farm labourer in Edmonton who was blinded by a flash of lightning when driving cattle in a field on a stormy night; the labourer regained his sight but it is not clear whether this could be attributed wholly to the medical attention given by Parkinson.

Rabies

Parkinson was also involved in at least two cases of possible rabies, then known as hydrophobia from the fear of convulsions when tasting water, a disease from which fortunately Britain has been free now for many years. In 1780 he treated a 28-year-old servant girl who worked for the Reverend Mr. Clare of Queens Row in Hoxton. Parkinson had not seen the disease previously. While awaiting Sir William Blizzard (1743–1835), Parkinson gently fed her water and calves'-feet jelly (and this was to remain a popular food for the sick until recently), noting that she was very distressed when the water touched her lips although when placed on her tongue she was able to swallow with some relief. She was transferred to The London Hospital at once but unfortunately died 48 hours later, the post-mortem examination being unremarkable. Although the patient did keep a dog, there is no evidence it was rabid. In fact, the illness may have been tetanus since it followed a compound fracture of the leg bones.

The second patient who might have had rabies was a 10-year-old boy whose hand had been severely savaged by a dog. Three weeks later Parkinson was summoned and he attended with his son. Dr. John Yelloly was also called in and advised the administration of lead but the boy died three days later. The dog survived and was apparently healthy two years later and so it is also difficult to incriminate that animal.

Treatment of this disorder was by means of general supportive measures; specific therapy was not available and so Parkinson had little therapeutic to offer. Since the first report of transmission of the disorder from a rabid dog was not reported until 1809 (by Zinke) and Parkinson might not have known of this immediately, he was astute in recalling his patient of 1780 and reported it in 1814 jointly with his son (Parkinson, 1814). Louis Pasteur (1822–1895) was to demonstrate the infective nature of the disorder in 1881.

Gout

From about 1790 Parkinson had been troubled with gout and he crystallised his observations in 1805 in his book *Observations on the Nature and Cure of the Gout* (Parkinson, 1805). He discussed the treatment of his own and his father's gout. He used to apply cold water to the gouty joints and described the stinging and burning pain of the swollen joints. He noted the fingers were affected after the feet had improved. Both interphalangeal joints of the right index finger were painful. Nodes were removed from his hand joints – the third finger of the right hand, the last joints of the first and second fingers, and the thumb. These problems did not seem to affect his flowery handwriting of which several examples survive. He also applied leeches and tried dietary treatment.

In his book he noted 'the proximate cause of gout' was 'a peculiar saline acrimony, existing in the blood, in such a proportion, as to irritate and excite to morbid action the minute terminations of the arteries in certain parts of the body' and he thought this was 'the acidifiable base of the uric acid' and so he very nearly realised that hyperuricaemia was the cause of gout although he confused the condition with Heberden's nodes, thinking them the same condition. A link with his interest in geology is provided in his monograph where he mentions that a dissection of the first joint of a gouty great toe was like a fossil shell.

Political Activist

In 1791 and 1792, Thomas Paine's *The Rights of Man* argued for proper representation of the people. The King should obey the law of the land, as should his subjects. In this setting, in 1792 Parkinson turned his gaze to politics and became a member of the *London Corresponding Society united for the reform of Parliamentary Representation*. This society, founded in 1792 by the shoemaker Thomas Hardy, aimed for universal suffrage and annual Parliaments. Five members planned to assassinate the King, George III, with a wind gun, an airgun consisting of a brass tube from which a barbed arrow could be fired. Dr. Robert Thomas Crossfield (1768–1802) was the principal instigator and the subsequent trial for High Treason was the fifth such since the suspension of Habeas Corpus in 1794. Parkinson wrote about the Society in 1794 (Parkinson, 1794a) and was a witness for the defence in the Pop Gun trial before the Privy Council in 1796. He also joined *The Society for

Constitutional Information, a reform society founded in 1780 to argue for proper representation of the people in Parliament.

Similar representational issues were rife in France, for this was the time of the French Revolution, a dangerous time, for in Paris more than 17,000 counter-revolutionaries were executed during the reign of terror from September 1793 to the summer of 1794. The new device of Dr. Joseph Ignace Guillotine (1738–1814), adopted in 1791, could remove the head of a victim in one two-hundredths of a second.

Parkinson argued that taxes should be levied in proportion to the ability of those expected to pay them, that the necessities of life including soap and candles should not be taxed, that a poor person should not be imprisoned for moving from his own parish to look for work and that workmen should not be imprisoned for uniting to obtain an increase in wages – perhaps an early attempt at trade unionism. The method of communication with many persons at once included the use of the pamphlet.

Pamphleteer

And so from 1793 to 1795 Parkinson became a prolific pamphleteer. Under the pseudonym *Old Hubert*, he published eleven political pamphlets.

In 1793 five pamphlets appeared. *The Budget of the People* (Parkinson, 1793a) consists of two parts, each costing one penny, which drew attention to the tyranny of the King and State. *The Village Association* (Parkinson, 1793b) describes Edley, a mythical small village in Yorkshire. Old Hubert and the statesman Edmund Burke (1729–1797), at that time MP for Malton in Yorkshire, are each featured in this story, the moral of which is that the authors of wicked and seditious pamphlets should be made known to the State. *Knave's-Acre Association* (Parkinson, 1793c) cost four pence and describes an imaginary association that actually supports the political abuses of the time; this publication was advertised in *The Times* (The Times, 1794). *An Address to the Hon. Edmund Burke from the Swinish Multitude* (Parkinson, 1793d) cost sixpence. This is a long and cynical pamphlet that applauds the term 'swinish'. The political attack is upon Burke, Leader of the Whig Party. Many of the pamphlets attack the Tory government in the same

manner. *Pearls cast before Swine by Edmund Burke scraped together by Old Hubert* was a penny pamphlet (Parkinson, 1793e) with a collection of aphorisms, published probably in 1793.

In 1794 two pamphlets appeared. *Revolutions without bloodshed* (Parkinson, 1794b) is a pair of pamphlets with further political aphorisms, and *Mast and Acorns* (Parkinson, 1794c) a two-penny collection of political writings.

In 1795 he published four pamphlets. *A Sketch by Old Hubert; Whilst the honest poor are wanting bread* (Parkinson, 1795a) is a single pamphlet in much the same vein as *The Budget of the People*. Next, *An account of some peculiar manners and customs of the people of Bull-land* (Parkinson, 1795b) bristles with political criticisms. *The Soldier's Tale* (Parkinson, 1795c) was priced at one penny and is a cautionary tale of politics and abuse. In *The Assassination of the King* (Parkinson, 1795d) he described his examination by the Privy Council for suspected involvement in The Pop Gun Plot that resulted in an attempt upon the life of George III, the name perhaps derived from the Pop-Gun public house in Portsmouth Street near Lincoln's Inn Fields – or perhaps the pub was so named afterwards.

Parkinson's transition principally from politician and political-pamphleteer to geologist occurred sometime after 1795, possibly influenced by the death of Burke in 1797 or, more likely, after his own inquisition by Pitt.

Churchman

This Parkinsonian idealist had strong religious convictions. He was Secretary of St. Leonard's Church Sunday School and remained a member of the vestry until his death. At some time after 1784 he became a Churchwarden. And so, philanthropic and well connected with the religious matters of the neighbourhood, in 1799 he was elected by the local parishioners of St. Leonard's to fill the post of Trustee for the Poor for the Liberty of Hoxton.

Parkinson's religious feelings were not really at variance with his scientific and medical opinions. To some extent he anticipated Charles Darwin in describing Creation as a constantly advancing process. Some of Parkinson's successors in science, especially in Victorian times, could not easily reconcile their views on religion and science when viewed in the light of Natural Selection.

Geologist

The beginning of the true science of palaeontology is attributed to the French biologists Lamarck and Cuvier. Lamarck is known best for his evolutionary theories and Cuvier for his classification of the animal kingdom. In England, Parkinson and the palaeontologist, actor and drawing-master William Martin can also claim fame. Martin's work relates to a study in Derbyshire dating from 1794 and published as a larger volume (*Petrificata Derbiensia*) in 1809. These four workers – Lamarck, Cuvier, Martin and James Parkinson were among the foremost founders of scientific palaeontology in the so-called 'heroic age of geology'. Parkinson made the greatest contribution to encouraging the hobby in England and it was popular especially away from London where access to cliffs and other geological features was easier.

Parkinson had started his geological collection in 1798 and bought specimens from W. Humphries in Rupert Street, from George Humphrey in Leicester Square, from Mr. Heslop in Finsbury Square and from John Mawe in the Strand.

He would have had leanings in this direction for a short time before this and the exact date upon which any of us transforms an idea into practice is not discernible but he seems to have collected voraciously. His collection was sold to several keen collectors after his death.

He was to become a Founder Member of the Geological Society in 1807 and to correspond with professional geologists of the day as well as with many amateurs, among them prominent medical men and clergymen including William Cunnington (1754–1810) of Heytesbury in Wiltshire (Gardner-Thorpe, 1986), William Buckland (1784–1856) of Oxford and Gideon Algernon Mantell (1790–1852) of Lewes in Sussex. Parkinson had already published Vol. I of his three volume magnum opus, *Organic Remains of a Former World,* in 1804. No doubt the influence of others would have modified Vols. II and III, published respectively in 1808 and 1811. Later at the age of 67, in 1822 he was to publish a more popular and concise textbook than *Organic Remains* (akin to his domestic medicine books) entitled *Outlines of Oryctology; or, An Introduction to the Study of Fossil Organic Remains.*

Domestic Medicine

Parkinson's interest in scientific matters probably started before the publication in 1799 of *The Chemical Pocket Book* (Parkinson, 1799, 1801a) that went into several editions between the years 1799 and 1809. *Chemical Essay* by Richard Watson (1737–1816) in 1781–1787 may have been responsible for arousing Parkinson's interest in addition to the influence of John Hunter. Watson was Professor of Chemistry at Trinity College, Cambridge in 1764. Parkinson's Preface noted: 'The following assemblage of chemical facts was formed, with the hope of rendering it an agreeable pocket companion for the lovers of Chemistry in general; and more particularly for those who may be just engaging in the study of this most useful and interesting science'.

Parkinson was one of the many authors of popular medical texts although most of these interesting volumes appeared in Victorian times. His was one of the earliest for in 1799 he published *Medical Admonitions addressed to families respecting the practice of domestic medicine and the prescription of health* (Parkinson, 1801b) that ran to two editions in that first year alone.

Times were not easy and diet largely unvaried. The so-called working and middle classes usually breakfasted on oatmeal porridge, often with milk or a piece of cheese. Dinner might consist of dumplings, boiled meat or broth. Sometimes potato pie was available but the content of beef or mutton was low. Supper tended to be similar to breakfast. As more machinery was introduced into factories, the health and happiness of the working classes seemed to deteriorate slowly. The enclosure of common and wastelands only added to the difficulties since the few animals possessed by each family were not able to graze adequately.

Manufacturing towns became notorious for overcrowding, unsanitary conditions and general squalor. The populations of most large towns doubled between 1800 and 1830. Thus all the help that could be gained from popular medicine texts was welcome. Parkinson was a pioneer of Community Health (Hertzberg, 1987).

In 1800 Parkinson published *The Hospital Pupil; or, An Essay intended to facilitate the Study of Medicine and Surgery* (Parkinson, 1800). This was an attempt to help medical students whose training was not well structured at that time. After James' death, his son John William Keys Parkinson (1785–1836) continued this interest and published *A scheme of a course of thirty lectures introductory to the study of medicine* – an *aide-memoire* of topics by which students might organise their studies. The copy at The Royal College of Surgeons is signed by John and was presented to Mr. Dickson. Many of Parkinson's medical interests were shared with his family – after all, there were at least three generations in medicine.

In 1804 another popular book followed: *The Villager's Friend and Physician* (Parkinson, 1804). Here he described how a village apothecary delivered a lecture to discourage unreasonable demands upon the doctor late at night for trivial complaints. Regular exercise and rest were advocated. *The Alehouse Sermon* was to discourage the drinking of alcohol – 'one shilling spent with the butcher is better than two with the publican' – 'spices should not be taken in large amounts'. A frontispiece depicts the apothecary delivering his talk and it is tempting to think that he may resemble James of whom a portrait has not yet been found – he lived too early for us to expect to find a photograph and the portrait allegedly of him was of a later dentist of the same name. Leeches were used for earache and toothache whereas bleeding, sweating and warm baths were proclaimed good for all illnesses. He advised that cancer of the breast should be treated early.

In 1802 *The Way to Health* was published (Parkinson, 1802a) as a first aid sheet to be hung by the cottage fire. *Hints for the Improvement of Trusses* appeared in 1802 (Parkinson, 1802b).

Parkinson wrote on child abuse (Currier & Currier, 1991) and might be considered a forerunner of the paediatrician (Pearn & Gardner-Thorpe, 2000, 2001). In 1807 *Observations on the Excessive Indulgence of Children* (Parkinson, 1807) appeared, followed by a further popular book, *Dangerous Sports, a tale addressed to children* in 1808 (Parkinson, 1808). He suggested that children should not be exposed to sudden changes of temperature, that every endeavour should be made to ensure their happiness and that they should be taught to read, especially the Bible. He encouraged the use of barley water with figs and raisins, and honey with oil and lemon juice, as good cough remedies. Honey and lemon are still popular for cough. He obviously felt that head injury was a real danger to children learning to walk and suggested that a quilted stuffed cap should be used for protection.

He recommended the use of Dr. James's Powders (Chenevix, 1802) – a popular cure-all in the eighteenth century. It contained antimony and bone ash but was not named after James Parkinson.

Conclusion

Parkinson held several medical appointments. He joined the Medical Committee of the Royal Jennerian Society in 1803. In 1805 he was a founder member of the Medical and Chirurgical Society of London along with Edward Jenner, Humphry Davy and others. Among his many further contributions, he argued that Mad Houses should be better regulated since for much of that time they were run as private institutions where the insane were incarcerated. In due course he was to write of the shaking palsy and he continued to publish into the first quarter of the nineteenth century – in an enlightened manner.

Parkinson was a polymath but this reflected the paucity of scientific information available in the late eighteenth and early nineteenth century when judged by the standards of today. When considered in relation to prior knowledge, Parkinson and his enquiring contemporaries add much and this was mainly by observation and somewhat by experiment. Parkinson was not an experimenter but a documenter of what he saw, deducing knowledge from his observations and applying this to the everyday care of his patients. He also applied the very important principle that publication is essential to record new knowledge in order to promote discussion and confirmation and to avoid repetition of that which has been proven. We should learn from this. Yet we remember him best for his single essay on a disorder little known in 1817 but he contributed a lot more.

James Parkinson was to continue to live nearly all his life at the family home but near the end he exchanged houses with his son, John, and went to live not far away, at 3 Pleasant Row, off the Kingsland Road where he died on 21 December 1824 after a stroke. He was buried at the nearby St. Leonard's Church where he had been baptized, married and been a churchwarden. The exact site in St. Leonard's of the grave of this polymath is not known; as a churchwarden, it might have been better marked.

References

Chenevix, R. (1802). Observations and experiments upon Dr. James's Powder, with a method of preparing, in the humid way, a familiar substance. *Nicholson's Journal* I, 22–26.

Currier, R. D., & Currier, M. M. (1991). James Parkinson, on child abuse and other things. *Archives of Neurology, 1991, 48*, 95–97.

Gardner-Thorpe, C. (1986). The Parkinson–Cunnington connection. *Wiltshire Archaeological and Natural History Magazine, 80*, 192–196.

Herzberg, L. (1987). Dr. James Parkinson. *Clinical and Experimental Neurology, 24*, 221–223.

Parkinson, J. (1780). *Observations on Dr. Hugh Smith's philosophy of Physics*. London.

Parkinson, J. (1789). Some account of the effects of lightning. *Memoirs of the Medical Society of London, 2*(193), 493–503.

Parkinson, J. (1793a). *The budget of the people, part I and part II*. London: DI Eaton.

Parkinson, J. (1793b). *The Village Association or the Politics of Edley containing The Soldier's Tale; the Headburough's Mistake; the Sailor's tale; the Curate's quotations; and Old Hubert's advice*. London: J Ridgway.

Parkinson, J. (1793c). *Knave's-Acre association. Resolutions adopted at a meeting of Placeman, Pensioners, etc, held at the Sign of the Crown Knaves Acre, for the purpose of forwarding the Designs of the Place and Pension Club latterly instituted in London. Faithfully copied from the original minutes of the Society, By Old Hubert*. London: T. Spence.

Parkinson, J. (1793d). *An address to the Hon. Edmund Burke, from the Swinish Multitude*. London: J Ridgway.

Parkinson, J. (1793e). *Pearls cast before swine, by Edmund Burke, scraped together by Old Hubert*. London: DI Eaton.

Parkinson, J. (1794a). *A Vindication of the London Corresponding Society*. London.

Parkinson, J. (1794b). *Revolutions without bloodshed; or, reformation preferable to revolt*. London: DI Eaton.

Parkinson, J. (1794c). *Mast and acorns: Collected by Old Hubert*. London: Daniel Isaac Eaton.

Parkinson, J. (1795a). *A sketch by Old Hubert: Whilst the honest poor are wanting Bread*. London: J Burks.

Parkinson, J. (1795b). *An account of some peculiar manners and customs of the people of bull-land, or the Island of contradictions; faithfully detailed by Old Hubert*. London: J Smith.

Parkinson, J. (1795c). *The soldier's tale, extracted from the village association: With two or three words of advice, by Old Hubert* (2nd ed.). London: DI Eaton.

Parkinson, James (1795d). *Assassination of the King: A letter to Mr. John Smith*. London.

Parkinson, J. (1799). *Letter to Wm Roscoe about chemical pocket book*. Roscoe's reply also preserved. Roscoe MSS 2876.7 in Liverpool reference library.

Parkinson, J. (1800). *The Hospital Pupil; or, An Essay intended to facilitate the Study of Medicine and Surgery. In four letters. I. On the qualifications necessary for a youth intended for the profession of medicine or surgery. II. On the education of a medical student, improved course of hospital studies, &c. III. Directions for the prosecution of hospital studies according to the present system of medical education. IV. Hints on entering into practice, on medical jurisprudence, &c*. London: Printed for HD Symonds.

Parkinson, J. (1801a). *The Chemical Pocket-Book; or Memoranda Chemical: Arranged in a compendium of chemistry: With tables of attractions, etc calculated as well for the occasional reference of the professional student as to supply others with a general knowledge of chemistry* (2nd ed.). London: Whittingham.

Parkinson, J. (1801b). *Medical admonitions to families, respecting the preservation of health, & the treatment of the sick*. London: HD Symonds.

Parkinson, J. (1802a). *The way to health. The villager's friend & physician*. London: HD Symonds.

Parkinson, J. (1802b). *Hints for the improvement of trusses*. London: HD Symonds.

Parkinson, J. (1804). *The villager's friend and physician; or a familiar address on the preservation of health, and the removal of disease on its first appearance; supposed to be delivered by a village apothecary: With cursory observations on the treatment of children, on sobriety, industry, etc, intended for the promotion of domestic happiness*. London: HD Symonds.

Parkinson, J. (1805). *Observations on the Nature and Cure of Gout; on nodes of the Joints; and on the influence of certain articles of diet, in gout, rheumatism, and gravel*. London: Whittingham, C, Dean Street.

Parkinson, J. (1807). *Observations on the excessive indulgence of Children*. London.

Parkinson, J. (1808). *Dangerous sports. A tale addressed to children*. London: HD Symonds.

Parkinson, J. (1814). *To the Editors of the London Medical Repository* (Vol. 1, pp. 289–292). London: Letter. London Medical Repository.

Parkinson, J. (1817). *An essay on the shaking palsy* (Vol. 2, pp. 964–997). Baltimore: Reprinted in Medical Classics, 1938.

Pearn, J., & Gardner-Thorpe, C. (2000). James Parkinson (1755–1824), a pioneer of child care. *Journal of Paediatrics and Child Health, 37,* 9–13.

Pearn, J., & Gardner-Thorpe, C. (2001). James Parkinson (1755–1824), an early paediatrician (abstract). *Journal of the Neurological Sciences, 187*(Suppl. 1), S155.

The Times. (1794). *Advertisement for book by Old Hubert and advertisement for the Village Association*. 15 December.

Section C
The Nervous System

Introduction

As we noted in Section B, it was not until the very end of the eighteenth century and indeed on into the nineteenth that real progress could be made in understanding the physiology of the nervous system. It was not until these later periods that techniques and instrumentation became adequate to the challenges presented by this, the most complex and mysterious of the body's systems. Nevertheless, much progress was made during the eighteenth century especially in clearing away the old time-worn ideas inherited from the seventeenth century and earlier. This section samples this progress. It examines six salient episodes in the long century, starting with a review of John Hunter's massive contributions to neuroscience and neurology and ending with two chapters detailing the beginnings of electrophysiology.

James Stone, James Goodrich and George Cybulski begin with an examination of the work of one of the late eighteenth-century's greatest figures – John Hunter (1728–1793). John was the younger brother of William who had established himself as anatomist and surgeon in London. In 1748, at the age of 20, John, tiring of Scotland, came south to assist his brother in his profession. John's mind had an intensely practical bent and he was soon dissecting not only human cadavers but many animals, including mammals, birds, reptiles, fish and invertebrates. He was also an avid collector, and his collections eventually formed the basis of the Hunterian Museum, now housed at the Royal College of Surgeons. This collection has been said to form John Hunter's 'great unwritten book'. Stone, Goodrich and Cybulski follow Hunter's own precept of allowing the facts to speak for themselves;

in this case the facts are Hunter's own writings on the nervous system and its diseases. They start by outlining his comparative anatomy, emphasising his attempt to use the nervous system to classify animals, and dwelling on his important contribution to the development of electrophysiology, that is, his investigation of the electric organs of specialised fish. After discussing Hunter's work on sensory systems, they review his take on the nervous system and its role in sensation, consciousness and the control of behaviour. Hunter also had forward-looking ideas on psychosomatic illnesses and on the action of voluntary and involuntary muscles. All of these investigations were put to good use in the clinic and operating theatre and the authors use his casebooks to outline his procedures. Indeed, John Hunter was not only 'the founder of scientific surgery' but also one of the most original of all eighteenth-century workers on the anatomy and physiology of the nervous system.

In the next chapter, Julius Rocca discusses the work of Robert Whytt (1714–1766) and William Cullen (1710–1790). Both Cullen and Whytt were major figures in the Scottish medical enlightenment and lived at a time when Edinburgh was the home of luminaries such as Adam Smith, David Hume, James Burnett (Lord Monboddo), Thomas Reid, James Hutton and many others. Both Cullen and Whytt, along with the Monros (Primus and Secundus), did much to establish Edinburgh as the leading medical school in the British Isles. Rocca shows that the ideas of Whytt and Cullen about 'animal economy' (physiology) were to an extent complementary. Whereas Whytt can be regarded as the 'finest experimental physiologist of his day',

Cullen was more concerned with systematising what was known and presenting it to others. Whilst Whytt was the first to recognise the pupillary light reflex and to gain some insight into the neural basis of reflex action in general, Cullen sought to teach his students (in English rather than Latin) a coherent system of medicine. Whereas Whytt struggled to understand the working of the nervous system at a fundamental level, distancing himself from both the animist (Stahlian) and mechanist (Cartesian) positions, Cullen sought to classify (a very eighteenth-century obsession) diseases, especially those involving the nervous system. Rocca examines these two professors of medicine (Cullen succeeded to the Chair when Whytt died), bringing out their differing strengths and showing how they contributed to the worldwide fame of the Edinburgh medical school.

In the next chapter, Lawrence Kruger and Larry Swanson review the life and work of Pourfour du Petit (1664–1714), who first served as an observant army physician and then an eye specialist. Petit found that the cervical sympathetic nerves do not originate from the cranium, and he also examined brain anatomy and functions. His animal experimentation, which constitutes an early example of what Kruger and Swanson call 'translational neuroscience', involved making lesions to follow up on some of his clinical observations and postmortem findings. Using dogs, he showed experimentally that limb movements originate on the opposite side of the brain and that paralyses follow destruction of the contralateral corpus striatum. Petit's underlying neurophysiology, however, remained within the more traditional theoretical framework: that is, he, like others at the time, still believed that the 'animal machine' worked by 'animal spirits'. Petit's experiments constitute an important landmark in the neurosciences, although he has remained, as Kruger and Swanson observe, a somewhat obscure and overlooked figure in the history of the neurosciences.

The famous debate between Albrecht von Haller (1708–1777) and Robert Whytt (1714–1766) on the nature of muscle contraction is addressed in the next chapter written by Eugenio Frixione. This issue had been the source of debate in the late seventeenth and into the eighteenth century. The old physiology, still to be found in Descartes *l'Homme*, envisaged animal spirit flowing through hollow nerves causing skeletal muscles to 'balloon' and thus shorten. William Croone in his *De ratione motus musculorum* (1667) disputed Descartes' theory, asking where are the cavities in muscles which can accept the flow of animal spirit from the motor nerves? As an alternative, he suggested that droplets of *succus nerveus* secreted from the endings of motor nerves interact with blood to give an 'ebullition' causing the muscle to shorten, once again by the ballooning of a multitude of tiny 'vesicles or globules' within the body of the muscle. The notable advance made by Haller was to insist that 'irritability' was an inherent property of all muscles, and that they would therefore contract when stimulated. Whytt, on the other hand, disagreed. He argued that the terminals of the nerve fibres transformed into muscle fibres, and that an 'active-sentient' principle ultimately identifiable with soul or spirit is responsible for contraction.

The next chapter brings us farther along in the eighteenth century. It deals with a great transition in our understanding of nervous physiology: from the time-worn concepts of subtle fluids and animal spirit to the first stirrings of an understanding of electrophysiology. Marco Piccolino traces research into some strange, specialized fish, some of which had been described in antiquity, fish that stunned their prey and those that dared to touch them with powerful, numbing jolts. This voyage of scientific discovery largely involves electric rays (the *Torpedo*) from the Atlantic and Mediterranean, and to some extent eels from South America. The story revolves around John Walsh (1725–1795), an English gentleman and member of the Royal Society of London with an obsessive interest in understanding how such creatures stunned other fish and even inquisitive experimenters, including himself. Walsh began his research in the 1770s, and ultimately he, along with John Hunter and Henry Cavendish, showed that these fish generate their numbing jolts from highly specialized, columnar organs with heavy nerve supplies. Their experiments showed that their discharges appeared to be similar to the electricity that was drawing the attention of physicists, philosophers and the general public at the time.

The final chapter discusses the contributions made by Luigi Galvani (1732–1798), who extended what was learned from the electric fish experiments to frog and mammalian neurophysiology.

Miriam Focaccia and Raffaella Simili review Galvani's life and many contributions, with emphasis on the neurosciences. They show how and why he deduced that the conductive principle in the frog sciatic nerve (and by generalisation, in all nerves) is similar to the electricity that Benjamin Franklin, Joseph Priestley and others were investigating in the inorganic world. Indeed, Galvani explicitly compared muscles and nerves to the Leyden jars and wires used by these experimentalists and now by him in his laboratory. The authors situate Galvani's work in the ongoing physiology of the time, showing how it also related to other fields, to the earlier work of Marcello Malpighi at Bologna, to the electrical investigations of Father Beccaria, to the wax modelling of Anna Morandi, as well to the research and theories of Felice Fontana, Albrecht von Haller and Robert Whytt. Luigi Galvani was in many ways a 'Renaissance man', an Italian patriot and, most importantly, a pioneer of modern neurophysiology.

The Editors

5
John Hunter's Contributions to Neuroscience

James L. Stone, James T. Goodrich, and George R. Cybulski

Introduction

John Hunter was a giant in the natural sciences and medicine (Fig. 1). His overall contributions to the basic and clinical neurosciences were substantial but are little known. One reason is because as a "naturalist" Hunter's underlying emphasis was upon the greater understanding of life itself, including paleontology and geology. His main interests were in the philosophy of life and nature, and he was one of the few in England at that time who took a really comprehensive view of these phenomena. Essentially a novel thinker rather than a studious scholar, he extensively utilized both inductive and deductive methods.

Hunter was never interested in the teaching of single systems or narrow confines within the "animal economy." This later term encompasses the broad fields of embryology, human and comparative anatomy, physiology, disease and repair, and pathology, all of which he pioneered and became the foremost eighteenth century British authority. Hunter performed roughly 2,000 cadaver or autopsy examinations early in his career, and dissected over 500 animal species. It is no surprise he foresaw the "emergence" and "divergence" of animal and vegetable forms over thousands of years, and his "heretical ideas" antedated Charles Darwin by more than one-half century (Hulke, 1895; Morris, 1909; Ottley, 1835; Paget, 1897; Pettigrew, 1839–1840).

Mr. Hunter was an anatomist and surgeon by occupation. Credited as the "Founder of Scientific Surgery," the surgical profession and Royal College of Surgeons (RCS) of England have taken the lead in paying him homage. Finding surgery a craft or trade, he left it a science and a philosophy. Nevertheless, his medical knowledge extended much more widely than the confines of surgery, and throughout we find scattered contributions to neuroscience. His deficiency in formal education proved beneficial in that he approached science objectively without preconception, and with remarkable originality. Still, his writings and lectures can be difficult to understand because of his peculiar style. Few contemporaries, and even later scholars, were fully aware of his immense contributions to the body of scientific knowledge. In addition, hundreds if not thousands of pages of observations, experiment notebooks, and writings were hoarded, plagiarized and burned 30 years after Hunter's death by a self-serving brother-in-law – Sir Everard Home (Dobson, 1969; Qvist, 1981). The surviving "casebooks" of John Hunter were finally given by Home to the RCS in the 1820s, but have only recently become available. The "casebooks" are an invaluable source of clinicopathological observations and deductions on hundreds of cases by the ever observant Mr. Hunter (Allen, Turk, & Murley, 1993).

We have now reviewed Hunter's various anatomical, medical, surgical, and pathological neuroscience observations. Some effort was made to place Hunter's neurological contributions within the context of what was then known or, accepted by future generations. In this chapter, we follow the Hunterian tradition of "originality," and we have accordingly largely allowed Hunter to speak for himself by including extensive quotation from his work.

FIGURE 1. John Hunter in 1788. From: Stephen Paget, "John Hunter" (1897)

Overview of Hunter's Life

Hunter was born in 1728 on a small farm in Scotland near Glasgow. The youngest of three surviving children, his parents of modest means imbued the children with high standards and aspirations. Their father died when John was 14 years of age, but before that he showed little interest in school despite the urgings of an indulgent mother. "He would do nothing but what he liked, and neither liked to be taught reading or writing nor any kind of learning, but rambling amongst the woods, braes, etc., looking after birds' – nests, comparing their eggs – number, size, marks, and other peculiarities." (Qvist, 1981: 4) Hunter later stated "When I was a boy, I wanted to know all about the clouds and the grasses. . . why the leaves changed colour in the autumn. . . watched the ants, bees, birds, tadpoles. . . I pestered people with questions about

what nobody knew or cared anything about." (Dobson, 1969: 13)

It is very possible John Hunter had a learning disability that limited his earlier education. At the age of 20 he joined his brother William, then an aspiring anatomist and obstetrician in London. John was set to the task of dissecting human cadavers, preparing dissections for the class, and permanent teaching specimens. He proved amazingly proficient at this work, followed the noted London surgeons William Cheselden and Percivall Pott for several years, and enrolled as a surgical pupil at St George's Hospital in 1754 (a medical degree as William had, was not required) (Dobson, 1969; Kobler, 1960; Ottley, 1835; Qvist, 1981).

John Hunter's quest for scientific knowledge soon drove him to understand function, as he believed all living matter had the power to move, digest, and feel. He devised many simple yet strikingly imaginative and ingenious experiments on man, animals, birds, fish, and a wide array of plants. In his anatomical and physiological work he made use of magnifying lenses and thermometers. His curiosity of disease and injury, and Nature's attempt at repair led him to explore the underlying pathological findings.

In 1760 John secured a position as an army surgeon for several years treating gunshot wounds and other injuries. He also found time to collect specimens of nature, which would later form the Hunterian Museum of the RCS in London (Dobson, 1969; Keith, 1919; Ottley, 1835). After his return to London he opened his own school of anatomy, began the practice of surgery, obtained appointment at St George's Hospital and presented a number of original natural science observations to the Royal Society where he was elected a fellow in 1767. The next year, he obtained a qualification from the Corporation of Surgeons (later renamed the RCS).

John Hunter's first book was on the anatomy and natural history of teeth, and it appeared in 1771. That same year, he married Ann Home, a charming, intelligent woman 14 years younger than himself and they had a son and daughter. Hunter, ever the experimenter, in 1775 enthusiastically wrote Jenner, of vaccination fame and his former house student, expressing the thought: "I think your solution is just; but why think? Why not try the experiment?" (Hunter, 1835: 56)

Hunter's private practice grew slowly but steadily. Most of the money he earned was spent lavishly on obtaining rare and unusual animal and human corpses for dissection, the upkeep of a suburban research farm, his London home with lecture and dissection rooms, a number of servants at all locations, and ever growing collection of specimens.

Dissatisfied with the prevailing teaching of surgery in England, he began a unique, once per year, course of lectures called the Principles of Surgery. Mr. Hunter's emphasis was on normal physiological principles, physiology and pathology of diseases and their surgery (Dobson, 1969; Hunter, 1835: 207–643; Qvist, 1981).

A major theme was that repair and restoration is effected by inherent powers in the living tissues of the patient. Specifically he believed the "coagulable lymph" component of blood united the opposite sides of a wound and was endowed with a "consciousness" or "wisdom" to form bone, tendon, nerve, vessels, etc. as the case might require. The surgeon can only help recovery by tending these powers, and thus "Nature is the master surgeon." (Hunter, 1794: 11–100; Jacyna, 1992; Keith, 1919) The course was supplemented by his Museum of over 13,000 pathological, comparative, and anatomical teaching specimens – highlighted by his background knowledge of each item (Abernethy, 1825: 142; Qvist, 1981: 67–75).[1]

Hunter emphasized an overall reluctance to perform surgery unless specific indications were met. One notable example was gunshot wounds, which usually did better if left alone, except for dressings. He did not teach anything that he had read or that had been traditionally passed on, and only taught what he had seen and learned to be true from his own observations. Hunter's students were made to feel like friends, and usually a number stayed at his house. With his rough Scottish accent, common

demeanor and fine but simple attire, he was devoid of the authoritarian, distant nature of most other anatomists and surgeons (Abernethy, 1825: 139–144; Dobson, 1969; Ottley, 1835).

Hunter's approach was so original and inspiring that despite his being a poor speaker, he attracted in excess of 1,000 students of surgery and medicine to his course or as pupils at St George's, a number who later became leaders throughout much of Europe and America. Hunter's influence changed medicine and surgery for all time and the Hunterian Museum has been described as his great "unwritten book."

In London, Hunter's immense skill and popularity did not go unnoticed and, after the death of Percivall Pott in 1788, he became the leading surgeon in England. He was appointed Surgeon General of the British Army, and Surgeon Extraordinary to the King. Commensurately, the pace of his teaching, hospital work, scientific studies, and writing continued unabated and he obtained about 4 h of sleep per night.

Around this time, John Hunter began to suffer increasingly frequent and more severe attacks of angina pectoris. Although the coronary arterial causation of angina was not yet known, Hunter recognized the role of emotional disturbances in initiating the episodes and said, "My life is in the hands any rascal who chooses to annoy and teaze (sic.) [me]." (Ottley, 1835: 119) These words were to be prophetic – in 1793, a quarrel over teaching issues with spiteful colleagues at St Georges Hospital led to a severe angina attack, unconsciousness, and death (Dobson, 1969; Ottley, 1835).

A short list of his individual contributions would include significant discoveries about the lymphatics, placental function, hemorrhage, shock, blood coagulation, inflammation, suppuration (abscess formation), the treatment of wounds, phlebitis, aneurysm, valvular heart disease, venereal disease,

---

[1] One of the more interesting pieces in the Hunterian Museum at the Royal College of Surgeons in London is the skeleton of an Irish Giant – Charles Byrne. Hunter obtained Byrne's body soon after death (1783) in a clandestine fashion at a cost of 500 pounds, did not perform an autopsy but quickly boiled off the soft tissue parts, and preserved the skeleton in secrecy for several years (Kobler 1960: 238–244; Dobson 1969: 262–264; Qvist 1981: 70). In 1912, X-ray films of the skull suggested evidence of acromegaly by thickened bone, large sinuses and a protuberant mandible (Allen 1974: 34–35). Some years later, Harvey Cushing the American neurosurgeon,

convinced then Museum curator Sir Arthur Keith to open the giant's skull for inspection. The sella turcica was grossly enlarged with evidence of a sizable intracranial extension of a pituitary tumor (Fulton 1946). After appreciating Hunter's in-depth gross autopsy examination of the brain including comments about the pituitary gland (Allen et al. 1993), there is no doubt had he performed an autopsy on Byrne, he would have come upon the tumor as the cause of giantism. It was to be more than 100 years after Byrne's death that a pituitary tumor or hyerplasia of certain pituitary cells was found to be the cause of giantism (acromegaly).

tissue transplantation, artificial respiration, and body temperature control (Dobson, 1969; Moore, 2005; Qvist, 1981). With this as background, we shall now attempt to detail his anatomical and experimental contributions to the neurosciences.

Comparative Anatomy of the Nervous System

Sir Richard Owen, a leading Victorian scientist, comparative anatomist, and Hunterian scholar, stated that Aristotle's knowledge and deductions on the classification of animals have "never been equaled by any other individual until the time of John Hunter... [whose]... arrangement of the Animal Kingdom more nearly corresponded with that of Aristotle and Nature, than the more artificial system of Linnaeus." (Owen, 1837/1992: 94–95) Hunter's published volume of work in this area exceeded 500 pages (Hunter, 1861b) and his total effort, much unpublished or lost, clearly eclipsed any other worker of his generation. Hunter's work on comparative anatomy was especially concerned with four aspects – the anatomical similarities of various animals, association of structure with function, classification from the most simple to complex on the basis of the above two factors, and finally the formulation of general laws (Qvist, 1981: 138–139).

Owen further stated – "Mr. Hunter's published writings on the nervous system bear no proportion to the extent of his anatomical investigations on this subject, especially as manifested in the Hunterian Museum preparations... [These ninety preparations]... trace ... the nervous system ... through its progressive stages of complication from the simple filaments of the Entozoon [simple internal parasitic animal], to the aggregated masses which distinguish the organization of man." (Owen, 1837/1992: 175–176)

The nervous system and specifically brain anatomy, in addition to the anatomy of the heart and digestive system, provided Hunter with a basis of classification of animals by increasing complex forms. He recognized six classes of animals according to visible aggregations of nervous substance or brain structure (Hunter, 1861a: 28–34; Ottley, 1835: 170–171; Owen, 1837, 1866: 281–282).

In the "first class," which includes the molluscs, the brain is in the form of an unprotected pulpy ring from which nerves radiate, and the esophagus passes through the ring. Hunter was of the opinion that this class had two senses, feeling and taste.

The "second class" has a pulpy brain in the head of the animal with two large nerves on each side of the esophagus joining ventrally at a knot and again dividing, and similarly, segmentally joining and dividing through the whole length of the animal, resembling a medulla spinalis (spinal cord) and great intercostal (sympathetic trunks) nerves. These include the leech, earthworm, insects, and lobster. The brain lacks a cranium or skull and is surrounded by soft parts. He believed this second class possessed the four senses – touch, taste, sight, and hearing (as in bees).

The following "four classes" included fish, amphibia/reptiles, birds, and quadrupeds with larger brains protected by a skull, special senses, and spinal cords encased within vertebrae. The fish brain (third class) was viewed as an irregular mass, inconsistent in its form and number of parts, yet in general similar to those of superior class. The skull is too large for the brain and the resulting space filled with an arachnoidal membrane.

In the amphibia/reptiles (fourth class), the parts of the brain are not compacted, but are relatively detached and follow one another, in a linear arrangement. Two anterior and middle cerebral lobes as well as a single cerebellar eminence have a cavity or ventricle within them. There are no convolutions on the external surface of the brain, and the smooth brain surface is covered by pia mater. The crocodile is of a higher degree within the fourth class having a more compacted brain with the same parts as others in that class.

The brain of the "fifth class," or birds, has greater relative size in proportion to the size of the animal, more superimposition of its component masses, and consists of a homogeneous pulpy substance. The cerebellum represents about one-sixth of the whole brain and is more convoluted than the cerebrum.

In the "sixth class," the quadrupeds, the brain is larger than in the preceding classes and the parts are more compacted into a globular form and the cerebral convolutions deeper. The cerebral cortical gray substance makes up the cerebral surface, and in the spinal cord this gray substance is enclosed within (Hunter, 1861a: 28–34).

Hunter's classification scheme thus approximated Cuvier's arrangement of invertebrates and vertebrates (Owen, 1837).

Hunter made very many neuroanatomical observations on animals, some of which had been noted earlier by others, yet which still seemed new to him. In his exquisitely detailed 1787 description of the whale brain, Hunter made particular note of the fibrous structure of the cerebral white matter, as well as the lack of olfactory nerves in the porpoise (Hunter, 1837a, Vol. 4: 373).

The phenomenon of electricity was an exciting and fascinating topic in the mid- to later-1700s (see other chapters in this volume), and John Hunter was presented two species of electric fish to study – the "torpedo" or "*electric ray*," and the Gymnotus or "*electric eel*." Hunter dissected both, and presented his findings in 1773 and 1775 to the Royal Society of London (Qvist, 1981: 128–129). In the Torpedo, the pair of electric organs formed large lateral masses made up of many gel-like parallel columns separated by thin fascia, and innervated by large prominent nerve trunks on each side originating from the lateral and posterior brain as part of the trigeminal complex. The nerve supply to the electric organs exceeded the innervation to any organ in any other known animal. Hunter believed the nerves subservient to the formation, collection and management of the electric fluid, and that the "will of the animal absolutely controlled the electric powers of the animal's body." (Hunter, 1837a, Vol. 4: 409–413; Owen, 1866: 350–354)

Although John Hunter utilized experimentation in the study of muscle and hearing, there is no hard evidence that Hunter experimented on the brain, nerves, or isolated reflexes. However, in search of the reason why patients with a high cervical spinal cord injury died suddenly of respiratory arrest, and low cervical cord injured patients could breathe using their diaphragm, Hunter enlisted a pupil to experiment with cervical spinal cord and phrenic nerve transections in the dog. This led to conclusive proof of a respiratory center in the upper cervical cord (Neuburger, 1897/1981: 201–202). Hunter referred to his "Book of Experiments Vol. 1" (Hunter, 1861a: 169) implying multiple volumes existed. But they were never recovered, and likely plagiarized and destroyed by Everard Home. As an example, some believe Home (likely John Hunter) may have been the first to describe brain tissue nerve cells ("globules or cell bodies") microscopically (Liddell, 1960: 3). Many of Hunter's deductions on neurophysiological principles have a modern ring and have been overlooked by nineteenth century and later scholars.

The Special Senses

The Huntarian Museum contained at least 400 preparations relating to the special senses (Ottley, 1835: 171–172). Hunter conducted some of the earliest anatomical studies on the organ of hearing in fish in 1760, and later presented his work to the Royal Society for publication (Hunter, 1837a, Vol. 4: 292–298). Over a number of years, in multiple animal forms, he examined the auditory apparatus and documented progression of complexity in higher forms and transitional forms between invertebrates and vertebrates (Owen, 1866: 343–344; Qvist, 1981: 127–128).

Hunter's evidence of hearing in fish was the following:

In the year 1762 . . . I observed in a nobleman's garden, near Lisbon, a small fish-pond full of different kinds of fish. Whilst I lay on the bank observing the fish swimming about, I desired a gentleman who was with me to take a loaded gun and fire it from behind the shrubs. . . . The moment the report was made the fish seemed to be all of one mind, for they vanished instantaneously, raising a cloud of mud from the bottom. In about five minutes afterwards they began to appear, and were seen swimming about as before. (Hunter, 1837a, Vol. 4: 292–298).

Today we know that although fish lack an acoustic opening they are nevertheless sensitive to pressure waves in the surrounding water and this could trigger reflexive escape. Fish belonging to the Suborder *Ostariophysi* (carp, goldfish, catfish, etc.) have, in addition to the usual lateral line system which detects low frequencies, an intricate system of Weberian ossicles connecting the swim bladder to the membranous labyrinth, and this gives them the ability to detect auditory frequencies up to 13 kHz. It may be that these were the fish Hunter observed in the nobleman's pond.

Regarding vision, Hunter wrote papers on the subjects of accommodation, eye color, color vision, eye movements, and visual adaptations in the animal kingdom. He rejected the then prevalent theory that the lens moved backward and forward within the eye, and that the eyeball altered its shape. He concluded the lens must vary its shape to provide accommodation (Hunter, 1837a, Vol. 4: 288).

He also described the muscularity of the iris, "The sphincter iridis of the eye contracts when there is too much light, but the radii contract when there is little or no light." (Hunter, 1837a, Vol. 3: 146) Hunter wrote on the direct and consensual pupillary responses secondary to light reaching the retina, presbyopia, and squint, which he divided into internal and external forms (Hunter, 1861a: 169–171). Hunter was also the first to describe the oblique eye muscles as responsible for the coordinated counter-rolling eye movements in man (Hunter, 1837a, Vol. 4: 274–276), and the 13 eye muscles associated with the crocodile eye (Owen, 1866: 341–342).

Hunter produced one of the earliest and likely the finest description of the human olfactory nerves (Hunter, 1837a, Vol. 4: 187–192) (Fig. 2). In 1754 after he had softened a skull in acidic solutions, he was able to trace the olfactory nerve distribution and follow the delicate nerve filaments through the narrow cribriform plate of the frontal skull base. He also observed that in animals, such as the crocodile, in which the nose is some distance from the brain, the olfactory nerves run a considerable way as "pulpy" (brain-like) (Dobson, 1969: 275–276; Hunter, 1861a: 163).

Hunter also dissected sensory branches of the human ophthalmic division of the trigeminal nerve

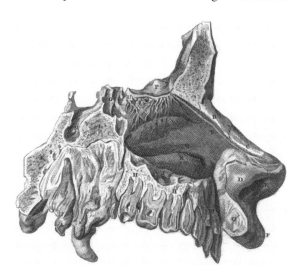

FIGURE 2. The nerves which supply the organ of smell in man. From Hunter (1837b), Plates from "The Works of John Hunter," p. 21, Plate XLII

in close proximity to the olfactory nerves, which he believed served common sensation. He deduced that nerves must be of different kinds to serve different functions, which could explain why branches of the first and fifth cranial nerves terminate within the same area of nasal mucous membrane (Hunter, 1837a, Vol. 4: 193–194).

John Hunter extensively discussed many interesting points regarding the special senses in animals and man, their utility in the animal kingdom, and progression from infancy to adulthood (Hunter, 1861a: 166–183).

The Brain and Nerves

Hunter was convinced that the development of a centralized brain mirrored or arose as a consequence of the presence of organs of sensation (Hunter, 1861a). He may be considered a "vitalist," in that he believed the blood in addition to the brain and nerves possessed the source of life he termed the living principle or "materia vitae," which was also diffused throughout each living creature (Bynum, 1996; Hunter, 1794: 89–90; Jacyna, 1992). An animal may live after a brain injury which takes away all sensation (consciousness), but when the power of the brain is taken away in animals or man "they waste... not... from the brain supplying those parts with nourishment... [rather] it arises from want of necessity to keep these parts in a state fit for action." (Hunter, 1835: 273)

Hunter believed the brain has a consciousness of the body, resulting in regulation of bodily motions dependent on life, and a general power to support simple life. "The actions of the brain towards the body are... of two kinds: one in consequence of the feelings or state of the mind at the time (actions of fear, courage, anger, love, etc.)... the other... the command of the will, called voluntary actions." (Hunter, 1835: 261)

Whatever properties brain size (in relation to body size) may have on an animal, are not employed upon the body, but rather about its own brain actions, as in a greater effort of the mind, and a greater scope of reason (Hunter, 1835: 264).

Hunter observed that the nerves are large in proportion to the size of the brain in lower animals, and believed the nerves nearer to the brain "...stronger their actions .... A larger brain does not require larger nerves to make impressions upon the senses, but less [sense organ size is generally inversely related to brain size]." (Hunter, 1861a: 163–164)

Although at that time many believed the nerves functioned as hollow tubes conveying a fluid, Hunter disagreed:

Nothing material is conveyed from the brain by the nerves, nor vice versa from the body to the brain; for if that was exactly the case, it would not be necessary for the nerves to be of the same materials with the brain; but as we find . . . [they are] . . . it is presumptive proof that they only continue the same action which they receive at either end. (Hunter, 1794: 89)

Moreover, "The nerves have but one mode of action. . .that of conveying impressions. . . from. . . the body towards the brain. . . the other, from the brain to the extremities (body), conveying the mandates of the will, etc." (Hunter, 1835: 260)

The nerve has done its business in communicating sensations or impressions to the mind, according to Hunter, and the mind, in turn, informs the seat of the senses (Hunter, 1861a: 164–165).

Additionally, Hunter believed "every nerve so affected to communicate sensation, in whatever part of the nerve the impression is made, always gives the same sensation . . . at the common seat of the sensation [in the brain] of that particular nerve." (Hunter, 1837a, Vol. 4: 190–191)

A blow on the eye often produces light, and on the ear sound. And besides, those senses are subject to diseases, where the sensation often arises without impression from without.
For distinct sensation two things are necessary, namely time and space . . . [and] every sensation appears to depend on the quantity of nerves acted upon, and in a given time (rate of stimulation) . . . a great quantity of nerves going to parts that are allotted for strong sensation. (Hunter, 1835: 263)

Hunter thus believed stimuli could be too brief or too gradual and prolonged to elicit sensations. Regarding the skin's organ of touch,

. . . distinguishing the different impressions, such as roughness, smoothness, heat, cold, etc. . . . we find the organ more perfect in those parts most sensible, as on the ends of the fingers, lips, etc. . . . [receptor structures] called villi, not of acute sensation, but of delicate. . . distinguishing sensation. . . . This structure for impression of touch, is perhaps perfectly mechanical, being adapted for the impressions of resistance. (Hunter, 1861a: 182–183)

Hunter also wrote:

The impression is made on the body, but the sensation is in the brain . . . referred. . . by the combination of other senses. . . in possession of . . . reference. . . . The reference produces a feeling in the mind receiving sensation, by joining it with the seat of impression. (Hunter, 1835: 261)

Higher Brain Functions

Hunter opined that the

brain and nerves . . . produce sensation, out of which arises mind and reason. . . without sensation the mind could not be formed, nor could we reason . . . it [is] about some object . . . that we reason . . . . objects affect our senses so as to form a peculiar state of mind: this I call mental impression [and] the mind can [then] reason and exercise volition respecting objects. (Hunter, 1835: 259)

Two pages later, he added:

Therefore a man should not feel that he exists but in thought; nor could he do that if he had never received impressions to think about, for thinking is no more than the memory of impressions put into order by the mind; nor should he have those [motor] actions which naturally arise from the brain till the brain makes the impression on the nerves. (Hunter, 1835: 261)

Consciousness to Hunter is

a conviction of the existence of one's self, or it is a feeling of itself . . .
. . . the reflection on one's own existence, both as to personal existence and the existence of the mind. . . . Consciousness is an act or impression of the mind which it cannot deny. . . . We often act without being conscious of it. . . called 'absence of mind'. . . . All animals may be said to have consciousness, but cannot be sensible of it. . . . To be conscious one has done a wrong thing, would appear to belong to some brutes: a dog when he has done a wrong thing shows signs of it.
[A] self consciousness not only regulates many of our natural actions when in health, [but] while under disease. . . . We have an internal monitor of our powers, and we use them accordingly. (Hunter, 1861a: 252–254).

As for the mind,

The mind is not only affected according to simple impression (as most probably is the case in brutes), but from experience and association of other impressions or ideas with the present. . . and this always produces a stronger effect than the simple impression. The effect arising from reflection is much stronger than that arising from simple sensation or impression. (Hunter, 1861a: 256)

Simple recollection of circumstances, or being in the same place or situation, may bring on a previous state of mind as when a memorable event happened. Different states of mind may compete within the same individual, the stronger state may drive out the weaker (Hunter, 1861a: 257).

The objects in the mind's eye, when we are young, are almost real. . . we then can hardly think without the object presenting itself strongly in the mind. If we connect a few of those ideas together so as to make a little train of thinking, it is almost like connecting real objects together. . . . What we think of when awake, we only see in the mind's eye; but what we think of when asleep appears to be an object immediately of the senses [dreams]. (Hunter, 1861a: 257).

As we become used to the acts of thinking and reasoning, we lose the strong impression of the objects on the mind, yet manipulate and connect these objects together so as to draw conclusions.

We can connect imaginary objects, almost without seeing them, in the mind, just as we can work in. . . any handicraft, almost without seeing or hearing what we are doing. . . at last, the hand seems to leave the mind, and appears almost to go on of itself.
    . . . the state of mind [may have] more difficulty in exciting the voluntary parts to action; for the will is often counteracting the actions that arise from mental emotion in voluntary parts, which produces an irregular action, as in trembling. (Hunter, 1861a: 258–259)

Regarding repetition,

The mind more readily goes into action the second time of an impression, though a considerable distance of time has taken place, but takes up the action with more ease, from merely collateral causes, from a recollection of the similarity, or often without any possible recollection whatever, as if the actions in consequence of the former impression were taking place in the brain again. (Hunter, 1835: 274)

A slight impression soon wears out, but those associated with strong emotions such as fear last longer. Additionally,

The more we have been. . . thinking on any object, the more readily does the train of thinking relating to that object recur. This principle is similar to inertia in matter,

for a motion begun is continued, [or remains at rest]. . . . [This] "does not allow men to think differently from what they have been accustomed to think. . . retards improvement and does not allow men to wander into novelty. It promotes improvement (in some), because it makes men perfect in what they have been long employed about." (Hunter, 1835: 276–277)

On memory Hunter would write:

The mind is improved by gathering sensations and has the power called `memory', of repeating those sensations without the original impressions and of combining those repetitions so as to form ideas and then of combining those ideas so as to form a complete action, story, or proposition of any kind. By habit the mind does these operations with ease, and often goes on doing them almost without being conscious of it. (Hunter, 1861a: 262–263)

Thinking is the forming of ideas, and a natural process like memory. Still, thinking and memory can be interfered with by a will prone to fear or anxiety. Reasoning is not a natural process and we can think without reasoning: "Reason is a kind of voluntary act; the mind brings itself to it." (Hunter, 1861a: 264–265) Reasoning is used to determine or prove, or disprove, some supposed fact, but it must be based on nonfallacious facts. Ideally we should not reason just on general principles but upon clearly established facts (Hunter, 1861a: 262–263).

In contrast, a "perfect belief" or "conviction," of some propositional truth arises from reasoning and is probably peculiar to the human species. Such "demonstrative" [scientific evidence] proof is absolute fact and stands as the first. . . order of evidence. (Hunter, 1861a: 253)

The "intellect or understanding" has an immediate connection with the senses, and the senses with the intellect. The senses are not equally proportioned in each person and some senses are capable of informing the intellect much more than others.

If it were possible to have more senses than what we have, it is probable we might lose by the gain. We are certainly not capable of managing more variety of sensations than what we have at present, and many are not capable of managing those. Take away a sense from some, they would be tolerably sensible. (Hunter, 1861a: 165–166)

Hunter also wrote about the mind and madness:

A disposition. . .when in health, always arises from some impression; but as both mind and body are capable of seemingly spontaneous actions, these may arise from diseased dispositions in both, producing madness in one case and strong disease in the other. (Hunter, 1835: 270)

Hunter goes on to discuss the influence of the mind upon the passions and described a number of psychological factors that predispose to loss of intimacy, virility or the inability to perform sexually, and relates several cases he cured by psychological means (Hunter, 1837a, Vol. 2: 306–308; Vol. 4: 214). He presents his ideas on sleep and additional psychological topics, such as anxiety, joy, grief, fear, superstition, deceit, age changes, heredity, sympathy, instinct, and custom (Hunter, 1835: 263–266; 1861a: 264–280). Hunter's analysis of man's higher brain functions is quite profound, clearly reasoned, and presents a framework not dissimilar from contemporary sources.

Psychosomatic Illness

As expressed many times in his writings and clinical cases, John Hunter was convinced of the very real influence of the mind over the body (Allen et al., 1993; Qvist, 1981: 126–127). The sympathetic trunks ("intercostal nerves") and plexuses, and vagal nerve innervations to the viscera had been known for some time, and Hunter believed an "intercourse between the will and the involuntary parts was necessary to maintain a harmoniously functioning system." (Atkinson, 1951)

Indeed there is not a natural action in the body, whether involuntary or voluntary, that may not be influenced by the peculiar state of mind at the time. . . some parts more readily influenced by it than others. . . . (Hunter, 1835: 359)

Regarding his own severe case of stress-induced chest pain or angina, Hunter stated

the spasm on my vital parts (heart) was very likely to be brought on by a state of mind anxious about any event. . . I have bees. . . and I once was anxious about their swarming lest it should not happen before I set off for town; this brought it on. . . . I saw a large cat. . . and was going into the house for a gun when I became anxious lest she should get away. . . this likewise brought on the spasm. (Hunter, 1835: 337)

Additionally,

I had the spasm in my heart upon the smallest exertion of the body. . . . yet I could tell a story that called up the finer feelings [likely his brother William's recent death], which I could not tell without crying, obliging me to stop several times in the narration, yet the spasms did not in the least take place. Therefore those [special] feelings of the mind we have for. . . people are totally different operations of the mind from. . . anxiety about events. . .. (Hunter, 1861a: 266)

Nevertheless, concerned about his recurring anginal problem and aware that his "imagination. . . worked up by attention to the part expected to be affected" was a significant part of the problem, Hunter sought assistance from a "magnetizer." "Convinced. . . [that] the apparatus. . . was calculated to affect the imagination," he trained himself to divert his attention from the magnetic feeling in the hand to his great toe. He found that he was soon able to create a distracting sensation in his great toe at will (Hunter, 1835: 337). This episode would appear to be an early example of self-therapeutic visual imagery or displacement.

Further regarding John Hunter's large belief in the influence of the mind over the body, it had been written of him "that to one patient in London (I think his own wife) he had prescribed common hot water, allowing it to be so, without any success; but that when he filled Bath bottles with the same water and pretended that they were received by the coach, she, after boiling them, derived all the benefit from them that she had expected from a course of the real waters drunk fresh at the Pump Room." (Qvist, 1981: 127)

Hunter's vast clinical experience also shed light on psychosomatic illnesses, hysteria, somatoform disorders, hypochondriasis, and hyperventilation (Allen et al., 1993: 65, 425, 489, 555). Regarding one such case in 1790, he dryly remarked:

I have to observe that our patient is sometimes affected along the inside of her arms; and at others a spasmodic cough and consequent hoarseness, very painful and troublesome. I forbear any physiological disquisitions respecting the seat of the disease; and shall leave to the more learned to point out the most probable means of relief. (Allen et al., 1993: 426)

Pain

Hunter believed that any sensory stimulation carried to excess could result in pain and when a large nerve is compressed the most acute sensation (pain or numbness) will be some distance below the compression (Hunter, 1835: 263, 363). He was aware that inflammatory pain arose from the nerves of the affected parts, and "Pain of the nervous kind arose from nerves themselves being affected – without the parts being affected that these nerves go to [referred pain]." (Hunter, 1861a: 165) Hunter was also familiar with hyperesthesia in relation to peritonitis and intestinal inflammation, noting that "I have seen complaints in the viscera of the abdomen produce a vast tenderness (soreness) in the skin of the abdomen." (Hunter, 1837a, Vol. 3: 291–292)

John Hunter presented an interesting theory on the mechanisms of referred pain (see other chapters in this volume). He wrote: "This delusion of the senses [and mind] makes disease seem where it really is not, from the different seat of the symptoms and of the diseased part. Thus, diseases of the liver are referred to the shoulder, of the testicles to the back, of the hip to the knee." (Hunter, 1835: 363)

Hunter disagreed with the usual view of the principle of "sympathy" – i.e., shoulder pain from disease in the liver "supposed to arise from the shoulder sympathizing with the liver. . . . But I believe it is a delusion in the mind. . . A delusion in the mind is an object appearing to be where it is not." (Hunter, 1835: 332)

Hunter then presents his theory of the mechanism of referred pain with the supposition that the seat of sensation in the brain splits part of the action of sensation in the brain (see Fig. 3). In Fig. 3, E represents the brain and A and B two portions of the brain, G and H two nerves which communicate by way of nerve F, and C and D two different parts of the body. Because of the connection between nerves, sensation may in part, be referred to D as well as C. But also the mind is aware of it, by way of brain part A having obtained sensation from C. But brain part B may also become aware of the sensation, by way of nerve F bringing the sensation to nerve H. Brain part B then refers it not to C, but to D, because that is the point that it has been accustomed to refer all of its sensation (Hunter, 1835: 332–333, Plate XVII, Fig. 3).

The above, may explain why we may feel both the disease and part of the sympathy, but in cases where the sympathizer (referral) of sensation takes on the whole action – all of the sensation passes by way of the communicating nerve (F), or alterna-

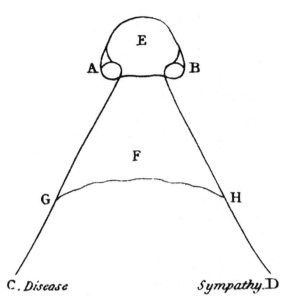

FIGURE 3. John Hunter's proposed mechanism of referred pain. From: Hunter (1837b), Plates from "The Works of John Hunter," p. 10, Plate XVII

tively by way of connections (between A and B) within the brain (Hunter, 1835: 333). In summary, he stated: "It is very remarkable that none of the sympathies can or ever are reversed, therefore they do not arise from the communication of the nerves, but [more likely] from the effect of the brain upon the nerves." (Hunter, 1861a: 275)

Hunter's theory of referred pain evoking the "mind" communicating with brain parts such as the "seats of the senses" clearly implies a nonholistic view of brain functions (see section "Apoplexy" below). A classic description of trigeminal neuralgia is given in a later section.

Muscular Function

Hunter extensively studied and experimented with the phenomena of motion in both plants and animals, and his observations were presented in six lectures to the Royal Society between 1776 and 1782. In accordance with his theory of constant adaptation of structure to function, Hunter stressed that the form and size of muscles are specifically adapted to their necessary usage and the nature of joints to be moved.

He detailed the structure and function of tendons and fascia, and their attachment to bone, muscles, and joints (Hunter, 1837a, Vol. 4: 228–241). He described the most simple muscle fiber in an animal as a bundle of fibers, distinct from end to end, and having one determined use. The muscles which move the globe of the eye came nearest to his idea of a distinct muscle (Hunter 1837a, Vol. 4: 224–225).

Hunter's experiments included water immersion tests on living animal muscles "To ascertain with as much precision as possible whether a muscle really alters in size or not when contracted." (Hunter, 1837a, Vol. 4: 258) His conclusions were that a muscle fiber is made up of [internal] "component parts" that undergo a change in their position during contraction (Hunter, 1837a, Vol. 4: 261). He concluded that "upon the whole, a muscle mass loses more of its length than it gains in thickness, unless the apparent difference arises from a universal approximation of all its parts." (Hunter, 1837a, Vol. 4: 260) Thus Hunter seems to have alluded to what would be later termed the sliding filament theory of muscle contraction.

Hunter believed relaxation of a muscle had to be an active process, or "a kind of negative action. . . a power as much depending on life as contraction." (Hunter, 1837a, Vol. 4: 250, 265) He even noted that, "Whatever becomes a stimulus to one set of muscles, becomes a cause of relaxation to those which act in a contrary direction." This line of thinking apparently led him to the principle of reciprocal innervation of antagonistic muscles (Hunter, 1794: 102; 1837a, Vol. 3: 146). This principle would later be developed and popularized by Charles Sherrington.

Muscles are of two types, wrote Hunter, one with a constant stimulus that does not tire (involuntary), and the other under the stimulus of the "will" (voluntary) that does tire. A mixed kind of muscle acts predominately by a natural stimulus, is predominately involuntary, and will not ordinarily tire, but can tire quickly if negatively exerted by the "will" (Hunter, 1837a, Vol. 4: 211; 1861a: 164).

Hunter commented that "one improvement voluntary muscles acquire from habit is the readiness with which they take up actions. . . . A man will learn one trade much more readily if he knows another, than if he knew no trade at all." (Hunter,

1837a, Vol. 4: 222) Furthermore the effect of habitual increased action on muscle (voluntary or involuntary) is often increase in size, and Hunter stated "whether this increase of the body of a muscle is a new addition of muscular fibers, or an increase in size of those already formed, is not easily determined, but I should be inclined to suppose the last." (Hunter, 1837a, Vol. 4: 223) Hunter was again correct in that assumption.

Sir Arthur Keith tells us Hunter regarded muscles as the most "educable" of all structures in the body, and stressed that a physician or surgeon must manage their treatment with an understanding of their patient's psychology (Keith, 1919). Hunter described patients with fractures and joint stiffness he brought back from the wheelchair to ambulation. This was done over months, beginning with passive motion and massage or "friction" (loosens and stretches muscle fibers) followed by active muscle strengthening (the patient must concentrate on the action) utilizing nongravity, gravity, and progressively increasing weights for resistance. Hunter also advocated gradual weight-bearing with crutches to heal nonuniting lower extremity fractures, thus utilizing the healing powers of Nature (Hunter, 1835: 511–513; Keith, 1919).

Involuntary Actions of Voluntary Parts

Hunter also described how cases of involuntary choreiform movements and torticollis, which seem to never tire, are worsened by anxiety and absent during sleep (Allen et al., 1993: 195; Hunter, 1835: 360; 1837a, Vol. 4: 212).

Muscle spasm arising from the muscle itself is usually attended with a painful sensation, but this is not always the case. "I saw a gentleman who had involuntary contractions of the orbicularis muscle of the left eye [blepharospasm], and the muscles of the angle of the mouth on the same side [hemifacial spasm], and there was no pain, not even in the muscles that acted." (Hunter, 1861a: 166) In that case, the antagonistic muscles [also] seemed to have lost their voluntary power of action, while the involuntary muscle spasms acted to some degree dependent on the "will" [worsened by anxiety] (Hunter, 1861a: 166).

Neurological Observations in Clinical Practice

Then, as today, the great majority of patients a surgeon such as John Hunter would see in his clinical practice did not require surgery – but consultation requested due to the difficult or critical nature of the case. In the preantiseptic era in which Hunter lived, surgical exposure of the abdominal or chest contents or incision of the dura and exposure of the brain were only rarely attempted, for it usually led to a fatal outcome secondary to infection.

Hunter utilized a number of common practices of the day, including blood-letting, gastrointestinal elimination or purging, rest, warm baths, sea bathing, alcoholic beverages, medicaments including opium, herbs, and the application of electricity. As noted, he was a reluctant and cautious surgeon, only performing an operation if he believed it were absolutely necessary. Hunter, who always questioned established principles, encouraged his pupils and fellow surgeons "to think of themselves as scientific professionals instead of merely as craftsmen." (Bynum, 1996)

Hunter's clinical-pathologic "casebooks," obtained by the RCS years after his death but only recently collated, well indexed, and published, disclose a careful history and thoughtful examination of hundreds of patients, many of whom had autopsies (Allen et al., 1993). The "casebooks" provide unique information on John Hunter's neurological views and clinicopathological findings. Hunter made use of an extensive, gross neuropathological technique, formulated well thought out deductions of a clinical–pathological nature, and often posed questions following each case.

Mr. Hunter's most famous and innovative operation was the cure of popliteal aneurysm by proximal ligation in the adductor (Hunter's) canal of the leg. In his day, carriage drivers wore high boots pressing into the popliteal fossa behind the knee during prolonged sitting, resulting in artery damage over years. The only treatment for such aneurysms, was amputation of the leg which carried a very high mortality. Hunter had experimented with deer and found that, if he ligated the external carotid arterial supply to one antler, it became cold, and he expected it would be shed. When that did not happen, he had the animal recaptured, felt the antler to

be normal in temperature, and had the animal sacrificed. To his surprise, the ligated external carotid artery had stimulated collateral arterial circulation and sustained the antler. This experiment led Hunter in 1785 to not amputate the leg of a patient with a popliteal aneurysm, but rather ligate the artery well proximal, to allow collateral circulation to develop above the ligature and sustain the leg. Patients were subsequently cured with this operation (Dobson, 1969: 259–261; Hunter, 1835: 548; Moore, 2005: 8; Qvist, 1981: 70, 109–110). "Hunterian Ligation" for various vascular lesions is still utilized today, often as adjuvant therapy for cerebral aneurysms, arteriovenous fistula, and other lesions in neurosurgery and vascular surgery.

Head Injury and Surgical/Pathological Observations

In John Hunter's time, as today, a surgeon would frequently be called to see a head injury patient, and consider the advisability of a trephination operation. Hunter was more conservative than his teacher Percivall Pott in the usage of the trephine and opening the dura mater (Flamm, 1992; Stone & Hockley, 2002), but made astute observations on concussion, brain compression, laceration (contusion), and skull factures (Ballance, 1922). Hunter reiterated the notions that a skull fracture in itself does not produce brain symptoms, and the trephine should be applied in all cases of depressed skull fracture.

"The dura mater must not, however, be perforated without good grounds; we should be. . . certain that there is fluid contained under the dura. . . ." (Hunter, 1835: 494–495) Hunter appears to have been the first to trephine with success in the posterior fossa (Hunter, 1835: 495). Hunter, like Pott, stated that if symptoms of brain dysfunction continue for 8 or 10 days or less, extravasation (blood clot) is expected, and if symptoms come on over 1 week later and especially 3–4 weeks, suppuration (pus) is more likely present (Hunter, 1835: 488).

In John Hunter's "casebooks," there are 12 patients trephined for complications secondary to head injury (Allen et al., 1993: 428). Eight of these patients died, especially if the dura was open (or opened), and three lived, including a child who had lost a "table-spoon-full" of brain tissue and had a cerebral fungus treated successfully with a poultice (dressings) applied (Allen et al.: 428).

An additional 28 head injury patients were examined, observed, and treated without operation. Hunter often made note of pupil size and reactivity, the level of consciousness, presence and quality of speech and behavioral responses, and motor weakness if present. In cases with autopsy (many of those who died), Hunter performed an elaborate gross brain examination technique (see section "Apoplexy"). Patients with spinal cord injuries are also included in Hunter's "casebooks." Hunter performed surgical excisions for orbital and skull tumors, nerve compression, tumors of the neck, face, and nasal cavities (Allen et al., 1993). Hunter was the first to excise a peripheral nerve tumor (Home, 1800). He treated a number of patients with eye problems and was very interested in their visual fields, having detected a loss in the upper visual field of his own left eye, distinct from the usual "blind spot" related to the optic nerve entrance into the globe (Allen et al., 1993: 70). He also was interested in ear and hearing problems, and carefully described the postmortem appearance of otitic brain abscess (Allen et al., 1993: 543).

Hunter examined a soldier with a solid tumor (not operated), which at autopsy was found to be situated largely in the dura. It had invaded the skull and deformed the underlying brain (likely a meningioma) (Hunter, 1837a, Vol. 3: 472–473). Another similar case was presented in the "casebooks" (Allen et al., 1993: 241).

Hunter described nine cases of hydrocephalus, eight with autopsies documenting the volume of ventricular fluid, thickness of cortical mantle, and other neuroanatomic features including the state of the optic nerves and pituitary gland. He also described spina bifida, anencephaly, and a two-headed calf (Allen et al., 1993).

He performed a number of autopsies on patients with brain abscesses, tuberculoma, and hydatid cysts (Allen et al., 1993). Of a large hydatid cyst, he stated: "This case. . . [shows] . . . that the brain can suffer very much mechanically before its uses and actions suffer; but when it comes to that period where the brain is. . . affected; why [deterioration] should be so quick [at times] is not easily accounted for." (Allen et al., 1993: 546) The problem of rapid decompensation from increased intracranial pressure related to structural brain lesions remains a significant challenge in surgical neurology even today (Fig. 4).[2]

Apoplexy and Other Neurological Disorders

Hunter treated many patients with apoplexy, and made a number of astute clinical and pathological observations. He often tried to correlate the clinical and pathological findings and made deductions or posed queries. Again, his "casebooks," which presented roughly 30 such cases, provide much of this information (Allen et al., 1993).

Hunter usually treated apoplexy or sudden conditions of paralysis with blood-letting, as much as the patient could tolerate, followed by other methods of the day (Cooke, 1820: 335). Furthermore, Mr. Hunter believed the blood-letting for apoplexy should be done from an area close to the head (jugular vein) or on the scalp (temporalis artery) if possible (Cooke: 306–308; 1821: 135–137). Hunter's clinical evaluation included behavioral observations, such as sensibility, memory, reasoning, comprehension, speech, visual, and sensory-motor functions.

In one of his cases, a man with a sudden left hemiplegia was also found to be speechless. The patient had a diminished level of consciousness, would open his mouth to command, but was unable to stick out his tongue. "When he died," he wrote, "his head was opened; and near half a pint of

[2] Dr. Matthew Baillie (1761–1823), the nephew and student of Dr. William and John Hunter, is credited with the first monograph on human pathology (Baillie, 1793). One chapter is devoted to the brain, but it does not go into the detail typified by John Hunter's "casebooks." Fascicle # 10 of Baillie's pathology atlas (Baillie, 1802) on the cranium and brain, includes 18 figures on eight plates. Specimens in the atlas from Mr. Hunter's Museum include: Plate 1, Fig. 2 a tumor from the orbit into the cranial cavity, likely a meningioma; Plate 2, Fig. 1 tertiary syphilis, Plate 3, Figs. 1 and 2 hydrocephalus patients; Plate 4, Fig. 1 chronic meningitis; Plate 6 a cerebral abscess; and Plate 7, Fig. 4 cystic choroid plexus lesions (Baillie, 1802; Compston, 1999).

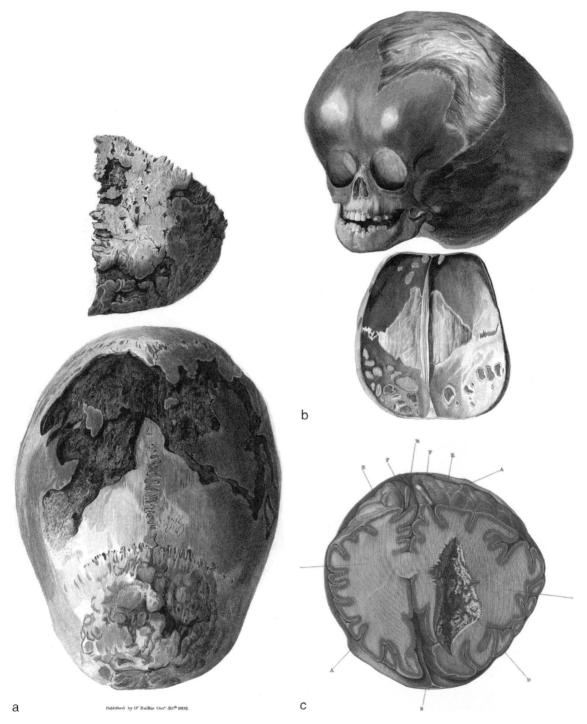

a

Published by Dr Baillie Oct. 30th 1802.

b

c

FIGURE 4. John Hunter's cases of (**a**) tertiary syphilis or cancer of the skull, (**b**) hydrocephalus, and (**c**) brain abscess. From Baillie (1802), A Series of Engravings to Illustrate the Morbid Anatomy of the Human Body, Plates 2, 3, and 6

coagulated blood was found lying on the left side, under the left anterior lobe of the brain; Observe, it was on the paralytic side." (Allen et al., 1993: 26–27)

In another patient with right side weakness, he wrote: "she recovered her faculty of discerning objects, distinguishing them particularly, and had a thorough knowledge of the questions put to her; but her articulation continued so bad, that her answers could only be known by the faint signs she made use of." Examination of the brain after death disclosed enlargement of the left lateral ventricle ". . . and in the anterior part of the brain appeared not so firm, but rather mashed. All other parts appeared quite sound." (Allen et al., 1993: 34–35)

Hunter also described patients with transient attacks and reversible problems. In his autopsy examination of the brain, Hunter occasionally referred to "ossification of arteries," such as the internal carotid or basilar (Allen et al., 1993: 385), in all likelihood referring to atherosclerosis. In the case of a woman, who was initially unconscious, vomited, developed a right hemiplegia and "talked in a rambling manner. . . not very distinctly, nor very reasonably," an autopsy revealed 3–4 oz of extravasated blood in the left middle (temporal) lobe with rupture into the ventricles. "It is most likely that the vomiting at the first attack was caused by the beginning of extravasation of blood; making first a slight pressure; not the rupture of the vessels by the vomiting as was supposed." (Allen et al., 1993: 345)

Hunter described cases of subarachnoid hemorrhage associated with violent head pain just before loss of consciousness (Allen et al. 1993: 348–349). He is also credited, in 1792, with the first clear pathologic description of a cerebral aneurysm, in a woman with bilateral cavernous carotid artery aneurysms (Blane, 1800; Bull, 1962). In 1757 Hunter described cardiac valvular vegetations (Allen et al.: 300). A 1793 case of probable rheumatic heart disease with hemorrhagic brain infarction secondary to emboli is also presented (Allen et al.: 528–529). In one apoplexy patient with such vegetations, Hunter observed the hemispheric cortex showing "red dots like cut ends of vessels, many of which were as large as a large pin's head, which were plainly extravasations." (Allen et al.: 358–359) The latter case (year unknown) likely represents one of the earliest descriptions of multiple small infarcts or bleeding points due to cerebral thrombosis or emboli from cardiac vegetations.

Several patients with apoplexy presented interesting visual problems. In one patient blindness followed a severe headache. For one such case, he wrote: "There was a kind of stare in his eye; and although he could not see, yet the pupils were not dilated; therefore I suspected that he did see; but from all the trials I could make it was plain that he did not; nor did . . . [the trial results] . . . alter by varying the light." (Allen et al., 1993: 527–528) Before his death about 8 weeks later, this patient recovered some sight gradually and he could describe what he saw. Autopsy showed extravasation of blood had occurred in the posterior lobe of the right hemisphere with extension into the ventricles.

John Hunter's father-in-law, Mr. Home, suddenly lost all sight, and almost an entire loss of memory.

It brought him in a great many circumstances to the state of a brute, or at least what I can conceive that state is, respecting reasoning. . . . The total loss of sight [and memory problems] produced a very curious effect: he lost entirely the remembrance of light, and did not annex any idea to light, although he would say that he had not seen you for some time, meaning you had not been there for some time. (Allen et al., 1993: 548–549)

Autopsy of the brain showed extensive bilateral destruction of the occipital lobes, as well as the undersurface of the medial posterior temporal lobes of both hemispheres adjacent to the tentorium. Today we would say the findings of pronounced visual and memory loss is compatible with occlusion of both posterior cerebral arteries, resulting in infarction of the occipital and posteromedial temporal lobes.

A case is described with mild memory loss, a curious faltering of pronunciation only when reading and not in regular speech, and difficulty concentrating on two things at the same time. Hunter explained these findings in terms of absolute sensations being the first and strongest of all impressions; and memory, reading, and simultaneous motor actions under the will being secondary and "rather an act of habit" are sooner lost "from a fault in the brain." (Allen et al., 1993: 193, 413–414) This description may be an example of reading apraxia.

In a most instructive case with right sided weakness, the patient

was perfectly sensible, but could not speak. As his attack was truly apoplectic; as he lived fourteen weeks, and recovered perfectly his sense; although not the uses of the parts, I was desirous of seeing what Nature had been doing those fourteen weeks towards restoring the parts injured. On removing the skull . . . above 4 table-spoonfulls of left sided subdural fluid, and slightly sunken and flattened left parietal convolutions were found. [This part the brain tissue] was irregular in its consistence, appearing ragged, or as if broken into a pulpy substance: there was no distinction between cortical and medullary. . . . In those who have died soon after an attack of apoplexy, I have often seen the extravasated blood mixed with the brain as if both were beaten up together. . . . I conceive this case that the extravasated blood had been absorbed as a first step; but the substance of the brain had not recovered it's natural texture. (Allen et al., 1993: 209–210)

This pathologic designation of nonhemorrhagic or "absorbed" apoplexy, with loss of the cortical gray/white border, may well represent the first adequate description of brain infarction. Another case seems to be a description of lacunar infarcts within the hemispheres – "these appeared to be (bean sized) cavities formed. . . [by] former [small] extravasations of blood. . . and were not lined with a membrane." (Allen et al., 1993: 551–552)

Hunter concludes:

To destroy the power of voluntary action by injuring a part only, and yet not the Sensorium, would shew (sic.) that they do not belong to the same parts in the brain. It seldom happens in these cases that sensation of the affected parts is destroyed; therefore while sensation can exist the Sensorium must be in some measure clear [separate]. (Allen et al., 1993: 210)

This and earlier quoted statement could well be interpreted that Hunter believed the brain is divided into autonomously functioning separate parts, at least involving motor, speech, sensory functions consciousness and the mind. Although a handful of cases with speech problems and left-sided brain pathology were described by Hunter, nowhere does he equate these findings.

Miscellaneous Neurological Disorders

Hunter described a number of children and adults with "fits" and "seizures," which were sometimes noted to be more common in sleep (Allen et al., 1993: 376, 419). He also described cases of infants and small children with seizures related to fever,

encephalitis, and postinoculation (Allen et al.: 154, 162, 175). Hunter believed a "convulsion" was an action that arose from a state of debility, and he often saw it as a futile struggle as death approached, though he did not elaborate further. (Hunter, 1835: 361)

One interesting case from 1783 was a woman aged 20 years with unusual spells often associated with vomiting

attended with [violent headache and] a kind of fit; but she hardly lost herself; heard them talk about her, but did not know what they said. Soon after, in some degree she lost the use of her left arm; it felt numb, and a vomiting succeeded, which continued till (sic.) next day. (Allen et al., 1993: 419–420)

Her headache lessened after Hunter removed 6 oz of venous blood, in an episode that would appear to have been a partial complex seizure or complicated migraine disorder. A somewhat similar case was also reported (Allen et al., 1993: 306).

Nervous problems and pain were of much interest to Hunter and, included in his 1778 treatise on diseases of the teeth, is an interesting chapter entitled "Nervous Pains in the Jaws."

There is one disease of the jaws which seems in reality to have no connexion with the teeth, but of which the teeth are generally suspected to be the cause . . . a tooth is . . . drawn out; but still the pain continues . . . it is then supposed . . . that the wrong tooth was extracted whereupon that in which the pain now seems to be is drawn, but with as little benefit. I have known cases of this kind where all the teeth of the affected side of the jaw have been drawn out, and the pain has continued in the jaw; in others . . . the sensation of pain has become more diffused, and has at last, attacked the corresponding side of the tongue. Hence is should appear that the pain in question [trigeminal neuralgia, tic douloureux] does not arise from any disease in the part, but is entirely a nervous affection. (Hunter, 1778: 61–63)

This early contribution to trigeminal neuralgia appears to have been largely overlooked in related reviews.

Conclusions

It had been stated that John Hunter's contributions to medicine and surgery have probably not been surpassed by any individual before or after his time, and he "was the greatest man in the combined character of physiologist and surgeon that the whole annals of medicine can furnish." (Ottley, 1835: viii) Every part of biological science was influenced by Hunter's physiological principles induced by tireless clinical and anatomic observation, inventive experimentation, and pathologic findings at autopsy. He laid the groundwork for a scientific, modern approach to medicine and surgery, and contributed significantly to the neurosciences.

John Hunter's inquisitive nature, empirical methods and carefully formulated principles would become adopted by future biological scientists who would examine all systems including the nervous system, with newer tools. A number of his neuroscience and neurological assumptions have now been validated or are still being debated today. Hunter made his mark by boldly exploring the body and seeing things with his own eyes, not by adhering to the speculative theories of the past. By casting aside traditional values he saw much more clearly than other contemporaries, captivated the brightest students with his free thinking approach, and opened the eyes of future generations.

Acknowledgment. The authors wish to thank Ms Tobi McCauley for assistance with preparation of the manuscript and illustrations, as well as communication with the RCS in London.

References

Abernethy, J. (1825). *Physiological lectures. A general view of Mr. Hunter's physiology and of his researches in comparative anatomy.* Hartford, England: Cooke & Co.

Allen, E. (1974). *Hunterian museum.* London: Royal College of Surgeons.

Allen, E., Turk, J. L., & Murley, R. (1993). The casebooks of John Hunter FRS. New York: Parthenon Press.

Atkinson, W. J. (1951). Surgery of the autonomic nervous system. In: Walker, A. E. (Ed.). *A history of neurological surgery* (pp. 428–450). Baltimore: Williams & Wilkins..

Baillie, M. (1793). *The morbid anatomy of some of the most important parts of the human body* (pp. 288–314). London: Johnson and Nicol..

Baillie, M. (1802). *A series of engravings with explanations to illustrate the morbid anatomy of some of the most important parts of the human body.* Fascicle # 10, Disease Appearances of the cranium, the brain and its membranes. London: Bulmer & Co.

Ballance, C. A. (1922). *A glimpse into the history of the surgery of the brain* (pp. 60–67). London: Macmillan & Co.

Blane, G. (1800). History of some cases of disease in the brain, with an account of the appearances upon examination after death, and some general observations on complaints of the head. *Transactions of the Society for Improvement of Medical and Chirurgical Knowledge,* 2: 192–212.

Bull, J. (1962). *A short history of intracranial aneurysms. London Clinical Medical Journal,* 3: 47–61.

Bynum, W. F. (1996). *Science and the practice of medicine in the nineteenth century* (pp. 5, 14). Cambridge, England: Cambridge University Press.

Compston, A. (1999). The convergence of neurological anatomy and pathology in the British Isles: 1800–1850. In: F. C. Rose (Ed.), *A short history of neurology: The British contribution 1660–1910* (pp. 36–57). Oxford: Butterworth-Heinemann.

Cooke, J. (1820). *A treatise on nervous diseases* (Vol. 1). *On apoplexy.* London: Longman, Hurst, Rees, Orme and Brown.

Cooke, J. (1821). *History and method of cure of the various species of palsy.* London: Longman, Hurst, Rees, Orme and Brown.

Dobson, J. (1969). *John Hunter.* Edinburgh: E & S Livingstone.

Flamm, E. S. (1992). Percivall Pott: An 18th century neurosurgeon. *Journal of Neurosurgery,* 76: 319–326.

Fulton, J. F. (1946). *Harvey Cushing. A biography* (p. 303). Springfield, IL: Charles C. Thomas.

Home, E. (1800). An account of an uncommon tumour formed in one of the axillary nerves. *Transactions of the Society for Improvement of Medical & Chirurgical Knowledge,* 2: 152–163.

Hulke, J. W. (1895). John Hunter the biologist. *British Medical Journal,* 1: 405–410.

Hunter, J. (1778). *A practical treatise on the diseases of the teeth – A supplement to the natural history of human teeth.* London: J. Johnson.

Hunter, J. (1794). *A treatise on the blood, inflammation, and gun-shot wounds.* London: John Richardson.

Hunter, J. (1835). In J. F. Palmer (Ed.), *The works of John Hunter F.R.S.* (Vol. I). London: Longman, Rees, Orme, Brown, Green and Longman.

Hunter, J. (1837a). In J. F. Palmer (Ed.), *The works of John Hunter F.R.S.* (Vol II–IV). London: Longman, Rees, Orme, Brown, Green and Longman.

Hunter, J. (1837b). In J. F. Palmer (Ed.), *The works of John Hunter F.R.S., PLATES.* London: Longman, Rees, Orme, Brown, Green and Longman.

Hunter, J. (1861a). In R. Owen (Ed.), *Essays and observations on natural history, anatomy, physiology, psychology, and geology* (Vol. I). London: John Van Voorst.

Hunter, J. (1861b). In R. Owen (Ed.), *Essays and observations on comparative anatomy* (Vol. II). London: John Van Voorst.

Jacyna, S. (1992). Physiological principles in the surgical writings of John Hunter. In C. Lawrence (Ed.), *Medical theory, surgical practice. studies in the history of surgery* (pp. 135–152). London: Routledge.

Keith, A. (1919). The orthopaedic principles of John Hunter. In A. Keith (Ed.), *Menders of the maimed* (pp. 1–17). London: Oxford Press.

Kobler, J. (1960). *The reluctant surgeon. A biography of John Hunter.* Garden City, NY: Doubleday & Co.

Liddell, E. G. T. (1960). *The discovery of reflexes.* Oxford: Clarendon Press.

Moore, W. (2005). *The knife man. The extraordinary life and times of John Hunter, father of modern surgery.* New York: Broadway Books.

Morris, H. (1909). John Hunter as a philosopher. *British Medical Journal, 1*: 445–451.

Neuburger, M. (1897/1981). In E. Clarke (Ed.), The historical development of experimental brain and spinal cord physiology before Flourens. 1981 reprint. Baltimore: Johns Hopkins University Press.

Ottley, D. (1835). The life of John Hunter. In J. F. Palmer (Ed.), *The works of John Hunter F.R.S.* (pp. xxi–xxiv, 1–188). London: Longman, Rees, Orme, Brown, Green and Longman.

Owen, R. (1837/1992). In P. R. Sloan (Ed.), *The Hunterian lectures in comparative anatomy.* May and June, 1837, 1992 reprint. Chicago: University of Chicago.

Owen, R. (1837). Preface to observations on certain parts of the animal oeconomy, by John Hunter F. R. S. In J. F. Palmer (Ed.), *The works of John Hunter F.R.S.* (Vol. IV, pp. xvi–xviii). London: Longman, Rees, Orme, Brown, Green, and Longman.

Owen, R. (1866). *On the anatomy of vertebrates.* Vol I. *Fishes and reptiles.* London: Longmans, Green, & Co.

Paget, S. (1897). *John Hunter. Man of science and surgeon 1728–1793.* London: T. Fisher Unwin.

Pettigrew, T. J. (1839–1840). *John Hunter, F.R.S., medical portrait gallery* (Vol. 2, pp. 1–14) London: Fisher Sons.

Qvist, G. (1981). *John Hunter 1728–1793.* London: Heinemann

Stone, J. L., & Hockley, A. D. (2002). Percivall Pott and the miners of Cornwall. *British Journal of Neurosurgery, 16*: 501–506.

6
William Cullen (1710–1790) and Robert Whytt (1714–1766) on the Nervous System

Julius Rocca

Introduction: Eighteenth Century Theories of Nervous Action

The long eighteenth century witnessed the full flowering of the Scottish Enlightenment, graced by figures such as Robert and James Adam, the architects; the chemist and physician Joseph Black, the discoverer of carbon dioxide ('fixed air'); Adam Ferguson, founder of the discipline of sociology; James Hutton, founder of modern geology; the philosopher David Hume and the political economist and moral philosopher Adam Smith. (Clayson, 1993). Their achievements amply demonstrated the characteristic pragmatic drive of the Scottish Enlightenment with its emphasis on the acquisition and promulgation of practical knowledge.

In such an atmosphere, where political patronage also played an important, albeit often underestimated role, it is little surprise that the Edinburgh Medical School, founded in 1726, eclipsed Leiden in fame and prestige, and would develop into one of the foremost institutions of its type in the world (Emerson, 2004; Phillipson, 1981). Its faculty numbered such eminent figures as Alexander Monro *primus* (1697–1767), the first and most famous holder of the Chair of Anatomy; his son, Alexander Monro *secundus* (1733–1817), who had a remarkable reputation as a teacher of surgery; and John Rutherford (1695–1779), professor of the practice of physic. Another key factor in the medical school's success was the clinical instruction provided by the Edinburgh Royal Infirmary. William Cullen, who counted Hume and Smith as colleagues and friends, played a signifi-

cant role in its reputation. But as well as promulgating clinical expertise, the Scottish Medical Enlightenment was deeply concerned with research into the function of the body, that is, with physiology, or, as it was then known, 'animal œconomy'. The three principal physiological theories of the eighteenth century which sought to account for nervous action were enumerated first by Herman Boerhaave (1668–1738), the dominant medical figure of the age, whose concept of a nervous fluid was in accord with his general principle that the human body is a complex hydraulic system regulated by a series of fluid dynamics in the vasculature and nervous systems. Health was due to the correct flow of these fluids; disease by their stoppage or interference. In this way, the key tenet of balance represented by the ancient humoural system "had thus been preserved but translated into mechanical and hydrostatic terms." Porter (1996, p. 162) Yet although Boerhaave's conception of the function of the body was not as overtly mechanistic as that of Descartes, the body was still conceived of as a machine.

Georg Ernst Stahl (1660–1734), founder of the Prussian medical school at Halle, reacted against such mechanistic theories. For Stahl, the soul (*anima*) "represented the vital principle (*principium vitae*) . . . the director of all movement [which] required no transmitting nerve fluid to act upon the body, which was thought of as a dead mechanism." Neuburger (1897, p. 114) The soul was not only the agent of consciousness but also of physiological function. Stahl's theory of *animism*, however, raised two main objections. The first was in what way could the soul, itself an immaterial entity, act on corporeal matter? The second, equally as pressing, was that in this scheme, the soul was a general principle that did not require the brain and nerves to have any separate, let alone defined, role. Stahl's system, based more on speculative philosophy than on experimentation, amounted to a gainsaying of the primacy of the nervous system. In contrast, the Swiss polymath,

Albrecht von Haller (1708–1777), who, like Whytt was a pupil of Boerhaave, sought to ground physiological function in experimentation. *Haller's Elementa Physiologiae Corporis Humani* (1757–1766) was a groundbreaking work. In it, Haller demonstrated the Dorset physician Francis Glisson's (1597–1677) hypothesis that all tissues, including muscle, contain an intrinsic contractility or 'irritability' (the *vis insita* or *vis nervosa*). Haller distinguished between nerve and muscle on the basis that the former exclusively possessed 'feeling' or sensibility, whilst the latter possessed the contractility or irritability (Garrison, 1929). Haller's work "differentiated organ structures according to their fibre composition, ascribing to them intrinsic sensitivities independent of any transcendental . . . soul." Porter (1996, p. 165).

The guiding spirit of Scottish research into the animal economy was governed in large measure by a reaction to these models of nervous action. It is no exaggeration to state that in the field of eighteenth century neurophysiology, Robert Whytt's name is paramount; his work on demonstrating spinal reflex action and the elucidation of the pupillary light reflex are sufficient to guarantee him a place as one of the most important figures in the history of the neurosciences, and as a founder of the discipline of neurophysiology (Rocca 2000, 2003, 2007). Whytt disagreed with Stahl's uncompromising animism, but he also viewed Haller's work as mechanistic in nature. Moreover, Haller maintained the *vis insita* to be independent of the nerves. As will be noted below, Whytt rejected overt animistic accounts of nervous action, denied the concept of animal spirits and argued forcefully against Haller's notion that muscle contractility was independent of the nerves.

William Cullen was also concerned with the nervous system, but he differed from Whytt in two ways. First, Cullen agreed with Whytt that Boerhaave's system of hydraulic fluids stood in need of correction. But unlike Whytt, Cullen accepted Haller's notion of irritability, eliding the difference Haller made between nerve and muscle, elaborating it to account for nervous function and nervous disease. Second, whilst Whytt's energies were devoted principally to physiological research, Cullen placed greater stress on medical classification, his guides here being, among others, Boissier de Sauvages and Linnaeus; on systematisation (principally in nosology); and, not least, on the promulgation of medical knowledge. That knowledge consisted of a broad understanding of the nervous system in health and disease. Both men, to be sure, never relegated the nervous system to a subservient role or subsidiary function. Although both were contemporaries, it was Cullen who succeeded Whytt as professor of the institutes of medicine on the latter's death, and Whytt's teaching was also a broad influence on Cullen's conception of the nervous system. Therefore, Whytt will be discussed first.

Robert Whytt

A Short Biographical Background

Robert Whytt was born in Edinburgh in September 1714, the second son of a Scottish lawyer who died before his son's birth. Whytt's mother died when he was six, and he was placed under the guardianship of his older brother, who died in 1728. He matriculated at Edinburgh University in 1729, and then studied medicine at Edinburgh in 1731–1732 and 1734, under Alexander Monro *primus*. Monro's scepticism regarding the existence of any sort of nervous fluid was surely an influence on Whytt's subsequent work on this subject. In 1734, with Monro's recommendation, Whytt moved to London to continue his studies under William Cheselden (1688–1752). He completed his education in Paris, attending lectures by Jacob Winslow (1669–1760), and in Leiden under Hermann Boerhaave and his student Bernhard Albinus (1697–1770). In 1736, Whytt took his MD at Rheims, and in 1737 was awarded an MD by the University of St. Andrews and granted licentiate of the Royal College of Physicians of Edinburgh. In 1738, Whytt was elected a Fellow of the College and commenced medical practice. Whytt's modification of an existing preparation for bladder calculi, published in 1743, made his reputation and gave him financial security in medical practice. It was Whytt's work on calculi that influenced Joseph Black's experiments on alkalies, and which led in 1754 to his discovery of carbon dioxide. In 1747, Whytt was appointed Professor of Theoretical Medicine (Medical Institutions) in the University of Edinburgh, which he held until his death. In 1752, Whytt was elected a Fellow of the Royal Society. In 1761, he was made First Physician to the King in Scotland, and in 1763 was elected President of the Royal College of Physicians of Edinburgh.

Whytt and Nervous Action

The experimental work of Jan Swammerdam (1637–1680), Stephen Hales (1677–1761) and Alexander Stuart (1673–1741), who repeated and published Hales' work in 1738, "paved the way for the evolution of thought which was to develop the concept of the spinal reflex arc." Pearn (2002, p. 24) Whytt's claim to greatness lies in his establishment that the spinal cord was absolutely crucial in mediating what Marshall Hall (1790–1857) would later describe as 'reflex action', for Whytt did not use the term 'reflex' himself. As Whytt stated in his *Physiological Essays* (1755):

After the destruction of the spinal marrow . . . the nerves distributed to the several parts of the body have no communication but at their termination in the brain or spinal marrow and that to this, perhaps alone, is owing the consent or sympathy observed between them.

For the first time, it was shown that only a discrete portion of the spinal cord was required to mediate the reflex arc. From these and similar experiments:

Whytt concluded that the nerves communicated only at their sites of origin (in the brain or spinal cord) and that possibly this communication accounted for the sympathy between the various parts of the body . . . Whytt here revealed a premonition of the true reflex concept by tracing the phenomena of sympathy, the phenomena of the consensus that played such an important role in early pathology and to some extent coincided with the reflex phenomena, back to a nerve impulse leaping from one nerve root to another. Neuburger (1897, p. 157)

Whytt was also the first to clearly and succinctly describe 'hydrocephalus internus', or tuberculous meningitis of children (*Observations on the Dropsy in the Brain*, published posthumously by his son in 1768). His findings, based on autopsy findings, are a masterpiece of meticulous description. A similar methodology enabled Whytt to elucidate the pupillary light reflex. This was based on the autopsy of a child with hydrocephalus whose pupils had been noted to have been non-reactive to light. The autopsy showed a cyst compressing the optic thalamus and Whytt deduced from this that this obstruction prevented the light reflex. Also noteworthy for their series of meticulous observations was Whytt's 1764 work, *Observations on the Nature, Causes and Cure of Those Diseases Which are Commonly Called Nervous, Hypochondriac or Hysteric*. This was based on his observations of patients at Edinburgh's Royal Infirmary, and was remarkable not only for its clinical expertise and diagnostic erudition, its accurate descriptions of referred pain, but also in part based on Whytt's dissatisfaction with the classification of those diseases described as 'flatulent', 'spasmodic', 'hypochondriac', 'hysteric' and 'nervous'. Whytt thought such descriptions merely reflected a physician's ignorance, and drew on his own work to try and explain that the nature of these diseases was rooted in the "sentient and sympathetic power of the nerves". Cited in Radbill (1976, p. 322)

Whytt's scientific reputation rests with his formidable experimental work on the physiology of the nervous system. His first study on this subject, *An Enquiry into the Causes which Promote the Circulation of Fluids in the Small Vessels of Animals* (1745/1746; in *Works*, p. 1768), consisted of an investigation into the 'sentient principle', which he believed governed voluntary and involuntary motions. This principle, also referred to by Whytt as 'the active power of an immaterial principle' or 'sentient principle', constituted the epistemological foundation of Whytt's work on the nervous system. This research was driven in part by Whytt's dissatisfaction with the several speculative theories outlined above, which sought to account for nervous action. Whytt disapproved both of their number and speculative nature, expressing himself forcefully in his 1751 *An Essay on the Vital and other Involuntary Motions of Animals*, as follows:

. . . whatever has been hitherto said on these subjects, is mere speculation; and to offer any new conjectures on matters so greatly involved in darkness, and where we have neither *data* nor *phænomena* to support us, is to load a science already labouring under *hypothesis* with a new burden. *An Essay* (1751, pp. 325–326)

This physiological scepticism extended to pathological conditions; in nervous disorders, according to Whytt, "faults may arise in the coats, the medullary substance or in the fluid of the nerves." *Works* (1768, p. 526)

Whytt denied unequivocally that muscular contraction was due to an influx of so-called animal spirits, the origins of which go back to Galen, or that there was a Newtonian *æther*, electrical fluid responsible. Nor were the muscles themselves responsible for contraction, independent of the nerves, as Haller claimed. That there was some form of *fluid* present in the nerves was admitted, albeit cautiously; according to Whytt, it was not possible to determine ". . . whether this fluid serves only for the support and nourishment of the nerves or whether it be not the medium by which all their actions are performed." *Works* (1768, p. 489) Whytt's own position was that there existed a 'superior principle', which depended on the presence of the soul which acted upon the body. This has resulted in Whytt being labelled a supporter of Stahlian animism. This, however, is not entirely accurate, for eighteenth century animism relegated the brain and the nerves to a subservient role, which Whytt never maintained.

By the end of the seventeenth century, there was a widely acknowledged concept of the nerves existing as channels for the passage of animal spirits. Yet how these spirits passed through the nerves was a matter of some dispute. Some thought they circulated in the body in a similar way to the blood (Clarke, 1978). A way of seeking to understand the physiology of the nerves was by an appeal to the wider authority claimed by chemistry and physics, as Willis did with his chemical explanation for nervous action. Giovanni Borelli (1608–1679) proposed a *succus nerveus*, to replace that of the animal spirits. This *succus*, however, acted in a similar way to Willis' nervous spirits. Yet its activity was strictly in accordance with then-current physical laws. To Borelli and his fellow iatrophysicists, "the laws of physics offered the key to the body's operations." Porter (1996, p. 160) That nerve spirits could be replaced by a *succus nerveus*, or by a *vis insita* or *vis nervosa*, was a deliberate echo of the *vis attractiva* of Newtonian physics, for Newton's

conception of an *æther* would prove a useful analogy to account for nervous action (French, 1981). Haller's *vis nervosa* was if nothing else a rephrasing of the concept of nervous spirits. Haller and Whytt's teacher, Boerhaave, believed the ventricles of the brain continuously exhaled a "thin Vapour or moist Dew", which was a "most subtle Fluid." Cited in Brazier (1973, pp. 199–200)

Another influence on Whytt's later cautious view of nervous fluid must have been his first teacher in Edinburgh, Alexander Monro *primus*, who noted that:

We are not sufficiently acquainted with the Properties of an Aether [or *electrical effluvia*] pervading every Thing, to apply it [them] justly in the animal Oeconomy, and it is as difficult to conceive how it [they] should be retained [or conducted] in a long nervous Cord as it is to have any Idea how it should act. These are Difficulties not to be surmounted. *Works* (1781, p. 333)

Whytt's physiology recognised the existence and importance of the soul, and the concept of a purposeful divine universe in which nothing happened without a cause. Nature operates in such a way that there are certain laws which the body and the soul follow in order to function. Whytt expresses this as follows:

Nature, as far as we can judge from the plan and scheme of things surrounding us, delights in simplicity and uniformity, and, by general laws applied to particular bodies, produces a vast variety of operations; nor is it at all improbable that an animal body is a system regulated much after the same manner. *An Essay* (1751, p. 4)

To Whytt, Nature operates by reasonably general laws from which derive a great number of physical activities. It is this simple, ineluctable fact which mechanists have failed to appreciate, since:

. . . all the spontaneous motions of animals are explicable upon the same principle, and owing to one general cause. How far some authors of great note have been unsuccessful in their inquiries into this matter, from their neglecting so obvious an analogy, and endeavouring to explain the vital motions of almost every organ, by a different theory, is left to the Reader to judge. *An Essay* (1751, p. 4)

Whytt's strong teleology is well conveyed in the conclusion to *An Essay on the Vital and other Involuntary Motions of Animals*, and his powerful stating of it is worth quoting at length for the insight it affords on his view of the nervous system:

If the human frame is considered as a mere CORPO-REAL system, which derives all its power and energy from matter and motion; it may, perhaps, be concluded, that the IMMENSE UNIVERSE itself is destitute of any higher principle: but if, as we have endeavoured to shew, the motions and actions of our small and inconsiderable bodies, are all to be referred to the active power of an IMMATERIAL principle; how much more necessary must it be, to acknowledge, as the Author, Sustainer, and Sovereign Ruler of the universal system, an INCORPO-REAL NATURE every where and always present, of infinite power, wisdom, and goodness; who conducts the motions of the whole, by the most consummate and unerring reason, without being prompted to it by any other impulse, than the original and eternal benevolence of his nature! ... The true Physiology, therefore, of the human body, not only serves to confute those Philosophers, who, rejecting the existence of IMMATERIAL BEINGS, ascribe all the *phænomena* and operations in nature to the powers of matter and motion; but, at last, like all other sound Philosophy, leads us up to the FIRST CAUSE and supreme AUTHOR of ALL, who is ever to be adored with the profoundest reverence by the reasonable part of his creation. *An Essay* (1751, pp. 391–392)

The agent responsible for nervous action, for 'the motions and actions of our small and inconsiderable bodies', is an immaterial active power or soul. The soul is coextensive with the body, which allows one to account for both voluntary and involuntary motions, since:

In every organ it (sc. the soul) exercised the faculties of life necessary to that part of the body. Only in the brain, at the *sensorium*, does the soul have the faculties of reason and consciousness; in the muscles it had the power of producing motion; in the nerves alone it exercised the ability to feel, and in the sense organs it was only sensitive to those physical or chemical stimuli for which the organ had been designed. French (1972, p. 45)

Whytt does not seek to further expand on or gloss this power, since "How or in what manner the will acts upon the voluntary muscles, so as to bring them into contraction, is a question wholly beyond the reach of our faculties." *An Essay* (1751, p. 229) Whytt acknowledges that there is an agent responsible for muscular contraction; although, how the immaterial soul can act on the physical body is not addressed by Whytt, nor did he feel the need to do so.

Once a general principle responsible for nervous action is defined, then coupled with the concept that the soul is coextensive with the body, it allows Whytt liberty to explore the ways in which nervous

action is manifest, and to deny the provenance of other, purely materialist, theories. Whytt prefaces his own view of nervous activity by reminding his readers of another fundamental axiom:

A certain power or influence proceeding originally from the brain and spinal marrow, lodged afterwards in the nerves, and by their means conveyed into the muscles, is either the immediate cause of their contraction, or at least necessary to it. *An Essay* (1751, p. 5)

By citing the role of the brain and the nerves in this way, Whytt attempts to distance himself from the animist position. The efficacy of this approach was recognised by the attack on Whytt by two of his Scottish colleagues, Thomas Simson (1696–1764) and William Porterfield (1696–1771), both strict animists. Simson denied that the brain and the nerves were "subservient to different faculties of soul or body", while Porterfield held sensation to be produced "either by the Mind itself, or by the power of God, or of some intelligent active Being acting under him." Cited in Mazzolini (1980, p. 67) Superficially, their position is similar to Whytt's, but animists rejected completely the role of the brain and the nerves in sensation or motion. The recognition of the importance of the nervous system is Whytt's main point of departure from the animist position. In the appendix to his *Physiological Essays*, Whytt stresses this importance:

The nerves are not to be considered merely as the excretory ducts of the brain and spinal marrow, but as real continuations or productions of the medullary substance, which are endowed with certain powers that they retain in a great measure, even after being divided from their origin. *Physiological Essays* (1761, pp. 251–252)

This encephalocentric position is virtually identical to Galen's, and to validate his view, Whytt notes similar experiments to that performed by Galen. For example, in his experiments, Whytt, like Galen, notes that "animals lose the power of moving their muscles, as soon as the nerve or nerves belonging to them are strongly compressed, cut through, or otherwise destroyed." *Physiological Essays* (1761, pp. 5–6) Further, deliberately recalling Galen's famous experiment on the function of the recurrent laryngeal nerve (Walsh, 1926), Whytt describes similar results when first one and then both nerves are ligated. *An Essay* (1751, p. 6) Whytt singles out Galen's experimental results to disprove those who hold that damage to a nerve does not result in muscular inaction and later atrophy, since "*Galen* informs us, that as often as a nerve has been quite cut through, the muscles to which it belonged were deprived both of sense and motion." *An Essay* (1751, p. 330) From all this, Whytt concludes, "The necessity therefore of the influence of the brain and nerves towards producing muscular motion, is not to be disproved." It is therefore crucial to understand "... the doctrine of the nerves being necessary to motion and sensation." *An Essay* (1751, pp. 8, 331) Whytt only describes an 'influence' or 'power' of nervous action. There is no mention of a 'nervous fluid' or 'animal spirits'. Here, Whytt stresses that he not be misconstrued:

The immediate cause of muscular contraction, which, from what has been said, appears evidently to be lodged in the brain and nerves, I choose to distinguish by the terms or *power* or *influence of the nerves*; and if, in compliance with custom, I shall at any time give it the name of *animal* or *vital spirits*, I desire it may be understood to be without any view of ascertaining its particular nature or manner of acting; it being sufficient for my purpose, that the existence of such a power is granted in general, though its peculiar nature and properties be unknown. *An Essay* (1751, p. 9)

Nearly a century earlier, in 1667, Nicolaus Steno (1638–1686) had reacted to Descartes physiological postulation of a blood-derived vapour passing through the alleged hollow nerves, lamenting that "Spiritus animales, finest component of blood, the vapour of blood and juice of nerves are names used by many; but they are only words without any meaning." (cited in Scherz, 1965, pp. 79–80) Alexander Monro *primus*, whose scepticism on the provenance of a nervous fluid has been noted above, amassed experimental evidence to refute the idea of a nerve fluid by, in one case, tying a nerve and watching for some form of swelling to take place, for this could be construed as evidence of a nerve fluid being impeded. Monro also observed the cut end of a nerve, to see if any sort of fluid gathered there. *Anatomy* (1746, p. 332) The problem of a nerve fluid, for Whytt, revolved about its physical essence as well as the nature of the nerve itself:

Although it seems probable that the nerves, which are continuations of the medullary substance of the brain and spinal marrow, derive from thence a fluid; yet the extreme smallness of the nervous tube and the subtility of that fluid which they contain, make us altogether ignorant of its peculiar nature and properties. Nor do we know certainly whether this fluid serves only for the support and nourishment of the nerves or whether it be not the medium by which all their actions are performed. *Works* (1768, p. 489)

Whilst this quotation shows Whytt as a traditionalist in maintaining the Galenic model of the origin of nerves, it also sees a distancing from the consequences of that model, namely that of the role of the 'nervous fluid' in transmission of the nervous impulse is being questioned. Whytt denies that muscular contraction is due to an influx of animal spirits. *Works* (1768, p. 37) Whytt's account of muscular activity is rooted in his teleological view. Thus, muscles are constructed "as to contract whenever a cause proper to excite their action is applied to them." *An Essay* (1751, p. 229) This activity may be initiated by an effort of the will (voluntary motion), or else it may be "vital and spontaneous" (involuntary motion). *An Essay* (1751, p. 229) Whytt denies that muscles themselves possess "an elastic power of their fibres" *An Essay* (1751, p. 230) which is the materialist standpoint, for if a muscle fibre is "an elastic body", then it "is no more than a piece of dead inactive matter." *An Essay* (1751, p. 231) The mechanistic view, "may seem to convey the idea of a mere inanimate machine, producing such motions purely by value of its mechanical construction: a notion of the animal frame too low and absurd to be embraced by any but the most minute philosophers." *An Essay* (1751, p. 1) Whytt has Haller firmly in his sights. Further, he notes that:

Others, giving scope to a lively imagination, have fancied the animal spirits lodged in the cavities of the muscular fibres, to consist of a number of little springs wound up, which, by the application of stimulating bodies, being put into voluntary motions, dilate these fibres, and so render the whole muscle shorter. *An Essay* (1751, p. 232)

This is "a refuge in ignorance" *An Essay* (1751, p. 233), for inasmuch as animal spirits used in this instance may indicate an acknowledgement of an animistic thesis, nevertheless, "these spirits must either act entirely as a mechanical power or not." *An Essay* (1751, p. 233) Still discussing the mechanistic hypothesis of muscular motion, Whytt now also dismisses contraction as due to:

some kind of explosion, ebullition or effervescence, occasioned by the mixture of the nervous and arterial fluids, or perhaps to the peculiar energy of some very subtile ethereal or electrical matter residing in the nerves . . . *An Essay* (1751, p. 234)

The problem with these stimuli lies in their effects, which, if granted, must be commensurate with their causative nature. Thus, if an explosive force is postulated to account for muscular contraction, then its most likely agent, gunpowder, will require fire; electricity in its turn requires a special medium to initiate it, and if the nervous fluid is, say, an alkaline agent, it will:

. . . only raise a commotion when mixed with acids; and no effervescences or sudden ebullitions can be produced, without the mixture of substances disagreeing in their qualities . . . If therefore muscular motion were owing to any of the causes . . . it might reasonably be expected that it would only follow upon the application of certain kinds of *stimuli* to the muscle fibres. *An Essay* (1751, p. 236)

All essentially mechanistic explanations are circumscribed by their own natures. Whytt, however, allows that a chemical or electrical agent, as well as a properly wrought metallic instrument, properly applied to a nerve, will elicit a similar contraction in the muscle concerned (*An Essay*, 1751, p. 236). But this is not the same as saying that the nature of the muscle reaction is chemical, electrical, or in any way mechanical. To Whytt, any hypothesis, which seeks to place in the muscle fibres "some latent power or property", is an appeal "to what, however modified or arranged, is yet no more than a system or mere matter." *An Essay* (1751, pp. 239–240) Whytt regarded matter as inert. For it to act in any way, the substance of which a muscle is formed must be "animated" with "an active sentient PRINCIPLE." *An Essay* (1751, p. 241) Whytt takes some pains to stress that this is not the same as granting to a muscle its own power of sensation and voluntary motion, thus legitimating the Hallerian thesis. For Whytt, any action a muscle makes is ultimately due to "an active sentient PRINCIPLE animating these fibres." *An Essay* (1751, p. 242) This principle, or first cause, is used by Whytt to undercut not only the mechanist position, but all other theories which describe a nervous fluid.

Upon the whole, as nature never multiplies causes in vain, it seems quite unphilosophical to ascribe the motions of the muscles of animals from a *stimulus* to any hidden property of their fibres, peculiar activity of the nervous fluid, or other unknown cause; when they are so easily and naturally accounted for, from the power and

energy of a known sentient PRINCIPLE. *An Essay* (1751, p. 265)

This teleological principle Whytt also refers to as a "nervous influence", and holds that its presence explains the experimental evidence that destruction of the cerebellum is not immediately followed by the absence of motion. That the heart is seen to beat in some animals for varying times after destruction of the cerebellum (or removal of the heart from the animal), constitutes for Whytt proof:

. . . of the nervous influence, lodged in the fibres of the heart and in the smaller filaments of the nerves, being sufficient to continue the motions of this muscle for some time, or to enable it to perform a great number of contractions. *An Essay* (1751, p. 332)

Whytt emphasises this in the *Physiological Essays,*

The plainest facts, therefore, prove that the irritability of the muscles depends on their nerves, and that the nerves preserve their power over the muscles, for a considerable time after they are separated from the brain or spinal marrow. (1755, pp. 249–250)

It may seem that all Whytt is doing is merely transferring the problem of the provenance of muscular motion to a convenient 'sentient principle'. Whytt is aware of this, and acknowledges that this transference is due to our ignorance of the ultimate nature of this principle, which he expresses thus:

. . . it must follow, by necessary consequence, that the motions of the heart, and other muscles of animals, after being separated from their bodies, are to be ascribed to this principle; and that any difficulties, which may appear in this matter, are owing to our ignorance of the nature of the soul, of the manner of its existence, and of its wonderful union with, and action upon the body. *An Essay* (1751, p. 390)

If this is an inadequate answer, then Whytt himself realised that his account of nervous action must remain hypothetical. Beyond a "sentient principle" one cannot probe further, due to the limits of our knowledge of the nervous system's fine structure: "It would be vain to enquire into this matter, unless we knew the minute structure and connections of the several parts of the brain . . ." *Works* (1768, p. 512) Whytt's objection to the notion of a nervous fluid was also grounded in what he regarded as the limits of physiological explanation to provide the complete answer to nervous transmission. However, the idea of a nerve fluid, in whatever guise, was persuasive, could be seen as a reasonable hypothesis and was therefore very difficult to dislodge. Whytt was unsuccessful in persuading his colleagues to abandon the idea of a nerve fluid, and the physics of the nineteenth century would apparently provide a justification for its existence, albeit clothed by the expression 'electrical or galvanic fluid'. Luigi Rolando (1773–1831), a pioneer in this field, held that the nerves served as conductors of this electrical fluid.

William Cullen

The Education of a Teacher of Medicine

William Cullen was born in Hamilton, Lanarkshire in April 1710. His father was factor (land agent) to the Duke of Hamilton. Educated at the local grammar school, he then studied mathematics at Glasgow University, and also became apprenticed to John Paisley, a Glasgow surgeon. After completing his apprenticeship, Cullen moved to London by the end of 1729, obtaining a post as surgeon to a merchant ship (whose master was a relative) trading between Britain and the West Indies. After a six months stay in Portobello, Cullen returned to London, becoming an apothecary's assistant. Following the death of his father and eldest brother, Cullen returned to Scotland in 1732, in order to support his younger siblings, and began medical practice.

In 1734, on receipt of a legacy, Cullen enrolled for 2 years in the Edinburgh Medical School, studying under Alexander Monro *primus*. Cullen's interest in chemistry was nurtured by Andrew Plummer, who had studied under Boerhaave. In 1736, Cullen returned to Hamilton, where, under the patronage of the Duke of Hamilton, he established a practice, and remained there for 8 years. From 1737 to 1740, William Hunter, elder brother of John, was his resident pupil. In 1740, Cullen took the MD at Glasgow, where he moved in 1744. He helped to organise Glasgow's fledgling medical school, where he lectured on botany, chemistry, materia medica and the theory and practice of medicine. In chemistry, one of Cullen's pupils was Joseph Black (1728–1799), who dedicated his treatise on 'fixed air' to him.

In 1751, Cullen became professor of medicine at Glasgow University. The medical school, however, did not develop as Cullen had anticipated; nor was his private practice a lucrative one. Consequently, on the advice of friends, he sought an appointment in Edinburgh, where in 1755 he was elected joint professor of chemistry, becoming sole professor the following year. From 1757, in addition to his popular lectures in chemistry, Cullen began clinical lectures in the Edinburgh Royal Infirmary, where he was also involved in patient care. These lectures, given in English rather than the customary Latin (a practice pioneered by Alexander Monro *primus*, and was also employed by Robert Whytt), were marked by careful preparation, the meticulous ordering of facts, practical experimentation and observations drawn from his own clinical practice. In 1760, at the request of his students, Cullen prepared and delivered a new course on materia medica (these were published in an unauthorised version in 1771; Cullen rewrote his lectures which were published in 1789). Although Cullen expected to be appointed to the chair of the practice of physic, this was denied him. Reluctantly, he accepted the position of chair of the theory of physic, following Robert Whytt's death in 1766. In 1773, on the death of its incumbent, Cullen succeeded to the chair of practice of physic. He held both positions until 1789, dying the following year. Cullen played a key role in the founding of the Royal Society of Edinburgh and the Edinburgh Royal Medical Society (Macarthur, 1993). He was president of the Edinburgh College of Physicians from 1773 to 1775.

Cullen's Edinburgh years saw him rise to become one of the most renowned teachers of medicine in the world. As a lecturer, he had few, if any rivals. His fame was such that he attracted many students from the American colonies, many of them carrying letters of introduction to Cullen by Benjamin Franklin (1706–1790). John Morgan (1753–1789), 'The Father of American Medicine', was one of the founders of the medical school of the College of Philadelphia, the first medical school in North America, which was established in 1765. Morgan used the medical school in Edinburgh as his paradigm, and discussed his plans with Cullen as well as with John Fothergill (1712–1780) and William Hunter (1718–1783). The first professors of the new American medical faculty – John Morgan, Adam Kuhn (1741–1817), Benjamin Rush (1745–1813) and William Shippen Jr.

(1736–1808) – were all Cullen's pupils and Edinburgh graduates. O'Donnell (1993)

Cullen's influence on British medicine was also significant. Apart from the chemist and physician Joseph Black, noted above, others that may be mentioned are Sir Gilbert Blane (1749–1834), a student of Cullen's, who later successfully persuaded the Royal Navy in 1795 to introduce citrus fruit to prevent scurvy; John Haygarth (1740–1827), pioneer epidemiologist, whose early draft of his ambitious 1793 work on the abolition of smallpox was given to Cullen for comment; John Lettsom (1744–1813), another student of Cullen, and the founder of the Medical Society of London and the London dispensary system; Robert Willan (1757–1812), founder of British dermatology, was also taught by Cullen; William Withering (1741–1799), whose 1785 *An Account of the Foxglove* Cullen recommended to every physician, also benefited from Cullen's teaching. It is no more than accurate to state that "Cullen's teaching . . . was possibly the most significant in eighteenth-century British medical education." Crellin (1971, p. 79). Cullen's teaching methodology as will be seen later, is the key to his conceptualisation of the nervous system.

Cullen's Concept of the Nervous System and Its Applications

William Cullen shared with Robert Whytt and his Edinburgh colleagues an abiding interest in the nervous system. As Bynum has cogently noted, "the long eighteenth century took the nervous system seriously as a cause of disease and, by implication, as an essential constituent of health." (1993, p. 152). However, Cullen, unlike Whytt, was not primarily a researcher of the nervous system and, in consequence, his reputation has suffered in comparison. The usual judgement has been that Cullen was more of a systematiser than an innovator. John Thomson, Cullen's first biographer, whose two-volume work remains unrivalled as a study on Cullen's life and times, notes that Cullen was the first in Britain to perceive "the value of the doctrine of the Nervous Pathology, introduced by Hoffmann." *Life I* (1859, p. 200) The 1882 *Dictionary of National Biography*, however, slightly damning with faint praise, refers to him as "distinguished for his clearness of perception and sound reasoning

and judgement rather than for epoch-making originality" (*DNB*, p. 280). Garrison's withering judgement was that "Sir William Hamilton was perilously near the truth when he said that 'Cullen did not add a single new fact to medical science'" (1929, p. 358). Cullen is not cited in Clarke and O'Malley's *Human Brain and Spinal Cord* (1968), and is given two pages (as opposed to 12 for Whytt) in Spillane's (1981) *Doctrine of the Nerves*. Bynum (1993, p. 153). It is futile to attempt to argue Cullen's case as an original researcher on the nervous system, and he undoubtedly would be the first to agree.

A more fruitful way to evaluate Cullen's contribution to the nervous system in the eighteenth century is to pay due regard to his undoubted skills as a lecturer. For it is in his role as a teacher with over forty years experience that Cullen articulated a systematic theory of health and disease based on the nerves, which, whilst not wholly original, nevertheless carried great integrative force and pedagogic rigour. This is seen in his *First Lines on the Practice of Physic* (1777), where Cullen disagreed with Boerhaave's hydraulic system. A word should be said about 'system'. Cullen often employed the term and it should be seen in the sense with which Cullen's students would be familiar. As Barfoot has pointed out:

Instead of thinking of Cullen's system as a totality of discourse, the students, were more familiar with another meaning of 'system' which Cullen inculcated in the classroom. 'System' in this most basic sense referred simply to an organised body of opinions on particular topics in the medical curriculum. For example, he might offer a 'short system of sympathy'. . . Thus 'system', although it implied subject matter of some kind, also referred to the way in which the contents were organised and presented in a general or theoretical way. (1993, p. 114)

Cullen's system was basically heuristic and reflected a very broad influence imparted by Adam Smith and David Hume. Risse (1986) Above all else, it was "simultaneously an expression of the necessary order of the human mind and a pedagogic strategy." Barfoot (1993, p. 124) This is seen to good effect in Cullen's *Lectures on the Institutions of Medicine* (1770). These are a compilation of lecture notes transcribed by a student, and comprise five volumes, the first three of which are devoted to physiology, with one each on pathology and therapeutics. They are invaluable, and "in many ways these verbatim accounts provide a better understanding of Cullen's thinking and teaching than his published work." Doig Ferguson, Milne, and Passmore (1993, p. 32) Thus, although not primarily an experimentalist (his chemical investigations aside), Cullen clearly recognised the importance of physiology in medical education. It is important "even considered as a piece of pure speculation with regard to the mechanisms of animal bodies; but, when considered as capable of a very useful application, it becomes a subject of the greatest importance, and this application is to explain the nature of the diseases of the body; and to explain the operation of remedies and thereby to lead to a more certain means of curing diseases, than we could otherwise obtain." *Lectures, IV* (1770, p. 1). In this context, Cullen's neural physiology "was used to convey knowledge of the gross functions of the human body, and in particular to illuminate patterns of disease." Stott (1987: 130). In this regard, then, it was the broad details that were more important. It was sufficient to note that the nerves "are merely channels of communication." *Lectures, I* (1770, p. 249). Thus, for Cullen:

Whether these nerves are solid strings, which vibrate from one extremity to another; or along which a fine elastic Aether moves; or if they are canals transmitting a fluid; hath long been, and still is a dispute which it is perhaps of little consequence to determine." Cited in Stott (1987, p. 130 n. 48)

How then, did Cullen conceive the nervous system? Broadly speaking, the nervous system functioned according to him as a type of "animated machine, as suited to perform a variety of motions, as fitted to have communication with the other parts of the universe, to be acted upon by external bodies, and to act upon these." *Lectures, III* (1770, pp. 8–9) The nervous system is one of three systems, the others being the simple solids and the fluids (animal functions) which make up the human being. Whilst each is apparently mutually dependent, Cullen prioritised the nervous system as follows:

Most powers acting upon living bodies do not act in the same manner or not at all upon dead Bodies, so that the Effects depend upon the powers of Life, upon sensibility and irritability in the whole or in the parts." *Lectures, V* (1770, pp. 9–10)

Cullen here shows the influence of Haller's concept of sensibility and irritability. However, he was no follower of Stahl either, since:

The Stahlians and some of the physiologists have taken a fancy that the soul has no seat, no particular part of the body with which it is connected more immediately than with the rest; they conceive that it is coextensive with the nervous system, perceives in the organs of sense, independently of any communication with the brain, and operates in the muscles without any motions being propagated to them from the brain. But in saying this, they cannot but be stumbled by the fact, that if a ligature or compression be applied to a nerve of sense, no sensation will be produced by any application between the ligature or compression, and the sentient extremity of the nerve; though, if we can go beyond the ligature, and apply a puncture between it and the brain, this will occasion a sensation. Since, therefore, the impulse or impression of bodies on the sentient extremities of a nerve does not occasion sensation, unless the nerve between the sentient extremity and the brain be free, we conclude that sensation, so far as it is connected with corporeal motions, is a function of the brain alone. *Life, I.* Thomson (1859, p. 273)

But Cullen was no blind adherent of Haller either. Cullen rejected the distinction Haller made between sensibility and irritability, since "for Cullen, there was no clear distinction between the nervous system and the muscles." Bynum (1993, p. 157) Cullen's nervous system consisted of four parts. The first was the 'medullary substance', the brain and spinal cord; the second are membranous nerves continuous with the medullary substance; third are sensory nerves; fourth are:

certain extremities of the nerves so framed as to be capable of a peculiar contractility, and, in consequence of their situation and attachments, to be, by their contraction, capable of moving most of the solid and fluid parts of the body. These we name the MOVING EXTREMITIES of the nerves: they are commonly named MOVING or MUSCULAR FIBRES. *Works, I.* Thomson (1827, pp. 15–16)

This statement is more in keeping with Galen than with an eighteenth century physician. However, it would seem that Cullen based this statement on the 1767 thesis of one Thomas Smith, one of his students. Smith apparently showed that:

. . . muscles could be stimulated to contract either by direct application of various mechanical or chemical means to muscle itself, or to its nerve; and that, when the muscle was no longer able to react, excitation of neither muscle nor its nerve was effective. They were thus functionally continuous and therefore, Cullen reasoned, part of the same system, i.e. the nervous system. Bynum (1993, p. 158)

Cullen's statement also indicates not so much a gainsaying of existing anatomical knowledge as of its relative unimportance in his overall system. Function, not form, counted for Cullen, especially the interlinked functional differentiae of his holistic conceptualisation of the body. It also allowed him, as will be noted below, "to broaden the class of diseases he called neuroses and to make literal sense of the traditional proposition that the principal functions of the nervous system are concerned with sense and motion". Bynum (1993, p. 158)

Cullen defined sensibility as:

[A] certain fitness to be acted upon by impressions, to be so moved by the impulse of external bodies, as that a motion may be propagated to the Brain and produce Sensation and its various consequences. . . this fitness of the Sentient Extremities may be called their Sensibility. *Lectures, IV* (1770, pp. 62–63)

Disease was defined as an excess or deficiency of sensibility, a very general definition which suited Cullen's nosology, as will be noted below. Irritability is not so easily defined. Although the term comes from Haller, for Cullen it seems to embrace muscle 'mobility' and 'vigour'. An excess of vigour produces an increase in muscle tone or strength, a decrease results in debility. Excess mobility causes irritability; a deficiency gives rise to torpor. Nervous diseases occur when there is any imbalance to sensibility and irritability. In this, the brain "had a peculiar function, in that it not only was acted upon by external powers, but was also classified as an external power itself, modifying the actions of the body and the actions of remedies." Stott (1987, p. 135)

Therapy was governed by what either stimulated or sedated the nervous system, since:

. . . the nervous power alone is capable of considerable and sudden changes, it is to this that our medicines should be chiefly directed. *Lectures on the Materia Medica* (1771, p. 20)

Stimulants formed the largest class of Cullen's therapeutic agents. They appeared to act to not only increase "the mobility of the nervous powers more generally", but also increased "the motion of the animal power in the brain . . . such [powers] as excite the action of moving fibres . . . and such as increase the motion of the blood and other fluids of the body." *Lectures, V* (1770, pp. 191–192) Therapy

could also have a beneficial affect on the "tone of mind", which is a disposition "on the one hand to joy, gaiety, and hope or on the other hand to sadness, seriousness and despair." *Lectures, IV* (1770, p. 174) Here, the doctor "must on occasion be the Moral Philosopher also, and he will sometimes practice with little success unless he can apply himself to the Mind." *Lectures, IV* (1770, p. 143)

Cullen's expansion of his concept of the nervous system reached its apogee in his study of disease. Cullen published his *Synopsis Nosologiae Methodicae* in 1769. This major aetiological work classified and systematised diseases into four principal classes ('Pyrexiae', 'Neuroses', 'Cachexiae and 'Locales'), 19 orders and 132 genera. Inspired by and based on the work of Boissier de Sauvages (1706–1767), especially his *Nosologia Methodica* (1768), whose disease classificatory system was itself inspired by the great Swedish botanical taxonomer, Karl von Linné's (Linnaeus: 1707–1778) *Systema Naturae* (1735), Cullen's work included a synopsis of de Sauvages, as well as reproducing the complete systems of Linné, and several others. It is the second of Cullen's classes which is of great interest. Here, Cullen coins the term 'neurosis' to describe:

All those preternatural affections of sense or motion which are without pyrexia . . . and all those which do not depend upon a topical affection of organs, but upon a more general affection of the nervous system, and of those powers of the system upon which sense and motion more especially depend. Cited in Kendell (1993, p. 223)

The four orders of the neuroses comprise 'Comata', 'Adynamiae', 'Spasmi' and 'Vesaniae'. Even without citing the subsequent genera and their divisions, it is apparent from these headings that Cullen's conception of the nervous system was considerably broader than our own. This becomes clearer when the genera are listed. Under 'Comata', defined as "A diminution of voluntary motion, with sleep, or a deprivation of the senses", are listed 19 genera of apoplexia and seven of paralysis. The second order, 'Adynamiae', is defined as "A diminution of the involuntary motions, whether vital or natural". Three genera listed here are concerned with descriptions of syncope, three with dyspepsia, and one each with hypochondriasis and chlorosis. 'Spasmi', the third order, is defined as "Irregular motions of the muscles or muscular fibres." The genera here comprise tetanus, trismus, two types of 'convulsio' ("an irregular clonic contraction of the muscles without sleep"), chorea, 'raphania' ("A spastic contraction of the joints, with convulsive agitations and most violent periodical pain"), three types of epilepsy ('cerebralis', 'sympathica' and 'occasionalis'), palpitatio, asthma, 11 types of dyspnoea, pertussis, 'pyrosis' ("burning pain in the epigastrium with plenty of aqueous humour, for the most part insipid, but sometimes acrid belching up"), seven types of colica, two of cholera, six of diarrhoea, diabetes, hysteria and hydrophobia. The fourth and final order, 'Vesaniae', is defined as "Disorders of the judgement without any pyrexia or coma". The genera here comprise one of amentia, eight of melancholia, three of mania ("universal mania"), and two of 'oneirodynia' ("A violent and troublesome imagination in time of sleep"). Cited in Kendell (1993, pp. 226–233)

Cullen's neurological nosology ambitiously abrogated vast swathes of what comprise the cardiovascular, gastrointestinal and endocrine systems, to say nothing of abnormal states of the mind – and one may note here that Cullen's concept of 'neurosis' was ameliorated first by Pinel and then by Freud. It was not "in any meaningful way exclusively psychiatric, or even neuropsychiatric in the modern sense" Bynum (1993, p. 159) Cullen's classificatory system offers a series of broad explananda of disease, but one which was meant to be considered a work in progress, "as a structure within which to study pathological anatomy and pursue nosography." Lawrence (1985, p. 172) In this structure, however, the role of the nervous system was paramount. To be sure, all the systems of the body may be considered to be under some degree of nervous control or influence, although such influence is not as far reaching or sustaining as the underlying physiological principle with which Cullen sought to imbue his classificatory system of disease and explain the absence and maintenance of health. This was his concept of 'nervous energy'. With this, "Cullen gradually convinced himself that excesses or deficiencies, local or general, of the hypothetical influence which interested him most, 'nervous energy', were the root cause of most diseases." Kendell (1993, pp. 223–224) Cullen's 'nervous energy' bears some comparison with Whytt's 'sentient principle', but the former possesses wider, pathological implications, as noted

by the sheer number of genera subsumed. While in his *Observations on the Dropsy in the Brain* (1768), Whytt echoed Cullen when he stated that:

All diseases may, in some sense, be called affections of the nervous system, because, in almost every disease, the nerves are more or less hurt; and, in consequence of this, various sensations, motions, and changes, are produced in the body. Cited in Kendell (1993, p. 224)

But Whytt's neurological nosology was much more restricted in number and scope. Whytt recognised three genera: hypochondriasis, hysteria and 'simply nervous', comprising convulsions, fainting and palpitations. Although Cullen's system of nosology barely survived his lifetime, its chief virtues were its coherence and scope, and the fact that it reflected the core principle of Cullen's conception of the nervous system, namely, sensibility and irritability. As Lawrence points out, Cullen went further still:

Developing Whytt's account of the sentient principle, Cullen used these concepts to create a model of the reactive organism, drawing attention on the one hand to the determining power of the environment and on the other to the original human constitution. These laws were used by him to construct a physiology, a psychology and then, in turn, an anthropology. They were also used as the foundation of pathology and, with natural history, to explain the geography of disease and the laws of hygiene (1985, p. 171).

These were, and remain, ambitious aims for any system of neurology.

Conclusion

Whytt and Cullen are key figures in the Scottish medical Enlightenment and eighteenth century neuroscience, though for different reasons. If Whytt represented originality in neurophysiological research, Cullen represented pedagogic excellence, which was a key factor in the international influence of the Edinburgh medical school. Whytt was perhaps the finest experimental physiologist of his day – the first to elucidate the pupillary light reflex and to experimentally determine that part of the spinal cord is required for reflex action ('sympathy', to use his term). Whytt's investigations into the nervous system were in large measure driven by an urgent need to try and make sense of the many theories of nervous action, as well as to attempt to unravel the often conflicting sets of sometimes frustratingly incomplete experimental results. Whytt decries the rash of speculative theories to account for the ultimate workings of the nervous system, and sought to bring order to the several competing theories of nervous action which characterised eighteenth century physiology. Whytt's answer to these linked problems was deceptively simple in its design, combining scepticism with experimental methodology. By stressing the need for the brain and nerves, together with an acknowledgement that the soul was also responsible for their function in ways beyond our understanding, Whytt hoped to undercut both the mechanist and animist positions. If he was not entirely successful, then his work paved the way for later investigators such as Charles Bell (1774–1842) and Georgius Procháska (1779–1820), and Whytt can with good reason be called one of the founders of neurophysiology.

William Cullen also sought to bring order to the nervous system, but in rather a different fashion. Cullen's physiological interests were of the clinical variety. Unlike Whytt, Cullen did not offer anything new as far as neurophysiologic theory was concerned. As Cullen famously remarked to his students, "no man can go much further than the state of science at his particular period allows him." *Works, I* Thomson (1827, p. 375). What Cullen did offer was a broad, coherent system of the nervous system in health and disease, which formed the core of his pedagogic enterprise, and gave him a widespread and justified fame. The fact that the very generality of Cullen's concept of the nervous system, which allowed for continuing speculation concerning the function of the nerves and their conflation with muscle fibres, also represented a backward step in eighteenth century neuroscience, and one which Whytt would have decisively rejected, was, it would appear, also a necessary consequence of William Cullen's broader systematisation, the latter nevertheless being as much a part of the Scottish Enlightenment as Robert Whytt's physiological research.

Acknowledgements. The author is grateful for the invaluable assistance provided by Gertie Johansson, Curator of the Hagströmer Medico-Historical Library, Karolinska Institute, Stockholm. I also wish to express my thanks to the editors for their courtesy and exemplary patience.

References

Barfoot, M. (1993). Philosophy and method in Cullen's medical teaching. In Doig et al. (Eds.), (pp. 110–132).

Brazier, M. A. B. (1973). *The evolution of concepts relating to the electrical activity of the nervous system. 1600 to 1800. In idem. The history and philosophy of knowledge of the brain and its functions.* Amsterdam, B. M. Israël, Amsterdam, Unchanged reprint of the 1958 Blackwell Scientific Publications Ltd. edition, Oxford.

Bynum, W. F. (1993). Cullen and the nervous system. In Doig et al. (Eds.), (pp. 152–162).

Bynum, W. F. & Porter, R. (Eds.), (1985). *William hunter and the eighteenth-century medical world.* Cambridge: Cambridge University Press.

Cantor, G. N. & Hodge, M. J. S. (Eds.). (1981). *Conceptions of ether. Studies in the history of ether theories 1740–1900* (pp. 111–134). Cambridge: Cambridge University Press.

Clarke, E. (1978). The neural circulation. The use of analogy in medicine. *Medical History, 22,* 291–307.

Clarke E., & O'Malley, C. D. (1968). *The human brain and spinal cord.* Berkeley: University of California Press.

Clayson, C. (1993). William Cullen in eighteenth century medicine. In Doig et al. (Eds.), (pp. 87–97). Edinburgh: Edinburgh University Press.

Crellin, J. K. (1971). William Cullen: His calibre as a teacher, and an unpublished introduction to his a treatise of the materia medica, London, 1773. *Medical History, 15,* 79–87.

Cullen, W. (1769). *Synopsis nosologiae methodicae.* Edionburgh: Royal College of Physicians of Edinburgh.

Cullen, W. (1770). *Lectures on physiology (=Lectures on the Institutions of Medicine)* (Vol. 5). Edinburgh: Royal College of Physicians of Edinburgh Cullen MSS #18.

Cullen, W. (1771). *Lectures on the materia medica.* London: (published without consent, 1789). *A treatise on the materia medica.* Edinburgh: Royal College of Physicians of Edinburgh.

Cullen, W. (1778–1784). *First lines on the practice of physic.* Edinburgh: Royal College of Physicians of Edinburgh.

Doig, A., Ferguson, J. P. S., Milne, I. A., & Passmore, R. (Eds.). (1993). *William Cullen and the eighteenth century medical world: A bicentenary exhibition and symposium Arranged by the Royal College of physicians of Edinburgh, 1990.* Edinburgh: University Press.

Emerson, R. L. (2004). The founding of the Edinburgh medical school. *Journal of the History of Medicine and Allied Sciences, 59,* 183–218.

French, R. (1969). *Robert Whytt, the soul, and medicine.* London: Wellcome.

French, R. (1972). Sauvages, Whytt and the motion of the heart: Aspects of eighteenth century animism. *Clio Medica, 7*, 33–54.

French, R. (1981). *Ether and physiology*. In G. N. Cantor & Hodge (Eds.), (pp. 111–134).

Garrison, F. (1929). *An introduction to the history of medicine* (4th ed.). Philadelphia: W.B. Saunders.

Kendell, R. E. (1993). *William Cullen's synopsis nosologiae methodicae*. In Doig et al. (Eds.), (pp. 216–233).

Lawrence, C. (1985). *Ornate physicians and learned artisans: Edinburgh medical men, 1726–1776*. In Bynum & Porter (Eds.), (pp. 153–176).

Macarthur, D. C. (1993). *The first forty years of the Royal Medical Society and the part William Cullen played in it*. In Doig et al. (Eds.), (pp. 247–251).

Mazzolini, R. G. (1980). *The Iris in Eighteenth-century physiology*. Bern: Hans Huber.

Monro, Alexander primus. (1746). *The anatomy of human bones and nerves*. Edinburgh.

Monro, Alexander primus. (1781). *The works of Alexander Monro*. Edinburgh: Published by his Son.

Neuburger, M. (1897/1981). *The historical development of experimental brain and spinal cord physiology before flourens*. Translated and Edited, with Additional Material by Edwin Clarke. Baltimore: Johns Hopkins University Press.

O'Donnell, J. M. (1993). Cullen's influence on American medicine'. In Doig et al. (Eds.), (pp. 234–246).

Pearn, J. (2002). *The final common path. Muscle action and the evolution of knowledge concerning neuromuscular disease*. Brisbane: Amphion Press.

Phillipson, N. (1981). The Scottish enlightenment. In R. Porter & M. Teich (Eds.), *The enlightenment in national context*. Cambridge: Cambridge university Press.

Porter, R. (1996). Medical Science. In R. Porter (Ed.), *The Cambridge illustrated history of medicine* (pp. 154–201). Cambridge: Cambridge University Press.

Radbill, S. X. (1976). Robert Whytt. In C. C. Gillespie (Ed.), *Dictionary of scientific biography* (Vol. XIV, pp. 319–324). New York: Charles Scribner's Sons.

Risse, G. B. (1986). *Hospital life in enlightenment Scotland: Care and teaching at the royal infirmary of Edinburgh*. Cambridge: Cambridge University Press.

Rocca, J. (2000). Robert Whytt (1714–1776): The sceptical neuroscientist. In F. C. Rose (Ed.), *A short history of neurology* (pp. 93–107). The British Contribution, 1660–1910. London: Butterworth-Heinemann.

Rocca, J. (2003). Robert Whytt. In M. J. Aminoff & R. B. Daroff (Eds.), *Encyclopedia of the neurological sciences*. (Vol. 4, pp. 754–755). USA: Elsevier Science.

Rocca, J. (forthcoming 2007). Robert Whytt. In H. Bynum & W. Bynum (Eds.), *Dictionary of medical biography*. Connecticut: Greenwood.

Scherz, G. (1965). *Nicolaus Steno's lecture on the anatomy of the brain*. Copenhagen: Arnold Busck.

Spillane, J. D. (1981). *The doctrine of the nerves*. Oxford: Oxford University Press.

Stephen, L., & Lee, S. (Eds.), (1882/1967–1968). *The dictionary of national biography. From the earliest times to 1900* (Vol. 5). Oxford: Oxford University Press.

Stott, R. (1987). Health and virtue: Or, how to keep out of harm's way. Lectures on pathology and therapeutics by William Cullen c. 1770. *Medical History, 31*, 123–142.

Thomson, J. (1859). *An account of the life, lectures and writings of William Cullen M.D.* (Vol. 2). Edinburgh: William Blackwood and Sons.

Thomson, J. (Ed.). (1827). *The works of William Cullen* (Vol. 2). Edinburgh: William Blackwood and Sons.

Walsh, J. (1926). Galen's discovery and promulgation of the function of the recurrent laryngeal nerve. *Annals of Medical History, 8*, 176–184.

Whytt, R. (1751). *An essay on the vital and other involuntary motions of animals*. Edinburgh: Hamilton, Balfour and Neill.

Whytt, R. (1761). *Physiological essays* (2nd ed.). Edinburgh, Hamilton, Balfour and Neill, Corrected and Enlarged, With an Appendix.

Whytt, R. (1764). *Observations on the nature, causes and cure of those diseases which are commonly called nervous, hypochondriac or hysteric*. Edinburgh.

Whytt, R. (1768). *Observations on the dropsy in the brain. To which are added his other treatises never hitherto*. Edinburgh: Published by Themselves.

Whytt, R. (1768). *The works of Robert Whytt, M.D.* (3rd ed.). Edinburgh: Balfour, Auld and Smellie.

7

1710: The Introduction of Experimental Nervous System Physiology and Anatomy by François Pourfour du Petit

Lawrence Kruger and Larry W. Swanson

Introduction

The beginnings of an experimental approach to brain function derived from the study of brain lesions can be traced to antiquity, but the emergence of a reasoned systematic methodology was surprisingly slow to mature in the early period of empiricism. The Oxford "virtuosi" associated with Willis (Lower, Wren, & others) in the late seventeenth century responded to the thrust of William Harvey's brilliant experiments, devised to explain the nature of the action of the heart and circulation of blood, by injecting various substances into the vasculature of dogs and observing their effects. Yet there was surprisingly little effort to add new empirical gains concerning brain function before the remarkable contributions of François Pourfour du Petit (1664–1741) in the early eighteenth century. His earliest and most important work, *Lettres d'un Medecin des Hôpitaux du Roy, a un Autre Medecin de Ses Amis*, was published in 1710 and contains an account of brain lesions as well as some experiments describing vascular infusion of acids and alkalis derived from the reports of Willis (1664). The treatise by Petit (as he was generally known) survives in but few copies and details of its contents remain astonishingly obscure to collectors, libraries, and historical accounts.

Although Petit's contributions to neuroscience and to ophthalmology are generally known and occasionally cited, it is puzzling that the historical context of his work remains largely unrecognized, for it constitutes a milestone in its descriptions of acute cerebral injuries in humans and in addition attempts to correlate motor signs with neuroanatomical findings that proved essentially correct; yet this work has remained shrouded and evidently had little impact in the eighteenth and nineteenth centuries. More remarkable, and most regrettable, has been the obscurity of Petit's pioneering efforts to employ ablation techniques successfully on experimental animals as a tool for developing a functional neuroanatomy. His later (1713) experiments (published in 1727), surgically interrupting the sympathetic trunk in the neck of dogs, is commonly credited as the earliest "discovery" of the cervical sympathetics and ascertaining their functional role, but the earlier (1710) and more difficult cerebral ablations in dogs, a pioneer effort derived from (or instigated by) clinical observations, remain largely unrecognized. In these two major reports, Petit established the impact of ablation as an experimental tool for a *functional* neuroanatomy from which he drew remarkably astute interpretations; advances enthusiastically credited with the introduction of a physiology derived from pathology in Max Neuberger's influential monograph of early neuroscience history (1897). Later historians of neuroscience generally acknowledge Petit's contributions in brief remarks, but details of his experiments remain largely unknown and there appears to be limited extant biographical material. The present chapter constitutes an attempt to correct these deficiencies and to place Petit's life and work in the broader context of its place in the development of neuroscience.

The significance of Pourfour du Petit's contribution initially lies in his analytical approach in interpreting morbid anatomy of wartime head wounds. It served as a springboard for devising animal experiments in which he was able, through remarkable skill, to perform somewhat comparable lesions to those he observed in his patients. His work probably constitutes the earliest example of what is currently designated "translational neuroscience" and represents the introduction of logically designed ablation experiments as a tool for analyzing the function of specific neural structures. This was achieved first in the series of lesions (both ablation and concussive compression) in dogs, involving specific regions of the cerebrum and cerebellum and published in his concise 1710 quarto monograph. During this period of military service Petit also applied the same strategy to uncovering the functional anatomy of the poorly understood cervical sympathetic trunk, then generally known as the "intercostal nerve" and believed to derive from the brain via the fifth and sixth cranial nerves, according to the widely accepted authoritative view published in 1664 by Thomas Willis and reiterated in a later dubious illustration by Vieussens (1684/1685) (see Rasmussen, 1947). Petit demonstrated the main features of sympathectomy in a rather elegant report to the academy, approaching this problem in functional neuroanatomy by cutting the "intercostal" nerve in the neck and observing that the ipsilateral eye filled with tears and the nictitating membrane extended over the cornea thus, unexpectedly demonstrating that the "animal spirits" passed *upward* into the head, rather than *from* it, as Willis (1664) had inferred. Further experiments, revealed that in addition to control of lacrimation and of the nictitating membrane, the pupillary

aperture and blood supply to the eye and facial skin were also modulated by the cervical sympathetic trunk (Best, 1969). This observation, published in 1727, was not generally recognized until the nineteenth century when it was rediscovered by Claude Bernard and became known eponymously as "Bernard-Horner's syndrome."[1]

The same basic methodology of resection was thus applied to both the central and peripheral nervous systems by Petit. Yet his surmise that the decussation of the pyramidal tract, which he also illustrated, might underlie the crossed motor deficits seen following cerebral lesions was not readily accepted in his lifetime and this milestone in functional neuroanatomy, although somewhat known, had curiously little impact. Perhaps this apparent oversight can be explained by the peculiarity of its publication in the form of three letters in a small quarto volume printed in Namur during wartime,[2] far from the mainstream of publishing. Without the author's name on the title page and in a run of only 200 copies, most of which were probably lost or were difficult to catalog without an author's name, this work was already a rarity at the time of Petit's death (as indicated in Mairan's eulogy, 1747). However, the obscurity of the small, unsigned edition of the 1710 three *Lettres d'un Medecin* . . . (Fig. 1) curiously does not adequately explain the seeming failed recognition and neglect of this landmark contribution.[3] But it is also likely that conditions in France could not have been as propitious for vivisection as was possible during wartime in Namur.

---

[1] In a touching, belated gesture attempting to rectify the injustice of depriving Petit of proper recognition, Italian neurologist-psychiatrist Serafino Biffi (1822–1899), a key supporter of Camillo Golgi's earliest experimental work, named the "syndrome of Pourfour du Petit" in his honor, although it refers to *stimulation* of the cervical sympathetics and thus the opposite of what now is generally known as "Horner's syndrome" and had been accurately described in dogs in 1727 by Petit (Ségura, Speeg-Schatz, Wagner, & Kern, 1998).

[2] The copy in the *Bibliothèque Nationale* upon which this account is based indicates authorship inserted with pen and ink as "*par Francois Petit*," a feature lacking in the London copy photographed by William Osler in assisting a scholarly report by H.M. Thomas (1910) on the pyramidal decussation. The requirement of obtaining

the royal imprimatur may have been difficult in wartime and perhaps Petit was intimidated by other factors that led him to avoid seeking the royal approval probably requisite for a military officer.

[3] The absence of the author's name in print on the title page is puzzling and probably constitutes a factor in archival cataloguing practice contributing to its rarity. Failure to obtain the royal imprimatur and financial support, for whatever reason (including the complexities of wartime) would suggest that Petit might have borne the costs of printing personally. The inclusion of illustrations also would have added significantly to the expense of this ostensible edition of 200 copies (according to Mairan's eulogy), and it is conceivable that this estimate might have been exaggerated for a privately financed publication of this size.

LETTRES
D'UN MEDECIN
DES HÔPITAUX
DU ROY.
A UN AUTRE MEDECIN DE SES AMIS.

LA PREMIERE LETTRE

Contient un nouveau Syſteme du Cerveau.

LA SECONDE LETTRE

Contient une Diſſertation ſur le ſentiment, & pluſieures experiences de Chimie contraires au Syſteme des Acides & des Alkalis

LA TROISIE'ME LETTRE

Contient une critique ſur les trois eſpeces de Chryſoſpleinium des Inſtituts de Mr. · Tournefort, trois nouveaux genres de Plantes & quelques nouvelles Eſpeces.

A NAMUR,

Chez CHARLES GERARD ALBERT, Impri-
meur du Roy. 1710.

FIGURE 1. Title page of the Pourfour du Petit *Lettres d'un Medecin*, etc. published in 1710 by the royal printer C.G. Albert in Namur, with the author's name appended by hand in this copy from the *Bibliothèque Nationale* in Paris

The French surgeon Antoine Louis (1723–1792) re-published the medical component of this publication together with the works of others (Louis, 1766, pp. 7–12), evidently serving as the source for most of the later references to its content. Jules Soury (1899) in his massive historical critique gives Petit short shrift, although he strangely cites Goltz's reference to Petit for describing the effects of cervical sympathectomy in the *horse* (Petit's (1727) paper published in the academy of science journal describes experiments on dogs). Remarkably, Soury followed the scholarly comprehensive history by Max Neuberger in 1897, where there already was an extensive account of Petit's 1710 studies of cerebral lesions. Neuberger suggested that Petit's treatise "fell into oblivion" because of its "opposition to the popular current of opinion" (Neuberger, 1981, p. 59), but the publications of other writers on the

subject of contralateral paralysis, discussed in some detail by Neuberger, include competitors for the academy's prize (see below) some of whom also performed animal experiments; notably Saucerotte, Chopart, and especially Jean Sabouraut, awarded the 1768 prize, "who is honored for his role in transmitting the results of [Petit's] research" (Neuberger, 1981, p. 60); but there is little apparent concrete evidence of Petit's results having fallen into disfavor.

Biographical

Our knowledge of François Pourfour du Petit's background and personal history is not extensive and the major sources of relevant information have been reviewed only briefly in modern accounts (Kruger, 1963; Zehnder, 1968). He was born in Paris on June 24, 1664, the only offspring of mercantile parents who died while he was still a child. During his early education in grammar, rhetoric, logic, and metaphysics, his teachers quickly discovered that he had difficulty in understanding and remembering didactic material, and that he took little interest in language and abstract philosophy, but his introduction to the mechanistic ideas of Descartes at the Collège de Beauvais revealed a greater receptivity to concrete experimentation that profoundly influenced his career. In addition to Cartesian physics, he acquired a preliminary education at Beauvais in natural history, chemistry, and anatomy, and upon completion of his studies there, he embarked upon extensive travels.

Despite a rather modest income, Petit visited most of the French provinces and a good deal of Flanders, seeking the company of natural philosophers and physicists, especially those of the Cartesian school – the only learned men he considered worthy of attention. At La Rochelle, he developed a close relationship with botanist Pierre Blondin (1682–1713), a practitioner of Cartesian science, who instructed Petit in anatomy and encouraged him to pursue a career as a medical doctor. Petit left for Montpellier at the end of 1687 and began his study of medicine there under Pierre Chirac (1650–1732), a member of the Royal Academy of Sciences. After receiving his doctorate in medicine in 1690, at the age of 26, Petit returned to Paris.

There he attended public lectures in anatomy at the *Jardin du Roi* offered by Joseph-Guichard DuVerney [also Duverney]. Duverney (1648–1730), sometimes regarded as the founder of otology, published the first comprehensive treatise on the anatomy, physiology, and diseases of the ear (1684), which contained ideas presaging the Helmholtz "place" theory of hearing. He was also considered an influential figure in establishing the principles of comparative anatomy (Cole, 1949). Upon Duverney's appointment as professor of anatomy at the *Jardin du Roi* in 1679[4] he advanced anatomical demonstration to a level that attracted large audiences, among them Petit. Although most of Duverney's work has not survived, some of his dissections were illustrated posthumously in beautiful color atlases by Gautier d'Agoty (1748). Petit also attended lectures in botany by Joseph Pitton de Tournefort (1656–1708), who had attended the faculty of medicine at Montpellier in 1679, where he acquired a fine reputation as a botanist (especially of the Pyrénées) and in 1683 was offered the post of Professor of Botany at the *Jardin du Roi* (later the *Jardin des Plantes*). In 1698, he received his medical degree in Paris and in 1700, Louis XIV assigned Tournefort the mission of studying and classifying the flora of the Levant as well as regions near the Black Sea and Asia minor and the Greek isles, reporting on more than 1350 specimens. On his return he was named professor of medicine at the *Collège de France* and Director of the *Jardin du Roi* until his death in 1708; a position in which Tournefort later was succeeded by the illustrious Carl Linnaeus. It may be presumed that Petit's contacts with Tournefort in Montpellier and later in Paris account for his intense interest in botany and the inclusion of an account of botanical taxonomy "concerning three species of chrysophyllum from the institute of M. Tournefort" in the third letter of Petit's 1710 treatise,[5] the rest of which is devoted to discussion of the nervous system.

Petit spent 6 months with Tournefort at the *Hôpital de la Charité* bandaging the wounded, and continued this work intermittently until 1692. He left Paris in April 1693 to serve in the dual capacity of physician and surgeon to Louis XIV's army in Flanders, and was widely admired for his zeal and skill working in the hospitals of Mons, Namur, and Dimant; serving until the Peace of Ryswick in 1697 and returning to Paris, only to set out once again the following year, to serve at Compiègne. His service for the military was resumed by the War of the Spanish succession, beginning in 1702. At Compiègne, where he worked with M. Prouvenza, *Médecin Inspecteur des hôpitaux*, who shared Petit's love of botany, and whom Petit would honor when he discovered an as-yet-unclassified plant along a riverbank and named it "*Provenzalia.*" Petit served admirably in Brussels and various towns along the "*pays-bas*" ("low countries"), and continued to instruct students. In 1710, he published his first work in Namur, where he was stationed – a collection of three essays in-quarto, under the title *Lettres d'un Médecin etc.* He served as an army surgeon in Namur during the War of the Spanish Succession (of Philip II) until the Peace of Utrecht in 1713. In addition to caring for the wounded French soldiers, Petit also established chemistry laboratories and anatomy demonstrations at the hospitals. He also collected and classified plants, teaching his students to do the same, as well as to use them as medical remedies. It was during this time that he began an *herbier*, which eventually grew to a total of 30 large folio volumes. The influence of Petit's interest in chemistry derives from Nicolas Lémery (1645–1715), who studied pharmacy in his native Rouen and then went to Montpellier where he lectured in chemistry and encountered Petit. He presumably influenced the portion of Petit's second letter in the 1710 treatise dealing with the influence of acids and alkalis on the nervous system. Lémery's "*Cours de*

---

[4] Petit followed Duverney in the Chair at the *Jardin du Roi* after his retirement in 1728. Conjecture concerning whether Petit recognized the anatomical arrangement of fiber decussation in the pyramidal tract from his teacher seems futile as Petit failed to cite learning what he claimed to have discovered independently and illustrates in his 1710 monograph. The posthumous publication of Duverney's dissections do

not depict the pyramidal decussation and were published long after Petit's death, without citing Petit's earlier "discovery" nor his barely persuasive drawing.
[5] The botanical letter may account for the survival of a rare copy of this work in the Plant Science Library at Oxford University (Sherard 163) with the name "Francois Petit" inscribed in pencil at an undetermined date.

Chymie" (1675) remained a standard work for about a century, appearing in 13 editions!

Returning to Paris after the war in 1713, Petit married (in 1717) and was now able to focus his attention upon scientific experiments in his *cabinet* and also present his candidacy for the *Académie Royale des Sciences*, to which he was elected in 1722, and where he succeeded his teacher, Duverney, as *pensionnaire anatomiste*, 3 years later. He published numerous works in experimental and theoretical physics, anatomy and surgery, but devoted himself principally to anatomical description and systematic measurements of the eye in humans and the visual system of various vertebrates, the mechanics of vision, and the causes and treatment of cataracts. The space between the anterior and posterior suspensory ligaments of the lens in the human eye remains known as the "canal of Petit." He invented an *"ophthalmomètre,"* an instrument for measuring the parts of the eye and quantifying astigmatism, as well as devising anatomical models and devices to aid in ophthalmologic procedures. Armed with the knowledge gained from his experiments and the instruments he developed, he met with some success practicing cataract surgery toward the end of his life, operating primarily on less affluent patients. His methods were attacked by Philippe Hecquet (1664–1737), the champion of a peculiar fad called "theological medicine," in his *Traité des maladies des yeux* and other works, to which Petit responded in a published letter of 1729. Little is known of Petit's distinguished late career aside from his numerous scientific reports; mostly published in the memoirs of the royal academy. On June 3, 1741, Petit fell ill from an intestinal hernia and died 2 weeks later, survived by one son, Etienne, who followed his father into medicine and who, upon his installation as dean of the faculty of medicine in

FIGURE 2. Portrait of François Pourfour du Petit at the age of 73 painted in Paris by Jean Restout, the younger (see footnote 6)

Paris in 1783, gave that institution the extraordinary portrait of his father (Fig. 2) by Jean Restout.[6]

Introduction of the Experimental Method

The *Lettres* are divided into three sections: The first letter "contains a new system of the brain" and deals principally with the nature of paralysis

---

[6] The portrait derives from an oil painting on canvas (93 × 74 cm) bearing an old ticket with the inscription "Restout 1737," and belongs to the *Musée d'Histoire de la Medicine, Paris* (Inv.7 458), given by his son Etienne Pourfour du Petit in 1783, the year he was elected dean of the faculty. A recent catalog containing criticism and interpretation as well as biographical comment and bibliographic material (Gouzi, 2000) describes the face of Pourfour as *"grave et décharné"* (p. 243) and *"le regard perçant et absorbé semblant évoqué une âme tormentée, presque mystique"* (p. 66); indeed, an austere, sallow and gaunt portrait of an aged (73-year-old) elegantly dressed man with a modestly subdued, seeming

glint in his eye. The painter, Jean Restout the younger (1692–1768), in his teens was a pupil of his famous uncle, Jean Jouvenet, the leading religious painter in Paris in his time. A traditional academic history painter, and from 1730 a professor of drawing at the *Académie*, Restout specialized in religious painting. Despite a good reputation as a leading history painter, his work did not succumb to the prevailing taste for ornate decorative painting and by the late eighteenth century, when the Italianate "grand manner" flourished, Restout's style became unfashionable, attracting little interest until art scholarship in the late twentieth century somewhat restored his reputation.

contralateral to cerebral injuries in soldiers with head wounds and in dogs in which experimental lesions were produced surgically; it comprises the main substance of interest for neuroscientists. It also constitutes the earliest successful attempts to achieve insight into cerebral localization by varying the site and nature of tissue disruption. The survival of the animals after vivisection without anesthesia served as testimonial to Petit's exceptional surgical skills-unmatched over the next century.

The second letter "contains a dissertation on consciousness and several chemical experiments which oppose the system of acids and alkalis." Much of this is a reply to the barely cited letter from a friend. The part concerning consciousness deals with issues of sensation and the flow of animal spirits based on head wounds and injections of various fluids in dogs, and obviously derives from the publication by Willis of the observations on blood transfusion performed earlier by Lower and his cohorts. It also contains an interesting account of producing a percussion injury of the head in dogs and of other neurological observations, including sensory observations and the effects of cerebellar lesions, but most of this discourse expounds early views about basic chemistry.

The third letter "contains a critique on the three species of chrysophyllum from the institute of M. Tournefort (Petit's teacher of botany); three new genera of plants and several new species." Although this letter is lacking in neurological interest, it should be noted that the prominent eighteenth-century medical biographical dictionary by Eloy (1778) discusses the botanical portions of Petit's monograph without even mentioning the neurological observations.

Letter One begins with an historical account from Bonet's Practical Anatomy (*Sepulchretum*, 1679), citing some authors from antiquity, specifically Cassius and Areteus, who believed that animal spirits passed from one side to the other, and notes, without specific references, that more contemporary authors (Prosper Martianus, Casalpin, & Hofman) also shared this belief for explaining paralysis on the side opposite of head wounds. This is followed by a discussion of the inflammation and blood extravasation resulting from head wounds as the cause of such *contrecoups* explanations of paralysis on the opposite side, but he notes with surprise that Bonet reports the work of

Diemerbroetil who had never seen a "*contrecoup*" in more than 200 soldiers wounded in the head, and that Fallopius saw more than 100 head wounds without observing an instance of contralateral effect (Petit, 1710, pp. 1–2).

Observation I describes an officer who received a sword wound injuring the right lower eyelid passing through the maxillary bone beneath the orbit; a small wound that healed after 4 days but resulted in a terrible headache on the same side of the wound on the second day, and also reported a slight pain in his left arm. Petit saw him a month later when blood was drawn (a common practice at that time) from the arm and the foot. The left arm became completely paralyzed and the left thigh was developing a paralysis, but there was no loss of vision and his judgment was sound to the end, when the officer died 3 months after the wound. Petit promptly performed an autopsy, starting with a dissection of the wound and proceeding to open the skull and the dura widely, detaching the brain from the base and opening the ventricle with a scalpel and found a large pus-filled abscess three inches (*pouces*) in length, two in width and at least two in thickness occupying the corpus striatum and internal capsule ("*le corps cannelez*") which had become softened.

Observation II was also a soldier who received a sword wound tearing the lower right eyelid and seen by Petit a week after the injury. Pain in the head was reported on the right side but he could not use his left arm and fingers, and there was no contralateral pain. The observation of the first case led him to suspect that inflammation of the *corps cannelez* might be involved and thus warranted treatment by blood-letting from the arm and foot several times; a treatment that was deemed efficacious when movement returned completely and the head pain lessened – a gratifying cure! But Petit openly demurs from offering an explanation of why the *corps cannelez* would be inflamed by a blow to the lower eyelid. He then cites some other cases that he did not see personally, including a wound to the right clavicle resulting in paralysis of the *left* arm and a wound to the right thigh resulting in paralysis of the *left arm*, but assures of the high integrity of those who reported these cases.

He next returns to his first observation of how the abscess on the right "furnished spirits" for the movement of the left part of the body and turns again to Bonet in search of similar observations, and finds an example of a girl who, while carrying a heavy burden on her head, felt something crack twice "as if something in her head had broken." Some months later her left side became paralyzed and displayed periodic convulsive movements. Postmortem examination revealed a right ventricle abscess the "size of a chicken egg" which Petit infers to have been in *les corps cannelez*. Two other cases cited from Bonet indicate contralateral paralysis and convulsive movements ipsilateral to the head wound. But he cites some other cases of wound-induced abscess resulting in crossed paralysis including a case from "*Job a Meckren dans la Chirurgie,* p. 86" in which a sharp instrument penetrating the right parietal region resulted in paralysis of the left arm that was preceded by paralysis of the middle finger. Although hundreds of cases of head wounds were known from the literature cited, there was little information concerning relevant morbid anatomy.

All of the above observations led Petit to proclaim "I no longer doubted that there was a passage by the animal spirits from one side to the other, and to assure myself still more of the belief, I carried out the following experiments on living dogs." He tied a dog belly-down on a table, exposed the *left* parietal bone, which he opened with a trephine and drove a penknife into the brain and made a very extensive lesion causing the brain to herniate out of the wound, which was prevented (without explaining how) and the wound dressed; it was now a "very weak dog" that was observed for the 76 h that it remained alive.

Both right extremities were devoid of movement although the left legs provided support and the next day his right forelimb regained some movement, but insufficient for walking. When the animal expired, he opened the skull and found much of the left side of the brain had been extruded. He repeated this experiment on other dogs with essentially similar results and tried to remove one entire hemisphere, but the animal died too quickly to be useful. He summarizes this group of experiments by concluding that if "*les corps cannelez*" has been cut or separated from the hemisphere, paralysis unfailingly ensues on the opposite side and never on the same side as the lesion.

Observation III provides an account of a 35-year-old cavalier brought to the hospital unexpectedly stricken by a unilateral paralysis after a mild episode of pleurisy from which he had recovered. He could not move his right arm and leg, and although his jaw was in normal position and able to open and close the mouth, he could not protrude the tongue nor pronounce words. There was no asymmetry or loss of somatic sensation, but a visual field defect was noted for the right "eye." His judgment remained sound to the end and he did not display convulsive movements, but he succumbed to scurvy and diarrhea after 6 weeks and Petit removed the brain and spinal cord for dissection. The cord and the right side of the brain revealed no abnormality, but the anterior protuberance containing the several components of the "*corps cannelez*" was essentially destroyed without obvious swelling, and there was no evident injury to the optic layers, nor the optic nerve. He concludes from this that the animal spirits causing limb movements pass down via "*les corps cannelez*" and that the animal spirits deriving from the cerebral hemisphere do not cause sensation. He also suggests that the visual field defect might have been caused by a loss of natural elasticity in the membranes. He ends noting that in his successful dog experiments, vision was lost "in the eye" contralateral to the side of ablation.

Observation IV was the case of a garrisoned soldier wounded by a stone of about two pounds that had fallen from 20 ft, "striking him on the superior and posterior aspect of the right parietal bone." There was a small (*trois lignes*-about 0.3 in.) superficial laceration, but the bone was not exposed, and although initially stunned by the blow he did not fall. Other than "general treatment" he was bled twice daily and in ensuing days the injured calvarium was uncovered and found unscathed. On the sixth day after the wound, he became feverish and on the eighth day his left arm and leg became paralyzed, but without sensory defect. He became delirious and died 11 days after his accident. Six hours after death, his skull was opened and although the parietal bone seemed intact, opening the dura revealed a layer of overlying pus extending laterally and posterior as well as into the midline down to the corpus callosum. The underlying inflammation was undoubtedly the result of the blow and was found restricted to the cortex with no disturbance in the medullary substance below, although two small abscesses the size of peas were found overlying the cortex opposite the lesion, but without disturbance in any other part of the brain.

Petit arrives at the conclusion that "the animal spirits which control movement of the arms and legs derive uniquely from the superior part of the cerebral hemispheres," but then relates an experiment he finds relevant that moved him to give pause ("*donne lieu d'en douter*").

He trepanned the middle of the left parietal bone of a dog and inserted a scalpel through the trephine hole, cutting the hemisphere horizontally from front to back and then dressed the wound with "spirits." He then observed that the legs opposite the lesion could be moved, but they were "so weak that, although the animal could support himself on them, he could not take two steps without falling on his right side, and during the time he lived, was never completely paralyzed."

The report of this experiment is strange on several counts; he neglects to tell how long the dog survived and apparently failed to perform a postmortem examination of the brain. He *is* noting the profound crossed motor deficit, but seems to be concerned that the affected limbs are not totally paralyzed and thus cannot securely attribute the dorsal portions of the cerebral hemisphere as the source (or at least, the sole source) of the "animal spirits" driving motor activity.

This puzzling terse account and unexplained comment is immediately followed by Observation V, a lengthy report on a soldier who received a violent saber blow to the superior and middle left parietal bone revealing a cavity containing cerebral substance and the *dura mater* embedded with several bone splinters resulting in a torpor state that compelled the surgeon-major to open the skull on the spot and remove the splinters. The wounded soldier soon regained consciousness but could move neither his right arm nor leg, but retained vivid sensation on both sides of the body. Three days later he was able to move his right arm and leg as easily as the left. He remained mentally sound until the tenth day, when he had convulsive movements on the left side and once again lost movement on the right, and finally succumbed on the 12th day after receiving his wound. Petit then opened the skull and found numerous bone splinters in the fractured area penetrating the dura and the cortical substance inflamed a bit "the size and thickness of a half-penny piece and the region of the wound less

than the width of a gold coin." Most astonishing to him was the absence of inflammation of the medullary substance below and "that such a small inflammation could have caused paralysis, and finally death," but he notes that the inflamed or compressed part of the cortex could compress tissue below and to the sides of it.

This again led him to experiment on another large dog and expose half an inch (*pouce*) of the middle part of the right parietal bone. On top of the bone he placed an iron rod whose end was less than half an inch ("*4 lignes*") and tapped the end with a hammer leaving an impression of the rod in the bone. At first the dog was "somewhat stunned"; an interesting way to describe what must have been one of the earliest, if not the first, description of experimental percussion injury.

The wound was dressed and they tried to make the dog walk, but he could not support himself on his left legs; he also seemed to have visual losses which Petit attributes to inflammation of the eyes. He ate and drank well for the next 3 days and on the fourth day seemed stronger and "was able to walk easily on all fours." On the eighth day, he would not eat, emitted some loud cries and died.

Opening the skull revealed adhesions and several splinters driven into and attached to the dura, "which was somewhat inflamed in the region of the blow and had suppurated to some degree," but there were no changes observed in the rest of the brain. Petit concludes "this, I believe, Monsieur, to be sufficiently convincing proof for the transfer of the animal spirits from one side to the other. It is presently a question of knowing how this change takes place. It is this which I believe to have discovered."

He then proceeds with his anatomical revelation "all of the cortical substance, which is located in the cerebral hemispheres, supplies the entire medullary portion" (the underlying white matter), which is only an accumulation of an infinite number of conduits, some of which form the corpus callosum, and others form the middle *corps cannelez* (sometimes translated simply as the corpus striatum but here "middle" presumably refers specifically to the internal capsule running through and splitting the corpus striatum). He then briefly offers the opinion that he can trace the medullary fibers from the *corps cannelez moyen* through the "annular process" (the middle cerebellar peduncles as they cross the base of the pons, shown in his Fig. 1, p. 11) to the inferior surface of the medulla to form "solely the pyramidal bodies" and notes that "fibers of the right side pass over to the left, and those of the left pass over to the right" recognizing the zone of "entanglement" comprising the pyramidal tract decussation of modern parlance (Fig. 3). Petit proceeds to explain his figure one illustration of the base of the pons and medulla offering an account of brain dissection and remarks consistent with the opinions of Willis and of Vieussens, managing some disapproving words for the latter's drawings, but there is nothing strikingly new and the account of the cranial nerves conforms to that illustrated by Willis. His Fig. 2 illustrates the spinal cord in transverse section where he indicates the position of the dorsal and ventral roots but also labels the location of the "transverse medullary fibers" as E, the approximate location of the pyramidal (corticospinal) tract determined experimentally in the late nineteenth century. Figure 3 is a longitudinal three-dimensional drawing of the spinal cord serving to illustrate the ventral location of transverse fibers.

Letter Two is largely devoted to issues of the chemistry of acids and alkalis in the context of the "animal spirits," but commences with advice to

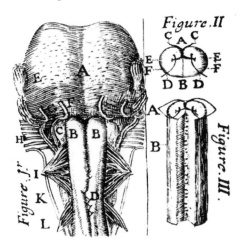

FIGURE 3. Anatomical Illustrations from the *Lettres* (Fig. I) of the crossing of medullary fibers of the "pyramidal bodies," a transverse section of the spinal cord (Fig. II) with the anterior (ventral) roots (C) on top and the dorsal roots (D) below and the transverse "medullary fibers" (E) and a three-dimensional segment of the spinal cord seen from below (Fig. III) illustrating the joining of the dorsal and ventral roots (A) and ventral transverse spinal cord fibers (B), from Petit, 1710, (p. 11)

his correspondent on the urgency of rapid dissection of the spinal cord if he is to confirm Petit's anatomical thesis for tracing fibers of the pyramidal tract decussation. It also addresses four matters raised in his friend's letter (p. 18). First, whether the cerebellum supplies the "spirits" that supply sensation or does this derive from other regions of the medulla, since Petit's work indicates that the brain proper supplies only the spirits for movement. Second, what is Petit's opinion of the nature and composition of the "animal spirits"? Third, what does he think of Willis's "Nerve sap"? And fourth, do the animal spirits "ferment" with some fraction of the blood to cause muscular contraction, and is this fraction acid and alkali in nature?

The first question deals with the part of the brain or brain stem concerned with sensation. He notes that in humans "the optic layers (diencephalon) are gray and composed of glandular substance and medullary fibers, but the majority of these fibers join the *corps cannelez moiens* (internal capsule) by a path which is opposite to that of the circulation of the spirits." He notes that the "*nates and testis*" (superior and inferior colliculi) are composed of gray and white matter, but it is not known whether or not it is glandular and thus he cannot conjecture whether sensation derives from animal spirits from the brain stem and opines that it is more probable that "sensation was produced by the spirits which filter through the cerebellum; but the following observation gives grounds to doubt this (p. 18)," and he proceeds to another case report (*observation*).

A soldier was brought to Petit's hospital 6 h after receiving a bullet wound in the neck while making his escape over a hedge, but the surgeon tried in vain to locate the bullet and detect its path except that it evidently passed from the bottom to the top of his neck, and he died 43 h after being wounded. Postmortem examination revealed the bullet had pierced the skull "to the left of the foramen from which the spinal cord passes" (foramen magnum), "passed through the left part of the cerebellum and penetrated the posterior lobe of the left cerebral hemisphere" but without damage to the brain stem (p. 19). During his 43 h of survival his judgment was good and he answered questions intelligibly but was more often delirious. Petit describes observations of his pulse, respiration, and autonomic function but ends noting, "if sensations were caused by spirits that filter through the cerebellum, it seems our wounded soldier would have had some sensory loss in the arm or leg . . ." (p. 19). This observation led to carrying out the following experiments.

He trepanned the left parietal bone of a dog and inserted a penknife through the hole into the cerebellum on that side and released the animal whose entire body bent to the left side into an arc "because of contraction of the cervical, thoracic and lumbar muscles on the left" with relaxation of these muscles on the right. The dog could support himself on his left legs but the right limbs were weak "although they did not lack the ability to move" (p. 19). Convulsions ensued (on both sides) and tests to determine a loss in any sensory modality proved equivocal and failed to satisfy Petit, but he proceeded with a postmortem examination and found that in cutting the posterior cerebral hemisphere he had opened the left ventricle (the entire ventricular system was filled with blood), the left side of the cerebellum was cut and he had damaged the "anterior part of the peduncle a bit" (p. 20). Not content with this experiment he decided to perform it in another manner.

This time he cut through a dog's right occipital bone "very close to the spine that divides at the midline" and then inserting a scalpel cut the cerebellum right to left. When the dog was untied it arched to the left and could not support himself on his right limbs and although weak, there was no sensory loss for the 10 days of survival; after which Petit opened the skull and found his cerebellectomy cut down to the "root of the peduncle." Similar results were obtained in repeated experiments enabling Petit to conclude that the cerebellum does not appear to "furnish the elements for sensation" (p. 20). He makes no comment on the observed *uncrossed* motor defect after cerebellectomy and focuses on the small amount of fluid transmitting the "animal spirits" required for sensation and movement.

The remainder of the letter is devoted to injecting various solutions into the jugular vein of living dogs; although not cited, he presumably was aware of such experiments by Lower and Wren, reported earlier by Willis. Petit recognizes the difficulties at the outset stating "as regards the nature of these animal spirits I feel it will never be known-it is not in any way tangible: it would be much simpler to learn the nature of pancreatic juice and of the stomach's secretion" (p. 21), but this doesn't deter him from trying, and although some animals died immediately, or shortly after various intravenous assaults with lethal doses of acids and alkalis, some substances (such as alcohol) proved less deleterious. In fact, wine seemed to revitalize the animal spirits (p. 24). In experiments in which the animal is opened up shortly after death, he was able to observe that pinching the distal end of the cut diaphragmatic nerve caused the diaphragm to contract and that different leg muscles contracted when different fascicles of the sciatic nerve were pinched. By contrast, pinching the intercostal (sympathetic) trunk) or the eighth pair of nerves (vagus), opposite the carotids, did not elicit contractions-not even of the heart or lower abdomen.

Petit struggles in vain with determining the nature of the animal spirits, to be distinguished from the soft and oily parts that circulate slowly in the nerves, a substance called nerve sap (*Suc Nerveux*) by Willis, with whom he agrees that another "ethereal substance can travel from the brain to the parts quickly to cause natural and unnatural movements and from the parts to the brain to cause sensations"(p. 24). Aware of inadequacies in determining the chemical nature of acids and bases as well of biological fluids, Petit modestly disclaims any original insight.

Petit in Historical Context

Neuberger's expansive account of Petit's contributions probably constitutes the principal source of information for over a century and some ambiguities in his account derive from possible misunderstanding of what Petit was describing anatomically. His interpretation of how Petit's observations provide insight into the role of the cerebral cortex is somewhat ambiguously stated, but proved tempting to infer from subsequent knowledge by later historians. Neuberger (1897) asserts that Petit was far ahead of his time "by hinting at the outstanding significance of the cerebral cortex" (Neuberger, 1981). This derives from the statement (in Petit, 1710, p. 7 Observation 3) "The animal spirits that cause the arms and legs to move come solely from the upper parts [cortex] of the hemisphere of the brain" (Neuberger, 1981), for it seems evident from Petit's subsequent statement that he traced the animal spirits from the white matter running across the *corps cannelez*, a term interpreted by Neuberger and Clarke as the "corpus striatum."[7] An English translation of key parts of Petit's 1710 monograph (Clarke & O'Malley, 1996) adds some confusion by translating the *corps cannelez moiens* as "the middle fluted bodies [corona radiata]" thereby confounding Petit's anatomical account. The "channel" where Petit's medullary fibers collect also could be the "internal capsule" of modern usage and the lesions observed by Petit in both human and dog brains were obviously too crude to distinguish between internal capsule fibers and striatal neuron involvement, but in any case, the wounds were not explicitly interpreted as lesions of the cerebral cortex *per se*. In the dog in which the lesion was largely in the dorsal portion of the hemisphere the motor defect was not described as limb paralysis, but rather of contralateral weakness in contrast to the profound paralysis he consistently observed following capsular lesions – a distinction not adequately understood until the twentieth century.

Petit also noted a contralateral visual field defect (pp. 7–8), but incorrectly concluded that animal spirits conveyed by the optic nerve "*passent par les corps cannelez*" (via capsular fibers), although he probably was able to trace the optic tract to the thalamus, and he explicitly (p. 7) states that the animal spirits from the cerebral hemispheres are not responsible for sensation. From his accounts it is apparent that none of the lesions were small enough to justify specific basal ganglia localization, but he

did get some clues of cerebral localization with different lesions consistent with a more caudal involvement for vision and a rostral localization of motor function. The concept of localization of function in the cerebral cortex did not emerge fully until the late nineteenth century.

Neuberger (1897, p. 58) assigns the failure of recognizing the importance of Petit's discovery concerning the cortex in Germany to the influential stand of Albrecht von Haller (1708–1777) against cerebral localization, but he also admonishes Petit for initially concluding that the cerebrum was not involved in sensation and turning to the cerebellum. However, the experiment (Letter Two, pp. 18–20) involving cerebellar damage failed to reveal a loss of sensation and actually caused what Neuberger interpreted as a sensory *enhancement* that was later confirmed by [Francois Gigot de] La Peyronie[8] (1678–1747) and [Francois] Chopart (~1750–1795). This is translated as "hyperesthesia and hyperalgesia" by Clarke (Neuberger, 1981) – a faithful stretch conveying Neuberger's exaggerated insight concerning Petit's observations and based on the erroneous notions of cerebellar function prevalent at the end of the nineteenth century, but not sustained in the subsequent century.

Neuberger provides an odd, but valuable account of the findings published after Petit's 1710 report, revealing a prize from the newly powerful *Académie Royale de Chirurgie* (in 1761 and again in 1768) for elucidating the "*contrecoup* phenomenon" (see footnote 8). Royal patronage prompted the prominent surgeon Antoine Louis to reproduce Petit's relevant text (1766) and led to further

---

[7] The "channel" where Petit's medullary fibers collect also could be the "internal capsule" of modern usage and the lesions observed by Petit in both human and dog brains were obviously too crude to distinguish between capsule fibers and striatal neuron involvement, but in any case, the wounds were not explicitly interpreted as lesions of the cerebral cortex *per se*. The translation of the term "*cannelez*" is a source of some ambiguity because it generally had been used in the sense of "channeled," undulated, or something grooved or fluted, as in a Doric column. But Petit also refers (p. 7) to "*la protuberance anterieure qui contient, les corps cannelez internes et superieures, les moiens et les externes ou inferieures*," which could also be interpreted as a description of the undulations of the rostral cerebral cortex rather than the *corpus*

striatum. Vieussens (1685), whose anatomy of the brain was an important cited source for Petit, apparently was following the Pordage translation of Willis (1681) in which the "*corps cannelez*" and "striatum" were used synonymously.

[8] La Peyronie is credited with establishing the *Académie Royale de Chirurgie* in 1731, and in 1761 the first academy prize for study of the *contrecoup* phenomenon, as well as its practical inferences (see Neuberger/Clarke, 1981, p. 171 ft. 5). Monetary prizes for the support of science were also recognized as an important incentive in England where in that same year of 1761, a huge prize for an accurate clock was established, resulting in John Harrison's landmark clockmaking achievement and a revolution in navigation and map-making (see Howse, 1980; Sobel, 1995).

experiments in competition for the prize. Military surgeon Louis Sebastian Saucerotte (1741–1814), vying for the 1768 prize, published results from experiments on 28 dogs in which brain injury or compression resulted in contralateral paralysis in those few animals that didn't bleed to death, but the lesions were far more extensive and crude than those of Petit's studies, damaging the thalamus and midbrain extensively. Neuberger (1897) credits Saucerotte with topographical localization of motor function in claiming that the hindlimb was innervated from the anterior part of the brain and the forelimbs from the posterior part, but with crude lesions, some extending to the base of the skull; this is hardly credible. Neuberger reveals his erroneous understanding of late nineteenth century experiments believing that the motor "center" for the posterior limbs lies medially and the anterior limbs laterally. Although not cited, Neuberger is evidently relying on the 1876 observations of David Ferrier (1843–1928) in monkeys and by J. Hughlings Jackson (1835–1911) in human Rolandic cortex, rather than maps of motor cortex in dogs; although he was clearly aware of Fritsch and Hitzig's (1870) experiments on the electrically excitable cortex in dogs. The eighteenth century investigators who followed Petit- Saucerotte, Sabouraut, and Chopart (all competitors for the royal prize and published in 1778; see Neuberger, 1981), were aware of Petit's work, but their experiments were poorly designed for physiological analysis, contributing more to the development of a French tradition in surgery than to stimulating sustained interest in the experimental approach devised by Petit (see Temkin, 1951).

The failure of Petit's work and methodological innovation to significantly alter the course of neuroscience research in the eighteenth century remains somewhat puzzling. Aside from the ingenuity and skill required to perform his experiments, he correctly assigned the contralateral paralysis following cerebral lesions to the decussation of the pyramidal tract and this alone should have been more broadly acknowledged. It can be argued that his surgical techniques were too crude to provide convincing evidence of functional localization, although his recognition that the region contributing to motor disturbance appeared to be different from that involved in visual field defects was an impressive start. His cerebellar lesion involved

other structures and animal survival was a serious problem, but recognition of the ipsilateral motor disturbance also constitutes a significant milestone. Petit's pioneer efforts in induction of concussive and compression injuries were followed with feeble attempts by other surgeons, but such studies have been performed in quantitatively controlled fashion only in modern times.

Petit's obscurity previously has been explained, in part, by the key publication being unsigned, and by its rarity, but the re-publication of the 1710 Namur monograph by Antoine Louis and citations by the several surgeons who later addressed the same problems with brain ablations suggest that he was not totally unknown in mid-eighteenth century France. Yet his 1727 paper describing the sequelae of cervical sympathectomy was widely recognized and generally credited as the functional "discovery" of the sympathetics in modern historical accounts. While several other surgeons attempted to expand on Petit's work competing for a prize, they were largely unsuccessful in maintaining the animals alive for long following surgery, and they provided few observations of consequence concerning cerebral localization; a concept that, as noted, did not advance significantly beyond Petit's modest accomplishments until the late nineteenth century. The eighteenth century neglect of Petit's accomplishments remains strangely unexplained. The biographical account in the standard medical biographical dictionary by Pierre Eloys (1778) contains a discussion of the all important 1710 monograph, but this is limited to the third letter devoted solely to botanical taxonomy without even mentioning the two letters of neuroscientific interest.

Conclusion

The religious wars, followed by the revolution in eighteenth century Europe, certainly cast a long shadow on the intellectual climate and the energies required for the birth of physiological sciences in general, and although the ablation technique was known and could be employed by a number of skilled surgeons, it was not until post-Napoleonic times that the experimental science which should have followed Harvey's empiricism and the experimentalists of the seventeenth century had the impact that might have been anticipated in a more conducive political climate. The brilliant force of Voltaire and the other *philosophes* was directed at a reorganization of society and a changing role for religion and the powers of royalty, but the scientific thrust of the "encyclopedists" led by Diderot's enormous achievement, focused efforts on collections, and indeed, the beautiful and elaborate *cabinets* that became an important feature of the *salon* life of the wealthy and educated, extending long beyond the revolution (Stafford & Terpak, 2001). Scientific taxonomy attracted the powerful organizing principles developed by Linnaeus, but experimental science remained practically moribund; perhaps a consequence of the dominance of Cartesian influence. While Pourfour du Petit's work was not totally unknown, it was sufficiently overlooked for the next 100 years to have had little impact on the blossoming of science. In a similar vein, this also might be said of his contemporary, Johann Sebastian Bach (1685–1750), indisputably the most significant figure in music in the early eighteenth century. While a few *cognescenti*, (notably Mozart & Beethoven), knew and were influenced by J.S. Bach's preludes and fugues, his work was little known and certainly less popular in his lifetime than that of his sons. It was not until after 1830 that Bach was rediscovered,[9] a consequence of the efforts of the young Felix Mendelsohn. There was no comparable proponent apparent in science and although Petit's work was known to Jules Soury, the latter's huge history in 1895 of what is now called "neuroscience" offers little appreciation of Petit's contribution or impact. Curiously, it was the German historian Max Neuberger who reawakened interest and lifted Petit from obscurity near the end of the nineteenth century (Neuberger, 1897). The relative inaccessibility of Petit's clinical and ablation studies may account for the failure of the nineteenth century "pioneers" performing selective ablations of cerebral cortex to fully acknowledge the impact of a work dating from 1710.

Historians of ophthalmology have been more astute in recognizing Petit's postwar contributions to the science of vision, especially his account of the sympathetic neural control of the eye and his work in physiological optics – most notably his method for measuring astigmatism.[10] Petit's true legacy lies in the profound importance of his systematic introduction of physiological animal experiments in neuroscience and linking these findings to clinical observations and morbid anatomy. His discovery that the decussation of the pyramidal tract can account for contralateral paralysis after cerebral injury and that the "intercostal nerve" provides autonomic innervation of the eye and facial skin also derive from the application of this methodology. Returning from a "war zone" to the refinements of a life in Paris imbued with medicine and science, it should be unsurprising that vivisection on conscious animals would be less appealing than studying the comparative anatomy of the eye, physiological optics or devising a practical technique for human cataract surgery. For all of these accomplishments, though long neglected, Petit stands as a towering transitional figure in the story of the thrust toward experimental science in an era that was ill prepared to receive it.

---

[9] Despite the earlier neglect of Bach, his re-discovery was enhanced by the efforts of the prominent German neurohistologist, Wilhelm His (1831–1904) in exhuming Bach's body for a detailed account of the remains of this musical giant (Barzun, 2001).

[10] A device for the purpose of achieving an accurate measure of astigmatism first described by Petit and not substantially improved upon until the efforts of Kohlrausch and Helmholtz two centuries later, enabled the construction of optical instruments of practical clinical utility based on the principles delineated by Petit, although this was largely unknown. An engaging account and diagram of Petit's ophthalmometer, which could only be used satisfactorily on excised eyes, was presented to the Optical Society in England a century ago (Sutcliffe, 1906). The principle employed was projection of a circular image with quadratic marks upon the spherical eye, and if the reflected image was oval instead of round, the difference between the length and breadth axes would constitute a measure of astigmatism. Knowing the size of the circular object, its distance from the convex eye and the size of the image, one could determine the curvature of the eye by measuring the dimensions of the image.

Acknowledgments. The microfilm of the Petit 1710 *Lettres* was obtained in 1958 at the request of the late Professor Denise Albe-Fessard (Institut Marey, Paris) through her friend, Mlle. Kleindienst, of the office of the Secretaire-Generale, M. Caen of the Bibliothèque Nationale, Paris. Professor Auguste Tournay, (the last Neurology resident trained at the Salpetrière Hospital by J.F.F. Babinski), found the Restout painting of Pourfour du Petit in (or through) the office of the dean of the faculty of medicine in Paris in 1958 and provided a photographic copy for an earlier publication (Kruger, 1963), as well as guidance in eighteenth century historiography. Nicolas Barker (British Library), Marcia Reed (Getty Research Institute), and Professor Roger Hahn (UC Berkeley) provided insights into the intricacies of eighteenth century publishing practices. Professor Marie-Françoise Chesselet was most helpful in interpreting early French anatomical nomenclature. Professor Peter Reill, Director of The UCLA Center for seventeenth and eighteenth Century Studies provided wise counsel and the expert student assistance of William T. Hendel and Diana Raesner. Gina Akopyan also assisted in final manuscript preparation. Russell Johnson and Cathy Donahue of the UCLA Louise Darling Biomedical Library provided helpful advice and the professional skills of trained reference librarians, without whom this work might have omitted many significant details.

References

Barzun, J. (2000). *From dawn to decadence*. New York: Harper Collins. 877pp.

Best, A. E. (1969). Pourfour du Petit's experiments on the origin of the sympathetic nerve. *Medical History*, 13, 154–174.

Bonet, T. (1679). Sepulchretum; sive, Anatomia practica: ex cadaveribus morbo denatis, proponens historias et observationes omnium pene humani corporis effectuum (4 Vols.). Geneva.

Clarke, E. (1981). *The historical development of experimental brain and spinal cord physiology before Flourens* (M. Neuberger, Trans., 1897). Baltimore, MA: Johns Hopkins University Press. 391pp.

Clarke, E., & O'Malley, C. D. (1996). Human brain and spinal cord: a historical study illustrated by writings from antiquity to the twentieth century. San Francisco: Norman. 951pp.

Cole, F. J. (1949). *A history of comparative anatomy*. London: Macmillan. 524pp.

Eloy, N. F. J. (1778). Dictionaire historique de la médecine ancienne et moderne . . . (4 Vols.).

Fritsch, G. T., & Hitzig, E. (1870). Über die elektrische Erregbarkeit des Grosshirns. Arch. Anat. Physiol. Wiss. Med, 300–332.

Gautier d'Agoty (1748). Anatomie de la tête, en tableaux imprimés, qui représentent au naturel le cerveau sous différentes coupes, la distribution des vaisseaux dans toutes les parties de la tête, les organes des sens, & une partie de la névrologie; d'aprés les piéces disséquées & préparées, par Duverney. Paris: Gautier.

Howse, D. (1980). *Greenwich Time and the discovery of the longitude*. Oxford/New York: Oxford University Press. 254pp.

Kruger, L. (1963). Francois Pourfour du Petit, 1664–1741. *Experimental Neurology*, 7, ii–v.

Louis, A. (1766). Recueil d'observations d'anatomie et de chirurgue; pour servir de base á la théorie des lesions de la tête, par contre-coupe. Paris: PG. Cavalier. 270pp.

Mairan, J. J., & Dortous de. (1747). Éloges des académiciens de l'Académie royale des Sciences: morts dans les anneés 1741, 1742, & 1743. Paris: Durand. 360pp.

Meyer, A. (1971). *Historical aspects of cerebral anatomy*. New York: Oxford University Press. 230pp.

Neuberger, M. (1897). Die historische entwicklung der experimentellen Gehirn-und Rückenmarksphysiologie vor Flourens. Stuttgart: Ferdinand Enke. 361pp.

Neuberger, M. (1981). *The historical development of experimental brain and spinal cord physiology before Flourens*/Max Neuburger; translated and edited with additional material, by Edwin Clarke. Baltimore: Johns Hopkins University Press.

Pordage, S. (1681). In W. Feindel (Ed.), *English translation of Willis's cerebri anatome* (1965). Montreal: McGill University Press.

Pourfour du Petit, F. (1710). *Lettres d'un medecin des hôpitaux du Roy, a un autre medecin de ses amis*. Namur: C. G. Albert. 38pp.

Pourfour du Petit, F. (1727). Mémoire dans lequel il est démontré que les nerfs intercostaux fournissent des rameaux qui portent des esprits dans les yeux. Mém. Acad. r. Sci.

Rasmussen, A. T. (1947). *Some trends in neuroanatomy*. Dubuque Iowa: W.C. Brown. 93pp.

Ségura, P., Speeg-Schatz, C., Wagner, J. M., & Kern, O. (1998). Le syndrome de Claude Bernard-Horner et son contraire, le syndrome de Pourfour du Petit, en anesthésie-réanimation. *Annales Françaises d'Anesthésie et de Réanimation*, 17, 709–724.

Sobel, D. (1995). Longitude: the true story of a lone genius who solved the greatest scientific problem of his time. New York: Walker. 184pp.

Soury, J. (1899). Le Système Nerveux Central: structure et functions histoire critique des théories et des doctrines (2 Vols.). Paris: Masson.

Stafford, B. M., & Terpak, F. (2001). *Devices of Wonder; from the world in a box to images on a screen.* Los Angeles: Getty Research Institute.

Sutcliffe, J. H. (1906). One-position ophthalmometry. *Transactions of Optical Society*, 8, 74–102.

Temkin, O. (1951). The role of surgery in the rise of modern medical thought. *Bulletin of the History of Medicine*, 25, 248–259.

Thomas, H. M. (1910). Decussation of the pyramids: An historical inquiry. *Bulletin of the Johns Hopkins Hospital*, 21, 304–311.

Willis, T. (1664). Cerebri Anatome, cui accessit nervorum descriptio et usus. London.

Vieussens, R. (1685). *Neurographia universalis.* Leyden.

Zehnder, E. (1968). Francois Pourfour du Petit (1664–1741) und seine experimentelle Forschung über das Nervensystem. (dissertation). Zürich: Juris-Verlag. 39pp.

8

Irritable Glue: The Haller–Whytt Controversy on the Mechanism of Muscle Contraction

Eugenio Frixione

Introduction

Animal motion has been one of the longest-lived great themes in neurophysiology. For the most part of that development the soul (*anima*), in any of its various versions – aerial, atomistic, or purely spiritual – was taken as the actual agent causing animation. Then, as it is well known, in the mid-seventeenth century this view was formally challenged by René Descartes (see Des Chene, 2001). The iatrophysical school he contributed to create maintained that the animal body is an automatic machine fully capable of executing and controlling all its operations, independently of any incorporeal assistance. Perhaps nowhere these two confronted positions – animist and mechanicist – clashed more loudly over a single specific topic than in the long dispute sustained, in the 1750s and early 1760s, between Albrecht von Haller and Robert Whytt about the persistence of irritability in isolated muscles.

The colorful Haller–Whytt debate, which was concerned also with the parallel physiological faculty of sensibility, has been thoroughly reviewed in stepwise detail (French, 1969; Miller, 1939). These excellent accounts of the subject, though, concentrate on the discussions about the relative irritability or sensibility of specific organs, the presence or absence of nerves in particular anatomical locations, or the spatial distribution of the soul in the animal body. Comparatively little attention has been focused on the radically opposed views of the two great authors regarding the basic mechanism of muscle contraction.

The present chapter examines this important aspect of the debate, starting from a brief background sketch of the theory of fibers as taught by Herman Boerhaave, under whom both Haller and Whytt studied. Then, following a summarized review of the main arguments on each side of the controversy, a few possible reasons are discussed of why and how was muscle the first province of the animal body to become permanently liberated from the sovereignty of the soul. A preliminary version of this paper has been published in abstract form (Frixione, 2004b).

Boerhaave's Teachings

Medical and chemical speculation in the eighteenth century was built over broad conceptual guidelines set largely by the revered systematizer of scientific knowledge and Dutch physician Herman Boerhaave (1668–1738). His anatomy was based on the all-comprehensive theory of fibers, which has a long history going back to classic Antiquity and was maintained, even validated, by the early microscopists in the seventeenth century (Frixione, 2004a; Grmek, 1970). It was apparently Antoni van Leeuwenhoek who first succeeded in directly visualizing that muscle substance is indeed made up of diminutive fibers, that are in turn constituted by "filaments" of a smaller caliber (Leeuwenhoek, 1682, 1948, pp. 393–397). Two decades later this notion had already been extended to encompass virtually all solid parts of animal bodies, thus giving rise to the "solidist" account of physiology

(Baglivi, 1702; see Bastholm, 1950) that gradually stripped the traditional doctrine of humors from its leading role in medical thought.

The foundations of the mature theory of fibers are discussed by Gerard van Swieten (1744), one of Boerhaave's most distinguished followers. Influenced by Giorgio Baglivi (see Hall, 1969), Boerhaave taught the whole body is built up of fibers in a wide scale of calibers: those discernible with the naked eye – like some muscle fibers – are composed of thinner threads clearly revealed by the microscope, which are themselves assemblies of still finer filaments at the limit of visual detection, and so on. Because the number of dimensional levels could not be infinite, a minimal category should exist of submicroscopic fibers which are not made up of smaller fibers. This basic anatomical unit or *fibra minima* (Boerhaave, 1737, pp. 4–5) must consist of inalterable elementary particles held together by a tenuous adhesive substance:

Those parts which being secreted from the fluid contained in the vessels, and applied to each other by the vital powers, and the assistance of a most fine aqueous or fat gluten, compose the least fibre, are very small, most simple, earthy, and scarce capable of being changed by any of those causes, which are found to act in us during life (Boerhaave, after Swieten, 1744, p. 39).

In his long commentary to this aphorism, Swieten explains that the elementary particles are necessarily of an earthy nature because observation shows that impalpable dry dust is the common ultimate remains of any thoroughly decomposed, desiccated or incinerated human or animal body. Such small particles are "adjoining to each other length-ways," he says, and "they cohere together […] by the interposition of aqueous or fat gluten, for water hath an incredible power to unite bodies together [… and because as long as some …] oil adheres to the parts of animals […] they still cohere; but this being expelled they fall asunder" (Swieten, 1744, pp. 39–42).

Above the minimal fiber level, organization of living matter was hierarchical. Minimal fibers would typically associate sideways so as to constitute delicate sheets or membranes, which could curl up into tiny cylinders. These tubules were actually fibers of the second order in the hierarchy, and could in turn either unite laterally or interweave to form membranes of the second order. As this sort of progressive arraying pattern continued,

macroscopic membranes and tubes of higher orders, like vessels or guts and eventually whole organs, were constructed. Accordingly, all kinds of flesh were really compact masses of fibers made up into a variety of ducts in widely different calibers where the bodily fluids were contained or flowed. Muscles were no exception, for their minutest visible fibers are very fine extensions of

the ultimate Extremities of the smallest Nerves, deprived of their Coats, hollow internally, […] being filled with Spirits or nervous Juice […] derived into the Nerve from its Origin or Fountain the Brain and Cerebellum, by the continual Force of the Heart (capitalized letters in the original; Boerhaave, 1744, p. 176).

Consideration of the various features characteristic of muscle contraction led Boerhaave to conclude that the cause of this phenomenon "can be no other than a very thin *fluid Body*, very easily or *quickly* moved, and that it must be *forcibly* thrust into or applied to the Muscle" (italics and capitalized letters in the original; Boerhaave, 1744, p. 217). These key qualities of that special fluid "are all found in the *nervous* Liquor, and in no *other* Humor; and therefore that Liquor is to be acknowledged for the true *Cause*; nor is it difficult to understand its Manner of Action" (italics and capitalized letters in the original; Boerhaave, 1744, p. 222).

Such an hypothesis of muscle contraction, dubbed "balloonistic" by contemporary critics, is essentially the same offered by Descartes (1664) and can be traced back to Erasistratus in the third century BC (cited by Galen, *De loc. aff.*, VI, K 429; see Siegel, 1976, p. 188). Variants of this model in which the thin fluid coming down the nerves from the brain, rather than filling tiny bladders within the muscles, would just initiate a local expansive reaction in them – i.e., intense agitation of particles or effervescence, even an explosion – by mixing up with "the spirits of the blood" or some other vital juice, were suggested by William Croone, John Mayow, Alfonso Borelli, and Thomas Willis in the seventeenth century (see Nayler, 2000; Wilson, 1961). Hence, Boerhaave must have felt reasonably safe in his basic position on this matter.

All of the above theories, however, were at odds with relatively recent evidence indicating that muscles do not swell at all, perhaps even shrink a little, upon contraction. A crude demonstration of this had been presented at a meeting of the Royal

Society (London) in 1669, where Jonathan Goddard showed that if a man clutched his fist while his whole arm was immersed in a narrow column of water, the level of liquid in a communicating slender glass pipe would descend instead of being raised (see Birch, 1756, 1968, pp. 411–412). A similar account of this fact was given a few years later by Francis Glisson (1677, pp. 166–167). Moreover, a vastly refined, and likely earlier, version of this experiment had been performed by Jan Swammerdam with an isolated muscle freshly dissected out from a frog, along with part of its associated nerve. This preparation was hanged inside a wide test tube terminated in an open capillary tube at its rounded end, the test tube being then tightly sealed with a piston plugged up into its mouth. Remote mechanical stimulation of the nerve from the outside was still possible by means of a silver wire inserted through the piston, and the muscle twitched in response to the stimuli applied to the nerve. Then Swammerdam noticed that when the muscle was thus induced to contract, a droplet of water intentionally lodged beforehand within the capillary tube sank slightly toward the muscle, thereby attesting a decrease in overall internal volume (pressure) within the glass apparatus. His conclusion, based on this and other studies, was categorical:

... I myself, relying on the propriety and certainty of the experiments I have proposed, can now, without any difficulty, maintain, that a muscle, at the time of its contraction, undergoes no inflation or tumefaction, from the afflux or effervescence of the supposed animal spirits; but that, on the contrary, it in this state becomes smaller, or collapses; or, to express my meaning more clearly, it takes up less room than it did before (Swammerdam, 1737, p. 128).

This work was published some 60–70 years after the experiments were actually carried out, i.e., when Boerhaave himself edited Swammerdam's posthumous book, but the surprising main findings became known by insiders since the 1660s. Nicolaus Steno (Stensen) heard about them directly from Swammerdam, since they were close friends and shared similar unorthodox views about muscle function (see Kardel, 1994). In fact Steno brilliantly demonstrated, through a strict physical and geometrical analysis, that muscles may indeed change in shape while keeping a constant volume (Steno, 1667). Among those convinced that, if anything, muscles become smaller during contraction was Richard Lower (1669, 1932, p. 77). In consequence, at least

Croone felt forced to revise his own theory (see Nayler, 2000; Wilson, 1961). Muscle inflation, a common belief held by many respected philosophers and physicians for two millennia, was suddenly thrown into serious doubt. Although the eclectic Boerhaave evidently chose to stay within the mainstream on this issue, his younger students may have preferred to look for possible alternatives, and a totally different mechanism would eventually be advanced by his most outstanding pupil.

Haller's Concept of Muscle

Albrecht von Haller (1708–1777), who after graduating as doctor of medicine at Leyden had visited other important medical centers in England and France before returning to his native Switzerland, was eventually appointed as professor at the recently founded University of Göttingen (see Frixione, 2006; King, 1966). Here he found the time, among many other duties and not a few quarrels, to write a textbook on general physiology "designed as a correction and improvement" of Boerhaave's treatises in the light of new findings, according to one of his later contemporary editors (Haller, 1747, 1786, p. v). Muscle contraction was one of those subjects needing revision, once the centuries-old theory of swelling by an influx of animal spirits could no longer be sustained.

Haller pointed out that muscle is endowed with at least three quite different kinds of forces, which are easily distinguishable from each other by their properties. Just like nearly all fibers, they present a "dead" force or elasticity which allows them to retract spontaneously after being distended, and which remains effective for a long time after death. Peculiar to muscle fibers, however, is a unique internal force – *vis insita* – capable of performing quicker and more ample motions than mere elasticity, and which disappears much sooner than the latter after death. This second force is excited into action by pricking, stretching, or applying acrid substances to the muscles, "but most powerfully of all by a torrent of electrical matter" (Haller, 1747, 1786, p. 232). The third force is the "nervous power" (Haller, pp. 215, 235), capable of activating the muscles although it does not reside in them like elasticity and the *vis insita*, and is therefore suppressed by tying or cutting the nerves connected to the muscles.

According to Haller, the *vis insita* characteristic of muscle "seems to be a more brisk attraction of the elementary parts of the fibre by which they mutually approach each other, and produce as it were little knots in the middle of the fibre" (Haller, 1747, 1786, p. 236). Somehow the nervous power promotes such approximation, but the actual motive force "seems not to be the soul, but a law derived immediately from God" (Haller, p. 237). This is evident from the fact that newly born animals already know how to execute compound movements, which would take a long time to calculate even for a well-trained soul. And it is still more certain that the *vis insita* is independent of the soul, because the heart and intestines or the organs of generation, for example, which are governed by the *vis insita*, move by themselves or in response to stimuli in a way that cannot be enhanced, reduced or modified in any form by the will. Indeed:

It is so certain that motion is produced by the body alone, that we cannot even suspect any motion to arise from a spiritual cause, besides that which we see is occasioned by the will; and, even in that motion which is occasioned by the will, a stimulus will occasion the greatest exertions, when the mind is very unwilling (Haller, 1747, 1786, p. 238).

It was precisely this kind of materialism that animist-bent physiologists like Robert Whytt found so unsatisfactory and disturbing. The climate was hardly favorable for such language since, by unfortunate coincidence for Haller, his textbook was published the same year as *L'Homme Machine*, one of the most scandalous works of the century, written by Julien Offray de La Mettrie (1709–1751) – also a former student of Boerhaave. Here matter was presented as endowed with sensibility (see Smith, 2002), and a general property closely similar to Haller's *vis insita* was taken as an incontestable fact: "chaque petite fibre ou partie des corps organisés se meut par un principe qui lui est propre et dont l'action ne dépend point des nerfs comme les mouvements volontaires"[1] (La Mettrie, 1747, 2004, pp. 71–72). Upon these and other considerations,

the author claimed, the time had come to accept that "l'âme n'est qu'un principe de mouvement, ou une partie matérielle sensible du cerveau qu'on peut, sans craindre l'erreur, regarder comme un ressort principal de toute la machine"[2] (La Mettrie, p. 75). Embarrassingly, moreover, La Mettrie's book was dedicated to Haller (La Mettrie, pp. 37–42), who deeply resented being involved with such impious contents mischievously related to his research and contrary to his devout personal feelings.

La Mettrie died unexpectedly much too soon (1751) for receiving a proper response (although the possibility has been argued that Haller might have been the author in disguise of another well-known work supposedly written by La Mettrie in which *L'Homme Machine* is largely refuted; see Saussure, 1949). Yet, after all of this, it was not materialism but spiritualism which would become more troublesome for Haller's physiological system, as an old and quite different "principle" also responsible for both sensation and motion was again put on the field.

Whytt's "Active Sentient Principle"

Robert Whytt (1714–1766) received first medical instruction at Edinburgh and then continued his professional education in London, Paris, and finally Leyden where, like Haller and La Mettrie, he studied under Boerhaave (see French, 1969; Radbill, 1981). After his return to Scotland, he combined a profitable practice as a physician with prestigious teaching and research, mostly on neurological matters. In 1744, "dissatisfied with the common theories of respiration and the heart's motion" (Whytt, 1751, 1768, p. v), he decided to write a critical review of the problem of animal movements in general. In all of this, we are assured, the author was "careful not to indulge his fancy, in wantonly framing *hypotheses*, but has rather endeavoured to proceed upon the surer foundations of experiment and observation." While perusing the various interpretations proposed to

---

[1] "Each small fiber or part of the organized bodies moves by a principle of its own and the action of which does not depend of nerves like the voluntary movements" (translated by the author).

[2] "The soul is but a principle of movement, or a sensitive material part of the brain that one can, without fear of error, regard as a principal spring of the whole machine" (translated by the author).

explain contractility, however, Whytt rejected all the purely mechanical models and endorsed the time-honored belief in an incorporeal agency controlling the behavior of animal bodies:

we cannot but acknowledge, that [the all-wise AUTHOR of nature] has animated all the muscles and fibres of animals, with an active sentient PRINCIPLE united to their bodies, and that, to the agency of this PRINCIPLE are owing the contractions of stimulated muscles (capitalized letters in the original; Whytt, 1751, 1768, p. 128).

Whytt could not understand just why, in an effort to explain the motions of the muscles, any intelligent colleague would resort to invoking mysterious attributes of the nervous fluid or the muscle fibers, when those phenomena were "so easily and naturally explained, from the power and agency of a known sentient PRINCIPLE" (Whytt, 1751, 1768, p. 140). He was ready to admit that little was actually known about this special principle, but saw no difficulty in accepting its existence as a scientific fact, supporting this claim with an analogy of obvious Newtonian overtones:

That there is such a thing as gravity, or attraction betwixt the parts of matter, is a thing not to be doubted of, because we see its effects, though its cause be unknown: And, if philosophers are allowed to make constant use of this power, in order to explain the *phænomena* of nature, why should it not be thought equally reasonable to have recourse, in accounting for the motions and actions of an animated body, to the power and agency of the mind, which we are sure is always present with it, and in numberless instances operates upon it? (Whytt, 1751, 1768, pp. 144–145).

Identifying the active sentient principle with both the soul and the mind seemed quite reasonable, for he was "inclined to think that the *anima* and *animus*, as they have been termed, or the sentient and rational soul, are only one and the same principle acting in different capacities" (italics in the original; Whytt, 1751, 1768, p. 148). Having equated the active sentient principle with the mind-soul, Whytt was obliged to offer some explanation for the occurrence of animal motions after death or in organs separated from the body. He provides brief descriptions and discussions of 35 instances of movements in headless animals or isolated body parts. Nevertheless, his main theory remained firm, declaring that:

the motions of the heart, and other muscles of animals, after being separated from their bodies, are to be ascribed

to this principle; and that any difficulties which may appear in this matter are owing to our ignorance of the nature of the soul, of the manner of its existence, and of its wonderful union with, and action upon the body (Whytt, 1751, 1768, p. 207).

Whytt was careful to stress that his position on this topic should not be confused with those of other animists like Godfried Leibniz and particularly Georg Stahl (e.g., Whytt, 1751, 1768, pp. 140–141), who was often ridiculed for pretending that the soul, as a rational agent, managed every nook of the body. Yet Whytt's essay found a cold reception in a book review authored by Haller, who wrote: "Nothing is more specious than to derive all motions from the soul" (Haller, 1752, no.1, part 3, p. 157).

Haller's "Irritability"

In the meantime Haller had also been busy studying sensibility and "irritability," a term and concept he borrowed from Francis Glisson (1677) although in fact it has a much older origin (see Temkin, 1964; Verworn, 1913), giving to it a precise definition: "I call that part of the human body irritable, which becomes shorter upon being touched; very irritable if it contracts upon a slight touch, and the contrary if by a violent touch it contracts but little" (Haller, 1753, 1755, 1936, p. 8). In collaboration with his student and friend Johann Zimmermann, Haller had set out to explore which parts of the animal body showed irritability and which exhibited sensibility, the latter being recognized by any signs of discomfort or pain in response to all kinds of nociceptive stimuli. The conclusion that "there is nothing irritable in the animal body but the muscular fibre" (Haller, p. 40) derived from a number of experiments performed from 1746 to 1752, using as many as 190 animals of various classes and ages just in 1751, and exercising "a species of cruelty for which [he] felt such a reluctance, as could only be overcome by the desire of contributing to the benefit of mankind, and excused by that motive which induces persons of the most humane temper, to eat every day the flesh of harmless animals without any scruple"(Haller, p. 7).

While discussing the nature of irritability, it was inevitable to mention the possible role of the soul in the phenomenon, especially after Whytt's recently published essay on the subject. Haller argued as follows:

a finger cut off from my hand, or a bit of flesh from my leg, has no connexion with me, I am not sensible of any of its changes, they can neither communicate to me idea nor sensation; wherefore it is not inhabited by my soul nor by any part of it; if it was, I should certainly be sensible of its changes. I am therefore not at all in that part that is cut off, it is intirely [*sic*] separated both from my soul, which remains as entire as ever, and from those of all other men. The amputation of it has not occasioned the least harm to my will, which remains quite entire, and my soul has lost nothing at all of its force, but it has no more command over that amputated part, which in the mean while continues still to be irritable. Irritability therefore is independent of the soul and the will (Haller, 1753, 1755, 1936, p. 28).

At the very beginning of this treatise Haller openly admitted ignorance about the ultimate explanations underlying both sensibility and irritability, and, like Whytt, offered to refrain from formulating any suppositions not based on direct observation:

For I am persuaded that the source of both lies concealed beyond the reach of the knife and microscope, beyond which I do not chuse [*sic*] to hazard many conjectures, as I have no desire of teaching what I am ignorant of myself. For the vanity of attempting to guide others in paths where we find ourselves in the dark, shews [*sic*], in my humble opinion, the last degree of arrogance and ignorance (Haller, 1753, 1755, 1936, p. 8).

Many pages ahead, however, following a detailed description of his numerous observations and coming to conclusions, little of the initial caution seems to remain. In discussing the locus of irritability, he after all hazards the following conjecture:

But the muscular fibres being composed of earthy particles and a glutinous mucus, it may be asked in which of these Irritability recides [*sic*]. It appears most probably to recide in the latter, because this when it is pulled endeavours to shorten itself; whereas on the contrary, dry earth never changes its figure of itself, and being extremely brittle, when its parts are separated, they constantly remain so (Haller, 1753, 1755, 1936, pp. 40–41).

The question then arises of how the gluten can become irritable, and this brings up again the possible involvement of the soul. Haller dismisses this idea on the grounds that:

Irritability acts without the soul being sensible of it, and that it is not subject to the command of the will, both which are proved by the example of the heart. To avoid the consequences of this argument, the anatomists are obliged to introduce an insensible sensation, and involuntary

acts of will, that is to say, to admit contradictory propositions (Haller, 1753, 1755, 1936, p. 42).

In the absence of a clear answer, Haller, also like Whytt, ventures to invoke basic general physics to defend the plausibility of his conjecture: "What therefore should hinder us from granting Irritability to be a property of the animal *gluten*, the same as we acknowledge attraction and gravity to be properties of matter in general, without being able to determine the cause of them" (italics in the original; Haller, 1753, 1755, 1936, p. 42).

Haller closes the original version of this paper with a last word bent back on metaphysics: "neither Irritability depends upon the soul, nor is the soul what we call Irritability in the body."

Further Exchanges

Whytt responded mentioning the Swiss physiologist by name in the very title of his *Observations on the Sensibility and Irritability of the Parts of Men and other Animals, occasioned by the celebrated M. de Haller's late treatise on those subjects* (1755, 1768). He complained for having been misrepresented by his colleague as a proponent of the divisibility of the soul, an opinion which he not only rejected but had also refuted with several reasons (Whytt, 1751, 1768, pp. 201–203). In regard to this point, Whytt states the dilemma:

we must either allow that both the head and body of a frog continue to be animated for some time after they are separated from each other; or else affirm, that the life, feeling, and active powers of animals, are merely properties of that kind of matter of which they are made. The former opinion is attended with some difficulties, which arise chiefly from our ignorance of the nature of immaterial beings; the latter seems to be inconsistent with all the known properties of matter (Whytt, 1755, 1768, p. 289).

Once again, ignorance on these questions was acknowledged but deemed not too different from that surrounding other mysteries in nature:

for, if we are far from understanding the communication of motion and other actions of matter upon matter, how shall we be able to comprehend the manner in which an immaterial principle acts upon it? But, as we can, from the little we know of matter, see that inactivity is one of its essential properties, we are hence convinced of the necessity of ascribing the life and motions of animals to the power of an incorporeal agent (Whytt, 1755, 1768, p. 292).

Having dealt with the defensive, Whytt turns to an overwhelming offensive against Haller's alternative conjecture. A general consideration is that "if irritability be a property of the muscular glue, why may not sensibility and intelligence be properties of the medullary substance of the brain?" (Whytt, 1755, 1768, p. 293). But there are more specific objections. The glue or gluten extracted from parts of animals may have elasticity, but absolutely no capacity for alternate contractions like the muscle fibers. Also, how is it that the skin, ligaments, and tendons contain more glue than the muscles, and yet they are not in the least irritable? Haller's analogy of irritability with gravity provided another wide flank for attack, for matter continues to possess gravity even after being altered by fire, solvents or otherwise, but the gluten removed from animal bodies "appears as inert and void of active powers as any other matter" (Whytt, p. 294). Moreover, even if allowed to remain in the muscles, the gluten loses its irritable power soon after the latter are separated from the body. Further, gravity had ultimately been ascribed to the active power of an incorporeal cause acting on matter, either directly or through some subtle elastic medium, and so it should be with irritability. Whytt concludes his reply with the following words:

It appears, therefore, after all that has been said to shew [sic] that the motions of irritated muscles are owing to a property of irritability in them or their glue, that we are at last obliged to refer them to the active power of an immaterial cause; unless we shall, contrary to sound philosophy, ascribe feeling and spontaneous activity to matter. And, as gravity must finally be resolved into the power of that BEING who upholds universal nature; so it is probable, that the irritability of the muscles of animals is owing to that living sentient principle which animates and enlivens their whole frame (Whytt, 1755, 1768, pp. 294–295).

Haller dismissed Whytt's objections as a marginal discussion that did not affect the general validity of his physiological system. In a further collection of papers on sensibility and irritability, the Swiss professor merely repeated the original consideration that, put to choose between glue and earth as the seat of irritability, and seeing the comparatively greater irritability of young or gelatinous animals, he leaned toward the first option (Haller, 1760, pp. 123–124). In turn, Whytt reviewed their whole controversy in an appendix to the second edition of his own essay on sensibility and irritability, where he points out that nothing new had been advanced in support of Haller's doctrine and regrets:

Strange, certainly it is, to find men of learning and abilities, misled by a few ill understood experiments and trusting to metaphysical notions, using every art to prove, that the irritable power of the muscles does not depend on their nerves, and ascribing it to a substance, the most unlikely to be possessed of it (Whytt, 1761, p. 312).

Undaunted, Haller discussed at length the key role of animal gluten in irritability in his monumental treatise on general physiology (Haller, 1762, pp. 464–466), and somewhat later he stated his position on the metaphysical aspect of the problem:

Religiously I recognize God as the mover of universal nature. But not for the same reason should the elasticity of expanding metal, or the weight of a stone, or the heat of an acid entering into reaction with an alkaline, or the contraction of a dissected muscle be attributed to incorporeal forces. God gave the contractile force to bodies, and he gave other forces which once having been received, are not due to any soul or spirit beneath God. I doubt about that spirit that is the cause of motion; I claim all from God (Haller, 1763, 1764, p. 27).

These last five words were apparently the only statement in which both Haller and Whytt could agree on the topic of irritability. By then not much time remained available for further debate anyway, as Whytt, though 6 years younger than Haller, died 11 years before his more durable adversary. The total discrepancy of two leading physiologists over such a clear-cut issue seemed to bring basic physiology, and thus the expectations of ever understanding in depth the human body, close to a hopeless end. But not for long, as the soul would soon be forced to yield its dominion over at least the voluntary muscles.

In Retrospect

Indeed, skeletal muscle was the first domain of the animal body to become permanently emancipated from the soul's rule. At the turn of the century, the body would no longer be a highly convoluted mesh of fibers, but either a composite of "tissues" (Bichat, 1801) or an assembly of living corpuscles (Oken, 1805). And, much like any other animal tissue, muscle exhibited the common properties of sensibility and contractility (Bichat, 1801). The role of the soul in muscle contraction, as debated by Haller and Whytt only 40 years earlier, is virtually absent from nineteenth century literature. If anything, its all-presiding aura is saluted in a prologue and then the text turns to explain physiology in materialistic fashion with the sole help of "vital forces." How could a millenary belief happen to disappear so fast from the scene?

It can be argued, of course, that in the mid-eighteenth century the field was ready for an imminent eclipse of the soul. Animism was in retreat following the evolution of iatrophysics to culminate in La Mettrie's assault with *L'Homme Machine*, and with the French Encyclopedists following immediately behind. Yet the case of muscle is unique because, in contrast to other body parts, it quite obviously possesses intrinsic capabilities for sensing stimuli and directly responding to them, even after being isolated from the rest of the organism. Whytt's hypothesis that the soul remains in the body for some time after death could perhaps have accounted for the persistence of irritability in recent corpses. The tougher fact to explain, however, was the persistence of irritability in fragments of a body, either dead or alive, because it implied the soul to be extensible or divisible; hence, Haller's conclusion that irritability was independent of the soul.

The conflict between the two authors was greatest in the case of muscle because, ironically, in this particular instance they were both right. In today's terms, irritability, as conceived by Haller and understood largely also by Whytt, refers to the coupled phenomena of excitability and contractility, which in macroscopic living systems occur together only in muscle. Here, irritability and the active-sentient – actually sentient-active – principle are one and the same.

As regards the immediate repercussion of their published arguments, it is perhaps fair to say that it was a tie. Whytt's orderly and more respectful style, apart from his sincere confession to the limits of our knowledge, produced an overall sense of confidence among his contemporaries. On the other hand, Haller's outstanding ability for experiment and perceived inclination to reductionism pleased the mechanically oriented theorists and foreshadowed the general scientific attitude of the following century. In the medium and long terms, however, there is no doubt that Haller can be declared the winner, as the soul was definitely expelled from muscle before the end of the century.

Several developments that took place almost simultaneously around 1790 may have contributed to this unexpectedly rapid outcome, beginning with the discovery of animal electricity. It was well known that electrical shocks are the most effective stimuli to irritate muscles; a quite different proposition, which led directly to the following major and noisy feud in the field, was that muscles might themselves function as electrical devices. Luigi Galvani's tentative comparison of "a muscle fibre to something like a small Leyden jar or to some other similar electrical body charged with a twofold and opposite electricity" (Galvani, 1791, 1953, p. 74) would be at the center of the confrontation between the brilliant Italian physiologist and a first-class physicist, Alessandro Volta (see Pera, 1992). The notion that all muscle activity might be a purely physical process was also likely fostered with the equally astonishing demonstration that respiration is, but combustion, analogous to the burning of oil in a lamp (Séguin & Lavoisier, 1789). On the other hand, the plausibleness of hypotheses concerning the soul as a physiological agency had largely diminished after Georg Stahl, another leading contender for animism in the eighteenth century, was discredited in relation to his very influential but misguided theory of phlogiston (Lavoisier, 1786). A last blow came from philosophy under the highly respected authority of Immanuel Kant:

if materialism is inadequate to explain the mode in which I exist, spiritualism is likewise insufficient; and the conclusion is that we are utterly unable to attain to any knowledge of the constitution of the soul, in so far as relates to the possibility of its existence apart from external objects (Kant, 1787, 1994, p. 126).

A closer examination of the possible link between the above (and other) factors and the sudden, inveterate twist in the concept of irritability toward soullessness should illuminate this key period in the history of muscle physiology.

References

Baglivi, G. (1702/1703). *Specimen quatuor librorum de fibra motrice et morbosa*. Roma/London: Buagni/Leigh and Midwinter.

Bastholm, E. (1950). *The history of muscle physiology. From the natural philosophers to Albrecht von Haller*. København: Ejnar Munskgaard.

Bichat, X. (1801). *Anatomie Générale Appliquée à la Physiologie et à la Médecine*. Paris: Brosson/Gabon.

Birch, Th. (1756/1968). *The history of the royal society of London for improving of natural knowledge from its first rise* (Vol. II). London: A. Millar in the Strand. (Reprint edition, New York Johnson Reprint Corporation.)

Boerhaave, H. (1737). *Aphorismi de cognoscendis et curandis morbis in usum doctrinae domesticae digesti*. Editio Leydensis quinta auctior. Lugduni Batavorum: Theodorum Haak, Samuelem Luchtmans, Joh. et Herm. Verbeek, et Rotterodami: Joh. Dan. Beman.

Boerhaave, H. (1744) *Dr. Boerhaave's academical lectures on the theory of physic, being a genuine translation of his institutes and explanatory comment, collated and adjusted to each other, as they were dictated to his students at the University of Leyden (Institutiones medicae)* (Vol. III). London: W. Innys.

Descartes, R. (1664/1986). L'Homme. In Ch. Adam & P. Tannery (Eds.), *Oeuvres de Descartes* (Vol. IX, pp. 119–202). Paris: Vrin.

Des Chene, D. (2001). *Spirits and clocks: Machine and organism in Descartes*. Ithaca: Cornell University Press.

French, R. K. (1969). *Robert Whytt, the soul, and medicine*. London: The Wellcome Institute of the History of Medicine.

Frixione, E. (2004a). Fibres (Théorie des). In D. Lecourt (Ed.) (Dir.), *Dictionnaire de la Pensée Médicale* (pp. 488–491). Paris: Presses Universitaires de France.

Frixione, E. (2004b). Irritable glue: The Haller–Whytt controversy over the mechanism of muscle contraction. *Journal of the History of the Neurosciences, 13*, 366.

Frixione, E. (2006). Albrecht von Haller (1708–1777). *Journal of Neurology, 253*, 265–266.

Galvani, L. (1791/1953). *Commentary on the effect of electricity on muscular motion*. Bononiæ. (*De viribus electricitatis in motu musculari. Commentarius*, M. G. Foley, Trans.; I. B. Cohen, Introd.; J. F. Fulton, M. E. Stanton, Bibliography). Norwalk: Conn. Burndy Library.

Glisson, F. (1677). *Tractatus de ventriculo et intestinis*. London: Brome; see also Fulton, J. F., & Wilson, L. G. (1966). *Selected readings in the history of physiology* (pp. 218–219). Springfield: Charles C. Thomas.

Grmek, M. D. (1970). La notion de fibre vivante chez les médecins de l'école iatrophysique. *Clio Medica, 5*, 297–318.

Hall, T. S. (1969). *History of general physiology 600 B.C. to A.D. 1900* (Vol. I). Chicago: The University of Chicago Press.

von Haller, A. (1747/1786/1966). *First lines of physiology (Primae lineae physiologiae)* "Translated from the correct Latin edition. Printed under the inspection of William Cullen, and compared with the edition published by H. A. Wrisberg. Edinburgh". (Reprint edition, with an introduction by Lester King. New York: Johnson Reprint Corporation.)

von Haller, A. (1752). *Relationes de novis libris*, no.1, part 3 (cited by French, 1969, p. 9).

von Haller, A. (1753/1755/1936). *A dissertation on the sensible and irritable parts of animals (De Partibus Corporis Humani Sensibilibus et Irritabilibus*, M. Tissot, Trans.) Republished with an introduction by Owsei Temkin. Baltimore: The Johns Hopkins Press.

von Haller, A. (1760). *Mémoires sur les Parties Sensibles et Irritables du Corps Animal* (Tome IV). Lausanne: Sigismond D'Arnay.

von Haller, A. (1762). *Elementa physiologiæ corporis humani*. Lausannæ: Francisci Grasset/Tomus Quartus/Liber Undecimus.

von Haller, A. (1763). *Ad Roberti Whyttii nuperum scriptum Apologia* (In pamphlet, 1764; cited by Miller, 1939, pp. 101–102).

Kant, I. (1787/1994). *The critique of pure reason* (Vol. 39) (*Critik der reinen Vernunft*, J. M. Meiklejohn, Trans., 2nd ed.). Chicago: Encyclopædia Britannica/Great Books of the Western World.

Kardel, T. (1994). Stensen's myology in historical perspective. In *Steno on muscles* (Vol. 84, Part 1, pp. 1–57). Philadelphia: American Philosophical Society, Transactions.

King, L. S. (1966). Introduction. In A. *von Haller* (1747/1786/1986), *First lines of Physiology (Primae lineae physiologiae)* "Translated from the correct Latin edition. Printed under the inspection of William Cullen, and compared with the edition published by H. A. Wrisberg. Edinburgh". (Reprint edition, with an introduction by Lester King. New York: Johnson Reprint Corporation.) pp. ix–lxxii.

de La Mettrie, J. O. (1747/2004). L'Homme machine. In J. P. Jackson (Ed.), *La Mettrie – Œuvres Philosophiques*. Vendôme: J. P. Jackson & Coda.

Lavoisier, A. L. (1786/1965). Réflexions sur le phlogistique. In J. B. Dumas (Ed.), *Oeuvres de Lavoisier publiées par les soins du Ministère de l'Instruction Publique et des Cultes* (Tome II, pp. 623–655). Paris: Imprimerie Impériale. (Reprint edition (1862–1893). New York: Johnson Reprint Corporation.)

van Leeuwenhoek, A. (1682/1948). Letter to Robert Hooke, March 3rd. In Committee of Dutch Scientists (Ed.) *The collected letters of Antoni van Leeuwenhoek*

(Vol. III, pp. 385–415). Amsterdam: Swets & Zeitlinger.

Lower, R. (1669/1932). *Treatise on the heart*. London: Allestry. (*De Corde*, K. J. Franklin, Trans.) In *Early science in Oxford* (Vol. 9). London: Oxford University Press.

Miller, G. (1939). *Albrecht von Haller's controversy with Robert Whytt*. Baltimore: Johns Hopkins University (MA Thesis, unpublished)

Nayler, M. (2000). Introduction. In W. Croone (Ed.) (1664/2000), *On the reason of the movement of the muscles* (*De ratione motus musculorum*, P. G. Maquet, Trans.) (Vol. 90, Part 1, pp. 2–53). Philadelphia: American Philosophical Society, Transactions.

Oken, L. (1805). *Die Zeugung*. Frankfurt bey Wesche.

Pera, M. (1992). *The ambiguous frog. The Galvani-Volta controversy on animal electricity* (J. Mandelbaum, Trans.). Princeton: Princeton University Press.

Radbill, S. X. (1981). Whytt, Robert. In Ch. C. Gillispie (Ed.), *Dictionary of scientific biography* (Vol. 14, pp. 319–324). New York: Charles Scribner's Sons.

de Saussure, Raymond (1949). Haller and La Mettrie. *Journal of the History of Medicine, 4*, 431–449.

Séguin, A., & Lavoisier, A. L. (1789). Premier mémoire sur la respiration des animaux. In J. B. Dumas (Ed.), *Oeuvres de Lavoisier publiées par les soins du Ministère de l'Instruction Publique et des Cultes* (Tome II, pp. 688–703). Paris: Imprimerie Impériale. (Reprint edition (1862–1893). New York: Johnson Reprint Corporation.)

Siegel, R. E. (1976). *Galen on the affected parts*. Translation from the Greek text with explanatory footnotes. Basel: Karger.

Smith, C. U. M. (2002). Julien Offray de la Mettrie (1709–1751). *Journal of the History of the Neurosciences, 11*, 110–124.

Steno, N. (1667/1994). Elementorum myologiae specimen. Florence. In T. Kardel (Ed.), *Steno on muscles* (Vol. 84, part 1, pp. 76–228). Philadelphia: American Philosophical Society, Transactions.

Swammerdam, J. (1737/1758/1978). *The book of nature* (*Bybel der Natuur*, Th. Flloyd, Trans., pp. 122–132). (Reprint edition, New York: Arno Press (Biologists of the World) II, pp. 122–132)

van Swieten, G. (1744). *The commentaries upon the aphorisms of Dr. Hermann Boerhaave, the late learned professor of physick in the University of Leyden, concerning the knowledge and cure of the several diseases incident to human bodies, translated into English* (Vol. I). London: John and Paul Knapton.

Temkin, O. (1964). The classical roots of Glisson's doctrine of irritation. *Bulletin of the History of Medicine, 38*, 297–328.

Verworn, M. (1913). *Irritability: A physiological analysis of the general effect of stimuli in living substance*. New Haven: Yale University Press.

Whytt, R. (1751/1768). An essay on the vital and other involuntary motions of animals. Edinburgh: Hamilton, Balfour, and Neill. In *The works of Robert Whytt, M. D., published by his son* (pp. 1–208). Edinburgh: Beckett, Hondt, and Balfour.

Whytt, R. (1755/1768). Observations on the sensibility and irritability of the parts of men and other animals, occasioned by the celebrated M. de Haller's late treatise on those subjects. Edinburgh: Balfour. In *The works of Robert Whytt, M. D., published by his son* (pp. 255–306). Edinburgh: Beckett, Hondt and Balfour.

Whytt, R. (1761). Appendix, containing a review of the controversy concerning the sensibility and moving power of the parts of men and other animals, in answer to M. de Haller's late remarks on these subjects in the 4th volume of the Mémoires sur les parties sensibles et irritables. *Physiological essays* (2nd ed. pp. 219–314). Edinburgh: Hamilton, Balfour, and Neill.

Wilson, L. G. (1961). William Croone's theory of muscular contraction. *Notes and Records of the Royal Society of London, 16*, 158–178.

9

The Taming of the Electric Ray: From a Wonderful and Dreadful "Art" to "Animal Electricity" and "Electric Battery"

Marco Piccolino

Introduction

The period spanning from the second half of the seventeenth century up to the end of the eighteenth century is marked by a truly paradigmatic episode of the transition between the classic science, still imbued with the themes of the wonderful and fantastic, and modern science based on experimentalism and objectivity. This episode concerns the study of strange fish capable of producing, at the simple contact of their body surface, a numbness or rather a violent shock. Two of these fishes were already known to the classical world since very ancient times (the torpedo and the Nile catfish). A third species, a singular eel of the rivers of tropical America, came to the attention of naturalists only in the second half of the seventeenth century, within the climate of the scientific revolution and of the interest for exotic countries.[1] In addition to representing a fundamental transition in the knowledge of the phenomena of the animated nature, the episode of these fish (which would be called electric) was important also because it opened the path to two of the most revolutionary episodes of the Enlightenment science: the demonstration of the electric nature of nervous conduction by Luigi Galvani and the invention of the battery by Alessandro Volta (see Piccolino & Bresadola, 2003).

---

[1] As to the historical aspects of the interest for the electric fish see for reviews of the old literature Kellaway (1946), Moller (1995), and Musitelli (2002).

Socrates, the Torpedo and the Nightmare

In the classic era there are frequent references, in both philosophic-naturalistic and literary texts to these singular fish, indicated by Greeks with the general names of ναρκη, and by Romans as torpedo, because of their power of "benumbing" or "paralysing" the fish or the persons coming in contact with them. These effects appeared at once wonderful and dreadful. Numerous are the artistic representations of these fish, sometimes dating to very old epochs. For instance, the Nile catfish are illustrated in some Egyptian bas-reliefs of the third millennium BC.

In the Greek literature a significant reference to the torpedo is in Plato's "Meno" where the protagonist compares Socrates to a torpedo, because of the "benumbing" effects of his conversation:

O Socrates, I used to be told, before I knew you, that you were always doubting yourself and making others doubt; and now you are casting your spells over me, and I am simply getting bewitched and enchanted, and am at my wits' end. And if I may venture to make a jest upon you, you seem to me both in your appearance and in your power over others to be very like the flat torpedo fish, who torpifies those who come near him and touch him, as you have now torpified me, I think. For my soul and my tongue are really torpid, and I do not know how to answer you; and though I have been delivered of an infinite variety of speeches about virtue before now, and to many persons - and very good ones they were, as I thought - at this moment I cannot even say what virtue is. And I think that you are very wise in not voyaging and going away from home, for if you did in other places as you do in Athens, you would be cast into prison as a magician.[2]

Another reference of literary value is found in the *Halieutica* of the Greek poet Oppian (I–II century AD), partially inspired by verses of Homer, in which the paralysing power of the torpedo is compared to dreadful images of nocturnal dreams:

It is like the dark images of the nocturnal dreams, when a man seized by terror would like to escape and his heart accelerates. His efforts notwithstanding, a tight tie blocks his kneels, in spite of his ardent wish to escape: they are powerful shackles that the torpedo brings about against other fish. (Oppianus, 1999, vv. 81–85)

In the Latin literature is of particular significance the poem *Torpedo* written by Claudian in the fourth century AD, and dedicated to the true torpedo (the flat torpedo of the sea), depicted as a small monster of the sea abysms ready to catch its ill-fated preys after benumbing them with its malefic power. The singular power of the fish is referred to as an "indomitable art": Claudian writes that the torpedo "scarcely does it mark the sand o'er which it crawls so sluggishly" ready to stun with its "cold venom" every creature happening to touch it; eventually it "greedily devours without fear the living limbs of its victim".[3]

Claudian's allusion to the "cold venom" ("*gelido veneno*") betrays the influence of an explanation initially suggested by Galen on the basis of the similarity between the benumbing power of the fish and the action of cold bodies[4]. Galen's hypothesis was indeed a first attempt to put forward a scientific interpretation of the fish shock, outside magic or demoniac conceptions, and well within the frame of the classic physiology. This was based on the well-known theory of the four qualities (dry, cold, moist and hot) in relation with the four elements (earth, water, air and fire) and the four humours (yellow bile, black bile, phlegm and blood).

In his poem Claudian makes allusion to the possibility that the dreadful power of the fish could act at distance, in the absence of an immediate contact. He narrates that if the torpedo:

carelessly swallow a piece of bait that hides a hook of bronze [. . .] it does not swim away nor seek to free itself by vainly biting at the line; but artfully approaches the dark line and, though a prisoner, forget not its skill, emitting from its poisonous veins an effluence which spreads far and wide through the water. The poison's bane leaves the sea and creeps up to the line; it will soon prove too much for the distant fisherman. The dread paralysing force rises above the water's level and climbing up to the drooping line, passes down the jointed rod, and congeals, e'ver he is aware of it, the blood of the fisherman's victorious hand.

The possibility that torpedo's effects might reach other fish (or persons) by the way of intermediate bodies was explicitly denied by some naturalists and philosophers but supported by others who based their contention mainly on the reports of fishermen. These claimed that the fish shock was normally transmitted through the water and the fishing nets. According to them, in order to detect torpedoes among the captured fish it sufficed to pour sea water over the net and a shock was immediately felt if live torpedoes were present. This was a clear indication that the shock could propagate through water.

Another important reference to torpedo in Latin texts is that of Scribonius Largus, a physician of the Roman Empire who advised the use of the torpedo shock for alleviating pain, particularly in the case of gout and headache:

For any type of gout a live black torpedo should, when the pain begins, be placed under the feet. The patient must stand on a moist shore washed by the sea, and he should stay like this until his whole foot and leg up to the knee is numb. This takes away present pain and prevents pain from coming on if it has not already arisen. In this way Anteros, a freedman of Tiberius, was cured.[5]
[. . .]

---

[2] Plato, "Meno" – English translation by Jowett in 'The Dialogues of Plato' 1892, p. 39, Sect. 80.

[3] I follow here with some modifications the English version of Barrie Hall in Claudianus C. 1985, poem XLIX; see also Claudianus C. 1922.

[4] See for instance Galenus (1533, p. 10; and 1541, p. 406). Galenus was, however, uncertain as to the true nature of the torpedo's shock. Extracts of torpedo's liver were used for their medical and/or magic power, and were

supposed to have anti-aphrodisiac effects (Trallianus, 1557, pp. 71–72, 115–116; Lacroix, 1868, p. 84; see, however, Redi, 1671, p. 41).

[5] The black torpedo mentioned by Scribonius is the *Torpedo nobiliana*, found in Atlantic Ocean but relatively rare in the Mediterranean sea. It is much larger than the common torpedoes (sometimes reaching the length of 180 cm) and its shock can be rather severe (more than 100 V).

Headache, even if it is chronic and unbearable, is taken away and remedied forever by a live torpedo placed on the spot which is in pain, until the pain ceases. As soon as the numbness has been felt the remedy should be removed lest the ability to feel be taken from the part. Moreover, several torpedo's of the same kind should be prepared because the cure, that is, the torpor which is a sign of betterment, is sometimes effective only after two or three.[6]

In addition to this analgesic use of the shock of live torpedoes (likely effective as a kind of electrotherapy *ante litteram*), parts of the fish body, and notably the liver, were employed to prepare potions or ointments, supposedly endowed with curative powers (and with other actions, as for instance anti-aphrodisiac effects and the capability to facilitate parturition). In order to be effective, the preparation should be made in a particular period of the year and with the moon in a particular phase. With time the rituality for these preparations would include magic elements, such as, for instance, the need to pronounce particular formulae or magic words (see for instance Rondelet, 1574, p. 1221; Trallianus, 1557, pp. 71, 115). This contributed to progressively situate the torpedo in the territories of the fantasy and oddness in the imagery of naturalists, together with other animals endowed with extraordinary powers: such were the remora (the small fish capable of slowing down the course of big ship), the unicorn, the basilisk and many other objects typical of Renaissance *Wunderkammern* (Daston & Park, 1998; Olmi, 1992).

From Redi to Réaumur: The Mechanical Shock

In the second half of the seventeenth century there was a great interest for electric fish, particularly among the members of the *Accademia del Cimento,* the prestigious cultural institution founded by Leopoldo dei Medici in 1657 with the aim to contribute to a scientific renewal in the spirit of the Galileian revolution. Particularly impressive is the number of scholars (more or less directly connected to the Cimento or somewhat related with the Medici) involved in the investigation of the mechanism of the torpedo's shock, all working between Pisa, Leghorn and Florence, on fish granted to them by the Granducal court. Among them Giovanni Alfonso Borelli (1608–1679), Francesco Redi (1626–1694), Nicola Stenone (Niels Steensen, 1638–1687), John Finch (1626–1682), Marcello Malpighi (1642–1694), Lorenzo Bellini (1643–1704), Holger Jacobaeus (1650–1701) and Stefano Lorenzini (1652–post 1700).[7]

This interest was part of the program carried out by the "new scientists" (*novatores*) to provide a physical explanation to unusual and singular phenomena, often attributed to magical forces or to animistic "virtues". It is not by chance that the first observations of Francesco Redi on the torpedo are published in a book entitled: *Experiences on several natural things and particularly on those coming from the India, written to the very Reverend Father Athanasius Chircher of the Company of Jesus* (Redi, 1671). In this book, where scientific rigour and argumentation ability go along with a fine polemic humour, Redi critically surmises the various beliefs on the therapeutic power or magical virtues of minerals, plants and animals. These were the typical objects of a *Wunderkammer*, as undoubtedly was the personal museum created in Rome by the German Jesuit.

On the basis of his personal observation, Redi suggested a muscular nature of the organs responsible for the shock ("musculi falcati") and put forward the hypothesis of a mechanical nature of the shock. This hypothesis was elaborated by Redi's pupil, Stefano Lorenzini, in a work published in 1678, *Osservazioni intorno alle torpedini* famous because it contains the first report of some nervous structures of ampullar shape ("Lorenzini's ampullae") now known to be part of receptorial system capable of detecting electric fields (and thus "electroreceptors").

---

[6] See Scribonius (1983), *Compositiones Medicae*, CXLII and XI.

[7] See Guerrini (1999). Some of the works of these authors dealing with electric fish are quoted in the bibliography of this essay.

As to the mechanism of the fish shock, Lorenzini postulated that it was due to a multitude of minute corpuscles ("*corpicciuoli*") emitted by the fish with great violence upon contraction of its "muscoli falcati". By penetrating deeply into the tissues of the prey (or of the experimenter) and hitting the nerves, these corpuscles would produce the commotion and the benumbing effect. As Lorenzini himself acknowledged, this hypothesis was inspired by the doctrine of the minute igneous bodies ("*ignicoli*") developed in 1623 by Galileo in the *Saggiatore* in order to account for the production of heat and for sensory processes.

A mechanical theory was also invoked by the French naturalist René-Antoine Ferchault de Réaumur (1683–1757), although it did not involve anything analogous to the Lorenzini's corpuscles. According to Réaumur the torpedo's shock was the consequence of a direct and rapid percussion of nerve trunks similar to that produced in the forearm by the action of a sharp body hitting the nerves in the region of the elbow: the shock would be produced by the fish at the moment when the dorsal surface of its body, normally flat or even concave in the preparatory phase, became suddenly convex as a consequence of a contraction of the *musculi falcati* that produced "a movement so prompt that even the most attentive eyes could not perceive it". (Réaumur, 1741, p. 351)

Réaumur developed his interpretation in 1714 on the basis of observations carried out on torpedoes caught on the coasts of Poitou, in France, near La Rochelle. It dominated the first part of the eighteenth century, and was considered a reference hypothesis also for the similar shock produced by the other species of fish that had come to the attention of the naturalists of the age.

Bancroft, Réaumur, the "Torporific eels" and the Scientific Challenge

On the wave of Galileian science and Descartes' philosophy, the principles of mechanics dominated the seventeenth century, and the law of motion and of the functioning of machines were invoked as references to account for the physiological processes, even those not immediately amenable to a mechanical interpretation (as for instance digestion, respiration and nervous conduction). New forces would attract the attention of the natural philosophers in the next century as long as the strictly mechanistic explanation reveals its inadequacies. Together with chemistry, whose development would culminate with Lavoisier's revolution, the other "force" dominating the eighteenth century was electricity. In addition to important conceptual advancements and discoveries, electricity attracted the attention for a series of practical achievements. Among them the invention of new and powerful electric machines capable of producing strong effects (see Heilbron, 1979).

Among the new instruments there was the Leyden jar, the first capacitor of the history, a device capable of accumulating huge quantities of electric charge which allowed people to easily experience the effects of strong electric shock. The similarity between these effects and the shock of the torpedo (and other singular fish) became thus rapidly apparent. It was first recognized by the French naturalist Michel Adanson in the Senegal in the years 1749–1753. Adanson investigated the shock produced by an African freshwater fish (the electric catfish) indicated as *trembleur* by the local Frenchmen because of its capacity to "produce not simply numbness as the Torpedo, but rather a trembling effect very painful in the arms of those that touch it". As to the comparison between the shock produced by this fish and by the Leyden jar, Adanson wrote:

Its effect which did not appear to me sensibly different from the electric commotion of the Leyden experiment that I had tried many times, is similarly communicated by a simple contact, with a stick or rod of iron five or six feet long, to such a point that one drops instantaneously all the things that he kept in his hands. (Adanson, 1757, p. 135)

Besides the correspondence of the effects produced on the experimenter, the possibility that the fish shock could be communicated through a metallic body also argued for the similarity with the shock produced by the Leyden jar.

Soon after the publication, in 1757, of the *Histoire naturelle du Sénégal*, containing Adanson's account of the shock produced by the *trembleur* fish, a series of observations were collected on the similar properties of another freshwater fish, similar to an eel, found in the tropical regions of America. These observations were prompted by the Dutch physicist Jean-Nicholas Sebastien Allamand, colleague of Pieter van Musschenbroek and one of the discoverers of the Leyden jar (Allamand, 1756; see Kellaway, 1946; van der Lott, 1762). Also in the case of this eel, the shock was found to be transmitted through metallic and humid bodies. These characteristics pointed to a possible electric nature of the phenomenon, and seemed therefore to contradict the mechanical interpretation advocated by Lorenzini and Réaumur for the torpedo's shock. In the case of the latter, the possibility that the shock could be transmitted to the experimenter by some intervening matter had been taken into account by Lorenzini, who, however, excluded it outright, on the basis of a series of experimental observations. According to Lorenzini "in order that the Torpedo might produce its effects, it was necessary to touch it [directly] in some part of its naked body". (Lorenzini, 1678, p. 111)

Lorenzini's conclusion fitted in with his hypothesis that the shock depended on minute corpuscles emitted by the animal and penetrating the body of the experimenter. It was unlikely that these corpuscles could pass through a long rod, or through the chords and threads of the fishing net. In his study of the torpedo, Réaumur found that the shock could be transmitted, though with great attenuation, through a rod, but he did not investigate whether the transmission's efficacy changed according to the material the rod was made of. Réaumur's view of the torpedo's shock as due to a kind of mechanical concussion or vibration could account for the partial transmission of the shock through a relatively rigid body.

To summarise, after the middle of the eighteenth century evidence was accumulating to suggest possibly the electric nature of the shock produced by some species of fish (notably the Nile catfish and the eel of Surinam), based on the similarity of the effects produced by these fishes to the shock of electric devices. However, the reference hypothesis on the nature of the fish shock still remained the mechanical one, based on Lorenzini's and Réaumur's studies of the torpedo.[8]

---

[8] On a mechanical explanation, inspired by Réaumur, was also based the account of the Torpedo given in 1765 by De Jaucourt in the *Encyclopedie* (article *Torpille*, Tome XVI, pp. 428–431).

Torpedo and "Torporific Eel" the Scientific Challenge of Edward Bancroft

The situation changed in 1769 when Edward Bancroft an American born doctor (who lately established himself in England), with an interest in natural history, published an account of observations made in equatorial America. Of particular relevance was Bancroft's description of a peculiar fish referred to as a "Torporific Eel", found "near the coasts and the rivers of Guiana". After a rather detailed description of the external morphology of the fish, which appears to be superficially similar to a Lamprey, Bancroft wrote:

But the most curious property of the Torporific Eel is, that when touched either by the naked hand or by a rod of iron, gold, silver, copper &c. held in the hand, or by a stick of some particular kinds of heavy *American* wood, it communicates a shock perfectly resembling that of Electricity, which is commonly so violent, that but few are willing to suffer it a second time. (p. 192)

The comparison between the shock of the eel and that of the torpedo as described by Réaumur inspired Bancroft to the following reflections:

What affinity there may be between the shock of Torporific Eel and that of Torpedo, I am unable to determine with certainty, having never felt the latter; but from all the particulars which I have been able to collect relatively thereto, I think it is pretty evident both are communicated in the same manner and by the same instruments.

Some years since the celebrated *Mons. De Réaumur* communicated to the Royal Academy of Sciences at Paris a Paper, in which he undertook to demonstrate, that the shock of the Torpedo was the effect of a stroke given by great quickness to the limb that touched it, by muscles of a peculiar structure. To this hypothesis all *Europe* beheld an implicit assent, and M. *De Réaumur* has either to enjoy the honour having developed the latent cause of this mysterious effect. But if we may be allowed to suppose, that is undoubtedly true, that the shock of the Torpedo and that of the Torporific Eel, are both communicated in a similar manner, and by similar means, it will be nowise difficult to demonstrate, that all of M. *De Réaumur*'s pretended discovery is perfect nonentity. You may, perhaps, think it an act of presumption in me, to dispute the authority of a man, whose literary merit is so universally acknowledged; but I am convinced, that an implicit faith, in whatever is honoured with the sanction of a great name, has proved a fruitful source of errors in philosophical research; and whilst I have sense and faculty of my own, am resolved to use them with the freedom for which they are given. Humanity is ever exposed to deception, and charm for novelty may perhaps have precipitated *M. De Réaumur* into an error. (pp. 195–196)

Bancroft continued by exposing those characteristics of the eel's shock that made it unlikely that it could be accounted for on a mechanical basis. He mentioned the transmission of the shock through the fishing line when the fish is caught by a hook, the transmission through an iron rod to a chain of people (up to 12) touching each other in a circle ("in a manner – he says – exactly similar to that of an electric machine"); and also the transmission of the shock through the water in which the fish is swimming. On the basis of these characteristics he assumed that the shock was produced by "an emission of torporific or electrical particles", depending on the life of the animal and under the control of its will. Finally, coming back to the torpedo and to *M. De Réaumur*, he concludes provocatively:

From whence it is self-evident that, either the mechanisms and properties of the Torpedo and those of the Torporific Eel are widely different, or that *Mons. De Réaumur* has amused the world with an imaginary hypothesis: and, from my own observations, as well as the information which I have been able to obtain on this subject, I am disposed to embrace the latter inference (pp. 198–199).

If one was inclined to endorse Bancroft's statement as a veritable scientific challenge, then all that remained to be done was to investigate the characteristics of the shock of the torpedo in order to see whether they corresponded to those of the eel; or, on the contrary, were so different, as to justify, in the case of the torpedo, a mechanical explanation apparently incompatible with the eel's shock.

This was just what John Walsh, a wealthy English gentleman with interest into natural history and "experimental philosophy", decided to pursue through a journey to France in the summer of 1772.[9] Walsh was indeed well acquainted with Bancroft's views on the subject. Both were members of the Royal Society and frequented a circle of natural philosophers in London interested in the study of electrical phenomena. This circle included scientists as eminent as Benjamin Franklin and Joseph Priestley.[10] The reason for travelling to France to investigate the torpedo's shock was due to the difficulty over finding live animals in England. Perhaps we could also be justified in assuming that a further motive might be the desire to perform experiments on torpedoes from the same region, the *Poitou*, where Réaumur had carried out his experiments more than half a century before. It was for these reasons that Walsh left London for France at the beginning of the Summer season of the year 1772 accompanied by his secretary David Davies and his nephew Arthur Fowke, as Walsh annotates at the beginning of his *Journal de voyage*:

---

[9] John Walsh was born in Fort St. George on 1 July 1726 and died in London on 9 March 1795. He was the cousin of Neville Maskelyne, the Royal Astronomer, and also of Lord Clive who was married to his mother's niece, Margaret Maskelyne. Walsh served in the East India Company since 1742 and, in 1757, he was appointed as private secretary to Clive. In 1759, Clive commissioned him to return to London in order to support his plans for reorganizing the administration of Bengal before the English Government. Walsh's richness (estimated to about £140,000) was mostly derived from the considerable fortune amassed as a share of war conquests in India. He was member of the Parliament from 1761 to 1780. He was elected Fellow of the Royal Society on 8 November 1770. In his certificate of election, Walsh is indicated as "a Gentleman well acquainted with philosophical, & polite literature, & particularly versed in the natural history and antiquities of India". A short but

accurate biographic sketch of John Walsh is contained in 'The papers of Benjamin', vol. 19, pp. 160–162 (see Franklin, 1959).

[10] As Walsh recognises in his 1773 article on the *Philosophical Transaction*, Franklin was influential in devising the plan of the experiments to be performed on the torpedoes at La Rochelle. Among the Franklin's papers there are two short notes undated but surely written before Walsh's journey to France, which attest the relation between Walsh, Franklin, and Bancroft. In the first it is written that Walsh asks Franklin for Bancroft's address in London saying that he "is very desirous of making some enquiries concerning the Torporific Eel" (p. 162, Franklin's papers, Vol. 19). From the other note we learn that Franklin had borrowed from Walsh the book of "Laurenzini]" (i.e. Lorenzini) and desired to have it back because "he wishes to look into it for some particulars" (see Franklin 1959, Vol. XIX, p. 163).

1772, June 8th. Whitsun Monday. Left London in the afternoon; Crowds of Men and Women returning from Greenwich Park and Maying. The hedges in very full Blow of May. In company with Mr. Davies, and my Nephew Arthur in Post Coach. Slept at Dartford 16 Miles.

Walsh in La Rochelle: The Taming of the Ray and the Transitions in the Eighteenth Century Science

We will provide but a rapid outline of the journey that brought Walsh to La Rochelle through a rapid *detour* and we will refer also rather briefly to the experiments he made in the small town of the French Atlantic coast and in the nearby island of the Isle de Ré. Together with other experiments that he would carry out 3 years later in London on electric eels imported from Guiana, these experiments led to the final demonstration of the electric nature of the shock of these singular fish. Walsh's journey and experiments have been diffusely dealt with in a recent book based on two unpublished manuscripts.[11] Here, we will content ourselves mostly with a discussion of some aspects of the way to do science in the second half of the eighteenth century, as they emerge from these manuscripts.

The first manuscript is the already mentioned *Journal de voyage* containing Walsh's observations on his passage from London to La Rochelle through Boulogne, Montreuil, Amiens, Breteuil, Clermont, Creil, Chantilly, Paris (where he stays for about a week in mid-June), and afterwards through the centre of France and the region of Poitou up to the Atlantic coast.

The second manuscript, *Experiments made in La Rochelle and Isle de Ré, June and July 1772*, conserved in the Library of the Royal Society of London, is of a particular relevance for tracing the events which brought Walsh to demonstrate the electric nature of the torpedo's shock. It bears the detailed registration of the experiments on the torpedo carried out in the region of La Rochelle, together with annotations and reflections. It also

contains the transcription of a letter sent by Walsh to Benjamin Franklin on 12 July 1772 which represents the first communication of the discovery of the electric nature of the torpedo shock made to a eminent member of the Republic of the Letters of the Enlightenment.[12]

Various elements emerge from these manuscripts useful to reconstruct the investigative attitude and the scientific psychology of a natural philosopher in a transition period as undoubtedly was the second half of the eighteenth century.

First of all there is what we could qualify in the case of Walsh as an "obsession for knowledge", an almost monomaniac interest for the problem underlying experimental investigation. The initial symptoms of this scientific compulsion appear in the first manuscript when Walsh betrays his anxiety for finding the torpedoes necessary to his experiments. In Calais, soon after the arrival in France, he visited the local fish market, and in the absence of torpedoes, he profited of the arrival of a boat unloading "6 or 700 Rayes of different sort", to carry out a dissection of some of these fish. This appears to be an anticipation of (and a preparation for) the dissection of torpedoes that he will carry out at La Rochelle. About a week later, on the way from "Boulogne to Montreuil", we find the first remark betraying explicitly Walsh's interest in the torpedo.

Spoke with Captain Palletti, and Captain Lobé, who commanded Coasting Vessels to La Rochelle and Bordeaux &c. They were acquainted with the Torpedo, and assured me that to their one knowledge, there were plenty of them at La Rochelle and La Tesle on the coast of Argenson. (WJ, p. 8)

During the period spent in Paris there is almost no mention of the beauty of the town, of his monuments or other interesting curiosities not related to torpedoes or to electricity. The day after the arrival Walsh visited Jean-Baptiste Le Roy, who he found engaged in electrical experiments, and discussed laboratory instruments with him. In the evening Walsh was at the *Jardin du Roy*, where he attended a lecture in chemistry and visited a naturalistic collection at the *Cabinet du Roy*:

---

[11] Piccolino (2003, 2005) and Piccolino and Bresadola (2002, 2003).

[12] Library of the Royal Society of London, Ms. 609. The quotation from this manuscript will be indicated by the notation WE followed by the page number.

a good collection of natural Curiosities, open to all three times a Week. Military mounted there, in Glass Cases, with Label of the name of each Article. (WJ, pp. 16–17)

Of the various "articles" on exhibition, his curiosity was first attracted by a "Male Torpedo, as called, dryed [*sic*] small about 9 Inches", and, moreover, by "two very small ones in Spirits mention'd from the Isle de Bourbon, two inches in Diameter". "No Anguille tremblante" – he wrote, very probably with some intention to deceive – after having noticed the absence of specimens of the electric eel. Afterwards he noted in the journal:

Visited M. Jussien [i.e. '*Jussieu*']; talked of the Torpedo, he never saw it or heard much of it; mentions his having heard that the Loadstone took off its Effects. One present mentioned the Eel to be in Salt water at Mauritius. M. Jussien named M. A Surgeon now dead to have collected some particulars of the Torpedo, and to have intended to have refuted M. De Reaumur. (WJ, p. 18)

On the next day Walsh endeavoured to visit several scientists, in particular the naturalist and botanist Michel Adanson, the same Adanson that 15 years before had reported on the similarity of the shock produced by the Leyden jar and the strange fish of Senegal.

In Walsh the compulsion for torpedoes and experimental science emerges, however, in a particularly clear way during the period spent in La Rochelle. In 27 days of experimental research in the journal are registered as many as 324 experiments, carried out on 76 torpedoes. Together with his collaborators (the nephew and the secretary, and sometimes members of the local upper class) Walsh made experiments even during the Sundays (five in all) and sometimes up to late in the evening.

In this obsessive attraction for the experimental investigation of a specific and well-defined problem, we can see a first element of transition in the way of doing science in the Enlightenment. In the eighteenth century the interest for science, and particularly for "experimental philosophy", was an aspect of the education and of the cultural taste of the members of the upper classes, and under certain respects Walsh's attraction for electric experiments and for the torpedo might correspond to this general interest for science of the cultivated members of his epoch.

This is, however, only partially true. People were interested into science as they were into music, art,

literature, for the pleasure to do it, and with the intention to carry out amusing and entertaining experiments. The "*amateurs*" of the Enlightenment "played"science in a relation of sane and harmonic equilibrium with their other interests and activities, as a form of spiritual recreation. The mental and physical discipline required by the experimental practice helped them to moderate the emotional excess brought about by art, poetry and music. This was not certainly the case with Walsh. His almost compulsive commitment to the investigation on the torpedo seems to anticipate the typical involvement of modern scientists in very specific and somewhat narrow research problems which often absorb in an exclusive manner all their physical and intellectual energies.

In Walsh's case the "transitional" features in the way of doing science concerned also other connotations, and particularly his interactions with people around him, both at the moment of doing experiments and when communicating his results.

The situation developed in La Rochelle with the arrival of the English gentleman interested in experiments seems to fit well in an almost idyllic frame of an old-fashioned "experimental philosophy" carried out in non-professional and non-institutional milieus and addressed to all peoples of high cultural and social rank, without any supposition of specific knowledge and competence. Walsh's studies attracted the attention of a literary party, comprising various personages typical of a provincial world of the *ancien régime*: the Bookseller, the Apothecary, the Hydrographer, the Naturalist, the Surgeon, many members of the local *Académie* and in particular the Director and the "Perpetual Secretary", who was also the Mayor of the town, the *Inspecteur des Fermes*, the Professor of Rhetoric at the Royal College, the military Governor and *Commandant en chef*, the *Ingenieur*, the *Avocat*, several Colonels and so on. Some of these personalities took an active part in the experiments, others discussed the effects of the torpedo or simply attended Walsh's demonstrations. However, after his return to London, Walsh succeeded in involving, in his electric fish studies, two of the most qualified English scientists of his time: Henry Cavendish, famous for his physical researches, and particularly for his measurement of the gravity constant of the Earth, and John Hunter, a leading anatomist and surgeon.

The collaboration among Walsh, Hunter and Cavendish was of great importance for the advancement of electric fish studies. Moreover, it is an indication of the transition towards a modern conception of scientific inquiry, based on the need of a tight collaboration among experts of different fields. Hunter provided an accurate study of the microscopic anatomy of both torpedo and electric eel. In particular he pointed to the great surface of the membranes of the elementary elements making up the columns of the electric organs and to the extreme abundance of nerves directed to these organs. As to the nerves, he concluded by saying that they are "subservient to the formation, collection, or management of the electric fluid", a statement which fitted within the framework of the hypothesis of the electric nature of nerve signals (Hunter, 1773, p. 487: On Hunter see Chapters 3 and 5 of the present volume). This hypothesis had appeared in the scientific literature in the first half of the century but since then had remained without any substantial experimental support (see Piccolino & Bresadola, 2003). Hunter's considerations were somewhat instrumental in confirming Luigi Galvani in the conclusion, developed in 1791 after an intense period of experimental investigation that nerve signals are due to the agency of an electric fluid. This fluid was indicated by Galvani as "animal electricity" (a phrase of which the first historical occurrence is in Walsh's journal of the experiments on page 143). In a memoir written in 1797, and concerning the result of his own research on torpedo shock, Galvani put a passage bearing an apparent correspondence with Hunter's consideration. In discussing his experiments aimed at ascertaining the source of the electric fluid involved in fish shock, he wrote that he wished to determine whether the brain "would be the elaborator and the collector of such electricity, and nerves the conductors" (Galvani, 1797, p. 65). According to Galvani animal electricity is accumulated in a condition of unbalance between the interior and the exterior of the muscle fibre: a nerve fibre penetrates inside it allowing, in both physiological or experimental conditions, for "the flow of an extremely tenuous nervous fluid [. . .] similar to the electric circuit which develops in a Leyden jar" (Galvani, 1791, p. 378). The neuromuscular complex was conceptualized by Galvani at a microscopic level as "minute animal Leyden jar", an image which developed the idea of torpedoes as "animate

phials" put forward by Walsh in his 1773 article (p. 477) and further developed by Cavendish.

Walsh–Cavendish collaboration in electric fish investigation was of particular relevance because it led to an immediate important development of broad physical interest. The problem particularly addressed by Cavendish concerned the reason why, in Walsh's experiments, a spark could not be obtained in association with torpedo's shock, even though the shock could be rather severe. This contrasted with what happened with Leyden jars. The discharge of Leyden jars were usually accompanied by visible sparks and by other perceptible phenomena (sounds, "electric winds" and odours). In the case of a strongly charged jar of small size, however, this occurred in spite of the fact that the shock perceived by the experimenter could be substantially weaker than that produced by torpedo (Cavendish, 1776).

To account for such differences Cavendish assumed, on the basis of mechanical-pneumatic analogy, that the action of electricity depended not only on the quantity of «electric fluid» but also on another factor indicated as "degree of electrification", measurable with the electrometer and relatively independent of electric fluid quantity. The sparks (and other perceptible phenomena) would be the effect of electricity present at a high degree of electrification. In the case of a strongly charged Leyden jar of small size, the shock might be weak in spite of the visible spark because of the small quantity of electric fluid involved. On the other hand, a great number of weakly charged large Leyden jars (49 in an actual experiment carried out by Cavendish), arranged in parallel, could give a severe shock (imitating in several respects that of the torpedo) in spite of the fact that the shock was not accompanied by any visible spark. It could be surmised that the shock of the torpedo is due to the discharge of a large amount of electric fluid present at a low degree of electrification. By developing an idea first advanced by Walsh, Cavendish surmised that the discs of the electric organ could behave as minute planar capacitors arranged in parallel. On the basis of Hunter's anatomical study he calculated for the ensemble of the membranes in the electrical organs a surface of about 30,500 in.[2] (which was of the order of magnitude of the total surface of his assembly of 49 large Leyden jars capable of giving severe shocks without sparks). In the elaboration of Cavendish to account for the torpedo's effects, there appear *in nuce* notions corresponding to those of "tension" (or potential) and "quantity of charge" and the laws of the capacitor, which would be afterwards developed by Volta and Coulomb. These notions were of fundamental importance for the advancement of electrical science in the eighteenth century (Heilbron, 1979).

In the case of Walsh, the interdisciplinary collaboration with expert scientists as Hunter and Cavendish went along, as already mentioned, with his interaction with amateurs and interested persons with no special commitment to science. It was in the circle of these local personages of La Rochelle, more or less directly involved in his experiments, that Walsh gave the first announcement of the success of his experiments in demonstrating "the Effect of the Torpedo to be Electrical". This occurred on 9 June 1772, during a dinner in Walsh's lodging. On that day Walsh could show that, similar to the "electric fluid" of a Leyden jar, the shock of the torpedo could circulate along a human chain, through a direct contact or via electric conductive bodies, whereas it was intercepted by insulating matters as glass and sealing wax.

It is worth quoting a rather long passage of the journal reporting the observations of this day, because this helps us in somewhat entering in Walsh's "room of experiments" at La Rochelle and in perceiving the intense and lively atmosphere pervading it. On that regard the journal of experiments is much more expressive that the published article (appeared in 1773 in the *Philosophical Transactions of the Royal Society* together with Hunter's anatomical study).

Walsh and his nephew Arthur Fowke touched with one hand the torpedo (one on its top and the other on the belly) and with the other hand established a mutual contact, directly or through conductive or insulating materials. In reading this passage we should take into account that the expression "signal" is used to indicate a small movement of the fish eye (usually in the form of a winking) which immediately precedes the shock. This "signal" is of particular relevance when it is necessary to ascertain the lack of transmission of the shock

Walsh and Arthur communicating with a Spoon, and touching one above, the other below the same flank; a Shock.

Communicating with Sealing wax, a Signal, felt nothing.

The Spoon again, immediately a Shock.

Sealing Wax twice; nothing.

Arthur put his Thumb upon Walsh' hand while the sealing wax was still in their hands; two signals – nothing.

Joined hands, immediately a smart Shock.

All insulated; sealing wax in hand; joining hands by thumbs, pressing harder; Arthur communicating with the under side; felt a very strong Shock. Repeated the same.

Changed Sides; Walsh, communicating with underside; felt nothing; but Arthur felt a Shock; plainly from being more sensible.

Joined hands; felt it immediately.

Communicating with Sealing wax: two signals, felt nothing.

Communicated by thumbs; placing ball of one upon the joint of the other thumb; two Signals; nothing.

Full hand within full hand; plain shock.

Holding Sealing wax in full hands joined; strong Shock.

Walsh and Arthur communicating by Spoons, and touching above and below the same flank with Spoons, felt it on touching hands. Repeated; the same effect.

Communicating with Glass, short, at an Inch distance, and touching with Spoons, two Signals, felt nothing.

Again communicating with Spoon, and touching with Spoons; felt it twice.

Communicating with Sealing wax, at half an Inch distance; touching with Spoons; two signals, felt nothing.

fish insulated; single person insulated, touched both flanks above with different hands, felt nothing.

Touched the same flank with different hands; nothing. In each of these experiments there were two winks.

Touched the upper and lower side of the same flank with Spoons; Shock, twice.

Repeated with Spoons; a Shock.

With Sealing Wax; nothing.

Repeated with Spoons; Six times.

With Sealing wax, twice; nothing.

Tried with 3 persons communicating with Spoons; too weak for the third person to feel the Shock.

Two persons communicating by a Spoon; felt it. (WE, pp. 49–53).

With regard to the page of the journal of the experiments in which Walsh registers his announcement of the electric nature of the torpedo (p. 54), it is also interesting to remark that it bears a marginal annotation, written for the second time by Walsh's proper hand and in a somewhat incorrect French: "*Je l'ai donté*". With this annotation Walsh wished to express his pride, by saying that he had "tamed" (*dompté*) with his research what Claudian had indicated as "the indomitable art of the marvellous torpedo".[13] In writing "*Je l'ai donté*", Walsh claimed in some way to have subjugated with the power of experimental science an elusive force which for many centuries had been felt at the same time wonderful and dreadful: a transition from the world of fantastic and weird to that of scientific knowledge and objectivity.

About 2 weeks later Walsh would present his results on the torpedo in somewhat more official situations, which are nonetheless expression of the typical cultural activities of a provincial town: on 22–23 July 1772 he gave two demonstrations for the members of the *Académie*, and on 24 July 1772 he also gave a third demonstration to the officials of the local garrison upon request of the Military Governor, the Baron of Montmorency.

These demonstrations were the occasion of pleasant entertainments and also gave birth to curious episodes. This was the case with the officer who appeared to be insensitive to the torpedo's shock because of the consequence of a severe war wound: he asked to try the effects of a Leyden jar, and, having well perceived the shock of the physical instrument, "went away satisfied of his sensibility" (WE, pp. 151–152). At the same time, the demonstrations were occasion for interesting discussions and remarks. A point particularly addressed by the members of the *Académie* concerned the transmissibility of the torpedo shock through the water, the natural habitat of the torpedo. It was not by chance that Walsh, who had been unable to obtain clear evidence on that regard in the first phase of his research, concentrated his experimentation on this point in his last days in La Rochelle.

---

[13] On the facing page Walsh quotes Claudian's verse by writing: "alluding to Claudian's epithet imdomitam applied to the Art of the Torpedo: Quis non indomitam mirae torpedinis artem audiit?"

Another Transition: Making Public Scientific Results in the Enlightenment

The experimental demonstration in a relatively official situation as in the case of the séances of the *Académie* of La Rochelle was for Walsh also a way to communicate in a public way the results of his experiments. This was in accordance to a custom typical of the eighteenth-century science when announcement made to the members of a learned society might sometimes be the only procedure of making scientific discoveries public. In the case of torpedo, however, Walsh supplemented these demonstrations with the publication of a printed document, appeared 1 year later in the *Philosophical Transactions* in the form of a letter addressed to Benjamin Franklin. On the other hand, in the case of the experiments made on the electric eels in London in 1775, in which he could obtain a spark from the fish, Walsh did not publish any account of his results and contented himself with demonstrations made to his colleagues of the Royal Society. He also wrote a letter to his French correspondent Jean-Baptiste Le Roy who published in 1776 an article entirely dedicated to Walsh's achievement bearing few lines of Walsh's communication:

C'est avec plaisir que je vous apprends qu'elles m'ont donné *une étincelle électrique*, perceptible dans son passage à travers une petite fente ou séparation pratiquée dans une feuille d'étain collée sur du verre. Ces poissons étoient dans l'air; car cette expérience n'a pas réussi dans l'eau; leur électricité est beaucoup plus forte que celle de la torpille, & il y a des différences considérable dans leur effets électriques.

This passage represents the only published record of this memorable experiment written by its author. It is somewhat ironical that only a few lines remain as personal attestation of an achievement which was considered at its time a crucial event in the progress of the eighteenth century science. It concluded the cycle of electric fish research started by Walsh in 1772 and proved the electric nature of fish shock, beyond any reasonable doubt for the standards of the epoch.

Walsh's achievements went far beyond the elucidation of a specific point of natural history concerning the curious property of some peculiar fishes, and prompted the interest in a potential role of electricity in other aspects of animal physiology, and particularly in nerve and muscle functions. They undermined *de facto* the relevance of some important objections against the possible electrical nature of nerve signals (and against other types of involvement of electricity in physiological processes). Since the tissues and "humours" of animal organisms are electrically conductive – it was argued – electricity would spread from nerve fibres to neighbouring tissues and it would not flow along the definite and restricted paths required by physiological necessities. Moreover, there could not be the stable unbalance of electrical fluid necessary for the conduction of electricity along nerve fibres between two distant sites of the animal body, because any such unbalance would be rapidly dissipated due to the conductive nature of living substances. All this notwithstanding, torpedo and the eel of Guiana were able to produce an electric shock, and it was therefore possible that electricity might be involved also in physiological processes of other animals (see Piccolino & Bresadola, 2003).

In spite of the absence of a direct publication in a printed form by the author, the news of Walsh's achievement with electric fish reached the "Republic of Letters" of the Enlightenment through the article of Le Roy (and through various *résumés* published in many journals of the time) and also through other channels of communication existing between the savants of the epoch. It renewed the interest for the possible involvement of electricity in neuromuscular function. Walsh's success with electric fish undoubtedly contributed to Galvani's decision to investigate the possible role of electricity in nerve function, and there is indeed some continuity, both logical and chronological, between Walsh's and Galvani's work. In this way it opened the path to one of the most revolutionary episodes in Enlightenment science, the demonstration of the electric nature of nervous conduction. The "animal spirits", the elusive entities considered in classical science as messengers of soul for sensation and will, were thus substituted forever by an electric fluid: this represented a really paradigmatic transition from old medicine to the new science that would dominate the next centuries.

Although Walsh's achievements with the electric eel did contribute to the advancement of the science of the epoch in spite of the lack of a direct

published record by their author, with time the absence of written documents contributed to the oblivion of Walsh's figure in scientific literature. Michael Faraday, who repeated the experiment of spark production from eel shock in 1839, could not find any clear written evidence of Walsh's experiment and considered its actual occurrence as doubtful (Piccolino, 2003).

The difference in Walsh's behaviour in making public the results of his experiments in the case of the torpedo and of electrical eel is another indication of the transition in the way of making science in the second half of the eighteenth century. In the case of the torpedo, private and public demonstration went along with the production of printed texts, whereas in the case of electrical eel Walsh contented himself with a demonstration to the members of an authoritative learned institution (as the Royal Society undoubtedly was).

The need of public demonstrations, which were customary in the experimental science of the Enlightenment, should not be considered simply as a way of communicating scientific results. It had also to do with the problem of the need of public credibility on both the real occurrence of the scientific events narrated and on the credibility and truthfulness of the results. These were particularly necessary for scientific results which, as in the case of Walsh's research, went against well consolidated paradigms and were difficult to account for within the framework of the physical principles of the epoch. In the paper published in 1773 Walsh comments on the need to give his public demonstrations at La Rochelle by saying that he wished "to give all possible notoriety to facts, which might otherwise be deemed improbable, perhaps by some of the first rank in science". "Even the Electrician – he wrote – might not readily listen to assertions, which seemed, in some respects, to combat the general principles of electricity" (p. 469). As to the experiments dealing with the "phenomena of the Torpedo", for which "Great authorities had given a sanction to other solutions", different from the electrical explanation, it was therefore necessary to have authoritative witnesses, because of their novelty and unexpectedness. Similar principles guided Walsh when, 3 years later, he was able to produce the spark from the eel's shock.

In modern times the ability to reproduce scientific results in a relatively easy way in various laboratories around the world limits the need of direct personal witnesses, and makes it necessary only to publish the experiments accurately in a written form, with appropriate specification on the methods, the type of preparation and the materials used. In the eighteenth century, science was not institutionalized and the possibility of checking the validity of experimental results was limited. This was particularly true in the case of the experiments, as those carried out by Walsh at La Rochelle, which depended on the availability of live animals, the torpedoes, which were difficult to obtain in the main centres of scientific inquiry of the time. There was an indirect allusion to this particular aspect of the problem in the last part of the discourse held by Walsh to the *Académie* of La Rochelle on 22 July 1772 and registered in the journal with these words:

That what we had been able to do had only opened the door to a curious and interesting enquire: That great points remained to be examined, both by the Electrician, and by the Anatomist.

That Nature had denied the animal to our Country and that the pursuit rested in a particular manner on them whose Shores abounded with it. (WE, pp. 144–145)

In addition to the public demonstration, Walsh made recourse to other strategies to attest the veracity and truthfulness of the results obtained in his work on the torpedo. These appear in a particularly evident light in the way he built up the article published in 1773 on the *Philosophical Transactions*. As already mentioned, this article was written in the form of a letter addressed to Benjamin Franklin, one of the most eminent authorities of the electric science of the eighteenth century. The importance of the addressee was by itself a warranty of the credibility of the results. But this is only one of the aspects of the more or less explicit concern for truthfulness in the article. After the initial paragraph, Walsh writes:

. . . I will request the favour of to lay before the [Royal] Society my letter from La Rochelle, of the 12th July 1772, and such part of the letter I afterwards wrote from Paris, as relates to this subject. Loose and imperfect as these informations are, for they were never intended for the public eye, they are still the most authentic, and so far the most satisfactory I can pretend at present offer, since the notes I made of the experiments themselves remain nearly, I am sorry to say it, in that crude and bulky state in which you have had the trouble to read it. (p. 461)

Walsh also speaks of his "deficiency" as to give to the Royal Society "a complete account of [his] experiments on the electricity of the torpedo". This deficiency, as well as his allusion to the imperfection of the letters on the subject sent to Franklin and to the "crude and bulky state" of the notes (likely his journal of experiments) should not be taken for their face value. They should instead be considered as a part of a communication strategy pointing to the veracity and to the "factuality" of the events described. Since the exordia of the modern experimental science, among the strategies proposed to reinforce the confidence in the facts narrated there was the tendency to a style overtly devoid of literary elaborations, leading to a non-systematic account, more akin to a minute verbatim: undoubtedly crude but «still the most authentic» as should clearly appear the letters written to Franklin in close temporal relation with the experiments at La Rochelle. The aim was to create a "virtual testimony" capable of attesting and making credible the events narrated. It was necessary to insist on the temporal occurrence of the events, either making recourse to a narrative type of style or presenting documents capable of certifying the time contingency of the facts (see Eamon, 1990; Shapin & Schaffer, 1985). In addition to the transcription of Franklin's letters, in Walsh's article there were other significant strategies of the same type. Walsh also reproduced the English translation of a short article that appeared on 30 October 1772 on the *Gazette de France* dealing with his demonstration at La Rochelle. This article was written (also in the form of a letter, sent to the publisher of the Gazette) by the Mayor of the town and "perpetual secretary" of the *Académie* "Sieur [Pierre-Henry] Seignette". It contained the description of all most salient facts concerning Walsh's demonstration and ended with the following conclusive remark: "the effects produced by the Torpedo resemble in every respect a weak electricity". (Walsh, 1773, p. 468)

The statement whereby Seignette's article is introduced by Walsh clearly betrays the testimonial character attributed to it:

As it [the article] came from a very respectable quarter. Not less so from the private character of the gentleman, than from the public offices he held, I must desire leave of the [Royal] Society to avail myself of such testimony to the facts I have advanced, by giving a translation of that narrative. (p. 466)

After reporting in its entirety Walsh declares:

This exhibition of the electric powers of the Torpedo, before the Academy of La Rochelle, was at a meeting, held for the purpose in my apartments, on the 22nd July 1772, and stands registered in the journals of the Academy. (p. 468)

We note here *en passant* that the *compte rendu* of Walsh's demonstration can still be found among the Registers of the *Académie* of La Rochelle (see Piccolino, 2003). Moreover, we cannot abstain from remarking that Walsh seems to have, at the time of his torpedo research, purposely left traces of his experiments, useful in a later time as a documentary and verifiable testimonies of their occurrence. Indeed the plan of Walsh's article on the *Philosophical Transactions* is, on that regard, rather elaborated: a letter to Franklin based on two previous letters to the same personage and containing an article in the form of letter sent by an eminent citizen of La Rochelle to the official French journal. We could also remark that the article of the Mayor of La Rochelle on the French Gazette makes in turn allusion to another article concerning Walsh's experiments also published on the *Gazette* (in the date of 14 August 1772) and also based on a communication by Seignette. All this contributes to give the impression of a Chinese box strategy of unquestionable efficacy in attesting the occurrence and verifiability of the events narrated, and thus capable of accentuating their "factuality".

Other Relevant Aspects of the Journal of Experiments

In recent years a great attention has been given by science historians to the manuscripts where scientists used to annotate the progress of their studies together with private reflections and remarks of various types (Grmek, 1973; Holmes, 1974; Holmes, Renn, & Rheinberger, 2003; Holmes, 2004). When preserved from the ravages of time, these texts appear to be useful both for the possibility they offer of tracing the discovery process and various aspects of the scientific psychology in addition to the published texts. The articles or memoirs written for the aim of publication, because of their communication scope, might tend to privilege the most successful and straightforward aspects of the investigation leaving aside the difficulties, the errors, the complex and tortuous

pathways leading to the discovery. These aspects appear more clearly in the experimental protocols manuscripts, particularly when, as in the case of Walsh's journal of the experiments, they present a precise day-by-day narrative of the research development together with various annotations and reflections. Here we can only survey rapidly some of these aspects that better emerge from the analysis of Walsh's manuscripts (in addition to those already discussed).

Among these aspects there is what we could call "didactics of the experiment", a form of learning acquired in the course of the investigation (and often not explicitly recognized by the scientist), by which he becomes progressively more capable of effectively pursuing his further research. There are various phases in which this emerges clearly in Walsh's experiments. One is when (attentive as he is to possible movements of the fish that could justify Réaumur's mechanical hypothesis) Walsh perceives a small motion of fish eye in the preparatory phase immediately preceding the shock. As already mentioned, this "winking" would henceforth represent a "signal" for ascertaining the production of fish shock, particularly useful in the condition in which the shock might not be perceived in other ways (as for instance when insulating matter is used to complete the discharge circle). Another interesting characteristic of the shock exploited by Walsh (even before he explicitly recognizes it in his annotations) is the fact that its intensity is significantly stronger when the fish is investigated "in the air" than when plunged in the water. Although Walsh recognized only in the late phase of his investigation (i.e. on 26 July 1772) that in water the shocks were "greatly weaker than in air, perhaps not a quarter part of the strength, at least to our sense of it" (WE, p. 170), nevertheless he started to examine more and more frequently the animal "in the Air" since the first casual observation made during the initial week of experiments (on 6 July 1772) of the stronger shock intensity with the animal outside the water.

The problem of the different intensities of the perceived shock in air and in water would be particularly addressed by Cavendish who succeeded in accounting for the difference on the basis of the shunting effect of the water and was also able to build an "artificial torpedo" capable of exactly reproducing the action of the natural one also under that point of view.

The mention of the subjective estimation of the intensity of the torpedo's shock brings us to consider another aspect of the transition in the way of doing science in the eighteenth century: the recourse to the experimenter's body as a measuring instrument that still persisted in spite of the development of objective physical measuring devices. Given the apparently crude and subjective character of these estimations of laboratory events, one may be astonished by the precision of the results obtained by the scientists of the *ancien régime*. Using a method introduced by the famous Italian electrician Giovan Battista Beccaria, Henry Cavendish succeeded in estimating, with impressive precision, the relative electric resistance of water, showing that it was much smaller than that of metals (about 400 million times). He did this by comparing the intensity of the shock perceived when an electric discharge was transmitted through a metallic wire of a given length and section, and tubes of various lengths and diameter, full of water (Cavendish, 1776; See Maxwell, 1879).

By estimating the taste sensation produced by a bimetallic arc applied to the tip of the tongue, Alessandro Volta was able to detect, in 1792, a feeble current, too weak to be measured by the physical electrometers of the age. Moreover, on the basis of the same physiological sensor he could establish the polarity of this current. These experiments were milestones in Volta's path to the discovery of the battery, a path that not only led to the invention of the epoch-making electrical device, but also resulted in an important series of physiological discoveries (Piccolino, 2000, 2003; Piccolino & Bresadola, 2003). Volta was able to show the eminently transient character of the responsiveness of the visual system, that contrasted with the tonic responses of the gustatory and nociceptive system. He anticipated, by about half a century, the fundamental idea of the functional organization of the nervous system, Johannes Müller's doctrine of "specific nervous energies". This stipulates that the physiological effects of nerve stimulation depend on the type of nerve stimulated, and not on the type of stimulus used to achieve the stimulation (Müller, 1826). Moreover, Volta showed that the effect of the electric stimulation of nerves depends on the polarity of the stimulus, again anticipating important laws of electrophysiology, to be fully clarified only in the nineteenth century. Finally, he measured with an impressive precision the time during which there was a partial time summation of the

physiological effects of a prolonged electric stimulation ("integration time" in modern words).

Among the other aspects of the way of doing research in the eighteenth century emerging from Walsh's work (and also from a comparison with the endeavour of other scientists of the age involved in a similar research, as Cavendish, Galvani and Volta) there is the interchangeable (and effective) way in which the investigator passed from the study of the animal preparation to that of physical instruments. Even though we cannot deal with this aspect, we cannot abstain from quoting a statement made by Walsh during his demonstration at La Rochelle:

as Artificial Electricity had led to a discovery of some of the operations of the Animal, the Animal if well consider'd would lead to a discovery of some truths in artificial Electricity which were at present unknown and perhaps unsuspected.

From Walsh to Volta

The passage just quoted (also reported in a slightly modified form in the 1773 article) might appear *a posteriori* somewhat prophetic, particularly if we consider the importance of the electric fish experiments for the research that led Alessandro Volta to the invention of the electric battery (Heilbron, 1978; Pancaldi, 1990, 2003; Piccolino, 2000; Piccolino & Bresadola, 2003). We have already alluded to the relevance of Walsh's studies in prompting the research of Luigi Galvani towards the demonstration of the electric nature of nerve signals. As already mentioned, in the case of Galvani, Walsh's demonstration of the electric nature of shock of the torpedo and of the electric eel undermined the significance of the physical objections against a role of electricity in physiological processes. Moreover, in the model of the "minute animal Leyden jar" elaborated by Galvani in order to account for neuromuscular function there is an echo of Walsh's conception of the torpedoes as "animate phials".

In the case of Volta the relevance of electric fish research was even more significant. From a reflection on the electric organs of the fish Volta derived in the years 1798–1799 the decisive inspiration in the attempt he was pursuing to obtain a strong electric action from the weak "electromotor" power of a bimetallic contact. Volta had a long interest in electric fish research. He had personally met Walsh and discussed with him on the subject in the occasion of a visit to London in 1782. However, in his writings the reference to the singular properties, very frequent in 1792 at the beginning of his research on animal electricity declined progressively in the following years. Suddenly, however, in 1798 electric fish started to dominate again Volta's elaboration during an intense period of the investigation on the electromotive action of bimetallic contacts. This renewed interest was motivated by the publication in 1797 of a memoir in which Galvani described his own experiments on torpedo and, moreover, by the appearance (also in 1797) of an article of William Nicholson proposing a physical model to account for the mechanism of fish shock (Pancaldi, 1990, 2003). In this article Nicholson suggested that the discs of the electric organs could behave like the plates of an "electrophore". The electrophore, an electric tool invented by Volta himself, consisted of a movable metallic plate and of another plate made of an insulating material (usually resin). Manifest electric effects were produced by the manipulation of the metallic plate that resulted in a change of its capacitance. Nicholson's model revived, in a somewhat modified form, the hypothesis formulated by Walsh, and elaborated by Cavendish, that electric organs could work as an assembly of minute planar capacitors.

Nicholson's article undoubtedly contributed to reorient Volta's attention on electric fish, but it acted somewhat "by contrast". As was the case with Walsh and Cavendish's model of the electric organ as composed by a multitude of planar capacitors, Nicholson's artificial electric organ was based on the interposition of insulating laminas between conductive plates. As a matter of fact, an insulating material was a constitutive element of all instruments of the age capable of maintaining an electric equilibrium (Leyden jars, Franklin's planar capacitors) and a proper insulation was also necessary in instruments capable of generating it (such as friction electric machines and electrophores). For Volta, however, the idea that an insulating matter could be a constituent of the electric organ was in evident contrast to the knowledge of "animal physics". This he will state in a clear-cut way in his 1800 letter to Banks on the invention of the battery, where, with reference to Nicholson's hypothesis, he wrote:

such a hypothesis falls entirely, these pellicles of the organ of the torpedo are not, and cannot be, in any manner insulating or susceptible of a real electric charge, and much less capable of retaining it. Every animal substance, as long as it is fresh, surrounded with juices, and more or less succulent of itself, is a very good conductor. I say more, instead of being as *cohibent* [i.e. *insulator*] as resins or talc, to leaves of which Mr. Nicholson has compared the pellicles in question, there is not, as I have assured myself, any living or fresh animal substance which is not a better *deferent* [i.e. *conductor*] than water, except only grease and some oily humours. But neither these humours nor grease especially semi-fluid or entirely fluid, as it is found in living animals, can receive an electric charge in the manner of insulating plates, and retain it: besides, we do not find that the pellicles and humours of the organ of the torpedo are greasy or oily. This organ therefore, composed entirely of conducting substances, cannot be compared either to the electrophore or condenser or to the Leyden flask, or any machine excitable by friction or by any other means capable of electrifying insulating bodies, which before my discoveries were always believed to be the only ones originally electric. (p. 430).

After the demonstration of the electromotive power of the contact of two different metals obtained in 1796, Volta had tried to multiply the weak electric effect of a single metallic contact by staking, one above the other, numerous bimetallic couples in a way that was similar to what happened with the discs of the fish electric organs. For a long time, these attempts were unsuccessful. Upon reflecting more deeply on the structure of the organs, Volta eventually decided to interpolate, between the bimetallic couples, a disc of paper or cloth soaked in a saline or acid solution which appeared to be more physically akin to the matter of the fish organs. This resulted in the invention of the electric battery, the realization beyond any hope of his scientific dream. In the letter to Banks on 20 March 1800 communicating the invention of this extraordinary device, Volta called it "*organe électrique artificiel*" in order to acknowledge – he wrote – that it was "similar at bottom", as he constructed it, "in its form to the natural electric organ of the Torpedo or electric eel" (p. 405).

According to Volta, in addition to an obvious similarity in the form, there was a deeper analogy between the natural and the artificial organs. As in the case of the newly invented tool, in the fish organ, the production and maintenance of a strong electric power occurred "by the mere contact of conducting substances of different species" (p. 419), in the absence of the interposition of any insulating matter, i.e. according to the new physical principle of the electromotive action he had discovered. The final passage of Volta's letter to Banks asserts this conception in a particularly expressive way:

To what electricity then, or to what instrument ought the organ of the torpedo or electric eel, &c. to be compared? To that which I have constructed according to the new principle of electricity discovered by me some years ago, and which my successive experiments, particularly those with which I am at present engaged, have so well confirmed, *viz.* that conductors are also, in certain cases, exciters of electricity in the case of the mutual contacts of those of different kinds, &c. in that apparatus which I have named the *artificial electric organ*, and which being at bottom the same as the natural organ of the torpedo, resembles it also in its form, as I have advanced. (p. 430)

The Battery and the Ancient Characters

Nobody could deny the importance of the battery for both the development of physics and for technological progress in the last two centuries. From the battery would originate electrochemistry and electromagnetism. On one hand, this will serve to orient physical investigation towards the study of the properties of ions and atoms, thus laying the grounds for the birth of modern physics. On the other hand, the battery would make possible discoveries and inventions rich in deep consequences for the progress of technology. Consider for instance Faraday's discovery of the reciprocity of electric and magnetic actions, which cortically depended on flow of constant electric current (feasible with the battery, but outside the reach of the friction electric machines of the age). On the reciprocity of electric and magnetic phenomena is based the possibility of producing great amount of electric energy (through the action of powerful alternators or dynamos), and thus the large-scale utilization of electric energy in everyday life. On it also depends the production of electromagnetic waves (and thus a large part of the science and technology of modern communications).

Compared to the historical importance of the invention of the electric battery, the pathway leading from Walsh to Galvani might appear less significant and less revolutionary. One could say many things on the importance of the electric hypothesis of nervous conduction which closed a millennial epoch of the science dominated by the doctrine of the "animal spirits" and opened the way to the development of modern electrophysiology (and thus to neurosciences). Through such paths scientists have succeeded in deciphering the nature of the signals that unceasingly flow along the circuits of our brain, the elements of the "electric storm" alluded to by Charles Scott Sherrington (1951) in a famous book. These signals allow us not only to see a distant castle, the visage of a nearby friend, but also to hear a voice, a beautiful music, to feel emotions, to speak and to think.

Even more than the "minute characters" (*caratteruzzi*) of the alphabet alluded to by Galileo in his *Dialogo sopra i massimi sistemi del mondo* (Galileo, 1632, p. 98), these electric signals represent the primordial characters underlying all that allow us to go beyond the more elementary level of animality. In the absence of these "electric characters" of our nerve cells, immensely more ancient of Galileo's *caratteruzzi*, writing itself would be impossible.

It has been said (Wu, 1984) that modern electrophysiology was born at the moment that Walsh produced for the first time a spark from the shock of an electric eel. This statement sounds a little paradoxical, but it is not totally devoid of foundation.

If one thinks of the continuity existing between Walsh's research on torpedoes at La Rochelle and the London experiment on the electric eel, we might perhaps say that modern electrophysiology initiated when Walsh wrote in his journal "*je l'ai donté*": I have tamed the "indomitable art of the wonderful torpedo".

References

Adanson, M. (1757). *Histoire Naturelle du Senegal*. Paris: J. B. Bauche.

Allamand, J. N. S. (1756). Kort verhaal van de uitwerkzelen, welke een Americanse veroosakt op de geenen die hem aanraaken. Hollandsche Maatschappij der Weetenschappen te Haarlem. *Verhandelingen, 2*, 372–379.

Bancroft, E. (1769). *An essay on the natural history of Guiana*. London: T. Becket and P. A. de Hondt.

Cavendish, H. (1776). An account of some attempts to imitate the effects of the torpedo by electricity. *Philosophical Transactions of the Royal Society (London), 66*, 196–225.

Claudianus, C. (1922). *Claudian. with an English translation by Maurice Platnauer*. London/New York: W. Heinemann & G. P. Putnam's Sons.

Claudianus, C. (1985). In J. Barrie Hall (Ed.), *Claudii Claudiani Carmina*. Leipzig: Teubner.

Daston, L. J., & Park, K. (1998). *Wonders and the order of nature: 1150–1750*. New York: Zone Books.

Eamon, W. (1990). *From the secrets of nature to public knowledge*. Cambridge: Cambridge University Press.

Faraday, M. (1839). *Experimental researches in electricity*. London: R. and J. E. Taylor.

Franklin, B. (1959). In L. W. Jr. Labaree, J. B. Jr. Whitfield, W. B. Willcox, & others (Eds.), *The papers of Benjamin Franklin* (37 Vols.). New Haven/London: Yale University Press.

Galenus, C. (1533). *De causis respirationis [. . .] Ianne Vasseo Meldensi interprete*. Parisiis: Apud Simonem Colinæum.

Galenus, C. (1541). *De causis morborum. Galeni operum omnium [. . .] Augustino Ricco authore*. Venetiis: ex officina Farrea.

Galilei, G. (1623). *Il Saggiatore*. Roma: Giacomo Mascardi.

Galilei, G. (1632). *Dialogo di Galileo Galilei sopra i due massimi sistemi del mondo*. Fiorenza: Gio. Batista Landini.

Galvani, L. (1791). De viribus electricitatis in motu musculari Commentarius. *De Bononiensi Scientiarum et Artium Instituto atque Academia Commentarii, 7*, 363–418.

Galvani, L. (1797). Memoria Quinta. In *Memorie sulla elettricità animale di Luigi Galvani [. . .] al celebre Abate Lazzaro Spallanzani . . .* (pp. 64–86) Bologna: Sassi.

Grmek, M. D. (1973). *Raisonnement expérimental et recherches toxicologiques chez Claude Bernard*. Genève: Droz.

Guerrini, L. (1999). Contributo critico alla biografia rediana, con uno studio su Stefano Lorenzini e le sue «Osservazioni intorno alla Torpedini». In W. Bernardi & L. Guerrini (Eds.), *Francesco Redi, un protagonista della scienza moderna* (pp. 47–69). Firenze: Olschki.

Heilbron, J. L. (1978). A. Volta's path to the battery. In G. Dubpernell & J. W. Westbrook (Eds.), *Proceedings of the symposium on selected topics in the history of electrochemistry*. Princeton N. J. : The Electrochemical Society Inc.

Heilbron, J. L. (1979). *Electricity in the 17th and 18th century*. Berkeley: University of California Press.

Holmes, F. L. (1974). *Claude Bernard and animal chemistry: the emergence of a scientist*. Cambridge (Mass.): Harvard University Press.

Holmes, F. L. (2004). *Investigative pathways: Patterns and stages in the careers of experimental scientists*. New Haven: Yale University Press.

Holmes, F. L., Renn, J., & Rheinberger, H. (Eds.) (2003). *Reworking the bench: Research notebooks in the history of science*. Dordrecht: Kluwer.

Hunter, J. (1773). Anatomical observations on the torpedo. *Philosophical Transactions of the Royal Society (London), 63*, 481–489.

Kellaway, P. (1946). The part played by electric fish in the early history of bioelectricity and electrotherapy. *Bulletin of the History of Medicine, 20*, 112–137.

Lacroix, P. (1868). *Secrets magiques pour l'amour: octante et trois charmes, conjurations, sortilèges et talismans/publiés. par un bibliomane*. Paris: Académie des bibliophiles.

Le Roy, J-B. (1776). Lettre adressée a L'auteur de ce Recueil par M. Le Roy. *Observations sur la Physique, 8*, 331–335.

Lorenzini, S. (1678). *Osservazioni intorno alle torpedini*. Firenze: Onofri.

Maxwell, J. C. (1879). *The electrical researches of Henry Cavendish*. Cambridge: Cambridge University Press.

Moller, P. (1995). *Electric fishes, history and behaviour*. London: Chapman & Hall.

Müller, J. (1826). *Zur vergleichenden Physiologie des Gesichtssinnes des Menschen und der Thiere neben einen Versuch über Bewegungen der Augen und über des menschlichen Blick*. Leipzig: Cnobloch.

Musitelli, S. (2002). *L'elettricità animale dalle origini alla polemica Galvani-Volta*. Pavia: La Goliardica Pavese.

Nicholson, W. (1797). Observations on the electrophore, tending to explain the means by which the torpedo and other fish communicate the electric shock. *Journal of Natural Philosophy, Chemistry and the Arts, 1*, 355–359.

Olmi, G. (1992). *L' inventario del mondo: catalogazione della natura e luoghi del sapere nella prima età moderna*. Bologna. Il Mulino.

Oppianus, A. (1999). In F. Fajen (Ed.), *Halieutica*. Stuttgart/Leipzig: Teubner.

Pancaldi, G. (1990). Electricity and life: Volta's path to the battery. *Historical Studies in the Physical Sciences, 21*, 123–160.

Pancaldi, G. (2003). *Volta: Science and culture in the age of Enlightenment*. Princeton: Princeton University Press.

Piccolino, M. (2000). The bicentennial of the voltaic battery 1800–2000: The artificial electric organ. *Trends in Neurosciences, 23*, 47–51.

Piccolino, M. (2003). *The taming of the ray: Electric fish researches in the Enlightenment, from John Walsh to Alessandro Volta*. Firenze: Olschki.

Piccolino, M. (2005). *Lo zufolo e la cicala: Divagazioni Galileiane tra la scienza e la sua storia*. Torino: Bollati-Boringhieri.

Piccolino, M., & Bresadola, M. (2002). Drawing a spark from darkness: John Walsh and electric fish. *Trends in Neurosciences, 25*, 51–57.

Piccolino, M., & Bresadola, M. (2003). *Rane, torpedini e scintlllle/Galvani, Volta e l'elettricità animale*. Torino: Bollati-Boringhieri.

Plato, (1892). *The dialogues of Plato/translated into English with analyses and introductions by B. Jowett* (3rd ed.). Oxford: Clarendon press.

Réaumur, R. A. F. (1741). Des effets que produit le poisson appelé an français torpille, ou trembleur, sur ceux qui le touchents; et de la cause dont ils dépendent. *Histoire de l'Académie Royale des Sciences pour l'Année 1714* (pp. 344–360).

Redi, F. (1671). *Esperienze intorno a diverse cose naturali e particolarmente intorno a quelle che ci son portate dall'Indie, scritte al Reverendissimo Padre Atanasio Chircher della Compagnia di Giesù*. Firenze: All'insegna della Nave.

Rondelet, G. (1574). *Gulielmi Rondeleti [. . .] methodus curandorum omnium morborum corporis humani in tres libros distincta; De dignoscendis morbis; De febribus*. Parisiis: apud Carolum Macaeum.

Scribonius, L. (1983). In S. Sconocchia (Ed.), *Compositiones*. Leipzig: B. G. Teubner.

Shapin, S., & Schaffer, S. (1985). *Leviathan and the air-pump: Hobbes, Boyle, and the experimental life*. Princeton: Princeton University Press.

Sherrington, C. S. (1951). *Man on his nature*. Cambridge: Cambridge University press.

Trallianus, A. (1557). *[L']onziesme livre d'Alexandre Trallian traittant des gouttes/trad. de grec en françois par M. Sébastien Colin*, . . . Poitiers: Enguilbert de Marnef.

van der Lott, F. (1762). Kort bericht van den conger-aal, afte drilvisch. *Hollandsche Maatschappij der Weetenschappen te Haarlem. Verhandelingen, 6*(Part 2), 87–93.

Volta, A. (1800). On the electricity excited by the mere contact of conducting substances of different species: Letter to Sir Joseph Banks, March 20, 1800. *Philosophical Transactions of the Royal Society (London), 90*, 403–431.

Walsh, J. (1772). *Experiments on the Torpedo or Electric Ray at La Rochelle and Isle de Ré - in June and July 1772*. Royal Society: MS 609.

Walsh, J. (1772). *Journey from London to Paris, begun 8th June 1772*. John Rylands Library Manchester: Rylands English Ms. 724.

Walsh, J. (1773). On the electric property of torpedo: in a letter to Ben. Franklin. *Philosophical Transactions of the Royal Society (London), 63*, 478–489.

Wu, C. H. (1984). Electric fish and the discovery of animal electricity. *American Scientist, 72*, 598–606.

10
Luigi Galvani, Physician, Surgeon, Physicist: From Animal Electricity to Electro-Physiology

Miriam Focaccia and Raffaella Simili*

Introduction

The two-hundredth anniversary of the death of Luigi Galvani (1732–1798), famed discoverer of animal electricity and professor of anatomy at the University of Bologna, provided an occasion to launch a rigorous and in-depth study of his work and multifarious scientific activity. Thanks to the many initiatives conducted in this direction, enriched by others held on the occasion of the anniversaries of the deaths of Lazzaro Spallanzani (1799) and Alessandro Volta (1799) it is now possible to render a reliable and detailed picture of Galvani's personality and theories.

Indeed, despite the widespread circulation of his work on animal electricity, Galvani's scientific endeavours have been largely misrepresented in the history of science. For a long time, he suffered the stereotypical image of a relatively unknown professor of anatomy who started his studies on electricity by chance, ignoring the physics of the time and claiming to have discovered, by means of his experiments on frogs, a new form of electricity, animal electricity. Officially presented in 1791 in his *De viribus electricitatis in motu muscolari Commentarius*, though his experiments dated back to at least 1780, his discovery provoked the harsh criticism of Alessandro Volta, physicist at the University of Pavia and future inventor of the battery.

It was the start of a scientific controversy – which would capture the attention of science historians for centuries – that has usually celebrated the physicist Volta as the winner and the physician Galvani as the loser. In fact, Galvani's approach to electrophysiology was revived in the nineteenth century (Nobili, Matteucci, du Bois-Reymond, von Helmholtz, Bernstein, Hermann), only to attain full legitimacy in research conducted in the twentieth century (Lucas, Adrian, Hodgkin, Huxley, Katz) (Piccolino, 1998).

Precisely on the occasion of the international symposium on Luigi Galvani, held in Bologna in 1998 as part of the Galvani celebrations, John Heilbron opened the meeting clarifying how the controversy between Galvani and Volta, the physician and the physicist, rested on the wrong bases, inasmuch as "boundaries between what we regard as distinct disciplines were indefinite and easily crossed." As a matter of fact, "physics signified a bookish study of the entire natural world, botany as well as chemistry and experimental physics, so animal electricity, like pneumatics, was an open field to which cultivators from different backgrounds brought various tools and approaches" (Heilbron, 1999, pp. 29–30).

Galvani was furthermore an "enterprising" scientist! Indeed, continued Heilbron, "he began his excitation on frogs thinking that their nerves contained ordinary electricity; but he did not hesitate to introduce a special nerveo-electrical fluid when his experiments seemed to him to require it" (Heilbron, 1999, p. 29).

In 1937, on the occasion of the two-hundredth anniversary of Galvani's birth, he was celebrated as

*The paragraphs 2, 3 are the work of Miriam Focaccia, whereas Raffaella Simili is responsible for the paragraphs 1, 4 and 5.

a "eroe della patria" according to the lines of fascist propaganda, while a great congress of physicists, biologists and radiobiologists (including nine Nobel prize winners) paid a special tribute to his prophetic insights as a physicist and physician.

Niels Bohr took the opportunity to enrol Galvani in the symbiosis of physics and physiology, then newly productive in both physics and biology: Galvani's work (said Bohr) demonstrated "an intimate combination of the laws of inanimate nature with the study of the properties of living organisms" (Bohr, 1937).

Two issues in particular marked his conceptual originality and experimental capability both in the fields of physics and biology. The Bolognese physicist Quirino Majorana affirmed that Galvani had not only discovered electricity but also electrodynamics, later called Galvanism. Furthermore, already in 1780, he had understood "the basic principles which today we find in wireless telegraphy" (Majorana, 1937, pp. 422–423). This opinion was formulated first of all by Augusto Righi in 1893 (Righi, 1893) and many years later reconfirmed in identifying Galvani the "experimenter" as the first to discover electromagnetic waves, which placed him in the gallery of the so-called fathers of Hertz (Süsskind, 1964).

For his part, Edgar Douglas Adrian, professor at Cambridge and 1932 Nobel prize winner, illustrated Galvani's merit in the field of physiology in a paper entitled: *Electrophysiology of the Sense Organs*.

[…] his discoveries were primarily physiological and their importance in this branch of knowledge was never greater than it is today. Electro-physiology has now spread into many fields and I cannot claim the honour of representing any one of them, but as one of the many electro-physiologists gathered for these celebrations, I shall try to describe a particular line of work which may illustrate the way in which "animal electricity" contributes to our knowledge of life.

Galvani's experiments have given us two methods of research in physiology. They showed that the current from a metallic couple would excite muscles and nerves and this has given us a means of producing an artificially controlled activity in the body; and they showed that active muscles and nerves produce electric currents, and this has given us a means by which various aspects of bodily activity can be analysed. By the first method, that of electrical stimulation, the nervous pathways have been explored and their function decided. As a modern example we have the use of a stimulating current by a neurosurgeon to explore the mechanism of the human brain, one might almost say

of the human mind. And on the other side, that of cell physiology, there are the precise studies of electrical excitation, such as those of Lapique and A. V. Hill, which aim at a mathematical formula for a vital process.

The second method of research, the analysis of the "animal electricity" produced by living cells, has been of constantly increasing value to medicine, as well as to physiology. Its value to medicine was most clearly shown when it became possible to study the activity of the human heart, normal or abnormal, by recording the electrocardiogram. We have now a fresh medical development in the electro-encephalogram and much has already been learnt from this discovery of Professor Berger that the electric currents from the human brain can be detected on the surface of the head (Adrian, 1937, p. 79).

Again in 1937, it was pointed out that Galvani was not only an able scientist and famed anatomist but also a physician, a practising doctor, a surgeon and a physician–obstetrician: aspects that proved crucial in constructing the theory of animal electricity. Galvani, then, was quite a special physician, for he cultivated a wide variety of interests with the idea of improving the art of medicine, whose subject was perceived to be the overall functioning of the human organism in sickness and in health, as well as related to the other living beings.

Galvani himself claimed membership of the stressful medical profession in hospitals and scientific institutions, referring almost everywhere, in published and unpublished writings, to his "many different kinds of work" at the various centres for the sick and in the governing bodies of the city health services. This he reiterated in the official orations delivered on the occasion of the degree ceremonies of his two nephews, Giovanni Aldini (Galvani, 1782, 1888) and Ludovico Galvani (Galvani (AASB, MSG, Cart. V, Plico XII, fasc. 6)), as well as in his best known publications, starting from *De Viribus*. The final part of the latter publication is entirely devoted to cures deriving from the intrinsic presence of animal electricity in living organisms, and hence, with reference to illnesses such as paralysis, epilepsy and tetanus, as well as serious rheumatic disturbances and sciatica. Seen in this context, Galvani's premise in *De Viribus* is quite clear: the ultimate aim of finding the "hidden properties" of nerves and muscles was precisely to heal diseases with greater certainty and to greater profit, as he and doctors in general wished to do (Galvani, 1791, 1998).

In 1782, in his oration to Aldini on graduating in physics, after underlining the usefulness of physics and its methods, as well as astronomy, nautical science and chemistry, Galvani recalled the specific uses of physics in medical practice. Thanks to them, physicians "may understand the human body better, and distinguish the causes of its hurts more correctly, for which, in short, it can conserve our health and defeat our infirmities with greater certainty" (Galvani, 1782, 1888).

On the occasion of the degree ceremony in medicine (December 16, 1797) of his youngest nephew Ludovico, Galvani pointed out several aspects of medical practice. First of all, practitioners must be competent and prudent, ready to fight against two obstacles, one concerning an uncertain diagnosis and the other an inappropriate therapy. And, secondly, practitioners must also exercise incessant study and deep meditation, but above all a great love of patients (Galvani (AASB, MSG, Cart. V, Plico XII, fasc. 6))!

This Chapter proposes to offer a unitary picture of the experimental research conducted by Galvani who sought to confirm his hypotheses of animal electricity not only through the usual experimental means (he was well-informed regarding the scientific matters of his times) but also by means of diverse and original research sectors (anatomy, medicine and surgery, obstetrics). Guided by the concept of a dynamic human organism, substantially dependent on the brain/nerves/muscles circuit, he elaborated animal electricity as a peculiar property of all living creatures.

Galvani and Iatrophysics

Halfway through the eighteenth century, chemistry, anatomy, natural history, physics, meteorology and astronomy were particularly thriving domains at the Istituto delle Scienze of Bologna. The Istituto delle Scienze was founded in 1711 by Luigi Ferdinando Marsili with the aim of promoting the experimental practice of scientific disciplines. The Istituto applied the experimental method to every branch of its activities: anatomy, physiology, chemistry and medicine, just as it frequently used electricity that had been very much welcomed in the city and had become extremely familiar to young students. Research was conducted on the bound-

aries of the different disciplines and, often, the most capable members of the Istituto practised more than one simultaneously (Angelini, 1993; Tega, 1984, pp. 65–108, 1986–1987).

These rather uncommon aspects for the time were joined by an original division of research along three lines: medicine, electrical physics and the study of the environment from the historical, naturalistic and meteorological viewpoints.

Medicine was closely tied to the teaching of Marcello Malpighi, for a long time in Bologna, and even more so to that of Giovan Battista Morgagni, one of the founding fathers of the Istituto (Bernardi, 1985). Their teaching sought to apply experimental practice also to the then delicate and uncertain sector of life sciences. The observation and anatomy of corpses gradually made way for physiology and pathological anatomy. The issue of anatomy was joined by wax modelling, surgery, obstetrics, comparative anatomy, the study of diseases and therapies, the taxonomy of diseases and epidemiology. In point of fact, the physician was also a chemist, an expert in electrical physics and familiar with natural history and meteorology (Simili II, 2001, p. 20).

The study of electricity, which soon embraced Franklin's theories, was strongly oriented towards diagnostics and medical therapy by virtue, on the one hand, of the influence exerted in Bologna by Newton's natural philosophy and, on the other hand, of the strong personality of Gianbattista Beccaria from Turin, an expert in electricity, who was in contact with Laura Bassi and Giuseppe Veratti in Bologna (Bernardi, 1992).

The experimental activity the two latter scientists conducted in their home-laboratory, made this part of physics an essential point of reference for medicine (Simili II, 2001, p. 20).

Practical medicine or, more precisely its theoretic aspect, was situated at the very centre of the Istituto's activity not only through the work of anatomists and surgeons but also through the construction of drawings, models and machines destined for the close-up and concrete study of the living, such as those of obstetrics proposed by surgeon obstetrician Giovanni Antonio Galli, and the anatomical ones realised by wax modellers Ercole Lelli, Giovanni Manzolini and Anna Morandi. Galvani, in fact, found himself part of the experimental tradition with a special eclecticism and a mixed education in medicine, chemistry and physics. Furthermore, he learnt the method of nearness between the various sectors of research and understood the reciprocal favours there can be in a system of knowledge in continuous evolution. Galvani completed his training in this labyrinth of relations, and this enabled him to comprehend and better define the contours and results of his programme.

As we leaf through scientific reviews of the time and examine Galvani's work, we are no longer surprised (with our present knowledge) to find traces of a journey that leaves no room to chance or improvisation. In physics and chemistry's laboratories, we find Galvani devoting himself to natural waters, phosphor and the composition of foods. Among naturalists, we find him dealing with classification and the problems of comparative anatomy. He was a habitual visitor to the home-laboratory of Laura Bassi and Giuseppe Veratti, while he assisted the physics professor and future secretary of the Istituto, Sebastiano Canterzani. He used the instruments and anatomical and wax models of the Istituto, as well as Galli's obstetric machines.

Galvani sought to explain the phenomena of the living state as part of natural science, so fascinating to European scientists of the time, insisting that metaphysical and theological hypotheses were otiose.

Here lies the profound sense of the project that medical studies helped mature in the course of the eighteenth century in which Galvani was a major player. The reconstruction of his scientific activity makes it possible to shed light on two facts of great importance. First of all, though an established physician, Galvani decided to focus all his attention and all his ability on constructing a research project susceptible to modification through time, as was the anatomy – physiology and physical – chemical study of the "life." Second, Galvani was convinced that his project required a lot of time in laboratories and a patient acquisition of the fundamental knowledge and techniques of the medical art. He considered this art an open system.

Galvani did not limit himself to frequenting the laboratories of Bolognese masters, but with great diligence also wrote his extraordinary lessons of anatomy and physiology (Galvani (AASB, MSG, Cart. IV)). He concentrated his attention on the living organism as a system of apparatuses and tissues that, in their stratification, are not considerably dif-ferent from those that make up the vegetable world. In this he was at one with Stephen Hales and Henri Louis Duhamel de Monceau, authors whom he read carefully. Moreover, it was by interacting with the natural scientist members of the Accademia, such as Reamur or Buffon, that he decided to follow the road of comparative anatomy rather than that of classificatory discussions (Simili II, 2001, p. 20). He became particularly keen on the relations between chemistry, physics and medicine.

In 1759, he graduated in medicine and philosophy, according to the university custom of the age. The most famous of Galvani's teachers were the anatomist, physiologist and physicist Jacopo Bartolomeo Beccari, who from 1737 occupied the first chair of chemistry in Italy; the anatomist Domenico Gusmano Galeazzi, practising doctor and experimental physicist; and the surgeon Giovanni Antonio Galli (Bresadola, 1998). After graduation, he immediately began to practise medicine and surgery in the city hospitals continuing without interruption until his death. Three years after obtaining his degree, in 1762 he became honorary lecturer, with a thesis *Delle ossa* (*On Bones*) (Galvani, 1762, 1998) in which he brilliantly described the bones in the human body, taking into account their genesis, their use and their diseases. In the same year, he married Lucia, the daughter of his teacher Galeazzi, bringing together family, academic and laboratory relationships, for his wife was also an active and intelligent partner in his experiments on frogs. In 1772, at only thirty-five years of age and only eleven years after his affiliation, he was elected President of the Accademia (Malagola, 1879).

He became Professor of Practical Anatomy at the University in 1775, having begun his training in 1763 as a lecturer in medicine, anatomy and surgery and he was often elected *Protomedico* and *Priore* of the *Collegio di Medicina* (official positions of responsibility for the public health services).

In 1766, he obtained the position of professor of anatomy at the Istituto delle Scienze, also assuming the post of *Custode e Ostensore delle Statue anatomiche* (Director and Demonstrator of the Institute's anatomical models), a position that before him had been occupied by wax modeller Ercole Lelli who died that same year. By virtue of this function, he was obliged to prepare a course in practical anatomy whose participants were not only students of medicine and surgery but also painters and sculptors of the city's Fine Arts Academy, known as Clementina, thus developing a profitable cooperation between the scientists and artists of the city. The lessons therefore concentrated on the description of the external structure and the function of muscles and bones; towards this end, he used the wax collections housed in the room of which he was the director (Simili, 1999: 33–63).

Dynamic anatomy, physiology, chemistry and medical practice also contributed a strong pragmatism in the choice of therapy, intended as the most functional response to a living organism, taken in its entirety and studied in relation to its natural environment.

In 1782, at the death of Galli, Galvani left the chair of anatomy at the Istituto delle Scienze to Carlo Mondini and opted for the chair in obstetrics, continuing, however, to hold the university chair in anatomy. He thus also began to directly use the *Suppellex Obstetricia*, the obstetric museum conceived by Galli who, from 1757, had held the first chair in obstetrics in Italy; Galvani was "Custode e Ostensore," of this museum, and of the wax models of the anatomical rooms.

Galvani was not spared the honour of the terrible annual Carnival public anatomic dissections of bones and corpses that were actually spectacular shows also for the general public. From these public exhibitions that he held four times in 1768, 1772, 1780 and 1786, Galvani learnt a lot because in the preparation of these events he successfully managed to combine anatomy and electricity.

Galvani had withdrawn from university and academic life a few months before his death, having refused to swear loyalty to the constitution of the Cisalpine Republic, according to the Napoleonic law of 26 *Ventose*. The strenuous efforts of his friend Spallanzani and nephew Giovanni Aldini (Malagola, 1879) to persuade him had failed, although the latter did manage to have him readmitted to the university as Emeritus Professor. But too late!

Galvani, the Bioscientist

It is no coincidence that the subject of many of the numerous dissertations on anatomy that he presented to the Accademia delle Scienze was the study of the structure, composition and function of the organs of specific animals. This is because, for Galvani, the study of animal and comparative anatomy was essential in order to understand human anatomy (Galassi & Giardina, 1956, pp. 464–464; Medici, 1857, pp. 365–366; Ruggeri, 1999, pp. 80–81).

Galvani's approach was of the functional–anatomical type: the study of structures was closely related to their functions.

The purpose was to be able to compare the information obtained by these studies with those available on man; studies that were not limited to a passive observation of corpses, but instead required an active intervention, as well as the application of knowledge acquired in the other spheres of natural philosophy. Confirmation of Galvani's innovative anatomical–physiological orientation is contained in the manuscripts of his anatomy lessons in the Archives of the Accademia delle Scienze (Galvani (AASB, MSG, Cart. IV, Plico I)).

Among these are the public dissections that traditionally ended with handling the bones. As we have seen, Galvani had been interested in this since 1762, enriching the anatomical part of his studies not only with reflections from physics and chemistry but primarily with observations on the functional meaning of bones against the background of the general morphology. In this respect, he compared the changeable vital course of bones with the life cycle of trees and plants; both seemed to obey harmonious and teleological criteria (Simili, 2001, p. 33).

As far as animal electricity was concerned, although he had already spoken of bioelectric phenomena as the cessation of the nervous fluid conductor of electricity in the 1780 lesson, it was at the public anatomy of 20 January 1786 that Galvani explicitly dealt with those "active principles," which as he himself remarked, had been described in part by botanists and physicists, but never comprehensively by physiologists and doctors (Galvani, 1762, 1998, pp. 5–14). Starting off from various references to physics and chemistry derived from the recent discoveries of Priestley and Cavendish, and according to "reason and experience," Galvani deduced in one great leap that there was a great quantity of electric fluid present in the bones, which would be distributed through them, and therefore throughout the body, like a "torrent." The development of the bones was produced by a metamorphosis linked to a mixture or proportion of the main chemical elements (fixed air, acid air, the inflammable or phlogistic principle considered also as hydrogen and oil earth), which varied according to the stage of development and the process of ageing. In the light of their chemical combination, and hence the prevalence or deficit of one of these "active principles," the more common bone diseases such as softening, rickets, caries and scurvy were analysed, while migraine, lumbago, rheumatism and odontalgia in which the nervous component prevails depended on the electric "energy" of the nervous fluid. Indeed, Galvani believed that diseases could be traced almost exclusively to modifications in the animal electric conductivity, without thereby denying electromagnetic forces of the atmosphere which depended not so much on the seasons as on the sharp atmospheric variations determined by unstable meteorological conditions (Simili, 2001, pp. 33–34).

Three further pieces of information on Galvani's investigative route can be drawn from this lesson (Galvani (AASB, MSG, Cart. IV, Plico I, 3, pp. 2–3)): the importance of chemistry, the meaning of electric force as a source of energy, anatomy as the study of life principles through examining the fluid and solid "parts," as well as through connecting these "parts" to the entire human organism:

They who, from this chair and with great wisdom, have dealt with the human body were accustomed to studying and examining with great diligence not only the naked body concealed in the form of a corpse, as it lies on this table covered with sad and gloomy veils, but also the active body, erect and, in a word, alive.

Indeed, anatomy is taught here not only so that the young may be instructed in the simple, naked knowledge of the parts of a human corpse, but so that once they are acquainted with these parts and their location, connection and structure, they may be more easily introduced to knowledge and a certain cognition of those found in the living man; hence they can better treat the suffering health of men with greater confidence and find remedies . . . (Galvani (AASB, MSG, Cart. IV, Plico I, 3, pp. 2–3)).

These words bring to mind the ones pronounced in 1777 on the occasion of the purchase of the collection of anatomical waxes of husband and wife Giovanni Manzolini and Anna Morandi by the Istituto delle Scienze that Galvani celebrated with his fine oration *De Manzoliniana suppellectili* (Galvani, 1777, 1998).

The tie with Anna Morandi (Focaccia (Forthcoming)), wax modeller of international fame,

becomes yet more explicit when the titles of Galvani's aforementioned Anatomy Lessons are examined. In addition to the transcriptions of his inaugural lessons on the occasion of public anatomy lessons, we find a series of lessons, *De Organis generationis in Viro, De Larynge et Pharinge, De Ore, Lingua et Naso, De Aure, De Oculis*, that bear no date but are presumably of 1768 (Galvani (AASB, MSG, Cart. IV, Plico I)). It is interesting to note that the last lesson is entitled *De cerebro*, a lesson in which Galvani highlighted the central connecting role of the brain and spinal marrow and, unlike his later pronouncements, explicitly spoke of the soul as its prime agent.

From the brain, the cerebellum and the spinal marrow depart all of the nerves [. . .] Let's turn our attention, however, to the extraordinary fluid secreted and elaborated in the brain. The noble function of this fluid is that of being, so to speak, the prime agent of the soul. From here, commanded by the soul, and increasing muscular contraction, it sets the parts and limbs into motion; it takes the sensations from the sensors to the brain so that some trace remains there in such a way that the soul, *which plausibly resides in the brain* [our italic], perceives the images of the presence of objects, as well as of those that it summons even in their absence. In the first action lies the imagination, in the other, memory: we have already dealt with how from the abundance of this fluid and the command of the soul imparted to the muscles, muscular movement is performed. [. . .] If one should therefore make conjectures, it is perhaps legitimate to presume that some sort of friction occurs on the ordered extremity of the sensor nerves, thereby provoking the greater strength of the electric fluid that accumulates at the extremity of the nerves, a fluid that concerns the sensor and brain; thus by the law of the electric fluid and a small change in it, we have a prompt and immediate irruption in the brain (Galvani (AASB, MSG, Cart. IV, Plico I)).

In the oration, he also underlined how Morandi's work presented innovative aspects in relation to the role of anatomy in empirical science, and with reference to the neurological circuit between brain and sense organs: a circuit that was at the basis of his theory of animal electricity (Focaccia, 2004, pp. 322–324).

Galvani exalted the importance of wax modelling for the study of anatomy in rendering the image of a living body and its organs and sheer elements, which, when ripped from corpses on the anatomy table, and "necessarily felt with the hands, lose their fluids (that no longer supply them), dry in contact with the air, wrinkle due to natural elasticity and, when they are laid on tables, flatten and are furthermore stretched by forceps and deformed when securing them with needles" (Galvani, 1777, 1998, p. 49).

The speech also expressed a special sensitivity for the intellectual capabilities of women; he indeed emphasised the greatness of Anna Morandi's work in her being a woman and, giving proof of surprising modernity, he wrote: ". . . what would you say if I declared that much more is added to this subject from the fact that it is the work of a woman? Would I perhaps not say the truth? It is certainly not rare that the Arts and Sciences are cultivated by men who seem predestined to this by nature. If it is a woman, though, to take interest in them and with utmost skill handle them, enlarge and I might even say take them to the extreme; a woman who seems born for wool and loom, isn't this fact so truly rare that it attracts the mind and eyes of everyone?" (Galvani, 1777, 1998, pp. 54–55).

We find the same sensitivity expressed in the *Elogio della moglie Lucia Galeazzi Galvani*, where the author celebrates, among other things, the "wonderfully acute intellect, a particular strength of mind superior to her age, and a special maturity of judgement" to the point that "she was very expert in orthography, so that her husband, a physician, anatomist and public professor, did not hesitate to entrust what he had written but would not re-read her correction, quite certain that no error of this type in his writings would have escaped his very cultured wife" (Galvani, 1790, 1937, p. 13). Lucia shared with her husband the laboratory and care of the sick, so much as to exhort Galvani himself "industriously and with particularly gentle means" to rush to their aid (Galvani, 1790, 1937, p. 23).

Moreover, in the course of his career, the "utility" of the medical art had gradually become more pressing for Galvani, as his clinical tasks were joined by the responsibility of childbirths and surgical interventions on new mothers in hospitals, often with serious risks for them and the newborn infants. Galvani indeed still worked in the city hospitals, in addition to performing his role as professor and researcher. On replacing Galli as surgeon at Sant'Orsola hospital, he was able to deepen his knowledge of the human body with its dysfunctions and possible cures; this aspect of his training was fundamental for his successive physiological research, as the surgical practice was naturally tied to manipulating the body and actively intervening on the organism.

Galvani's concept of obstetrics doubtlessly falls within his general vision of the medical art and illustrates the profound connections between anatomy, physiology and surgery. While on one hand, it in fact returned directly to the Bolognese experimental tradition, on the other, it used clinical and surgical experience acquired in practising medicine in hospitals under the guidance of Galli and Gaetano Tacconi.

In obstetric research, Galvani used a microscope to render a minute description of the embryo or of the placenta; in teaching he used Galli's *Suppellex Obstetricia* (described in part in his notes), continuously referring to the many clinical and functional aspects that concern the performance of the obstetric art.

Testimony to Galvani's obstetric thought remains in his autograph notes used as outlines for the lessons that, as of 1782, he held at the Istituto delle Scienze. These outlines refer to the years 1782, 1785 and 1791 (Galvani (AASB, MSG, Cart. IV, Plico IV); Galvani, 1965).

Reading these original and well constructed lessons, which were open not only to future physicians and obstetric surgeons but also to midwives (by way of example, see the lesson specifically entitled, *Degli aiuti e differenze che deve recare la mammana alla donna partoriente (About help and differences that the midwife must render to the parturient)*), it emerges how, resuming his studies on animal electricity and the consequent attention devoted to the brain – nerves – muscles circuit, Galvani invoked the primary role of the nervous system both in the functioning of the human organism and also in the context of childbirth, a context till then neglected and little known (Focaccia, 2006).

In one of the latter, Galvani gave lengthy considerations to the nature of uterine contractions and on the force of muscular contraction, distinguishing two different types of characteristic forces, on the one hand contractility and elasticity and on the other irritability. The forces of contraction depend on an extrinsic principle like nervous force, or else on an intrinsic principle, like irritability. Of these two, only the extrinsic principle, in Galvani's opinion, had been demonstrated to exist, and hence the forces of contraction depended exclusively on nervous force. As proof of this, he recalled his recent experiments in which he provoked the contraction by pricking the nerve. Of course, by stimulating the nerve a contraction could be obtained, but in this case it was more difficult to establish the cause, given the complexity of the organ and its intimate connections with the nerve. Still concerning the muscles, he then recalled their motory function, and their being equipped with nerves, as well as tendons and vessels, "for nutrition and not for sensation" (Galvani (AASB, MSG, Cart. IV, Plico IV); Giardina, 1965, pp. 14–16).

In lesson XXII, where he dealt with the phenomenology of childbirth, Galvani came out against the interpretation of Levret, Baudelocque and Stein, then current, that childbirth was a mechanical operation determined exclusively by the muscular forces of the uterus and the abdomen, arguing for the existence of an additional, decisive factor: that of the nerve (Galvani (AASB, MSG, Cart. IV, Plico IV); Giardina, 1965, pp. 126–128). Indeed, if pregnancy is considered, as it was then, to be the product of an equilibrium between forces, what set off the uterine force at the moment of childbirth was the result of an upset equilibrium produced both by the contractions of the muscular fibres and by the distortion of the nervous fibres, as well as by the full-grown placenta. It was precisely during labour that the delicate circuit comprising muscles, nerves and brain came into play – powerfully – set in motion by the distortion of the nerve fibres, which in their turn, in so far as they were strained, produced the contraction, – "in fact there is evidence that they [the nervous fibres] are mainly responsible for provoking the contractions, beginning (labour) with their pains (Galvani (AASB, MSG, Cart. IV, Plico IV); Giardina, 1965, p. 128)."

Galvani and Bioelectricity

September 1786 is usually taken as the crucial time that convinced Galvani of the existence of a specific animal electricity, even if the organisation of the experiments on Christmas Day 1780, and those that followed up to 1782, already pointed the way to later developments. Galvani argued that muscles contain electricity, comparing them to Leyden bottles, whose external surfaces are charged with negative electricity and whose internal surfaces are charged with positive electricity, according to a metaphor drawn from physics. This metaphor had already been employed by Walsh in his work on electric fish (Piccolino, 2003). The nerve is the conductor of this bottle and together with the blood vessels provides the muscles with electricity. Galvani's interpretation of the phenomenon was that the organism was not just an electroscope that indicated an action within a particular circuit, but *the originator* of the observed reaction.

The question whether animal irritability, and more specifically nerves, were conductors of a kind of "nervous fluid," which was analogous or equal to the electric fluid, was a long-standing issue when Galvani approached it in 1780.

In a famous treatise on the effects of viper poison (1781), Felice Fontana, a former collaborator with Leopoldo Caldani at Bologna, summed up the debate, and called attention to recently completed anatomical work on the torpedo and other electric fish. He expressed the belief that the nervous fluid was most likely a form of electricity; yet, he also invited colleagues "to assure by certain experiments, whether there is really an electrical principle in the contracting muscles; we must determine the laws that this fluid observes in the human body" (Fontana, 1781). Hence, in Galvani's era, there was great animation around the unresolved question of electric fluid, nervous fluid and muscular irritability.

Several years had passed since the famous dispute involving the younger and not so young scientists of the University, which had quickly given rise to two opposing sides: for and against Hallerian irritability. It is obvious that Galvani followed the various stages of the controversy, because it took place mainly in the years just before and after he graduated. Indeed, he had had to present his graduating thesis to Paolo Battista Balbi, who with Laghi and Fabri could be numbered among the enemies of Haller. However, he must have known the Hallerian, Leopoldo Caldani, who taught anatomy at Bologna, and who in the public anatomy lesson of 1760 presented the new theses on irritability, to the great scandal of the more conservative doctors. And he must have read the three weighty volumes *Sulla insensività ed irritabilità halleriana* published in 1757–1759, edited by G. B. Fabri, who gathered together, without distinction, the various interventions for or against Hallerian irritability (Fabri, 1757–1759).

The debate on irritability at Bologna had a significant effect, without a doubt weighing on Galvani and his future investigations, in which among other things he made use of Caldani's method of preparing frogs for experiments.

On 9 April 1772, the year in which he was elected President of the Accademia dell'Istituto delle Scienze, whose secretary was the physicist Sebastiano Canterzani, his assiduous adviser in the course of various experiments and professor of his nephew Aldini, Galvani produced a paper for his colleagues of the Istituto called *Sull'irritabilità halleriana* (*On Hallerian Irritability*), followed in 1773 by *Sul moto muscolare nelle rane* (*On the Muscular Movement of Frogs*) and in 1774 another on the *Azione dell'oppio nei nervi delle rane* (*The Action of Opium on the Nerves of Frogs*). None of these papers were ever published in the *Commentarii* of the Istituto.

Among the questions left unresolved, one above all was decisive. This was the lack of further detailed investigation into the nature of irritability, although Haller had generally characterised it as a force intrinsic to the muscles and separate from sensibility. It being internal was a problem. Galvani, as can be seen from both published and unpublished writings prior to the publication of *De Viribus*, kept in touch with Fontana's experiments, repeating in his turn those concerning the use of opium as a stimulant on the frogs, and following very carefully the investigations on viper's poison. He certainly learnt a lot from Fontana's *traité* on this subject, in addition to the *Ricerche filosofiche sopra la fisica animale* of 1775 (Fontana, 1775, 1781; Knoefel, 1984; Zanobio, 1959, pp. 307–320), in which a good deal of space was given to the anti-Haller positions of the Scottish doctor Robert Whytt, who believed the seat of energy and nervous power was the brain/mind, on which the entire functioning of the living organ depended. Whytt's pupil, William Cullen, also proceeded in this direction, and the long and frequent reviews of his works in the issues of the Bolognese *Memorie Enciclopediche* cannot have escaped Galvani's notice. In 1772, the year of Galvani's dissertation on irritability, Caldani's *Institutiones pathologicae* had appeared (Caldani, 1772, 1773, 1787), in which the concept of irritability was analysed and applied to various parts of the body.

Once Galvani had discarded sensibility, which was just a hindrance to an enquiry into the nature of irritability, he developed a special idea of irritability as an inherent force that *was* present in the muscles as a force capable of provoking both contractions and movements, but was not thereby independent of the nerves, and had its seat only in the gluten (Piccolino, 1997, pp. 443–448). The point is that Galvani, faithful to Malpighi's view of the anatomy of movement and to Newton's idea of the attraction of particles, believed that an organism was a dynamic system of communication. It was not a coincidence that animal electricity, often defined in terms of irritability (above all in the *Trattato sull'arco conduttore*) (Aldini, 1792, pp. 221–237), corresponded to a state of lack of equilibrium, ready to move in response to external stimuli or external influences. What was problematic, however, was managing to pick up the mechanism of functional communication relative to this process, whether this meant sending out electric signals, or stimuli to the cellular membranes, as Fontana had pointed out.

On the other hand, if as his experiments proved, the contractions seemed to be caused not by an extrinsic electric influence but instead by some internal force of the animal, how was it that Galvani did not follow the direction opened up by irritability? This, after all, was understood as the reaction to an external stimulus on the basis of an internal organisation of the organism, independently of the specific nature of the external influence. Why, also, could he not be persuaded, on the basis of the fact, that the electric stimulus excited the contractions but did not cause them, so that there existed an electric force in the animals themselves, generated by a communicative circuit existing between the exciting forces and the muscular fibre?

By mid-1780 Galvani had set up his own laboratory to conduct experiments on the electrical stimulation of frog preparations. He intended to study the power of irritability of muscles and nerves, hoping to prove the strong analogy between the electrical and the nervous fluid. A series of observations, followed by patient, repeated experimentation, convinced him that there was indeed an electrical fluid active in the muscles. Muscles, as we said before, were, for Galvani, biological Leyden jars that collected the electrical fluid produced by the brain, and conveyed to the periphery by the nerves. Galvani was struck by the facts that the contractions occurred at the discharge of the machine, when the nerves of the preparation were touched with a knife and grounded through the body of the experimenter. His investigations continued with experiments on the effect of electrical discharges in the atmosphere on frog preparations attached to the iron railings of his balcony.

The further unexpected result Galvani achieved was the production of contractions indoors, when the nerves and the muscles were touched with the ends of an arc made of two metals, and without any possibility of chance electrostatic induction. He further varied the metals used to build the arc, and rightly noted that the intensity of the contractions depended on the kind of metals used. Galvani triumphantly announced to the world that he was now prepared to offer various, complex and repeatable experiments showing the presence of electricity in frog preparations. He felt he had provided concrete answers to the question Fontana had posed. The sensation was enormous, worldwide.

In the last part of the *De Viribus* Galvani claimed

Now inasmuch as we have already shown that electric fluid is carried through the nerves of the muscles, it must therefore be transmitted through all of the nerves. Furthermore all these nerves must draw it from a single common source, namely the cerebrum. Otherwise [. . .] they do not seem to be adapted to activating and secreting one and the same fluid. We believe, therefore, that the electric fluid is produced by the activity of the cerebrum, that it is extracted in all probability from the blood, and that it enters the nerves and circulates within them in the event that they are hollow and empty, or, as seems more likely, they are carriers for a very fine lymph or other similarly subtle fluid which is secreted from the cortical substance of the brain, as many believe. If this be the case, perhaps at last the nature of animal spirits, which has been hidden and vainly sought after for so long, will be brought to light with clarity. But however this may be, I think no one in the future will have doubts concerning their electrical nature in view of our experiments (Galvani, 1791, 1998, pp. 108–110).

This, however, meant extending these characteristics to living species in general, so that "our observations may be made use of, provided that we can transfer this information of ours from animals and particularly from warm-blooded animals to man (Galvani, 1791, 1998, p. 110)." And Galvani added,

as if to correct the almost materialistic form the theory seemed inexorably to be taking on:

In the case of voluntary motion, it is possible that the mind is able, with its extraordinary power, residing either outside of the brain, or as is easier to believe, inside it, to stimulate whatever nerve it chooses (Galvani, 1791, 1998, p. 114).

In this respect, while Galvani never forgot the connection between electric fluid and the phenomenon of life itself, he had theorised at least ten years beforehand ("where finally that most noble electric fluid to which the movements, the sensations, the circulation of the blood, even life itself seemed to be entrusted" – these are his words at the Carnival anatomy lesson of 1780) – he also did not forget to emphasize those interesting aspects of his theory that could be applied to his therapeutic practice. In the present case his theory could be applied to those diseases such as, for example, paralysis, epilepsy and rheumatism, in which a neurological circuit was at work.

Yet, in the course of six years, Alessandro Volta was able to convince colleagues in Italy, France and England that Galvani had not discovered animal electricity, but had unknowingly observed the action of the electrical currents produced by dissimilar metals forming a circuit through the medium of moist materials, the frog preparations of his experiments.

Galvani repeated his experiments, and added new ones. He also started a series of anatomical investigations on torpedoes. He devised, but gave little publicity to, an experiment showing that frog crural nerves, applied to wounds in the muscles of the limbs of the animal, provoked contractions, every care having been taken to prevent electrically charging the preparation through induction. In other words, he proved the existence of contractions even when a purely biological arc was created.

It is hardly surprising that hardly anyone took any notice, for Galvani chose to publish these extremely important findings in an anonymous pamphlet, *Trattato dell'arco conduttore*, to which was added an anonymous *Supplemento* (1794) (Galvani, 1794, 1998, pp. 153–297)! A year later he went as far as Rimini, and the nearby Senigallia, to carry out experiments on electric torpedoes, whose findings he set down in a little book, the *Taccuino*, not published until 1937 (Galvani, 1937, pp. 177–198, 1998), that considerably help us to understand his evolution of his thought from *De Viribus* to the *Trattato*, and thence to the *Supplemento* and the *Memorie*. These experiments were published in 1797 in a more refined and finished version with the title *Memorie sull' elettricità animale* (Galvani, 1797, 1998, pp. 300–343) addressed to Lazzaro Spallanzani. (Di Pietro, 1976: 115–146). In these publications, Galvani went more deeply into the nature of animal electricity, correlating it with work on the torpedo, in which he had singled out two kinds of electricity, both animal and that which was characteristic of the electric ray itself. It is curious that the *Saggio sulla forza nervea* written in 1782 began precisely with reference to the electric organ of the torpedo.

From 1792 to 1800, Volta worked at hundreds of experiments on arcs made of two different metals, and in 1800 he able to show his own home-made torpedo, the electric pile. It is significant that one of the early names he used to describe his invention was "organ electrique artificiel," artificial electric organ.

On July 7, 1798 the indefatigable Galvani, who over the years had continued to provide papers for the *Accademia*, as well as to show his students the electricity intrinsic to animal organisms and carry out experiments on frogs using opium, wrote a note summing up his electro-physiological ideas:

Animal electricity seems to be located in two separate places. One is in the muscles, the other in the brain and spinal medulla. That located situated in the muscles operates with the law of the circuit, either alone, or mainly. I have not been able to demonstrate that when the nerve is irritated the contractions occur by means of this circuit by experiment, although it is likely. The same can be said of the contractions that arise, although to a lesser extent, when the muscle is cut. What is certain is that such contractions, well known from the animal electricity in the muscles, do not arise from a simple, very slight contact either of the muscles or the nerves, but require to be stimulated either by alteration or division of substance, or at least by some change in the disposition and union of the component parts of the nerve and the muscle, and such changes must be greater in the muscle than in the nerve.

It is not too far-fetched to conjecture, indeed it seems quite in conformity with experiment and observation, that the contractions arising through similar changes induced in the muscles also depend on changes made in the nerve, since the nerves are distributed throughout the entire muscle and also form an integral part of the same (Galvani (AASB, MSG, Cart. V, Plico IX)).

On the appearance of *De Viribus*, Felice Fontana had written to his friend Caldani in May 16th, 1792:

I have read Galvani's work on animal electricity, and repeated most of his experiments, which are true and surprising, and one can no longer doubt that a fluid in animals analogous, at least, to electric fluid exists, discovered through the muscular movement in Galvani's experiments. The subject is new, and may become of really great interest. However, I find great anomalies, difficult to reconcile with the known laws of ordinary electricity, which leads me to think it cannot possibly be the same principle. What I have found, which seems to me new and important, is that this principle perishes in animals when the sentient principle perishes, and reappears when the same sentient principle awakes; in short, I found it a true animal movement (Fontana, 1980, p. 332).

Paradoxically, Volta used Galvani's tools to start a new era in physics, opening a new electro-dynamic perspective composed of electricity, chemistry and electromagnetism, known as "Galvanism." It began to grow rapidly and make converts everywhere: the stages of this development, numbering famous names such as Fabroni, Oersted, Davy, Wollaston and Faraday, had the existence of galvanism substantially accepted as a specific sector of scientific enquiry.

The second wave of galvanism took other directions, assuming an important role in various forms including culture in general (Engelhardt von, 1992). Schelling's, Hegel's and also Schopenhauer's philosophy of nature contain significant reflections on galvanism. Similarly, Ritter discussed galvanism from the point of view of a philosophy of nature (Simili, 2005: 145–160). Even Oersted based his discovery of electromagnetism on dynamic reflections drawn from the philosophy of nature. True galvanism, as it was at the outset, connected to animal electricity, medical therapy, electro-neural-physiology, was taken up at different times, amidst controversies and difficulties, as we said, from the nineteenth to the twentieth century.

Conclusion

At the end of 1798 Galvani died, and thanks to the splendid success of the battery in 1800, and the mere ripple of interest aroused by Aldini's 1804 *Essay on Galvanism* (Aldini, 1804), animal electricity disappeared from the official scientific scene for many years. It took the "current of the frog" – as Leopoldo Nobili called it in 1811 – to bring up the problems of electrophysiology once more, without realising its true mechanism, however.

Carlo Matteucci and Émil Du Bois Reymond confronted these problems from a new scientific perspective. Their experiments, on the basis of their discoveries on muscular tissues, rehabilitated an unambiguous version of the theory of galvanism, affirming that the measured current was truly biological in origin, and confirming Galvani's supposition that an electrical disequilibrium exists in muscle tissue.

In his work on animal electricity of 1848–1849, *Untersuchungen über tierische Elekrizität* (Du Bois Reymond, 1848–1849), Émil Du Bois Reymond wrote a historical introduction that clarified Galvani's experiments. According to Du Bois Reymond, this introduction turned out to be important, first because the history of animal electricity had never been taken into consideration from any other perspective than that of Voltaic electricity, and second because precisely at that time, several unpublished writings of Galvani were published, which testified to the large number of successful experiments in the field of animal electricity well before the appearance of *De Viribus*. These manuscripts were those donated to the Accademia delle Scienze after Giovanni Aldini's death, as well as those from the family of Camillo Galvani, a lesser-known nephew who had been a doctor, and the most assiduous collaborator in Galvani's experiments, especially before *De Viribus*.

The manuscripts quoted by Du Bois Reymond were edited in 1841 by the Bolognese physicist Silvestro Gherardi; they included the *Memorie sull'elettricità animale al celebre abate Lazzaro Spallanzani* (Galvani, 1797, 1998) where Galvani defined animal electricity as muscular electricity common to all living beings.

After the experiments on the mechanisms of electrical nerve and muscle excitability of other representatives of the German School (von Helmholtz, Bernstein and Hermann), the modern phase of electrophysiological studies in excitable cells began in Cambridge around 1934, thanks to Keith Lucas and Edgar Douglas Adrian.

Alan Hodgkin published the results of his first experiments in the "Journal of Physiology" in 1937, just when at the University of Bologna his predecessor Adrian, and Majorana and their illustrious colleagues recounted in detail Galvani's later developments in various branches of research.

Within his lifetime, Galvani the physician, surgeon and physicist progressed from his early experiments to explore a magical domain, creating a new kind of electricity essentially linked to life, animal electricity. Others found the keys to it, but without his prophetic insights they would have had to engage a far longer struggle.

References

Adrian, E.D. (1937). The electro-Physiology of the sense organs. In: *Celebrazione del secondo centenario della nascita di Luigi Galvani (Bologna, 18-21 ottobre 1937)* (pp. 79-85). Bologna: Tip. Luigi Parma.

Aldini, G. (1792) *De animale electricae theoriae ortu atque incrementis*. Bononiae.

Aldini G. (1804) *Essai théorique et expérimental sur le galvanisme*. Paris: Fournier et fils.

Angelini A. (Ed.) (1993) *Anatomie Accademiche* vol. 3. Bologna: Il Mulino.

Bernardi, W. (1985) *Le metafisiche dell'embrione: scienze della vita e filosofia da Malpighi a Spallanzani, 1672-1793*. Firenze: Olschki.

Bernardi, W. (1992) *I fluidi della vita: alle origini della controversia sull'elettricità animale*. Firenze: Olschki.

Bohr, N. (1937) Biologia e fisica. In: *Celebrazione del secondo centenario della nascita di Luigi Galvani (Bologna, 18-21 ottobre 1937)*. Bologna: Tip. Luigi Parma.

Bresadola, M. (1998) Medicine and Science in the life of Luigi Galvani (1737-1798). *Brain Research bulletin* 46, 5: 367-380.

Caldani, L. (1772) *Institutiones pathologicae*. Patavini.

Caldani, L. (1773) *Institutiones physiologicae*. Patavini.

Caldani, L. (1787) *Institutiones anatomicae*. Venetiis.

Di Pietro, P. (Ed.) (1976) Carteggio di Lazzaro Spallanzani con Luigi Galvani e Giovanni Aldini. *Atti e Memorie dell'Accademia nazionale delle Scienze, Lettere e Arti di Modena* 6th ser. 18: 115-146.

Du Bois-Reymond, É. (1849) *Untersuchungen über tierische Elektrizität*. Berlin: Reimer.

Engelhardt von, D. (1992) Filosofia e pratica del Galvanismo tra Settecento e Ottocento. In: Verra V. (Ed.) *Il problema del vivente tra Settecento e Ottocento*. Roma, Istituto della Enciclopedia italiana.

Fabri, G. B. (Ed.) (1757-1759) *Sulla insensitività ed irritabilità halleriana. Opuscoli di vari autori*. Bologna: per Girolamo Corciolani.

Focaccia, M. (2004) Le cere anatomiche di Anna Morandi Manzolini. In Olmi, G. (Ed.) *Rappresentare il corpo. Arte e anatomia da Leonardo all'Illuminismo* (pp. 322-324). Bologna: Bologna University Press.

Focaccia, M. (2006) *Una nuova pratica sperimentale: arte, anatomia e ostetricia a Bologna nel XVIII secolo.* PhD Thesis: University of Bologna.

Focaccia, M. (forthcoming) *Anna Morandi Manzolini. Una donna fra arte e scienza.* Firenze: Olschki.

Fontana, F. (1775) *Ricerche filosofiche sopra la fisica animale.* Firenze Repr.: (1996) Barsanti, G. (Ed.) Firenze.

Fontana, F. (1781) *Traité sur le venin de la vipère, sur les poisons americains, sur le laurier-cerise et sur quelques autres poisons vegetaux.* Firenze.

Fontana, F. (1980) *Epistolario. Carteggio con Marc'Antonio Caldani, 1758-1794.* Mazzolini, R. and Ongaro, G. (Eds.). Trento: Società di studi trentini di Scienze Storiche.

Gallassi, A. and Giardina, B. (1956) L'opera medica di Luigi Galvani. *Studi e memorie per la storia dell'Università di Bologna* n.s. 1: 461-478.

Galvani, L. AASB (Archivio dell'Antica Accademia delle Scienze), MSG (Manoscritti Galvani), Cart. IV, Plico I, Lezioni di anatomia.

Galvani, L. AASB, MSG, Cart. IV, Plico IV, Scheletri di lezioni di Ostetricia.

Galvani, L. AASB, MSG, Cart. V, Plico IX.

Galvani, L. AASB, MSG, Cart. V, Plico XII, fasc. 6.

Galvani, L. (1762) *De ossibus. Theses physico-medico-chirurgicae.* Bononiae, S. Thomas Aquinatis. Repr.: (1966) Pantaleoni M. (Ed.) De ossibus. Lectiones quattuor. Bologna, Compositori. Repr.: (1998) Bologna, Arnaldo Forni editore.

Galvani, L. (1777) De Manzoliniana suppellectili. Repr.: (1841) Gherardi S. (Ed.) *Opere edite ed inedite di Luigi Galvani.* Bologna, Tipografia di Emidio dall'Olmo. pp. 41-60. Repr.: (1998) Bologna, Arnaldo Forni editore.

Galvani, L. (1791) De viribus electricitatis in motu musculari Commentarius. *De Bononiensi Scientiarum et Artium Instituto atque Academia Commentarii* 7: 363-418. Repr.: (1841) Gherardi S. (Ed.) *Opere edite ed inedite di Luigi Galvani.* Bologna, Tipografia di Emidio dall'Olmo. pp. 61-135. Repr.: (1998) Bologna, Arnaldo Forni editore.

Galvani, L. (1794) *Dell'uso e dell'attività dell'arco conduttore nelle contrazioni de' muscoli.* Bologna, S. Tommaso d'Aquino. Repr.: (1841) Gherardi S. (Ed.) *Opere edite ed inedite di Luigi Galvani* (pp. 155-278). Bologna, Tipografia di Emidio dall'Olmo. Repr.: (1998) Bologna, Arnaldo Forni editore.

Galvani, L. (1794) *Supplemento al Trattato dell'arco conduttore.* Bologna, S. Tommaso d'Aquino. Repr.: (1841) Gherardi S. (Ed.) Opere edite ed inedite di Luigi Galvani. Bologna, Tipografia di Emidio dall'Olmo. pp. 279-299. Repr.: (1998) Bologna, Arnaldo Forni editore.

Galvani, L. (1797) *Memorie sull'elettricità animale al celebre abate Lazzaro Spallanzani.* Bologna, Sassi. Repr.: (1841) Gherardi S. (Ed.) Opere edite ed inedite di Luigi Galvani (pp. 300-438). Bologna, Tipografia di Emidio dall'Olmo. Repr.: (1998) Bologna, Arnaldo Forni editore.

Galvani, L. (1937) Taccuino delle esperienze sulla torpedine fatte a Senigaglia ed a Rimini nel 1795 (pp. 177-198). Bologna, Zanichelli. Repr.: (1998) *Il "taccuino" di Luigi Galvani.* Fabriano, Arti grafiche "Gentile".

Galvani, L. (1888) *Orazione, letta nel 25 novembre 1782, per la laurea del nipote Giovanni Aldini.* Bologna, Monti.

Galvani, L. (1790/1937) *Elogio della moglie Lucia Galeazzi Galvani.* Bologna, Azzoguidi.

Galvani, L. (1965) Giardina L. (Ed.) *Lezioni inedite di Ostetricia di Luigi Galvani.* Bologna, Clueb.

Gherardi S. (Ed.) (1841) *Opere edite ed inedite di Luigi Galvani.* Bologna, Tipografia di Emidio dall'Olmo. Repr.: (1998) Bologna, Arnaldo Forni editore.

Heilbron, J. L. (1999) Galvani, Volta and the uses of centennials. In: Bresadola M. and Pancaldi G. (Eds.) *Luigi Galvani. International workshop* (pp. 17-32). Bologna, University of Bologna.

Knoefel, P. K. (1984) *Felice Fontana: life and works.* Trento, Società di studi trentini di scienze storiche.

Majorana Q. (1937) Commemorazione di Luigi Galvani. In: *Celebrazione del secondo centenario della nascita di Luigi Galvani (Bologna, 18-21 ottobre 1937)* (pp. 36-39). Bologna, Tip. Luigi Parma.

Malagola, C. (1879) *Luigi Galvani nell'Università, nell'Istituto e nell'Accademia delle Scienze di Bologna. Documenti.* Bologna, Gaetano Romagnoli editore.

Medici, M. (1857) *Compendio storico della scuola anatomica di Bologna dal rinascimento delle scienze e delle lettere a tutto il secolo XVIII.* Bologna, Tipografia della Volpe e del Sassi.

Piccolino, M. (1997) Luigi Galvani and animal electricity: two centuries after the foundations of electrophysiology. *Trends in Neurosciences* 20: 443-448.

Piccolino, M. (1998) Animal electricity and the birth of electrophysiology: The legacy of Luigi Galvani. *Brain Research bulletin* 46, 5: 381-407.

Piccolino, M., (2003) *The taming of the ray. Electric Fish Research in the Enlightenment, from John Walsh to Alessandro Volta.* Firenze: Olschki.

Righi, A. (1893) Su alcune disposizioni sperimentali per la dimostrazione e lo studio delle ondulazioni elettriche di Hertz. *Atti della R. Accademia dei Lincei. Rendiconti della classe di scienze fisiche, matematiche e naturali* 2: 333-337.

Ruggeri F. (1999) Galvani medico. In: *Discorsi e scritti in onore di Luigi Galvani. Nel bicentenario della morte, 1798-1998* (pp. 77-84). Bologna: Accademia delle Scienze dell'Istituto di Bologna, Arnaldo Forni Editore, 1999.

Simili, R. (1999) Luigi Galvani. In: Bresadola, M. and Pancaldi, G. (Eds.) *Luigi Galvani. International workshop* (pp. 33-63). Bologna, University of Bologna.

Simili, R. (2001) Luigi Galvani. Animal electricity and medical therapy. In: Beretta, M. and Grandin, K. (Eds) *A Galvanized Network. Italian-Swedish scientific relations from Galvani to Nobel* (pp. 13–46). Stockholm: The Royal Swedish Academy of Sciences.

Simili, R. II (2001) Luigi Galvani, médecin et savant. *La Revue* 33 : 16-25.

Simili, R. (2005) Erasmus Darwin and Luigi Galvani. Two Special Doctors. In: Smith, Ch. and Arnott, R. (Eds) *The Genius of Erasmus Darwin* (pp. 145-160). Aldershot: Ashgate.

Süsskind Ch. (1964) Observations of electromagnetic wave radiation before Hertz. *Isis* LV/179.

Tega, W. (1984) Mens agitat molem - L'Accademia delle Scienze di Bologna (1711-1804). In: Cremante R. and Tega W. (Eds.) *Scienza e letteratura italiana nella cultura italiana del Settecento* (pp. 65-108). Bologna: Il Mulino.

Tega, W. (Ed.) (1986, 1987) *Anatomie accademiche* vols. 1-2. Bologna: Il Mulino.

Zanobio, B. (1959) Le osservazioni microscopiche di Felice Fontana sulla struttura dei nervi. *Physis* 1: 307-20.

Section D
Brain and Behaviour

Introduction

From earliest times, medical writers have recognised that the brain is the seat of thought, sensation and willed behaviour. Hippocrates in the fifth-century BC was quite clear about this in his well-known tract *On the Sacred Disease*. With the famous exception of Aristotle, who was not a physician although the son of a physician, most subsequent writers have agreed. But exactly how the mushy 'quagmire', as Samuel Johnson described it, within our skulls supports the processes of thought remained for most in the eighteenth century an enigma. This section reviews the work of four eighteenth-century thinkers who concerned themselves with this problem.

First Nicholas Wade provides an account of the somewhat obscure Scottish physician, William Porterfield (c. 1696–1771). Although, as Wade points out, Porterfield was not a great experimentalist; he made important theoretical contributions to both ophthalmology and the study of phantom limbs. In the first case, he was most concerned with physiological optics, discussing not only the way in which light was focused by the crystalline lens, but also the way the eye moves in a series of jerky motions that we now call saccades. Porterfield also concerned himself with the visual pathways, estimating the magnitude of the nerve endings in the retina with remarkable accuracy and theorising on the way in which images from the two retinae could be fused in the 'sensorium' into a unity. In his analysis of saccadic movement, Porterfield concluded that we do not 'see' the image on the retina (which is continually moving) but instead the whole scene clearly and at rest. The brain is therefore something more than a simple camera. This thought also appears in his analysis of phantom limbs. Indeed, Porterfield, who had had a leg amputated, was the first physician to write about phantom limbs from personal experience. Unlike Horatio Nelson, who believed that the phantom experience of his fingers digging into the palm of his lost hand was evidence for the existence of an immortal soul, Porterfield contended that the experiences arose from the brain's interpretation of messages from the stump. In many ways, Porterfield anticipated the neuroscience of later years, and Wade helps to bring him out of an undeserved obscurity.

In the second chapter, Robert Glassman and Hugh Buckingham discuss the theoretical constructs of Englishman David Hartley (1705–1757), who was also interested in brain and behaviour. Hartley's major work, *Observations on Man*, was published in 1749, and in it he combined his theory of intra-neuronal vibrations with the tenets of association psychology. He based his vibration theory on certain speculations in Newton's *Principia* and *Opticks*, and his associationistic ideas from the British empiricist tradition, especially John Locke. Glassman and Buckingham review Hartley's work and explain how it made a break with earlier 'fluidist' theories of nerve physiology. They also seek to show that Hartley's ideas foreshadow the thought-world (zeitgeist) of much later times. Waves and vibrations still figure large in the scientific understanding of both the microcosm (man) and the macrocosm (the world).

Next, Harry Whitaker and Yves Turgeon examine the thoughts of Charles Bonnet (1720–1793), the Swiss naturalist and later, because of failing eyesight, theoretical "neuroscientist." In some ways, Bonnet's neuroscientific ideas resemble those of David Hartley. Both contended that the physical correlate of mind was to be found in the activity of the brain, and both believed that this activity was to be traced to the nerve fibres of which it was constituted. This is a very different view from that held by Descartes and his followers in the previous century. Both Bonnet and Hartley lived in a post-Newtonian world and, in d'Alembert's phrase, are attempting to create 'an experimental physics of the mind'. Bonnet, like Hartley, conceived that nerve fibres form a complex web within the brain, and that the mosaic of activity within this web underlies the operations of the mind. Like Hartley, also, he seems to anticipate some of the concepts normally attributed to a later age: localisation of function, specific nerve energies, a 'molecular' basis for memory, etc. Whitaker and Turgeon conclude their account of Bonnet's cognitive science with a discussion of what is now known as 'Charles Bonnet syndrome'. Bonnet observed this derangement in his long-lived grandfather who, in old age, suffered from conscious hallucinations. He knew he was hallucinating but was unable to escape the experience.

Finally, Ulf Norrsell shows us that Emanuel Swedenborg (1688–1772) made perceptive though little appreciated contributions to the brain sciences. Most people associate Swedenborg with mystical theology and remember him in the context of the Swedenborgian (New Jerusalem) Church. In fact, his theological writings and mys-

tical experiences were features of the second part of his remarkable life. In the first part, he devoted himself to many more earthly scientific studies, especially metallurgy, biology and medicine, hoping to become another Newton. In the 1740s, he planned a zoological work of some seventeen volumes (*Regnum animale*), in which the relation of soul and body would be elucidated. Norrsell describes these early engineering and scientific endeavours and then focuses on what we would now call his neurophysiology. We learn that Swedenborg preferred to synthesise, as objectively as possible, the observations of others skilled in medicine. Because of his philosophical/theological conviction that the natural (physical) world and the divine (spiritual) world are related by 'correspondences', he concluded that the mind must somehow be related to the brain's structure. Thus, by looking at patient records, he suggested that certain cerebral areas regulate the voluntary movements of different parts (of the opposite side) of the body. He also had interesting ideas about habitual behaviours; notions that seem to foreshadow those of Herbert Spencer a century later. Norrsell discusses Swedenborg's insights in the light of subsequent investigations of localisation of brain function, especially those of Fritsch and Hitzig on the motor cortex in 1871. He concludes, and we agree, that Swedenborg's physiological ideas, perhaps because of his theology, have been unjustly neglected.

The Editors

11
The Vision of William Porterfield

Nicholas J. Wade

Introduction

In eighteenth-century Britain, research on vision was conducted in the context of either optics or medicine, and both were influenced by philosophy. These threads were woven together by William Porterfield (ca. 1696–1771) in his essays on eye movements and in his treatise on the eye and vision. The scene for investigating vision was set by Isaac Newton (1642–1727) in the first decade of the century with his *Opticks* (Newton, 1704). The Newtonian mould was loosened by Thomas Young (1773–1829) in the last decade with his initial observations on vision (Young, 1793). Newton and Young adopted contrasting theories of light; Newton's (1704) theory was based on its corpuscular properties whereas Young (1800, 1802) provided further evidence (mainly from studies of interference) for its action as a wave. Despite the controversies in physical optics, their studies of visual optics had much in common (Wade, 1998). They examined the image forming properties of the eye similarly and their analyses of errors of refraction were in accord.

Newton made many astute comments about vision and his optics were extended further in the visual domain by Jean Théophile Desaguliers (1683–1744), Robert Smith (1689–1768) and Joseph Harris (1702–1764). The medical dimension was represented by William Cheselden (1688–1752), John Taylor (1708–1772) and Erasmus Darwin (1731–1802). James Jurin (1684–1750), Porterfield, William Charles Wells (1757–1817) and Young combined optics and medicine with a flavouring of philosophy. Certain visual problems were under examination throughout the century, whereas others were addressed seriously for the first time during that period. Colour vision, visual direction, eye movements, accommodation and binocular vision belong to the first category and visual vertigo to the second. Porterfield assessed all of these areas of vision, as well as many other aspects of the senses.

Relatively little is known about Porterfield's life despite the fact that he was Librarian and Secretary of the Royal College of Physicians of Edinburgh and later its President. Much more is known about his contributions to visual neuroscience. Between holding these posts he wrote two long articles on eye movements in the *Edinburgh Medical Essays and Observations* of 1737 and 1738. The first essay was addressed to what he called the external motions of the eye; he described the scanning and binocular eye movements. The second was directed to its internal motions; he coined the term "accommodation", examined it in an aphakic (lensless) individual, and invented the optometer. However, the scope of the essays was broader than motions of the eye, since much more space was devoted to perception than to anatomy.

Twenty years later he published his two-volume *Treatise on the Eye, the Manner and Phænomena of Vision*. Volume 1 contains accounts of the gross anatomy of the eye and its attendant structures, the properties of light and image formation, theories of accommodation, and his experiments with the optometer. Volume 2 is on vision: it commences with further reflections on accommodation and progresses to myopia and presbyopia, variations of pupil size, and eye movements; it ends with the phenomena of vision – binocular single vision, colour, size, distance, shape and motion perception. The *Treatise* presented a survey of the then contemporary knowledge of vision and the eye, and placed them in comparative and historical contexts. Between writing the essays and the *Treatise*, Porterfield had a leg amputated and provided an account of his phantom limb experiences and gave them a theoretical interpretation (Porterfield, 1759a). This was the first self-report of the phenomenon by a physician.

William Porterfield

It might be expected that someone who had held an esteemed post, like President of the Royal College of Physicians of Edinburgh, would be celebrated in word and image. Such is not the case for Porterfield – even his birth date is not accurately recorded, nor is there a known portrait of him. Thus, despite his prominence in Scottish medicine, very little is known about his life (see Chance, 1936; Comrie, 1932; Hirschberg, 1911; James, 1937). He does not appear in many of the standard biographical dictionaries, although there is an entry for him in the *Dictionary of Eighteenth Century British Philosophers* (Price & Yolton, 1999). What little is known of his early life was presented by Bower:

Dr. Porterfield was a native of the shire of Ayr, descended of a reputable family, and was a man of considerable private fortune. . . . It is probable that he was educated at the University of Glasgow, where the mathematics especially were much cultivated. He very early made proficiency in this fascinating study; and has employed it in all the works of which he has acknowledged himself to be the author. (1817, pp. 201–202)

Porterfield studied medicine under Herman Boerhaave (1668–1738) at Leyden and was awarded his medical degree from Rheims in 1717. He was elected a Fellow of the Royal College of Physicians of Edinburgh in 1721, was its Librarian and Secretary from 1722 to 1725 and President from 1748 to 1752. In 1723, the members of the College "signed a recommendation in favour of Dr. William Porterfield, to teach the Institutes and Practice of Medicine . . . to such of the inhabitants

of the Good Toun of Edinburgh, as have sons who are to follow Medicine" (Ritchie, 1899, p. 295). He was appointed Professor of Medicine by the town council of Edinburgh in 1724. He held the post for 2 years although there is no record confirming that he actually delivered any lectures:

It is a singular circumstance, that no documents are known to exist, by which it can be proved whether Dr. Porterfield ever delivered a course of lectures or not in the university of Edinburgh. For my own part, I am inclined to believe that he never did. The lectures would necessarily require a considerable time in composing; and it is well known, in this part of the country, that he was a man of a peculiar temper. I do not blame him for unsteadiness, or for shrinking from the task which he had proposed to perform. There was no member of the college of physicians better qualified than he was. I rather conjecture that some of his colleagues had the intention of beginning courses of medical lectures about this very time, and that he voluntarily declined interfering with their plans. (Bower, 1817, p. 203)

The Edinburgh Medical School was founded by Drs. John Rutherford, Andrew Sinclair, Andrew Plummer and John Innes (see Comrie, 1932; Craig, 1976). Porterfield played a role in its foundation, but as with so many other aspects of his life, its precise nature remains uncertain:

In the Year 1724 they [Drs. Rutherford, Sinclair, Plummer, and Innes] obtained the use of the ground from the Town Council, and delivered lectures on the above and other branches of Physic with considerable success. On the 9th of February, 1726 having apparently made some arrangement with Dr PORTERFIELD, who had been appointed two years earlier to the medical chair but who probably did not perform any active duty, they were promoted from the situation of private teachers to that of Professors in the College, and in the subsequent October were united with their illustrious predecessor MONRO into a Medical Faculty, the first that ever existed in the University. From this period the commencement of the Medical School of Edinburgh, which has since enjoyed so uninterrupted a prosperity, seems properly to be dated. (*List of members, laws, and library-catalogue of the medical society of Edinburgh*, 1820, pp. vii–viii)

The reasons for Porterfield's "arrangement" remain obscure. Indeed, it has been remarked that "Porterfield was not a success as professor and in a way which has mystified historians quietly disappeared from the academic scene" (Craig, 1976, p. 392). In keeping with his life, the nature of his death and interment is mysterious; he died in Sancerre, France.

Publications

Porterfield was an active member of the Edinburgh Society for the Improvement of Medical Knowledge, founded in 1731, and he contributed two long articles on eye movements to their journal. In his initial essay (Porterfield, 1737), he described the external movements of the eyes by the actions of the straight and oblique muscles. This essay can be regarded as providing the impetus for more detailed examination of eye movements later in the century (see Tatler & Wade, 2003; Wade, 2000a; Wade & Tatler, 2005; Wade, Tatler, & Heller, 2003). In the second essay, Porterfield (1738) discussed the internal movements of the eye – principally those involved in accommodation. He described the first optometer, for measuring the near and far points of vision. Moreover, his studies of an aphakic man (lacking crystalline lenses) who was unable to accommodate established the lenticular basis for accommodation – a term he coined: "That our Eyes change their Conformation, and accommodate themselves to the various Distances of Objects, will be evident to every body, who but reflects on the Manner and most obvious Phænomena of Vision" (1738, p. 126).

Vision was vital to the neuroscience of the day, as estimates of the sizes of nerves derived from measurements of visual acuity. Neither nerve cells nor receptors were known about, and it was assumed that the sizes of nerves serving vision were defined by the minimum spatial resolution. That is, two points were thought to be seen as separate if the light from them fell on neighbouring nerves. Porterfield's (1737) estimate of this was one 7,200th part of an inch, or around 3.5 µm, which he gave as the estimate of the diameter of nerves.

The ideas presented in the two essays provided the foundation for his two-volume *Treatise* which was widely read and remained in use for many years. When Wells was commenting on an aspect of it, he remarked that "his Treatise is in every body's hands" (1792, p. 132). Now the *Treatise* is rare and the few volumes that still exist would seem to be seldom consulted. Nonetheless, it is important because it heralds the application of experiment and observation in the analysis of the spatial dimensions of vision. It should also be borne in mind that Porterfield was writing at a time when quack oculists were rampant, and ophthalmology as a distinct discipline had not been established in Britain (see James, 1933).

Optics and the Eye

Porterfield had a thorough knowledge of the optical properties of the eye, and he was familiar with the contemporary literature regarding its anatomy and function (Fig. 1, upper left). The overarching analogy that had been applied to the eye was that of the camera obscura with a lens (Wade & Finger, 2001). While the analogy assisted an understanding of image formation, it led to the assumption that there was a picture in the eye. Porterfield was aware of the problems that this could raise, and argued against it: "I have said, that according as the Pictures upon the *Retina* are perfect or imperfect, the Objects are seen perfectly or imperfectly: But we are not from thence to imagine, that the Mind sees or perceives any Pictures *in the Retina*, or that it judges of Objects from what it observes in these Pictures. This is a vulgar Error" (1759a, p. 361).

Understanding the optics of the eye both clarified the manner in which images were formed on the retina, and it introduced the problem of how the eye adjusted its focus to objects at different distances. When Hermann Helmholtz (1821–1894) looked back on his part in resolving this puzzle, he remarked: "The mechanism by which this is accomplished ... was one of the greatest riddles of the physiology of the eye since the time of Kepler ... No problem in optics has given rise to so many contradictory theories as this" (1873, p. 205).

The camera with a lens provided the principal source of potential solutions to this problem, and the most commonly adopted view was that the lens moved forwards and backwards in the eye itself. Porterfield provided not only the name for the process, but also some vital evidence regarding its basis. He established the involvement of the lens by examining vision in a person who had the

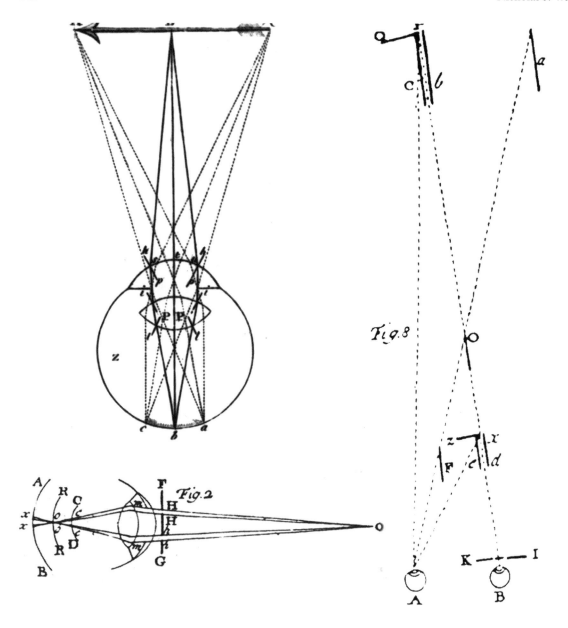

FIGURE 1. Upper left, Porterfield's (1759a) diagram of image formation in the eye. Lower left, his interpretation of Scheiner's experiment (from Porterfield, 1738). Right, Porterfield's (1738) optometer for determining the near and far points of distinct vision

crystalline lenses removed from both eyes in surgery for cataracts:

A Man having a Cataract in both Eyes, which intirely deprived him of Sight, committed himself to an Oculist, who finding them ripe, performed the Operation, and couched the Cataracts with all the Success could be desired; but after they were couched, he could not see

Objects distinctly, even at an ordinary Distance, without the Help of a very convex *Lens*; which is what every body has observed to be necessary to all those who have had a cataract couched: Neither is the Reason thereof difficult; for as a Cataract is not a Philm swimming in the *aqueous Humour*, as has been generally believed, till of late, but an Opacity in the Crystalline itself, and as the couching of a Cataract consists in introducing a Needle into the Eye, and turning down the opaque Humour below the Pupil, it is evident that the Crystalline cannot be displaced and turned down to the under part of the Eye, but the vitreous Humour must, in giving way to it, be pushed into its Place. ... But this is not all that happens after the Depression of the Cataract; for it was observed, that the same *Lens* was not equally useful for seeing all Objects distinctly, but that he was obliged, for seeing them distinctly, to use Glasses of different Degrees of convexity, still the more convex the nearer the Object. ... Seeing that nothing happens in the Eye, in couching the Cataract, but that the Crystalline is depress'd, it follows that the Change made in our Eyes, according to the Distance of Objects, must be attributed to this Humour (1738, pp. 182–184, 186).

Porterfield concluded that since elongation of the eye was still possible for such a person, the crystalline lens must be involved in accommodation, although he remained unsure of the manner in which it functioned. The involvement of the ciliary process was acknowledged, but its location and attachment to the lens led him to the conclusion that its action moved the lens forward and backward in the eye itself. Eight reasons for supporting his view were stated and Porterfield concluded: "As to the *efficient Cause* it has already been demonstrated, that this lies in the *Ligamentum ciliare*, which being muscular, does by its Contraction change the Situation of the *Crystalline*, according as Objects are nearer or further off" (1738, pp. 212–213). Porterfield did not entertain changes in the curvature of the lens because no muscle fibres could be found in the lens itself, and it was believed that all motion was a consequence of muscular contraction.

Young (1793, 1801), on the other hand, did consider that the lens could change in curvature and advanced a theory of accommodation based on this idea. He was able to confirm Porterfield's results with an aphakic person, and provide negative evidence against alternative theories. Young's speculations were given substance by Cramer (1853) and Helmholtz (1855), who described the elasticity of the lens and the varying tension that could be applied to it to achieve changes in its curvature.

Evidence for the changes in the optical power of the eye derived from an experiment Christoph Scheiner (1571–1650) had described: "Make a number of perforations with a small needle in a piece of pasteboard, not more distant from one another than the diameter of the pupil of the eye ... if it is held close to one eye, while the other is shut, as many images of a distant object will be seen as there are holes in the pasteboard ... at a certain distance, objects do not appear multiplied when they are viewed in this manner" (Scheiner, 1619, p. 38). This has become called "Scheiner's experiment", and an accurate interpretation of it was provided by Porterfield (1738; Fig. 1, lower left). He described it as follows:

Now it is certain, that if the Rays of Light that come from each Point of the Object are exactly united in a corresponding Point of the *Retina*, the Object will always appear single, though it be viewed through several small Holes, for the luminous Cones, OHH, O*hh* which have for their Apex or Top a Point of the Object, O, and for their Basis the little Holes in the Card, HH, *hh*, will also have all their opposite Tops *o*, *o* in one and the same Point *o*, of the *Retina*, RR, which must needs make the Object appear single: But if the Eye have not that Conformation, which is necessary to unite these Rays in a Point in the *Retina*, each of these Cones will be cut by the *Retina*, either before or after their Reunion; and therefore each Point of the Object shall, by its Rays, touch the *Retina* in as many Places as their are Holes in the Card, and consequently the Object will appear multiplied, according to the Number of Holes. (Porterfield, 1738, pp. 140–141)

Porterfield (1738) utilized the phenomenon in his optometer (Fig. 1, right). It consisted of a metal plate with two narrow and close vertical slits in it so that when it was held close to the eye their separation was less than the pupil diameter. The stimulus viewed was a vertical slit illuminated by a candle in a lamp-housing; he also tried a black line on white card, and a white line on a black card, but these were not so appropriate for distant viewing. The position of the vertical light line from the vertical slits could be changed so that the near and far points of vision could be measured; that is, the nearest and farthest positions at which the line could be seen as single; variations in pupil diameter were not controlled.

Young stated that "Dr. Porterfield has employed an experiment, first made by Scheiner, to the determination of the focal distance of the eye; and has described, under the name of an optometer, a very excellent instrument founded on the principle of the phenomenon" (1801, pp. 33–34). Young improved on Porterfield's optometer by incorporating a lens and a graduated scale so that corrective prescriptions for myopia and presbyopia could be made. Porterfield was able to use the optometer to establish the link between accommodation and convergence, which he believed was a consequence of learning: "This Change in our Eyes, whereby they are fitted for seeing distinctly at different Distances, does always follow a similar Motion in the Axes of Vision with which it has been connected by Use and Custom" (1738, p. 164).

Eye Movements and Motion Perception

Porterfield's first essay was concerned with the external motions of the eyes. These were considered to be important because the range of high spatial resolution was restricted to a small region around the visual axis. He realized that this fact conflicted with the impression that the whole scene is seen with similar clarity, and he sought to resolve this conflict by scanning the scene rapidly:

Now, though it is certain that only a very small Part of any Object can at once be clearly and distinctly seen, namely, that whose Image on the *Retina* is in the *Axis* of the Eye; and that the other Parts of the Object, which have their Images painted at some Distance from this same *Axis*, are but faintly and obscurely perceived, and yet we are seldom sensible of this Defect; and, in viewing any large Body, we are ready to imagine that we see at the same Time all its Parts equally distinct and clear: But this is a vulgar Error, and we are led into it from the quick and almost continual Motion of the Eye, whereby it is successively directed towards all the Parts of the Object in an Instant of Time. (1737, pp. 185–186)

The quick movements of the eye were not given a name, but they are now called saccades, nor were they open to measurement other than by observation of another's eyes or the awareness of one's own eye movements (see Wade & Tatler, 2005). Porterfield also related the pattern of eye movements to reading:

Thus in viewing any Word, such as MEDICINE, if the Eye be directed to the first Letter M, and keep itself fixed thereon for observing it accurately, the other Letters will not appear clear or distinct. ... Hence it is that to view any Object, and thence to receive the strongest and most

lively Impressions, it is always necessary we turn our Eyes directly towards it, that its Picture may fall precisely upon this most delicate and sensible Part of the Organ, which is naturally in the *Axis* of the Eye. (1737, pp. 184–185)

Wells (1792) was later to distinguish between the optical and visual axes of the eye: when a candle flame is aligned with its reflection in a mirror, the reflected corneal image is not in the centre of the cornea.

Porterfield was far-sighted in indicating the importance of eye movements in scanning the scene, but he was less successful at measuring them. Indeed, he did not measure them at all in the situations mentioned above, but assumed that they must be occurring on theoretical grounds. The views expressed in his essay of 1737 were repeated in the *Treatise* of 1759. However, he did add a new motion phenomenon for which he attempted to determine how his eyes moved. This was visual vertigo following body rotation. The visual motion of the world following body rotation was clearly described in antiquity (see Wade, 1998, 2000a, b), but Porterfield (1759b) added an eye movement dimension to it. In fact he denied the existence of eye movements following rotation because he was not aware of feeling his eyes moving. That is, the index of eye movement he used was the conscious experience of it:

If a Person turns swiftly round, without changing his Place, all Objects about will seem to move in a Circle to the contrary Way, and the Deception continues, not only when the Person himself moves round, but, which is more surprising, it also continues for some time after he stops moving, when the Eye, as well as the Objects, are at absolute Rest. (Porterfield, 1759b, p. 425)

Porterfield sought to accommodate this visual vertigo within his broad analysis of visual motion (Fig. 2, left). His analysis of visual motion was sophisticated for his time, and this can be appreciated with regard to the figure he presented:

Let *abcd* ... be the Eye moving round, according to the Order of the Letters *a*, *b*, *c*, *d*, and let A be an Object at rest seen by that Eye: When the Eye moves from *a* to *b*, if we attend to, and are sensible of the Velocity with which it moves, we will attribute the change of Situation of the Object intirely to the Motion of the Eye, and will therefore conceive the Object at rest, as it truly is: But if we are entirely unattentive to the Motion of the Eye, and

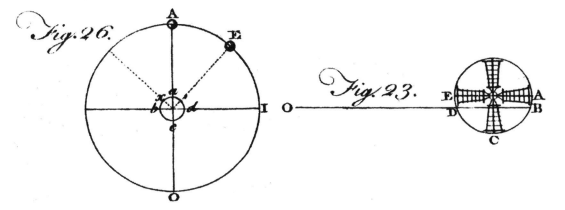

FIGURE 2. Left, Porterfield's interpretation of motion seen during body rotation, and right, his illustration of the ambiguous rotation of a windmill seen from afar (both from Porterfield, 1759b)

imagine that it is absolutely at rest, the Object A will then seem to move the contrary Way, with a Velocity proportional to its Distance, so as to describe an Arch similar to that described by the Eye in the same time, that is, in the Time the Eye moves round from *a* to *b*, the Object will seem to move from A to I. . . . But if, when the Eye is absolutely at rest, it be imagined to turn round from *b* to *x*, with the Velocity *bx*, all Objects, as A, will appear to move the same Way in the Order of the Letters AEI; and consequently will appear to continue their Motion for some Time after the Eye ceases to move. (Porterfield, 1759b, pp. 427–428)

Porterfield applied this analysis to vertigo and proposed that the post-rotational visual motion "proceeds from a Mistake we are in, with respect to the Eye; which, tho' it be absolutely at rest, we nevertheless conceive it as moving the contrary way to that in which it moved before: From which Mistake with respect to the Motion of the Eye, the Objects at rest will appear to move in the same way, which the Eye is imagined to move in, and consequently will seem to continue their Motion for some Time after the Eye is at rest" (1759b, p. 426). In modern terminology he was suggesting that it was the signals for eye movements, rather than the eye movements themselves, that generated the visual motion following body rotation. Porterfield's analysis, though incorrect, was the stimulus for others to examine vertigo and for Wells (1792) to provide the correct analysis of it (Wade, 2000b, 2003a). Wells formed an afterimage (which acted as a stabilized image) before rotation so that its apparent

motion could be compared to that of an unstabilized image when rotation ceased. The direction of the consequent slow separation of the two images and their rapid return (nystagmus) was dependent on the orientation of the head and the direction of body rotation. This represented the first attempt at recording the pattern of eye movements (Wade & Tatler, 2005).

Vertigo was the final phenomenon mentioned by Porterfield in his *Treatise*, but it was preceded by accounts of several others, of which visually induced motion was one. The apparent motion of the moon when clouds pass over it was then long known, but Porterfield provided an elegant generalization of the situation: "If two or more Objects move with the same Velocity, and a third remain at rest, the Moveables will appear fixed, and the Quiescent in Motion the contrary Way. Thus, Clouds moving very swiftly, their Parts seem to preserve their Situation, and the Moon to move the contrary Way" (1759b, p. 424). As with most other aspects of his vision, Porterfield was content to describe rather than to determine the characteristics of phenomena. This also applied to his report of the ambiguous rotation of a distantly observed windmill (Fig. 2, right):

Another *Phænomenon* of this Kind is, that of a Wind-Mill seen at a great Distance; for, by taking the nearest End of the Sail for the most remote, we sometimes Mistake the Course of its circular Motion. This *Phænomenon* may easily be accounted for: For, if a Spectator at O (Fig. 23.) situated nearly in the Plan of the Sails produced, imagines the furthest End A of the Sail AE to be the nearest, and the real Motion of the Sails be in the Order of the Letters ABCDE; when A is moved to B, and the Line BO is drawn cutting the Circle ABCDE in D; since he at first imagined the End A to be at E, he will not now conceive it at B, but at D; and so will imagine the Course of the Motion to be from E to D; which is contrary to the real Motion from A to B. The Uncertainty we sometimes find in the Course of the Motion of a Branch or Hoop of lighted Candles turned round at a Distance, is owing to the same Cause; and also that we mistake a convex for a concave Surface sometimes with the naked Eye, but more frequently in viewing Seals and Impressions with a convex Glass or a double Microscope. (Porterfield, 1759b, p. 384)

Porterfield had described, but had not illustrated, this phenomenon in his essay of 1737. It was also reported, and illustrated, by Smith (1738) in his *Opticks*, and it is frequently attributed to Smith rather than to Porterfield.

Porterfield's analysis of motion perception was insightful. He argued that perceived motion could not be equated with motion over the retina, and he appreciated the relativities involved in determining what was seen: "we can never know the absolute Magnitude or Celerity of their [Bodies] Motions, but only the Proportion that these Motions bear to one another" (1759b, p. 417). Despite his training in medicine, he did not advance physiological interpretations of his vision.

Visual Pathways and Binocular Vision

Porterfield's interests in eye movements were fuelled by his analysis of visual resolution. Since only a small part of an object can be seen in detail, with increasing obscurity towards the periphery, some means of linking the distinctly seen parts was required, and eye movements performed this function. The limits of visual resolution were taken as 1 min of arc, following the value determined by Robert Hooke (1635–1703; 1705). Porterfield calculated the dimension on the retina that would correspond to this small angle and took it as the dimensions of the nerve endings in the retina (Fig. 3, lower left):

And here by the way it may not be improper to observe, that this Experiment of *Dr. Hook*'s, serving to determine the *minimum visibile*, affords pretty certain Proof of the Magnitude of our nervous Fibres: For if *ao* (Fig. 12.) be the End of one single Fibre, the small Object IE, which is here supposed to be bright and luminous, will by means of its Picture on the *Retina ie*, move the whole Fibre, and the Appearance of the Object will be the same as if its Picture were extended over the whole End of the Fibre *ao*; and therefore if from the extreme Points of the Fibre *a* and *o*, the right Lines *ax*A, *ox*O are drawn thro' the Center of the Eye *x*, these Lines will be perpendicular to the *Retina* at the Points *a* and *o*, and consequently the small Object IE will be seen under the Angle O*x*A; which Angle being given, the Angle *oxa*, which is equal to it (both being Angles at the Vertex *x*) will also be known, from which the Diameter of the nervous Fibre *ao* may easily be found. Thus if the Angle O*x*A be one Minute, as *Dr. Hook* found it in most Eyes, though there were some that could see to the third of a Minute, the Angle *oxa* will also be one Minute, which is the 60 Part of a Degree, or the 21,600 Part of a Circle: Whence if the Eye be supposed to be one Inch Diameter, or thee Inches in Circumference, the Diameter of the Nervous Fibre *ao* will be the 21,600 Part of three Inches, or the 7200 Part of one Inch (pp. 250–251).

Thus, the dimensions of fibres in the optic nerve were considered to be similarly small. Although the assumptions upon which the indirect estimates were based were flawed, the values derived were reasonable approximations to direct measures obtained a century later and with modern microscopic values of receptor dimensions (see Wade, 2004, 2005).

Porterfield's diagram of the visual pathways (Fig. 3, upper) signified his concerns with perception as well as anatomy: he related the correspondences in the two retinas with binocular single vision. He was well aware of Newton's (1704) description of partial decussation, and reprinted Query XV of the *Opticks* in full. While concluding that "This is indeed the most beautiful and ingenious Explication of the Manner how an Object appears single from the Coalition of the Optick Nerves that ever appeared" (1737, p. 197). However, he rejected it because he believed it was based on speculation rather than anatomy, and he was more content to rely on the authority of anatomists (see Wade, 1987, 2000a). He wrote:

Suppose, as in Fig. 1. the Nerves composed of five Fibres, whose Extremities in the right Eye are A, B, C, D, E, and in the other Eye, *a*, *b*, *c*, *d*, *e*. The corresponding Fibres A*a*, B*b*, C*c*, D*d*, and E*e*, are supposed to meet in the *Sensorium* S, in the Points α, β, χ, δ, ε.

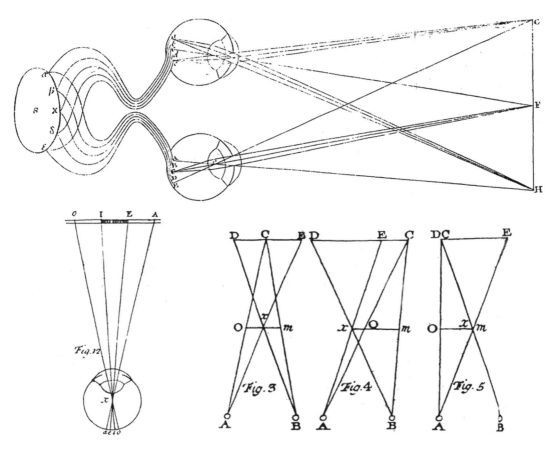

FIGURE 3. Lower left, Porterfield's (1737) diagram illustrating the relationship between the minimum visible and dimensions of retinal nerve fibres, described in the following words: Upper, Porterfield's (1737) diagram of the visual pathways. Lower right, Porterfield's (1738) diagrams demonstrating double vision

Hence if both Eyes are directed to F, its Image will fall on the *Retina* at the Optick *Axes*, and there strike the sympathizing Fibres C and *c*; which motion being propagated to the single Point of the *Sensorium* χ, must there make but one Species or Picture. In like manner the Eyes retaining the same Direction, the Image of the Point G will fall upon the right Side of both Eyes; and by striking the correspondent Fibres E and *e*, will, in the *Sensorium*, make but one Impression at ε, where these Fibres terminate; and the Image of the Point H, by striking the corresponding Fibres A and *a*, will, in the *Sensorium*, make but one Impression at α: And thus, though both Eyes receive the same Impressions from Objects, yet they are not seen double, because of these two Impressions or Images, one is only formed in the *Sensorium*. (Porterfield, 1737, pp. 201–202)

Binocular vision represents one of the oldest issues examined by students of vision. It was commented upon by Aristotle and it has been a source of constant interest and experiment since his time, not least because it could be related to the common condition of strabismus or squint. The aspects of both single and double binocular vision were assessed by Porterfield. Single vision with two eyes was related to stimulation of corresponding points on each retina, as is indicated in Fig. 3, lower right, and these in turn were related to the concept of visual direction (see Wade, 2000a). Double vision (diplopia) followed if non-corresponding points were stimulated. This could be demonstrated in normal vision by viewing a ruler extended outward from the nose. Double vision also occurred under other conditions, like restraining the motion of the eyes or as a consequence of inebriation:

When our Eyes are restrained from moving uniformly, all Objects are seen double. Neither is it to be doubted, but, when the same *Phænomenon* occurs in drunk or maniac Persons, it proceeds from the same Cause; the uniform Motion of our Eyes requiring an easy and regular Motion of the Spirits, which frequently is wanting in such Cases. . . . The same Way of Reasoning applied to Objects in all Manner of Situations, will shew that all of them appear double, when placed out of the Plan of the *Horopter*; all which is exactly agreeable to Experience: And this is also the Reason why a double Appearance will be seen when the End of a long Ruler is placed between the Eye-brows, and extended directly forward with its flat Sides respecting Right and Left; for, by directing the Eyes to a remote Object, the right Side of the Ruler seen by the right Eye, will appear on the left Hand, and the left Side on the right Hand. (Porterfield, 1737, pp. 194, 238)

As with the windmill phenomena mentioned in section "Eye Movements and Motion Perception", the ruler demonstration was both described and illustrated by Smith (1738), to whom it is usually attributed.

Double vision was experienced after the onset of strabismus, and Porterfield provided a lengthy description of the possible causes of this condition:

First, This Disease may proceed from Custom and Habit, while in the Eye itself or in its Muscles, nothing is preternatural or defective. . . . *Secondly*, The *Strabismus* may proceed from a Fault in the first Conformation, by which the most delicate and sensible Part of the *Retina* is removed from its natural Situation, which is directly opposite to the Pupil, and is placed a little to a side of the *Axis* of the Eye, which obliges them to turn the Eye away from the Object they would view, that its Picture may fall on this most sensible Part of the Organ. . . . *Thirdly*, This Disease may proceed from an oblique Position of the Crystalline. . . . *Fourthly*, This Disease may arise from an oblique Position of the *Cornea*. . . . *Fifthly*, This Want of uniformity in the Motions of our Eyes may arise from a Defect, or any great Weakness or Imperfection in the Sight of both, or either of the Eyes. . . . *Sixthly*, Another Cause from which the *Strabismus* may proceed, lyes in the Muscles that move the Eye. When any of those Muscles are too short or too long, too tense or too lax, or are seized with a *Spasm* or *Paralysis*, their Equilibrium will be destroyed, and the Eye will be turned towards, or from that Side where the Muscles are faulty. From all that has been said on this Head laid together and duly considered, we may clearly deduce this Inference: the double Appearance of Objects that happens when either of the Eyes is, from a *Spasm* or *Paralysis* of any of their

Muscles, or from any other Cause, restrained from following the Motions of the other, does not prove, that to see Objects single, it is absolutely requisite that both Eyes be directed to the same Object, and that this is one of the final Causes of their uniform Motion. (Porterfield, 1737, pp. 239–254)

Most analyses of strabismus proposed in the eighteenth century focused on one of these eight possible causes; Porterfield displayed a modern appreciation of the multiple reasons for the occurrence of the condition.

In the course of his analysis of binocular vision, Porterfield (1738) distinguished between viewing objects that are either farther from or nearer to the eyes than the point of bifixation. Thus, if the object is more distant, then covering one eye will result in the object appearing on the same side as the open eye, whereas the opposite occurs if the object is closer than the fixation distance. The situation was illustrated both in the essay of 1738 and in the *Treatise* (see Fig. 3, lower right). The condition of particular interest is his Fig. 5, where the distant object (C) is aligned with the fixation point for the right eye but not for the left. He wrote:

To illustrate this, see Figs. 3, 4 and 5. where A and B are the Eyes, x the Object, which is at a smaller Distance than the Point C, to which both Eyes are directed; it is evident, that while the Eyes continue to be directed to C, the Object x must be seen in two different Places, which, with respect to the *Horopter*, to which all Objects are referred, will be D and E; for being seen by the Right Eye B, in the Direction of the visual Line BxD, it must, at D, hide a Part of the *Horopter* DCE; and being seen by the left Eye A, in the Direction of the visual Line AxE, it must hide a Part of the *Horopter* at E, and therefore, with respect to the *Horopter* on which the Eyes are fixed at C, the Object x must appear to the Right Eye B, as at D, and to the Left Eye A as at E; and in covering either of the Eyes, the Appearance that is on the contrary Side will be made to vanish. In like manner, if the Eyes are directed to x, the Object C, which is further off the x, will be seen by the Right Eye B, in the Direction of the visual Line BmC; and by the Left Eye A, it will be seen in the Direction of the visual Line AoC; and therefore, with respect to the *Horopter* mxo, to which all Objects are referred, it must appear double, as at m and o; and in covering the Right Eye B, the Appearance that is on the Right Side towards m will vanish; and in covering the Left Eye A, the Appearance that is on the Left Side towards o will vanish; all which is exactly agreeable to Experience. (Porterfield, 1738, pp. 156–158)

The situation displayed in Porterfield's Fig. 5 was later studied by Peter Ludvig Panum (1820–1885, p. 1858) with the aid of a stereoscope. He presented two vertical lines to one eye and a single vertical line to the other and described the apparent depth. Panum does not seem to have been aware of Porterfield's early examination of similar conditions (without a stereoscope), as he is not cited in Panum's monograph, nor has it been cited by subsequent writers on binocular vision. The situation is now referred to as Panum's limiting case.

Phantom Limb

Porterfield's agenda was philosophical as well as physiological. He adopted a nativist view of vision and he argued vehemently against empiricist theories such as the one advanced by Berkeley (1709). For example, Porterfield disputed Berkeley's proposition that touch could guide vision in the perception of the third dimension of space: "for the tangible Ideas are as much present with the Mind as the visible Ideas, and, on that Account, must be equally incapable of introducing the Idea of any Thing external" (Porterfield, 1759b, p. 307). He used essentially the same arguments to contradict Locke's (1694) answer to a question posed by William Molyneux (1656–1698) regarding perception in someone who was born blind but had their vision restored. This has become known as Molyneux's Question, and it has stimulated considerable interest and speculation ever since (see Fine et al., 2003; Morgan, 1977; von Senden, 1960). Porterfield addressed the question in the second volume of his *Treatise*:

To this Question, both these profound Philosophers [Molyneux and Locke] pronounce in the negative … and yet, notwithstanding the great Deference I have for the Opinion of so able Judges, I cannot help thinking that they are mistaken; for, I have already demonstrated, that the Judgments we form of the Situation and Distance of visual Objects depend not on Custom and Experience, but on an original, connate and immutable Law, to which our Minds have been subjected from the Time they were first united to our Bodies; and therefore the blind Person, immediately upon receiving his Sight, must, by virtue of this Law, by his Eyes alone, without any Assistance from his other Senses, immediately judge of the Situation of all Parts of the Globe and Cube. (Porterfield, 1759b, pp. 414–415)

In the period between writing his essays and his *Treatise*, Porterfield had a leg amputated; in the first volume of the *Treatise* he used his experiences of a phantom limb to support the projective features of perception generally (Wade, 2000a, 2000b, 2003b). He was specifically attacking Berkeley's theory that a pictorial image existed on the retina, and that this was perceived by the mind. He described his own experiences of a phantom limb in this same general context:

It is therefore evident, that, did the Mind perceive Pictures in the Retina, it behoved to be there present: And for the same Reason, did it perceive in the other Organs of Sense, it behoved also to be present to all the Parts of the Body; because the Sense of Feeling is diffused thro' all the Body: Nay, in some Cases it behoved to be extended beyond the Body itself, as in the Case of Amputations, where the Person, after Loss of his Limb, has the same Perception of Pain, Itching, &c. as before, and feels them as if they were in some Part of his Limb, tho' it has long been amputated, and removed from the Place where the Mind places the Sensation. Having had this Misfortune myself, I can the better vouch the Truth of this Fact from my own Experience; for I sometimes still feel Pains and Itchings, as if in my Toes, Heel or Ancle, &c. tho' it be several Years since my Leg was taken off. Nay, these Itchings have sometimes been so strong and lively, that, in spite of all my Reason and Philosophy, I could scarce forbear attempting to scratch the Part, tho' I well knew there was nothing there in the Place where I felt the Itching. And however strange this may appear to some, it is nevertheless no way miraculous or extraordinary, but very agreeable to the usual Course and Tenor of Nature; for, tho' all our Sensations are Passions or Perceptions produced in the Mind itself, yet the Mind never considers them as such, but, by an irresistible Law of our Nature, it is always made to refer them to something external, and at a Distance from the Mind; for it always considers them as belonging either to the Object, the Organs, or both, but never as belonging to the Mind itself, in which they truely are; and therefore, when the nervous Fibres in the Stump are affected in the same Manner as they used to be by Objects acting on their Extremities in the Toes, Heel or Ancle, the same Notice or Information must be carried to the Mind, and the Mind must have the same Sensation, and form the same Judgment concerning it, viz, that it is at a Distance from it, as if in the Toes, Heel or Ancle, tho' these have long ago been taken off and removed from that Place where the Mind places the Sensation. If this should prove hard to be conceived, it may be illustrated by what happens in the

Sensation of Colours; for tho' the Colours we perceive are present with the Mind, and in the Sensorium, yet we judge them at a Distance from us, and in the Objects we look at; and it is not more difficult to conceive how Pain may be felt at a Distance from us, than how Colours are seen at a Distance from us. (Porterfield, 1759a, pp. 363–365)

Porterfield displayed considerable sophistication in the analysis of his phantom limb, by associating the projective features of the experience with other aspects of perception. He was well-versed in Newtonian colour theory, and cited Newton many times. Indeed, he gave a quotation from Newton's *Opticks* on the title page of volume 1 of his *Treatise*. The reference to colour relates to Newton's statement that the rays are not coloured, but that the experience of colour is subjective and projected externally to objects. Porterfield extended this subjectivity of sensation to phantom limbs, and incorporated the sensations into the body of perceptual theory.

The term "phantom limb" was coined by Silas Weir Mitchell (1829–1914) following his treatment of injuries sustained by soldiers during the American Civil War; these included the sensations that amputees experienced in their lost limbs (Mitchell, 1871). Reporting experiences from amputated parts has a much longer history, but it remains remarkably short considering the incidence of the condition. Medical interest started with Ambroise Paré (1510–1590, p. 1551) and continued to fascinate philosophers in the seventeenth century (see Finger & Hustwit, 2003; Price & Twombly, 1972, 1976). Porterfield's account was the first self-report by a physician (Wade & Finger, 2003). There were many subsequent reports of phantom limbs in the eighteenth century, but unlike Porterfield's, all had to rely on descriptions provided by the amputees. He did not regard the experiences of the lost limb as phantoms, but as a natural consequence of stimulating the brain in a manner similar to that which existed prior to amputation. He integrated the phantom limb experiences with a general theory of perception.

This position was generally accepted by physicians in the eighteenth century. For example, George Fordyce (1736–1802, p. 1771), in his text on medicine, related phantom sensations to the normal functioning of the nervous system. A similar sentiment,

voiced again with primary reference to the nerves and their pathways, was written in the next decade by John Hunter (1728–1793, p. 1786). In his book *Observations on certain Parts of the Animal Œconomy*, he described two cases of phantom sensations in the missing penis. Such examples were of particular significance as both Fordyce and Hunter considered that all senses responded to touch and pain, in addition to their specific sensations. In his papers (unpublished during his lifetime), Hunter (1861) expressed it thus: "Touch is probably the only sense that is cognizable by another sense besides the immediate sensation" (p. 7). That is, if touch alone was experienced as a phantom sensation, then it might reflect the central operation of common sensitivity. If the specific sensations associated with a particular body part could be experienced after amputation then that was stronger evidence for the localization of sensation in the brain.

It is clear that Porterfield's lost limb provided a stimulus for the new found desire to localize perceptual experience in the brain rather than in the sense organs themselves. In the following century, the phenomenon of phantom limbs was employed by both Charles Bell (1774–1842, p. 1811) and Johannes Müller (1801–1858, p. 1826) to support their notions of the specific actions of the sensory nerves (Wade, 2003b). Porterfield foresaw these developments in his analysis of vision:

Tho' there was a Picture in the *Retina* in that vulgar gross Sense that so many imagine, yet it is impossible that the Mind could perceive it there; because all the Sensations or Perceptions of the Mind are present within it and in the *Sensorium*: I appeal to every one's Experience, if he ever sees or observes any Pictures or any Thing else in the *Retina*. And to say we see, observe or perceive Pictures there, without being sensible or conscious of it is absurd and ridiculous. The Mind or sentient Principle does not at all perceive in the *Retina*, but in the *Sensorium* where it is present. . . . All Things perceived must therefore be present with the Mind and in the *Sensorium*, where the Mind resides; and that not only virtually but substantially. (1759a, pp. 362–363)

Thus, his whole approach to perception, including the experiences of missing limbs, was based upon his analysis of visual phenomena. Porterfield's vision extended far beyond the modality of sight.

Conclusion

Porterfield remains a shadowy figure in the history of neuroscience although his position in its evolution should be more securely grounded. He was an acute observer and theorist, but these were not matched by his experimental skills. Indeed, his contributions to vision were those of extending theoretically the experiments of his predecessors: he set the scene for others to explore phenomena that he had placed in a more secure theoretical light. In one area he was able to give a unique interpretation of perceptual experience – in the descriptions he provided of sensations apparently arising from his amputated leg. Rather than treating these as illusions, as others had, he incorporated them into the body of the extant neuroscience. This was achieved because of his prior analyses of visual phenomena.

References

Bell, C. (1811). *Idea of a new anatomy of the brain; submitted for the observations of his friends.* London: Published by the author; printed by Strahan and Preston. (Reprinted in Wade, N. J. (Ed.). (2000). *The emergence of neuroscience in the nineteenth century* (Vol. 1). London, Routledge: Thoemmes Press.)

Berkeley, G. (1709). *An essay towards a New Theory of Vision.* Dublin: Pepyat.

Bower, A. (1817). *The history of the University of Edinburgh* (Vol. 2). Edinburgh: Oliphant, Waugh and Innes.

Chance, B. (1936). William Porterfield, M.D. An almost forgotten opticophysiologist. *Archives of Ophthalmology, 16,* 197–207.

Comrie, J. D. (1932). *History of Scottish medicine* (2nd ed.). London: The Wellcome Historical Medical Museum.

Craig, W. S. (1976). *History of the Royal College of Physicians of Edinburgh.* Oxford: Blackwell Scientific Publications.

Cramer, A. (1853). *Het accommodatie-vermogen, physiologisch toegelicht.* Haarlem: Loosjes.

Fine, I., Wade, A. R., Brewer, A. A., May, M. G., Goodman, D. F., Boynton, G. M., et al. (2003). Long-term deprivation affects visual perception and cortex. *Nature Neuroscience, 6,* 909–910.

Finger, S., & Hustwit, M. P. (2003). Five early accounts of phantom limb in context: Paré, Descartes, Lomas, Bell and Mitchell. *Neurosurgery, 52,* 675–686.

Fordyce, G. (1771). *Elements of the practice of physic, in two parts* (3rd ed.). London: Johnson.

Helmholtz, H. (1855). Ueber die Accommodation des Auges. *Archiv für Ophthalmologie, 1,* 1–74.

Helmholtz, H. (1873). *Popular lectures on scientific subjects.* Trans. E. Atkinson. London: Longmans, Green.

Hirschberg, J. (1911). Geschichte der Augenheilkunde. In Graefe-Saemisch (Ed.), *Handbuch der gesamten Augenheilkunde* (Vol. 14). Leipzig: Engelmann.

Hooke, R. (1705). *The posthumous works of Robert Hooke, M.D.* London: Waller.

Hunter, J. (1786). *Observations on certain parts of the animal œconomy.* London: Published by the author.

Hunter, J. (1861). *Essays and observations on natural history* (Vol. 2). London: Voorst.

James, R. R. (1933). *Studies in the history of ophthalmology in England prior to the Year 1800.* Cambridge: Cambridge University Press.

James, R. R. (1937). William Porterfield, M.D. *British Journal of Ophthalmology, 21,* 472–477.

List of members, laws, and library-catalogue of the medical society of Edinburgh. (1820). Edinburgh: Aitken.

Locke, J. (1694). *An essay concerning humane understanding* (2nd ed.). London: Awnsham, Churchill, and Manship.

Mitchell, S. W. (1871). Phantom limbs. *Lippincott's Magazine, 8,* 563–569.

Morgan, M. J. (1977). *Molyneux's question. Vision, touch and the philosophy of perception.* Cambridge: Cambridge University Press.

Müller, J. (1826). *Zur vergleichenden Physiologie des Gesichtssinnes des Menschen und der Thiere, nebst einen Versuch über die Bewegung der Augen und über den menschlichen Blick.* Leipzig: Cnobloch.

Newton, I. (1704). *Opticks: Or, a treatise of the reflections, refractions, inflections and colours of light.* London: Smith and Walford.

Paré, A. (1551). *La méthode de traicter les playes faictes tant par hacquebutes que par fleches.* Paris: de Brie.

Porterfield, W. (1737). An essay concerning the motions of our eyes. Part 1. Of their external motions. *Edinburgh Medical Essays and Observations, 3,* 160–263.

Porterfield, W. (1738). An essay concerning the motions of our eyes. Part 2. Of their internal motions. *Edinburgh Medical Essays and Observations, 4,* 124–294.

Porterfield, W. (1759a). *A treatise on the eye, the manner and phænomena of vision* (Vol. 1). Edinburgh: Hamilton and Balfour.

Porterfield, W. (1759b). *A treatise on the eye, the manner and phænomena of vision* (Vol. 2). Edinburgh: Hamilton and Balfour.

Price, D. B., & Twombly, S. J. (1972). *The phantom limb: An 18th century Latin dissertation text and translation, with a medical-historical and linguistic commentary.* Washington, DC: Georgetown University Press.

Price, D. B., & Twombly, S. J. (1976). *The phantom limb phenomenon. A medical, folkloric, and historical study*. Washington, DC: Georgetown University Press.

Price, J. V., & Yolton, J. W. (Eds.). (1999). *The dictionary of eighteenth century British philosophers* (Vol. 2, pp. 706–707). Bristol: Thoemmes.

Ritchie, R. P. (1899). *The early days of the Royall Colledge of Phisitians, Edinburgh*. Edinburgh: Johnston.

Scheiner, C. (1619). *Oculus, hoc est Fundamentum Opticum*. Innsbruck: Agricola.

von Senden, M. (1960). *Space and sight*. Trans. P. Heath. London: Methuen.

Smith, R. (1738). *A compleat system of opticks in four books*. Cambridge: Published by the author.

Tatler, B. W., & Wade, N. J. (2003). On nystagmus, saccades, and fixations. *Perception, 32*, 167–184.

Wade, N. J. (1987). On the late invention of the stereoscope. *Perception, 16*, 785–818.

Wade, N. J. (1998). *A natural history of vision*. Cambridge MA: MIT Press.

Wade, N. J. (2000a). Porterfield and Wells on the motions of our eyes. *Perception, 29*, 221–239.

Wade, N. J. (2000b). William Charles Wells (1757–1817) and vestibular research before Purkinje and Flourens. *Journal of Vestibular Research, 10*, 127–137.

Wade, N. J. (2003a). *Destined for distinguished oblivion: The scientific vision of William Charles Wells (1757–1817)*. New York: Kluwer/Plenum.

Wade, N. J. (2003b). The legacy of phantom limbs. *Perception, 32*, 517–524.

Wade, N. J. (2004). Visual neuroscience before the neuron. *Perception, 33*, 869–889.

Wade, N. J. (2005). *Perception and illusion. Historical perspectives*. New York: Springer.

Wade, N. J., & Finger, S. (2001). The eye as an optical instrument. From *camera obscura* to Helmholtz's perspective. *Perception, 30*, 1157–1177.

Wade, N. J., & Finger, S. (2003). William Porterfield (ca. 1696–1771) and his phantom limb: An overlooked first self-report by a man of medicine. *Neurosurgery, 52*, 1196–1199.

Wade, N. J., & Tatler, B. W. (2005). *The moving tablet of the eye: The origins of modern eye movement research*. Oxford: Oxford University Press.

Wade, N. J., Tatler, B. W., & Heller, D. (2003). Dodge-ing the issue: Dodge, Javal, Hering, and the measurement of saccades in eye movement research. *Perception, 32*, 793–804.

Wells, W. C. (1792). *An essay upon single vision with two eyes: Together with experiments and observations on several other subjects in optics*. London: Cadell (Reprinted in Wade, 2003a).

Young, T. (1793). Observations on vision. *Philosophical Transactions of the Royal Society, 83*, 169–181.

Young, T. (1800). Outlines of experiments and enquiries respecting sound and light. *Philosophical Transactions of the Royal Society, 90*, 106–150.

Young T. (1801). On the mechanism of the eye. *Philosophical Transactions of the Royal Society, 91*, 23–88.

Young, T. (1802). On the theory of lights and colours. *Philosophical Transactions of the Royal Society, 92*, 12–48.

12
David Hartley's Neural Vibrations and Psychological Associations

Robert B. Glassman and Hugh W. Buckingham

David Hartley's Conjecture

In the mid-eighteenth century David Hartley published a treatise that combined ideas about the psychology of mental associations with conjectures drawn broadly from neuroanatomy, mechanics, optics and electricity. Recognizing that a complete mechanistic theory must consider not only causally related mental associations but also their physiological substrates, Hartley conceived of neural activity as vibrations, suggested earlier by Isaac Newton, Thomas Willis, Pierre Gassendi and others (Buckingham & Finger, 1997; Finger, 1994; Glassman & Spadafora, 1997; Robinson, 1995; Smith, 1987; Wallace, 2003). Hartley approached the mind/brain issue by bridging causal concepts, not localizing psychological functions except to imply that the input vibrations of distinct sensory modalities would likely terminate in different parts of the sensorium (Aubert & Whitaker, 1996).

Hartley's Life

David Hartley was born on 21 June 1705 in Halifax, Yorkshire. He was admitted to Jesus College, Cambridge, where he earned his BA in 1726 and MA in 1729. Although he had been educated for the Church, his conscience would not allow him to agree with certain ecclesiastical dogmas and he turned instead to medicine. He began practicing "the art of physick" in 1730 in Newark, although he never took a medical degree.

In the 1730s he wrote on the benefits of inoculation for small pox, praised Joanna Stephens' "cure" for "the stone," and, in a series of letters to the Reverend John Lister, commented on "... benevolence and the attainment of universal happiness through reason and scripture" (Allen, 1999, p. xix; Hartley, 1746). It was during the 1730s that he had become aware of the claims made by the Reverend Mr. Gay of Sidney Sussex College that it was possible to deduce all intellectual pleasures and pains from association principles (Young, 1990: 95–96). His initial aversion to certain church dogma lifted and by 1735 he claimed that he had rid himself of his early doubts as to the "truth" of religion.

At this time Hartley began writing his extension of Gay's thoughts on the principles of association, which would ultimately expand into his *magnum opus*, *Observations on Man (OM)* of 1749. Three years before *OM*, Hartley published a short "trial balloon" monograph, *Various Conjectures (VC)* in Latin. It comprised much of the previous correspondence that he had had with Lister, various unpublished papers on associationism and religion, and a section on Joanna Stephens' cure for stone (Strohl, 1963, p. 509). "Conjectures" was written in the Newtonian style with propositions, corollaries and scholia (a coda-like summary). This monograph thus focused on Hartley's science – not his religion. After the publication of *VC* and *OM*, Hartley started corresponding regularly with the experimental natural philopshers Stephen Hales and Joseph Priestley. A mere 8 years after the publication of *OM*, Hartley died at the age of 52 at

Bath on 28 August 1757, "… actively practicing medicine to the time of this death" (Allen, 1999, p. xxiii).

Hartley borrowed from Newton the idea of vibrations, as seen in *Optiks*, "… excited in the bottom of the Eye by the Rays of light are of a lasting nature, and [thus] are they not of a vibrating nature?" (pp. 347–348). What distinguished Hartley's vibrations was that he extended them to his psychology and his overall account of human nature, its moral codes, ethics and belief in a creator. This was no small feat, since Hartley incorporated the deep religiosity of a belief in God and that, in turn, allowed him a very useful level of causation that helped him steer around the unwanted charges of materialism.

Association psychology has a long history, going back at least to Aristotle (Warren, 1921). A masterful history of associations as memory traces in the work of Descartes through Hartley, continuing on to modern connectionist modeling, is found in Sutton (1998). Locke and Hobbes were major influences on Hartley's associationism, which he then linked to Newton's vibrations.

Observations on Man

Hartley's *OM* is divided into two parts, the first of which runs 512 pages, and the second, 455 pages. Part I, "Containing OBSERVATIONS on the frame of the human body and mind, and their mutual connexions and influences," is an extensive expansion of his Conjectures but followed its outline. Part II, "Containing OBSERVATIONS on the duty and expectations of mankind," is a discourse in the spirit of what the Rev. Gay had written 15 or 20 years earlier on the ontology of moral thought along lines laid down by Christianity. Allen (1999) has brought to light the fascinating connections between Hartley's stimulus–response psychology and stimulus–response theology. Hartley incorporated Newton's conception of vibrations as a neurophysiological basis for associative information processing. The vibratory medium was hypothesized to process associative linkings for motor-sensory information, as well as for the ultimate scaffolding of an architecture of moral thought. The seeds of a Pavlovian model of behavior were being sown in *OM*.

Hartley's Apologists and Detractors

Hartley had supporters and critics of all parts of his psychophysical model, in which associative information was processed through exceedingly small "subtle" vibratory motions (vibratiuncles), postulated to be the medium of nerve transmission. During the late-eighteenth and early-nineteenth centuries, his ideas were promulgated in abridgements of his *Observations on Man* by his long-term admirer, Joseph Priestley, who omitted from his editions of 1775 and 1790 the sections of the book that explored vibrations, on the grounds that this was technical, difficult and dubious (Hatch, 1975, pp. 548–550; Spadafora, 1990). Priestley thus lopped off the physical side of Hartley's model and popularized the sections on association psychology.

Later, Coleridge criticized Priestley for his non-material remake of Hartley, claiming that without half the castle, the building would collapse. However, Coleridge at the same time "smiled" with others at the puzzling notion of nerves vibrating (Smith, 1987, p. 132). Ultimately this admirer (and father of David Hartley Coleridge) became a strong critic of Hartley's model (Smith, 1987, p. 133).

Bower (1881) shared Priestley's skepticism over Hartley's medium as well as his associative model of the acquisition and use of moral dictates. Bower (1881, p. 248) wrote of Priestley's 1775 edition of *OM* that his re-release of it, "… has the advantage of omitting the vibration theory, and the theological speculations."

Herman Boerhaave also chided Hartley for proposing that animal nerves vibrate like plucked tense chords, since the nerves were too "soft, pulpy and flaccid" for such vibrations. What seemed lost on Boerhaave and some of Hartley's other critics was that he never really had the vibrations of musical strings in mind, as will be seen below.

Nevertheless, Hartley strongly influenced William Carpenter, James Mill, Alexander Bain and Benjamin Rush. As for Erasmus Darwin, one of the most respected physician–philosophers of the century, he referred to "the ingenious Dr. Hartley" (Smith, 1987, p. 135).

Most of these natural philosophers abided by Hartley's psychophysical stance, but concentrated on its associationism while adopting the reductionist article of faith that there must be somewhere, somehow, a physical and/or an immediate cause for the processing of psychological information. Hartley's coexistent science, religion and natural philosophy may have appeared as a logical contradiction, but there were and are areas of legitimate intellectual unity among these domains of inquiry (Barbour, 1997; Campbell, 1975; Glassman, 1996; Hefner, 2006).

Hartley's Vibrations

Hartley's physiological psychology of neural vibrations presupposes that a physical system is the substrate for mental activities. It was well known in the eighteenth century that the communicative properties of sound involved vibrations (Smith, 1987). For example, Newton (1723, 1995, 1730, 1979) had mathematically analyzed mechanical oscillations and described wavelike properties of light. Jean Phillipe Rameau's exquisitely detailed treatise on the mathematics of musical harmonic ratios had appeared in 1722 (Rameau, 1722, 1971). Thus, there was scientific promise in hypothesizing "vibrations" as a candidate for opaque neural communicative processes. Hartley's coupling of Aristotelian associationism and Newtonian vibration forged together two promisingly robust constructs. Each had a balance of specificity and openness that could be submitted to further scientific inquiry. Quoting Hartley:

And as a vibratory Motion is more suitable to the Nature of Sensation than any other Species of Motion, so does it seem also more suitable to the Powers of generating Ideas, and raising them by Association. However, these Powers are evident independently, as just now observed; so that the Doctrine of Association may be laid down as a certain Foundation, and a Clue to direct our future Inquiries, whatever becomes of that of vibrations. (*OM*, Pt.I, Prop. 11, p. 72)

The seventeenth- and eighteenth-century Enlightenment had witnessed an upsurge of scientific activity in mechanics, optics, electricity and chemistry. Hartley cited findings of Boerhaave and Hales, and was aware of his theory's speculative and metaphorical aspects. He understood that scientific progress requires reasoning from wholes to parts, as well as from parts to wholes, and his comments were often appropriately qualified, as he considered the possible relevance of oscillatory phenomena (his prudence and awareness of the

vicissitudes of work at the mental-physical juncture are examined by Spadafora, 1990). Hartley's model, in addition to suggesting a specific mechanism of neural communication, was also a sort of "binding hypothesis" relevant to the neural basis of cognition.

Vibrations and Waves

Waves involve repetitions and transitions, both essential aspects of communication and seemingly appropriate as a neurodynamic underlying psychological associations. Hartley had similar intuitions, while acknowledging his limited understanding of the work of Isaac Newton. At the same time, Hartley wisely warned that his "vibrations" should not simply be construed literally as vibrations of musical strings (*VC*, Prop. 4, p. 3).

Hartley skirted between the Scylla of the physical and the Charybdis of the mental by postulating distinct levels of causation (*OM*, Pt. 1). He postulated an "original" cause, by which he meant a kind of knowledge of the "rational" soul, cast at the level of "will" or "volition," much as those terms are used today. Here, Hartley placed the mental forces of reflection, reason, judgment, memory and imagination. Memory as a function of the volitional system was at a clear remove from the motor and sensory systems.

Another level of causation in Hartley's scheme was referred to as the "efficient" cause, otherwise labeled the "immediate" cause. If a soul was involved at this level, it was the "sensitive" soul and was subject to the workings of "immediate instruments," such as the white medullary substance of the cerebrum, the spinal marrow, and the nerves proceeding from them. At this causative level, the instruments could be directly examined in healthy organisms by the physiological techniques of stimulation (or irritations) and ablation. They could also be investigated by disease-causing pathologies, where "ideas may fail to cohere and where motions were disrupted" (*OM*, Pt. 1).

Closer to the periphery was another level of causation, also referred to as "efficient." Here Hartley holds that all sensory perception is derived from motion and corporal impressions. At this level, muscle contraction also causes movement – a physical causation through physical force. Hartley's preferred medium for this was perhaps electrical within the muscle (see below), but at the "immediate" level he incorporated vibrations, as suggested by Newton.

God remained Hartley's overriding, or ultimate, causal level. Hartley's theology is at work here and it far transcends his psychology. In fact, Hartley's *OM* is, according to Allen (1999), three quarters theology. Allen rebukes academic psychologists for concentrating on only a quarter of *OM* (Buckingham, 2001).

Evidence for Hartley's Irresoluteness on the Theory of Vibrations

A close reading of relevant passages in Hartley and Newton reveals that these authors often displayed caution in balancing observation, theory, and conjecture. They personified eighteenth-century philosophical sophistication and were well aware of what they could and could not directly observe. They knew the possibilities of their available empirical methods, while recognizing an important set of problems outside that range. Hartley was concerned that his logical reasoning was incomplete, and that any challenge to theism rested on uncertain ground. His opening sentences in *OM* are both forthright and circumspect:

Man consists of Two Parts, Body and Mind. The First is subjected to our Senses and Inquiries, in the same manner as the other Parts of the external material World. The Last is that Substance, Agent, Principle, &c. to which we refer the Sensations, Ideas, Pleasures, Pains, and Voluntary Motions.

I shall not be able to execute, with any Accuracy, what the Reader might expect of this kind, in respect of the Doctrines of Vibrations and Association, and their general Laws, on account of the great Intricacy, Extensiveness, and Novelty of the Subject. However, I will attempt a Sketch in the best manner I can, for the Service of future Inquirers. (*OM*, p. 6)

Buckingham and Finger (1997) argue that Hartley's balancing act between associationism and religious commitments was not the psychophysical parallelism inferred by Aubert and Whitaker (1996), Robinson (1995, p. 250), Smith (1987), and Young (1990, p. 96), but rather served as a convenient "tightrope walk," or yet another form of mind–brain obfuscation. Hartley's general cautiousness and

careful consideration of the scientific literature and the "art of physick" (Webb, 1989) contributed to the success of his theory.

The Observable and the Conjectural in Hartley's Model

Hartley expressed the hope that his intended science of the mind would have the power to "analyse all that vast Variety of complex Ideas … into their simple compounding Parts, i.e., into the simple Ideas of Sensation, of which they consist." His scientific sophistication in dealing with the mental, as with hypothesized underlying physical constructs, is evident in his discussion of Newton's controversial hypothesis that there is an "aether." Following an analogical lead-in, Hartley invoked two principles, the first a version of operationalization and the second a version of falsifiability, akin to the mathematical technique of assumption and contradiction. He wrote:

The Emission of odoriferous Particles, Light, magnetical and electrical Effluvia, may also be some Presumption in favour of the Existence of the Æther. … Lastly, Let us suppose the Existence of the Æther, with these its Properties, to be destitute of all direct Evidence, still, if it serves to explain and account for a great Variety of Phaenomena, it will have an indirect Evidence in its favour by this means. … The rule of False affords an obvious and strong Instance of the Possibility of being led, with Precision and Certainty, to a true Conclusion from a false Position; and it is of the very Essence of Algebra to proceed in the way of Supposition. (*OM*, pp. 15–16)

Although the aether hypothesis misled some subsequent inquirers, his underlying intuition that communication requires a medium was consistent with the inference that information somehow travels through nerve fibers. This idea has ancient roots (Clarke, 1968, p. 123).[1]

---

[1] Clarke's paper also describes the creative array of metaphorical conjectures about neural "porosity" in the seventeenth and eighteenth centuries, including a citation of Willis, who in his 1664 book may have been the first to refer to the microscopic structure of nerves, and who likened it to the macroscopic structure of sugar cane. Wallace (2003) as well alludes to the sugar cane analogy of Willis (also see, this volume, chapter by Ford, on microscopy).

The Communicating Medium of Nerves

Boerhaave, committed to the Galenic "fluidist" view of transmission through nerves (Brazier 1959; see Koehler, this volume), may have been disingenuous in assuming literalness and thereby criticizing neural vibration theory on the basis that nerves are too soft and flaccid by comparison with musical strings. As pointed out, Hartley carefully qualified his references to vibrations, as had Newton earlier, in a way that suggests the quality of metaphor. It was not until many years after its initial publication that Newton had augmented the *Principia* with the cautiously speculative "General Scholium" – in which he raised the possibility of neural vibrations and their possible electric character. Hartley, intrigued and puzzled (Webb, 1988), commented that

It seemed credible to Newton that a very subtle and elastic fluid, and hence very suitable for reception and communication of vibrations both lies hid in gross bodies and is diffused through the open spaces that are void of gross matter. ... I remain somewhat doubtful that I have sufficiently understood his views. (*VC*, p. 4)

Hartley followed this comment with a hope that more might turn up in Newton's posthumous writings. Hartley wrote of the relevant characteristics of nerve fibers, with clear caution and awareness of the partial nature of his concept of vibrations:

These vibrations are Motions backwards and forwards of the small Particles; of the same kind with the Oscillations of Pendulums; and the Tremblings of the Particles of sounding Bodies. They must be conceived to be exceedingly short and small, so as not to have the least Efficacy to disturb or move the whole Bodies of the Nerves or Brain. For that Nerves themselves should vibrate like musical Strings is highly absurd; nor was it ever asserted by Sir Isaac Newton, or any of those who have embraced his Notion of the Performance of Sensation and Motion, by means of Vibrations. (*OM*, pp. 11–12)

In this passage, Hartley generalizes the idea of vibrations across scales of time and space by reference to the gross behavior of pendulums and the finer oscillations of sound. He then extrapolates to the idea that there may be a still finer level of oscillation, one with a substrate and dynamics that are

as yet unclear. He also recognizes that oscillatory dynamics may have abstract properties that are common to different material substrates:

Vibrations descend along the motory nerves ... in some such manner as sound runs along the surfaces of rivers, or an electrical virtue along hempen strings ... (*OM*, Pt. 1, Prop. 11, p. 88)

Two historical observations, cited by Robinson (1995) and Brazier (1959, 1984), help to put Hartley's vibrations in perspective. The first concerns Petrus van Musschenbroek, a pupil of Boerhaave and inventor of the Leiden jar, who warned against metaphysics and against the dangers of argument by analogy (Brazier, 1959, pp. 9–10). The second concerns Croone's seventeenth-century criticism of Descartes' animal spirits. Croone reasoned that there is a liquid, or rather a "most subtle, active and highly volatile liquor of the nerves, in the same way as we speak of spirit of wine or salt ..." (Brazier, 1984, p. 60). Anticipating later developments in the theory of vibrations, Croone additionally commented:

For at the same time as the fibrils of the taut nerve are struck in the brain, immediately these droplets of liquor exude from all its branchlets ... this liquor creates an effervescence in the blood in an instant. And at the same instant blood flows through the artery ... like water from an opened pipe. (Brazier, 1984, p. 61)

Associationism and the Doctrine of Vibrations

As others have pointed out, the idea of association has roots in Aristotle's writing and has been passed down the centuries from Epicurus through Hobbes to Hartley and ultimately, via twentieth-century behaviorism, to modern day connectionism (Buckingham, 2002; Buckingham and Finger, 1997; Warren, 1921). Hartley writes:

Any sensations A, B, C, &c. by being associated with one another a sufficient Number of Times, get such a Power over the corresponding Ideas a, b, c, &c. that any one of the Sensations A, when impressed alone, shall be able to excite in the Mind b, c, and c. the Ideas of the rest.

Other Instances of the Power of Association may be taken from compound visible and audible Impressions. Thus the Sight of Part of a large Building suggests the Idea of the rest instantaneously; and the Sound of the

Words which begin a familiar Sentence, brings the remaining Part to our Memories in Order, the Association of the Parts being synchronous in the first Case, and successive in the last. (*OM*, pp. 65–66)

In addition,

Simple Ideas will run into complex ones, by means of Association. (*OM* Prop. 12, p. 73)

Scientific and Intuitive Prodromes of the Doctrine of Vibrations

We now focus more closely on what induced Hartley to utilize a conception of vibrations as the body's way of physically instantiating the centuries old conception of associative processes. As already noted, Hartley cited the General Scholium, at the end of Newton's *Principia*. In this often-quoted paragraph, Newton conjectured that there had to be something that had properties of electricity, vibrations, and particle-like elements, which would serve as a medium:

And now we might add something concerning a certain most subtle Spirit which pervades and lies hid in all gross bodies; by the force and action of which Spirit the particles of bodies mutually attract one another at near distances, and cohere, if contiguous; and electric bodies operate to greater distances, as well repelling as attracting the neighbouring corpuscles; and light is emitted, reflected, refracted, inflected, and heats bodies; and all sensation is excited, and the members of animal bodies move at the command of the will, namely by the vibrations of this Spirit, mutually propagated along the solid filaments of the nerves, from the outward organs of sense to the brain, and from the brain into the muscles. But these are things that cannot be explained in few words, nor are we furnished with that sufficiency of experiments which is required to do an accurate determination and demonstration of the laws by which this electric and elastic Spirit operates. (443)

Brazier (1984, pp. 60, 172) examined the pre-Newtonian work of Borelli, Baglivi and Croone, each of whom hypothesized that there is relatedness between vibrations and electricity, and she concluded that the concept of neural vibration reached a zenith with Hartley (1749). Moreover, much was made of a notion of subtlety in early speculations on the medium of cognitive transport. The word "sub-

tle" often modified the hypothetical conceptions of neural gases, liquids, and vibrations. It appeared above in Hartley's quotation from the *Principia* (*VC*, p. 4) and in Croone (quoted in Brazier, 1984, p. 60).

Returning to the early-eighteenth century and specifically Newton's *Opticks*, the following observation appears in the form of a query:

When a Man in the dark presses either corner of his Eye with his Finger, and turns his Eye away from his Finger, he will see a Circle of Colours like those in the Feather of a Peacock's Tail. If the Eye and the Finger remain quiet these Colours vanish in a second Minute of Time, but if the Finger be moved with a quavering Motion they appear again. (*Qu.* 16, p. 347)

As Newton continued, we see a cautious but justified desire to go beyond existing knowledge, to grasp what we take to be transductions in the forms of energy. Newton makes reference to essential properties of waves that makes them useful for short-term storage of information and communication. He writes:

Do not these Colours arise from such Motions excited in the bottom of the Eye by the Pressure and Motion of the Finger, as, at other times are excited there by Light for causing Vision? And do not the Motions once excited continue about a Second of Time before they cease? And when a Man by a stroke upon his Eye sees a flash of Light, are not the like Motions excited in the Retina by the stroke? And when a Coal of Fire moved nimbly in the circumference of a Circle, makes the whole circumference appear like a Circle of Fire; is it not because the Motions excited in the bottom of the Eye by the Rays of light are of a lasting nature, and [thus] are they not of a vibrating nature?

If a stone be thrown into stagnating Water, the Waves excited thereby continue some time to arise in the place where the Stone fell into the Water, and are propagated from thence … (*Qu.* 17, p. 348)

Emerging Knowledge of Electrical Conduction

Electricity was often thought of as a form of matter, for example by Boyle in his 1675 treatise on electricity (Roller & Roller, 1967, p. 16). By the 1740s, entertaining public demonstrations were being held of lights, sounds and stimulations in the form of electric shocks (see Bertucci, this volume).

In 1746, The Abbé Nollet reported Musschenbroek's experiment using a gun barrel as a mass to collect an electric charge:

I am going to tell you about a new but terrible experiment which I advise you not to try for yourself. ... I attempted to draw sparks from the gun barrel. Suddenly my right hand was struck so violently that all my body was affected as if it had been struck by lightning ... I thought it was all up with me ... (Roller & Roller, 1967, p. 52; Boynton, 1948, pp. 309–312)

Hartley had read several treatises reporting the recent findings on electricity. For instance, he writes:

The Impulse, Attraction, or whatever else be the Action of the Object, affects both the Nerves and the Æther. ... And the Result of these Actions, upon the Whole, may be supposed such a Compression or Increase of Density in the Æther, as must agitate its Particles with Vibrations analogous to those which are excited in the Air by the Discharge of Guns, by Thunder-claps, or by any other Method of causing a sudden and violent Compression in it. (*OM*, p. 21)

The sequence of eighteenth-century developments on electrical conduction included Henry Cavendish's 1773 psychophysical comparison of his own resistance with that of metal wires, salt water and fresh water, in a series circuit with an electric fish (Heilbron, 1979, 1999, p. 487; chapters in this volume by Piccolino and Focaccia and Simili). C. A. Hausen, in 1743, had held exhibitions in Leipzig, in which a small boy suspended with insulating cords transmitted a spark to an observer from his feet, which were touching a spinning glass globe. Also in Leipzig, G. M. Bose cleverly replaced the small boy with an attractive young woman; members of the audience, invited to kiss her, received an electric shock. In 1746, Gralath, in Danzig, discharged a number of von Kleist's bottles (Leyden jars) along chains of up to 20 people holding hands or connected by wires. In an experiment with two long parallel wires connected to a Leyden jar, LeMonnier found that the time interval from the visible spark to the shock was inappreciable (Roller & Roller, 1967; Wolf, 1952, 1961). Hartley was quite likely aware of this popular research into electricity, news of which motivated Benjamin Franklin's research in America (See Finger, 2006, for an in depth treatment of Franklin's electrical investigations as related to medicine; also chapters by Finger, Locke and Finger, and Bertucci in this volume).

Hartley's Grasp of Logical Requirements for a Material Substrate of Mind

During the early-eighteenth century, the idea that mind was "in" brain was sufficiently current to appear in arch satirical form in Swift's (1735) chapters on Gulliver in Laputa, where refractory political debates were said to be resolved as follows:

You take a hundred leaders of each party; you dispose them into couples of such whose heads are nearest of a size; then let two nice operators saw off the occiput of each couple at the same time, in such a manner that the brain may be equally divided. Let the occiputs thus cut off be interchanged, applying each to the head of his opposite party man. ... [T]he two half brains being left to debate the matter between themselves within the space of one skull, would soon come to a good understanding. (211–212)

Swift was vague about details. In contrast, Hartley, Newton and other scientists of the time grappled with the imperative that a true understanding of mind–brain relatedness required the assignment of specific mechanistic properties to the material system. In the following six numbered sections we offer additional quotations as support for this contention.

1. Coding: Transduction of Dynamic Patterns Across Substrates. Two and three centuries ago, techniques for studying the functions of neural tissue by stimulation, such as pricking and applying acids and alkalis, complemented ablation techniques (Walker, 1957). A precursor of the concept of transduction was the "irritability" of neural tissue, a concept on which Glisson published in 1677, and on which Haller elaborated in 1739 and more fully in 1752 (Brazier, 1959, p. 12–14; see Frixione, this volume).

The growing "solidist" opposition to the Galenic emphasis on humors, which is amply covered in this volume, is one of the most intriguing developments of the eighteenth century. At the very least, it accords with the concept of vibratory movement. In turn, this promoted conceptualizations of neural substrates of mind that are phenomenologically very different from "fluid" dynamics. Hartley provides a closer approximation to the idea of neural transduction than Newton when he claims that

We are to conceive, that the Vibrations thus excited in the Æther will agitate the small Particles of the medullary Substance of the sensory Nerves with synchronous Vibrations, in the same manner as the Vibrations of the Air in Sounds agitate many regular Bodies with corresponding Vibrations or Tremblings. (*OM*, pp. 21–22)

2. Differentiation. The brain has to provide a rich medium for impressions, transforms and storage of representations, which must be kept distinct. Hartley developed the idea of differentiation, while noting that there can be a distinction between the microstructure of a signal and its effect on cognition. For example, he writes that

The same Continuance of the Sensations is also evident in the Ear. For the Sounds which we hear, are reflected by the neighbouring Bodies; and therefore consist of a Variety of Sounds, succeeding each other at different Distances of Time, according to the Distances of the several reflecting Bodies; which yet causes no Confusion, or apparent Complexity of Sound, unless the Distance of the reflecting Bodies be very considerable, as in spacious Buildings. Much less are we able to distinguish the successive Pulses of the Air, even in the gravest [lowest pitch] Sounds. (*OM*, p. 10)

Hartley observed that a brain needs access to multiple means of coding. In his earlier book, *Various Conjectures*, he postulated three kinds of vibration: degree, kind and place (10), modifying this somewhat in *Observations* by adding a peculiar fourth "line of direction" or a sort of trajectory code. Here Hartley suggested,

Since the Vibrations, or reciprocal Motions, of the small Particles of each Nerve are made in the same Line of Direction with the Nerve, they must enter the Brain in that Direction, and may preserve some small Regard to this Direction at considerable Distances within the Brain; especially if this be favoured by the Structure of the nervous Fibrils in the Brain. Hence the same internal Parts of the Brain may be made to vibrate in different Directions, according to the Different Directions of the Nerves by which the Vibrations enter. (24)

An additional question about differentiation is how an adequate substrate can handle the sheer volume of our long-term memories. Hartley's approach to this challenge was "miniaturization" (analogous to "subtle"). He believed that this would allow the fine differentiating control that

nerves must have over our diverse muscular movements. In *OM* he states:

Sensory Vibrations, by being often repeated, beget in the medullary Substance of the Brain, a Disposition to diminutive Vibrations, which may also be called Vibratiuncles and Miniatures, corresponding to themselves respectively. This Correspondence of the diminutive Vibrations to the original sensory ones, consists in this, that they agree in Kind, Place, and Line of Direction; and differ only in being more feeble, i.e. in Degree. (*OM*, p. 58, Prop. 9)

He continues, employing a rough antecedent to present notions of coding:

For since Sensations, by being often repeated, beget Ideas, it cannot but be that those Vibrations, which accompany Sensations, should beget something which may accompany Ideas in like manner. (*OM*, p. 58)

He struggles conceptually when he writes that some sort of neural competition and summation must be involved in hedonic states:

If the Vibrations go beyond the common Limit of Pleasure and Pain in one Part of the Brain, at the same time that they fall short of it in the others, the Result will be a Pleasure or Pain, according as this or that Sort of Vibrations prevails. (*OM*, p. 40)

Although not in contemporary terms, Hartley seems cognizant of the necessities both for encoding and some sort of neural localization. (This is tightly linked with Hartley's modular vantage point regarding dimensionality; item 6, below)

3. Communication Lines. Information about events in the world must enter the mind, and information structuring willed movements must go out to the muscles. Anticipating fundamental concerns of modern communications theory, Hartley looked carefully for properties of the nervous system that might be conducive to low distortion ("pellucidity" of nerve fibers), maintenance of a sustained signal for an appropriate amount of time and continuity of connections. An earlier statement about neural communication lines can be found in Newton, who provided a correct diagram of the anatomy of the optic chiasm inferred from his knowledge of optics (including his reading of the 1687 *Theory of Vision* of ophthalmologist William Briggs; Brazier, 1984, pp. 92–93). Quoting Newton:

the Fibres on the right side of both Nerves ... going thence into the Brain in the Nerve which is on the right side of the head, and the Fibres on the left side of both Nerves uniting in the same place, … meeting in the Brain in such a manner that their Fibres make but one entire Species or Picture, half of which on the right side of the Sensorium comes from the right side of both Eyes. (*Opticks*, pp. 346–347)

Commenting on correlations of behavior and physiology, Hartley used the logic of the ablation method (see Glassman, 1978; Gregory, 1961; Walker, 1957) to evaluate claims that the fibrous nature of the white matter implies a route for communication of signals. He wrote:

Sensibility, and the Power of Motion, seem to be conveyed to all the Parts, in their natural State, from the Brain and spinal Marrow, along the Nerves. These arise from the medullary, not the cortical Part, every-where, and are themselves of a white medullary Substance. When the Nerves of any Part are cut, tied, or compressed in any considerable Degree, the Functions of that Part are either intirely destroyed, or much impaired. When the spinal Marrow is compressed by a Dislocation of the Vertebrae of the Back, all the Parts, whose Nerves arise below the Place of Dislocation, become paralytic. (*OM*, pp. 7–8)

Subsequently, he cited "evidence … from the writings of physicians and anatomists" that deficits are seen most surely when the white matter is involved, concluding that:

The white medullary Substance of the Brain is also the immediate Instrument, by which Ideas are presented to the Mind: Or, in other Words, whatever Changes are made in this Substance, corresponding Changes are made in our Ideas; and vice versa. (*OM*, Prop. 2, p. 8)

Following Newton's 24th query in the *Opticks*, Hartley opted for continuity and "pellucidity" of nerve fibers. He sometimes seemed to equivocate, however, on the significance of neural fibers as communication lines, perhaps assigning too much emphasis to volume conduction. Hartley compared the heterogeneity of the structure of glands to that of brains, suggesting that the apparent homogeneity of white matter provides the needed medium for transmission of vibratory signals. He even alluded to the role of the brain's blood supply:

We come next to consider the Uniformity and Continuity of the white medullary Substance of the Brain, spinal Marrow, and Nerves. Now these are evident to the Eye, as far as that can be a Judge of them. The white medullary Substance appears to be every-where uniform and similar to itself throughout the whole Brain, spinal

Marrow, and Nerves; and tho' the cortical Substance be mixed with the medullary in the Brain, and spinal Marrow, and perhaps in the Ganglions and Plexuses, yet it does not appear, that the Communication of any one Part of the medullary Substance with every other, is cut off any-where by the Intervention of the cortical. There is no Part of the medullary Substance separated from the rest, but all make one continuous white Body; so that if we suppose Vibrations apt to run freely along this Body from its Uniformity, they must pervade the Whole, in whatever Part they are first excited, from its Continuity. … If we admit the foregoing Account of the uniform, continuous Texture of the medullary Substance, it will follow, that the Nerves are rather solid Capillaments, according to Sir Isaac Newton, than small Tubuli, according to Boerhaave. And the same Conclusion arises from admitting the Doctrine of Vibrations. The Vibrations hereafter to be described may more easily be conceived to be propagated along solid Capillaments, so uniform in their Texture as to be pellucid when singly taken, than along hollow Tubuli. For the same Reasons, the Doctrine of Vibrations will scarce permit us to suppose the Brain to be a Gland properly so called; since the Difformity of Texture required in a Gland, appears inconsistent with the free Propagation of Vibrations. Neither can we conclude the Brain to be a Gland, from the great Quantity of Blood fed to it by the Heart. It is probable indeed, that this is required on account of the important Functions of Accretion, Nutrition. (*OM*, pp. 16–18)

Hartley's foregoing comments on communicative continuity are followed up by a remark suggesting the importance of integration coexisting with differentiation: "It is reasonable also to think, that the Nerves of different Parts have innumerable Communications with each other in the Brain. …" (*OM*, p. 19).

One might speculate about a variety of possible historical metaphorical sources for the notion that line-like entities are literally needed for neural communication – on analogy with the need for a line of sight for visual signaling, or sequences of voices and ears in a rumor. One old line-like device for communicating an event is a bell-rope. Dramatic empirical observations of conduction along long objects, and then along long lines of "packthread," further emerged with the study of electricity in the eighteenth century (Roller & Roller, 1967, pp. 29–48). Such observations were easily related to the line-like peripheral nerves, the long spinal cord and the fibrous white matter of the brain.

Giving such possibilities serious consideration, Monro Secundus, along with Haller, wondered whether the nerves are insulated. In a carefully qualified comment Monro remarked:

We are not sufficiently acquainted with the properties of aether or electrical effluvia pervading everything, to apply them justly in the animal oeconomy; and it is difficult to conceive how they should be retained or conducted in a long nervous cord. (Brazier, 1959, p. 14)

This type of scientific thinking led to the diagrams of the classical aphasiologists of the middle and late-nineteenth century (c.f., Lichtheim, 1885) and ultimately to more modern models of disconnection syndromes (Geschwind, 1965). Stephen Gray and others in the 1720s had discovered that electric charges could be communicated along elongated objects, eventually including fishing rods, then rods with lines attached, and even through threads 650 ft long (Roller & Roller, 1967, pp. 29–37). These findings were also harbingers of the concepts of communication inherent in the connectionist models of present day neuropsychology.

4. Organization.
Hartley struggled with the metaphysics of connectionist theorizing, as scientists do today, sometimes confronting it explicitly, at other times more indirectly. How does brain organization provide a foundation for coherent memory and personality, for the origin of new ideas and for reasonable intentions? How might ethical and moral sensitivity arise in the natural course of materialistic events, as individuals' histories accumulate in their memories?

Allen (1999) has examined Hartley's associationist/behavioral theology, and more recently Gazzaniga

(2006) has written of what he calls "the ethical brain" and a kind of "science" of our "moral" dilemmas. We do not yet know whether simple axioms of association - such as the "Hebbian neuron" or Pavlovian conditioning and learning via reinforcement of operants – genuinely attain the power to generate such robust social epiphenomena, although psychology textbooks are replete with examples of social ramifications purportedly derived from basic learning principles, which in turn presumably depend upon brain function.

5. Emergent Properties.
In moving beyond Newtonian speculations about vibrations in visual perception to a deeper and more general level of cognition, Hartley wagered that his physiology of associations would provide the most fruitful account of those deeper levels.

The Doctrine of Vibrations may appear at first Sight to have no Connexion with that of Association; however, if these Doctrines be found in fact to contain the Laws of the Bodily and Mental Powers respectfully, they must be related to each other, since the Body and Mind are. One might expect, that Vibrations should infer Association as their Effect, and Association point to Vibrations as its cause. (*OM*, p. 6)

Here Hartley confronted the same challenge that still exists in contemporary mind–brain science, which strives to identify, unify, or otherwise align neuroanatomy, neurophysiology, and neurochemistry with the psychological sciences (Kovac, 2006; Miller, 2003).[2]

6. Dimensionality/Modularity.
The organismic brain is a substrate rich with structural, electrical and chemical attributes, with dynamics that vary

---

[2] Does knowledge of particular physical phenomena – through ingenious experimentation or clever inference – explain the psychological phenomenon? Is our understanding of elementary parts fully explanatory of the whole? Understanding behavioral architectures per se is often necessary for establishing the boundary conditions within which the assumed supporting biological processes operate. In the investigation of complex systems, bottom-up inferences from the parts to the "emergent whole," even when those parts and their computations are well understood, can still be highly underdetermined. Kovac (2006, p. 565) has suggested that "Chemistry, more than any other science, abounds in emergences … The brain … is … not a computer with

hardware and software. … Perceptual and emotional qualia, and even … self-consciousness, lose much of their mystery if we think of them as emergences … in which myriads of teleonomic chemical interactions – molecular cognitions – are occurring all the time." Throughout the historical study of the human mind–brain, the so-called "chemical analogy" has been used time and again to help explain what an epiphenomenon might be. Here, Kovac's claim for the chemical analogy has more far reaching consequences for the mind–brain puzzle of emergent systems, although his rejection of complementary computer metaphors of emergent complexity may be excessive (see Simon, 1996).

somewhat independently of each other. Our brains far exceed computers in density of multivariate properties and perhaps, even in this era of amazing microchips (e.g. Burr-Brown, 2005) in density of connections among tiny nodes (Glassman, 2002). Hartley sought his own examples of what we now call modularity, as he discussed "the Ganglions" and "Plexuses" (19), and hypothesized about the four kinds of vibratory codes previously noted.

Aubert and Whitaker (1996) and Glassman and Spadafora (1996) have written about Hartley's seemingly innovative idea that specific sensory nerve functions terminate not in some equipotential or mass action-like *sensorium commune*, but rather in specifically localized regions in the higher nervous system. Although parts of the above quotations from *OM* (pp. 24, 16–18) may be read to suggest Hartley considered the possibility of neural information transmission sometimes escaping from the anatomical lines of direction in white matter fibrous tissue, his emphasis in those quotations seems to be on the adherence of information flow to those lines of direction. In a broad sense, Hartley's notions seem to prefigure later claims of brain modularity, such as Johannes Mueller's nineteenth century notion of "specific nerve energies."

Modularity has to entail emergence of higher-level phenomena from a rich tapestry of underlying energetic interactions (potential organizations of information). Hartley suggested that the vibrations are sustained by

Active Powers of the medullary Substance [which] may serve to explain or evince the Vibrations of the medullary Particles, … Want of active Powers in these Particles would suffer the excited Motions to die away prematurely. … [T]he Vibrations of the Æther must be conceived as regulating and supporting the Vibrations of the Particles, not as exciting them originally. (*OM*, pp. 21–22)

Because Hartley was making these claims without access to the two parsimonious laws of thermodynamics, he might be suspected of vitalism. Upon further reflection, however, this attempt to explain emergent agency has a mechanistic flavor that seems to be modeled after Newton's laws of motion. As noted earlier, these active powers of the medullary substance represent Hartley's "efficient" or "immediate" levels of causation. Moreover, the following quotation from Hartley, which follows a comment on the relatedness of the

different disciplines of science, illustrates general insights into the existence of what we now understand to be biochemical processes that liberate and transform energy.

Electricity may also extend, without being excited by Friction or otherwise, to small Distances, … The Effervescence which attends the Mixture of Acids and Alcali's, and the Solution of certain Bodies in Menstruums, Fermentation, and Putrefaction, are all general Principles of very extensive Influence, nearly related to each other, and to the forementioned mutual Attractions and Repulsions, and are possessed of the same unlimited Power of propagating themselves, which belongs to several Specieses [sic] of Plants and Animals. (*OM*, p. 29)

The idea of potential energy triggered into kinetic energy was earlier part of fluidist notions of neural action. For example, Croone hypothesized that "spiritous liquid … nourished the juice of muscle," and Thomas Willis put forth the notion of a flame in the vital fluid of the blood interacting with light in the nervous juice. Although these tropes were mocked by Danish physician Niels Stensen as "mere words," as metaphors they were valuable words. In his own way, and with the available knowledge of his time, Hartley seemed to anticipate what we now know about the effect of motor neuron action potentials on muscle.

Looking Back at Hartley

Our modest claim is to suggest that Newton's foresight developed further by Hartley, that communication at any scale might take advantage of an energetic repetitiveness, which both spoke of as "vibration," remains valid. Indeed, we see evidence that Newton's and Hartley's thinking contained a partial anticipation, of today's knowledge that neural activity is replete with oscillations on different scales. Smith (1987) however cautions that there is such a large conceptual and empirical gap between the science of the mid-eighteenth and mid-twentieth centuries, that there is little sense in crediting Hartley for nervous vibratory theory as a forerunner of the discovery of electrical oscillations in the cortex. Smith additionally observes that nowhere is

Hartley referred to in the early papers of Robert Caton and Hans Berger.

Yet we see a need for caution about this caution. Sources are often not explicitly recalled. Such, for instance, is true of the learning and memory of human individuals during their long natural lives (e.g Linton, 1982, 2000; Reder & Schunn, 1996). Michael Polanyi (1964) famously referred to "the tacit component" of human knowledge. In the spirit of our subtext, Smith writes, "One cannot help suspecting, however, that if Hartley had known of the discovery of, for instance, the evoked potential, he would have felt that his speculations had at long last been justified" (1987, 133).

In our chapter, we have extended beyond specific investigator citation of Hartley to a subtext where we highlight the proliferation of vibratory phenomenology in physics and mathematics. Newton and Hartley were beginning to consider waves and vibrations in their search for the physical mechanisms of information processing and communication in the mind. Newton had initially focused upon the ethereal vibrations of the universe of space outside the human body (Descartes' view), but in his *Opticks* he suggested that not only were there external vibrations in nature associated with sound, light, smell, and touch, but also that these were transduced at the sensory periphery of animal bodies and carried throughout the nervous system by means of different, yet somehow analogous, vibrations. Hartley, not a physicist-mathematician, focused exclusively upon these latter phenomena as the principal internal medium for information processing within the nervous system.

Waves and vibrations are still regarded as basic features of nature, both inside and outside the human body. Hartley's eighteenth-century vibratory model is not operative in its details today, but nevertheless seems to us close to the presuppositions of modern physical and mathematical treatments of the oscillatory communications underlying information processing. Our text for this chapter has been the specific contributions and their historical assessments by others of Hartley's ideas. Our subtext has been the ubiquity of vibration in nature and its scientific investigation, which can be traced to Hartley's Newtonian neuropsychology.

Acknowledgments. We thank David Spadafora for introducing Dr. Glassman to Hartley's work and for many illuminating conversations about that work and other aspects of Enlightenment history. We also thank Robert M. Young (1970) for introducing Dr. Buckingham to Hartley's work. We thank anonymous reviewers for excellent critical reviews and additional references on Hartley. Wes Wallace's stimulating paper in Brain and Cognition (2003) on early mechanical theories of the conductive capacity of vibrating nerve impulses in the work of Newton, Willis and Gassendi greatly helped us formulate many of our arguments concerning Hartley's theories of the same.

References

Allen, R. C. (1999). *David Hartley on human nature.* Albany, NY: State University of New York Press.

Aubert, D., & Whitaker, H. (1996). David Hartley's model of vibratiuncles as a contribution to localization theory of brain functions, with a side note on short-term memory. *History and Philosophy of Psychology Bulletin, 8,* 14–16.

Barbour, I. (1997). *Religion and science.* New York: HarperCollins.

Bower, G. S. (1881). *Hartley and James Mill.* London.

Boynton, H. (1948). *The beginnings of modern science. Scientific writings of the 16th, 17th and 18th centuries.* Roslyn, NY: Walter J. Black, Inc.

Brazier, M. A. B. (1959). The historical development of neurophysiology. In J. Field, H. W. Magoun, & V. E. Hall (Eds.), *Handbook of physiology. Section 1: Neurophysiology* (Vol. 1, pp. 1–58). Washington, D.C.: American Physiological Society.

Brazier, M. A. B. (1984). *A history of neurophysiology in the 17th and 18th centuries.* New York: Raven Press.

Buckingham, H. W. (2001). Review of Richard C. Allen's *David Hartley on human nature. Journal of the History of the Neurosciences, 10,* 328–330.

Buckingham, H. W. (2002). The roots and amalgams of connectionism. In R. G. Daniloff (Ed.), *Connectionist approaches to clinical problems in the clinic.* Mahwah, NJ: Lawrence Erlbaum.

Buckingham, H. W., & Finger, S. (1997). David Hartley's psychobiological associationism and the legacy of Aristotle. *Journal of the History of the Neurosciences, 6,* 21–37.

Burr-Brown Products from Texas Instruments (2005). INA128/INA129/Precision low power instrumentation amplifiers. Data sheets, 18 pp., downloaded December 2005 from http://www.ti.com.

Campbell, D. T. (1975). On the conflicts between biological and social evolution and between psychology and moral tradition. *American Psychologist, 30,* 1103–1126.

Clarke, E. (1968). The doctrine of the hollow nerve in the seventeenth and eighteenth centuries. In L. Stevenson & R. Multhaui (Eds.), *Medicine, science and culture: Historical essays in honor of Owsei Temkin* (pp. 123–141). Baltimore, MD: Johns Hopkins University Press.

Finger, S. (1994). *Origins of neuroscience.* New York: Oxford University Press.

Finger, S. (2006). *Dr. Franklin's medicine.* Philadelphia, PA: The University of Pennsylvania Press.

Gazzaniga, M. (2006). *The ethical brain: The science of our moral dilemmas.* New York: Harper-Perennial.

Geschwind, N. (1965). Disconnection syndromes in man and animal. *Brain, 88,* 237–294, 585–644.

Glassman, R. B. (1978). The logic of the lesion method and its role in the neural sciences. In S. Finger (Ed.), *Recovery from brain damage: Research and theory* (pp. 3–31). New York: Plenum Press.

Glassman, R. B. (1996). Cognitive theism: Sources of accommodation between secularism and religion. *Zygon, Journal of Religion and Science, 31,* 157–207.

Glassman, R. B. (2002). "Miles within millimeters" and other awe-inspiring facts about our "mortarboard" human cortex. *Zygon, 37,* 255–277.

Glassman, R. B., & Spadafora, D. (1996). David Hartley's (1749) doctrine of neural vibrations: Enlightenment or accident. *Society for Neuroscience Abstracts, 22* (Part 1), 243.

Gregory, R. L. (1961). The brain as an engineering problem. In W. H. Thorpe & O. L. Zangwill (Eds.), *Current problems in animal behavior* (pp. 307–330). Cambridge: University Press.

Hartley, D. (1746). *Various conjectures on the perception, motion, and generation of ideas.* (R. E. A. Palmer & M. Kallich (1959) Eds. and Trans.). Los Angeles: The Augustan Reprint Society, Publication #77–78, William Andrews Clark Memorial Library, University of California.

Hartley, D. (1749). *Observations on man, his frame, his duties, and his expectations,* (2 Vols.). London: S. Richardson.

Hatch, R. B. (1975). Joseph Priestley: An addition to Hartley's observations. *Journal of the History of Ideas, 36,* 548–550.

Hefner, P. (2006). Editorial: The mythic grounding of religion and science. *Zygon, 41,* 231–234.

Heilbron, J. L. (1979/1999). *Electricity in the 17th and 18th centuries: A study in early modern physics.* Mineola, NY: Dover Publications.

Kovac, L. (2006). Life, chemistry and cognition. *European Molecular Biology Organization Reports, 7,* 562–566.

Lichtheim, L. (1885). On aphasia. *Brain, 7,* 433–484.

Linton, M. (1982/2000). Transformations of memory in everyday life. In U. Neisser & I. E. Hyman, Jr. (Eds.), *Memory observed.* New York: Worth Publishers.

Miller, G. A. (2003). The cognitive revolution in historical perspective. *Trends in Cognitive Sciences, 7,* 141–144.

Newton, I. (1723/1995). *The principia* (3rd ed.). Trans. Motte A. London/Amherst, NY: Prometheus Books.

Newton, I. (1730/1979). *Opticks* (4th ed.). Eds. A. Einstein, I. B. Cohen, E. Whittaker, & D. H. D. Roller. London/New York: Dover Publications.

Polanyi, M. (1964). *Personal knowledge: Towards a post-critical philosophy.* New York: Harper Torchbooks.

Rameau, J.-P. (1722). *Treatise on Harmony* (P. Gossett, Trans. (1971)). New York: Dover Publications, Inc.

Reder, L. M., & Schunn, C. D. (1996). Metacognition does not imply awareness: Strategy choice is governed by implicit learning and memory. In L. M. Reder (Ed.), *Implicit memory and metacognition.* Mahwah, NJ: Lawrence Erlbaum Associates.

Robinson, D. N. (1995). *Intellectual history of psychology* (3rd ed.). Madison: University of Wisconsin Press.

Roller, D., & Roller, D. H. D. (1967). *The development of the concept of the electric charge.* Cambridge: Harvard University Press.

Simon, H. A. (1996). *The sciences of the artificial* (3rd ed.). Cambridge, MA: MIT Press.

Smith, C. U. M. (1987). David Hartley's Newtonian neuropsychology. *Journal of the History of the Behavioral Sciences, 23,* 123–136.

Spadafora, D. (1990). *The idea of progress in eighteenth-century Britain* (pp. 135–178). New Haven: Yale University Press.

Strohl, E. L. (1963). Parliament Hoodwinked by Johanna Stephens. *Surgery, Gynaecology & Obstetrics, 116,* 509–511.

Sutton, J. (1998). *Philosophy of memory traces: Descartes to connectionism.* Cambridge, UK: Cambridge University Press.

Swift, J. (1735). *Gulliver's Travels.* Windermere Series (1912). Chicago: Rand McNally & Company.

Walker, A. E. (1957). Stimulation and ablation: Their role in the history of cerebral physiology. *Journal of Neurophysiology, 20,* 435–449.

Wallace, W. (2003). The vibrating nerve impulse in Newton, Willis, and Gassendi: First steps in a mechanical theory of communication. *Brain and Cognition, 51*, 66–94.

Warren, H. C. (1921). *A history of the association psychology*. New York: Charles Scribners & Sons.

Webb, M. E. (1988). A new history of Hartley's observations on man. *Journal of the History of the Behavioral Sciences, 24*, 202–211.

Webb, M. E. (1989). The early medical studies and practice of Dr. David Hartley. *Bulletin of the History of Medicine, 63*, 618–636.

Wolf, A. (1952/1961). *A history of science, technology, & philosophy in the 18th century* (Vol. 1). New York: Macmillan/Harper Torchbooks.

Young, R. M. (1970/1990). *Mind, brain, and adaptation in the Nineteenth Century*. London: Oxford University Press.

13
Charles Bonnet's Neurophilosophy

Harry A. Whitaker and Yves Turgeon

Isaac Newton's concept that auditory and visual images were transmitted by vibrations and that these vibrations were transduced to vibrations in nerves, built upon earlier ideas of vibrations in nerves by Pierre Gassendi and Thomas Willis, likely in contrast to Cartesian "hydraulic" models of mechanical pressure (Wallace, 2003). Newtonian "vibrations" strongly influenced David Hartley's neuropsychology (Glassman & Buckingham, 2007; Smith, 1987); less often discussed is the movement or "vibration" model of Charles Bonnet, a rather more elaborated mid-eighteenth century model of the internal representation of ideas. Bonnet, a Swiss naturalist and philosopher, proposed in an original fashion that an understanding of animal and human behavior requires, first, knowledge of how the nervous system functions.

Life and Career

Charles Bonnet was born on March 13, 1720, in Geneva, Switzerland; his family, originally French, had evidently left France in the sixteenth century on account of religious persecution. It appears that he did not travel outside of Switzerland, dying May 20, 1793, at his country home, Genthod, Lac Lemond, near Geneva. Bonnet's hearing and vision were problematic most of his life; while still a child he became partially deaf and from his mid-twenties to his forties he gradually lost his vision, forcing him to abandon natural science for philosophical speculations in cognitive neuroscience. While still a teenager, he read Noël-Antoine Pluche's (1688–1761) *Spectacle de la Nature*

and René-Antoine Ferchault de Réaumur's (1683–1757) *Mémoires pour servir à l'Histoire des Insects*, inspiring him to turn away from law, which he had first studied, to biology. In 1740, at the age of 20, he presented a paper to the Parisian Académie des Sciences comprising his famous study demonstrating parthenogenesis in aphids. In the following year, the Academy made him its youngest "corresponding" member. By 1742, Bonnet had identified the pores by which butterflies and caterpillars breathe; these studies earned him membership in London's Royal Society in 1743. These distinguished awards put him in communication with some of the best scientific and philosophic minds of the mid-eighteenth century, including Réaumur (mentioned above), Albrecht von Haller, Carl Linneaus, and Georges-Louis Leclerc, the Count of Buffon. Bonnet's *Traité d'Insectologie* (1745) and *Recherches sur l'Usage des Feuilles dans les Plantes* (1754) cemented his reputation in the history of biology; the latter studies on plants included his original ideas that plants exhibited sensation and powers of discernment. Failing eyesight after mid-century restricted his later research to theoretical questions in biology, psychology (Nicolas, 2006) and what we would now call neuroscience. Bonnet was likely the first person to use the term "evolution" as a biological concept in *La Palingénésie Philosophique* (1770), a work in which he developed a catastrophe theory of evolution that would influence Erasmus Darwin. Bonnet's catastrophe theory proposed that changes in species occur after great natural disasters; this view contrasted with Cuvier's view of a continuous goal-driven ascent toward perfection and Lamarck's view that evolution occurred through mutation. Bonnet's philosophical studies in cognitive neuroscience are less well known. The *Essai de psychologie* (1754), published anonymously in London in 1755, (reissued in 2006, edited by Serge Nicolas) and the *Essai analytique sur les facultés de l'âme* (1760), published in Copenhagen, not only present his ideas on the brain representation of mental activity, but also introduce for the first time a discussion of the conscious hallucinations experienced by his grandfather, Charles Lullin, to be called the Charles Bonnet Syndrome in the twentieth century.

Philosophy

The eighteenth century is remembered for some of the great contributions to rationalist (Leibniz, Wolff, Kant) and empiricist philosophy (Berkeley, Hume, Condillac), as well as bridging viewpoints such as the commonsense philosophy of Thomas Reid. Bonnet, too, should be viewed as one who borrowed major viewpoints of both empiricism and rationalism, but unlike Reid, was clearly interested in brain functions. Following Locke and Condillac, Bonnet believed that psychological phenomena are dependent on sense experience, the result of external objects acting upon bodily organs. As did Condillac, he attached great importance to the faculty of smell, but unlike the empiricists Bonnet argued that sensations coalesce in the mind according to natural dispositions of the brain, that is, innate dispositions that allow sensations and ideas to associate according to constant laws.

In the tradition of Pierre Gassendi, Thomas Willis, and Isaac Newton (see Wallace, 2003), and parallel with but evidently independently of David Hartley, Bonnet directly correlated psychological phenomena with the functioning of the brain. The psychological life rests upon nerve fibers; vibrations in nerve fibers are the substrate for sensations, for the association of ideas and for habits. Unlike many eighteenth-century empiricists, the mind of Bonnet's "statue" has active, that is to say, rationalist powers that make it an independent entity. The mind of each sensitive being, animal and human, is the internal power of action; the mind is both acted upon by external objects and in turn acts on the body, i.e., the brain, by producing movement in nerve fibers. The mind's "power," *la force motrice de l'ame*,[1] is the cause of psychological experiences such as sensation, memory, abstraction, judgment, and reasoning, modifying psychological life through attention and thought.

For Bonnet, the subject that empirical psychology can study is the material evidence found in the brain. Although Bonnet agreed that the soul is immaterial, he believed that its functioning is not; the working of the mind, the mind's powers acting on the nerve fibers belongs to the body and can be scientifically observed. The classical, traditional (e.g., Cartesian) concept of soul is not a psychological phenomenon and thus is of no help in explaining the faculties of the mind.[2] Bonnet thus distinguished metaphysical concerns from psychological concerns, concluding that, since the problem of the soul is out of the reach of science, one should concentrate on the functions of the brain. He defined his cognitive neuroscience as the science of psychological phenomenon in relation to the activity of nerve fibers.

---

[1] Throughout the Essai, Bonnet uses the word *l'ame* and not *l'esprit*. The former is usually translated as "soul" and the latter as "mind". We translate *l'ame* as "mind" for the following reasons: First, Bonnet was severely visually handicapped when he wrote the Essai de Psychologie and the Essai Analytique; he wrote in a sparse, almost spartan style, transcribed by an aide who may or may not have followed his line of thinking. Second, Bonnet makes a point of distinguishing the spiritual soul from the material soul; it is the material soul and its mental images that he writes about and that empirical psychology may investigate. He is everywhere materialistic: mental images are equated with changes in the structure of nerve fibers (a psycho-physical parallelism). In contemporary English, the word "soul" conveys the spiritual sense much more than the material or physical sense, that which, in the eighteenth century, would have been of concern to the Church, something that would have been beyond evidentiary, empirical testing. Given these reasons, using "mind" conveys to the modern reader a more accurate sense of what Bonnet intended, even though it clearly is not what he wrote. At the very least, the issue is debatable.

[2] The phrenologist Johann Caspar Spurzheim was later to identify Bonnet as one of the scientists prior to Gall who believed that the mind had multiple faculties, physically represented in the brain.

Fibers

Bonnet (1755) was first interested in the facts of consciousness and in their links to the activity of the nervous system. How does consciousness arise? How is the psychological life possible? How does the brain function to make a person feel, move, and will? Bonnet represented the brain as "the instrument of the operations of the mind" and viewed it as a machine (e.g., a harpsichord, an organ, a clock) as did many of the so-called French sensationalists. He fabricated a physiological psychology somewhat different from, independent of and, in psychological respects, more detailed than what Hartley did in the same epoch. For Bonnet, all of psychological life is equated to physiological activity; all intellectual and affective psychological phenomena have this biological substrate, the nervous system, a machine made of nerve fibers which are the loci of mind's activity.

Here are springs intended to move the head; there are those that move the limbs. Higher are the movements of the senses. Below are those of breathing and of speech, etc. And what numbers, what harmony, what variety in the parts that make up these springs and these movements! The mind is the Musician who performs, on this (brain) machine, different tunes, who judges those being performed and who repeats them. Each nerve fiber is a sort of key intended to render a certain pitch. Whether the keys are moved by (perceived) objects, or the movements are created [internally in the nerve fibers] by the mind's capability to move, the play is the same; it can only vary in duration and in intensity. (Bonnet, 1755, pp. 13–14)[3]

He asked:

If all our ideas, even the most spiritual, depend upon movements in the Brain, it is fair to ask if each idea has its own particular fiber intended to produce that same idea, or if that same fiber when moved differently will produce different ideas? (Bonnet, 1755, p. 51)

To answer that question, Bonnet proposed that each of the five senses has its distinct physiological substrate, in part anticipating Johannes Mueller's (1801–1858) "Law of Specific Nerve Energies," a mainstay of nineteenth-century physiological psychology. The logic of Bonnet's

reasoning developed thus: first, there is no feeling where there are no nerves, next there is a vast diversity of sensations and feelings, and last, all nerves take their origin from the brain. Consistent with the prevailing anatomical notions of his time, he believed nerve fibers to be "white, homogenous, solid bodies." Although observationally nerves are similar in physical nature, in Bonnet's view they vary according to their internal structure; the variety of structures of nerves represents the functions they perform, e.g., transmitting to the mind the impressions of external objects. Thus, employing Aristotelian teleology, nerve fibers have as their purpose the transmission to the brain, by "movement," not precisely by overt physical vibrations, the impressions external objects make on the sense organs themselves (as in taste, for example) via "corpuscular emanations." These same nerve fibers also transmit movements from the brain to the body via the "action" of the mind.

After analyzing each of the five senses, Bonnet concluded that nerve fibers have specific characteristics of action or energies, e.g., specific characteristics that define specific sensory experiences. In the *Essai analytique* (1760), he clarified and generalized his theory of nerve fiber specificity, arguing that, however impressions are made, "it is certain that [nerve] fibers are moved and that they cannot be moved other than if a change occurs in the actual or primitive state of their molecules or in the state of their essential elements" (Bonnet, 1760, p. 56). An object perceived by the senses imparts a particular movement to a specific nerve fiber; the movement once imparted, the impression once received, then the fiber acquires and conserves, by repetition, a tendency to transmit that particular movement. "A natural consequence of this change is a tendency toward reproducing that same movement, or a disposition to execute that same movement" (Bonnet, 1760, p. 57).

Thus, the structure of a nerve fiber is malleable or elastic; starting with a primitive arrangement at birth, nerve fibers can be reorganized or rearranged under the influence of either the internal context or the external world. Each nerve fiber has a distinct molecular organization specific to the appropriate impressions it is destined to receive, e.g., visual images. The structure of a nerve fiber differs according to the arrangement of the *molecules* from which it was built. "A fiber can essentially be

---

[3] Translations from the *Essai de psychologie* and the *Essai analytique* are by the senior author.

different from another [fiber] because of the nature and the arrangements of its basic elements" (Bonnet, 1760, p. 437). Hence, the brain's putative complexity rests on the hypothesis of specific nerve energies. Using the metaphor of the statue originally popularized by Condillac, Bonnet explained:

The facts lead us to assume that the diversity of sensations do not depend on the diversity of movements caused by objects in identical fibers; [. . .] Thus, we are compelled to acknowledge that there is in each sense, fibers designed for diverse sorts of sensations that the sense can generate in the mind; that there is, for example, in the organ of smell, fibers designed for the particular set of corpuscles that emanate from the rose . . . (1760, p. 52)

As mentioned earlier, Bonnet observed that sensations vary in their intensity. The intensity of a sensation depends not only on the objects' activity upon specific nerves, but also on the number of fibers moved and on the ability of its basic elements to be moved.

The Object is a composite of corpuscles: the Organ (brain) is a composite of fibers. As these corpuscles are nearly similar or identical, so are the fibers. Each corpuscle, each fiber, each of the smallest fibers, therefore produces the same essential effect. It is infinitely small forces that, because of their gathering, contribute to produce a Sensation at a particular level of intensity. The Sensation is essentially the same in each of even the smallest fibers. If there was only one very small fiber that was affected, the Sensation would be infinitely weak. Therefore it is of the identity and of the simultaneity of the fibers' action that results in the intensity of the Impression. From these aspects of the impression results the simplicity and the intensity of Sensations. (Bonnet, 1760, pp. 151–152)

At the beginning of *Essai analytique* (1760, p. 23), Bonnet argued that the degree of "liveliness" of the impressions and consequently the level of the mind's activity, are directly correlated with the degree of movement in the nerve fibers. Since the cerebral matter is the medium for sensations, it follows that the greater or lesser mobility of the fibers of the brain will depend upon the degree of liveliness of impressions. In eighteenth- and nineteenth-century parlance, every sensitive, moral and intellectual faculty is connected to a particular bundle of fibers in the brain; each has its own set of mechanical rules which link it to other faculties and determine its mode of action. Not only does each faculty have its special fasciculus of fibers, but also each word has its own fiber.

Memory

Bonnet's ideas on memory are consistent with his theory of nerve fibers. There exists an anatomical substrate for memories and as well for lost memories. He was among the first to suggest that the biological basis for memory is a "collection" of nerve fibers. The logic of his argument about the nature of memory starts with the evident truth that the brain maintains sensations, and the movements they correspond to, in nerve fibers; movements, induced by an object, of nerve fibers change the primitive state of these fibers both in their structure and in their constitutive elements. He wrote that "the action of objects on the sense fibers changes, up to a certain point, the primitive state of these fibers, and then it imprints on them various dispositions that they did not have beforehand" (Bonnet, 1760, p. 87). According to Bonnet the nerve's capacity to receive these "dispositions," or "determinations," depends upon its particular structure, specific to each different neural structure. A simple nerve fiber is constituted of molecules ("corpuscles"), elementary parts of which the arrangement determines "the type or construction of the fiber," that is, the functional purpose of a fiber, such as memory, recall, or recognition.

Internal or external objects lead to two different consequences in nerve fibers: impressions lead to a change in the molecular structure of fibers or to a change in the molecular position in a fiber of a selected group or collection of fibers. The brain preserves something of impressions. It is a kind of trace but not the sort of trace that it is possible to make on solid matter. Bonnet believed that the brain preserved of objects only their types, or "determinations" and he thought these to be expressed anatomically in structural, molecular modifications and rearrangements of fibers. Bonnet's memory traces seem to bear no direct resemblance to the original object and consequent impressions of the object on the sense organs, except for the object's action on a nerve fiber. It would appear that Bonnet understands that the movements in fibers are some sort of a neural code, arguably foreshadowing modern views. This clearly distinguishes his ideas from the classic conception of memory traces as an object's direct impression on a wax-like brain as well as that of his contemporaries, many of whom still considered memory to be "souvenirs," miniature objects somehow inscribed on the cerebral material in the form of spirits or fluids.

For Bonnet, memory is at the core of psychological life; memory places ideas in the brain, memory simplifies and makes coherent one's life experiences. He conceived memory to be closely linked to what we might call attentional resources (the intensity of a memory is related to the amount of attention given to it), to what we might call habit (memory without attention can also lead to a strong trace through repetition as well by means of associations with other traces), and to the association of ideas (a memory is stronger when integrated into an ensemble of memories, remembered in a specific order). Attention, habits, and associations act from memorization to recall and then to recognition (1760, p. 88). Bonnet, like others before him, distinguished the two different processes, recall and recognition; the latter he called "reminiscence." He wrote that "one can distinguish two things in memory: the first is the operation by which one or several ideas are recalled to mind; the second is the operation by which the mind recognizes that these ideas are those which have already been presented to it" (Bonnet, 1760, pp. 59–60).

According to Bonnet, recall rests upon the associations of ideas and the mechanics of nerve fibers. He admitted that science knows little about "the distribution or arrangement of the diverse orders of fibers in each sense," the particular organization of a specific neural structure in each sense organ and correspondent brain area. But he knew that, for recall to be possible, one must recognize that nerve fibers are interrelated so that they may communicate with each other in some way. Bonnet (1760) thus underlined the importance of the "relationships between" each fiber, and that each group of fibers has with other groups. Communication between nerve fibers in memory is the cornerstone of our understanding of psychological life. To quote:

we will never succeed in satisfying ourselves regarding the linking and reproduction of our ideas while we ignore the relationship between the fibers to which one of the ideas are attached. All that we can catch a glimpse of on this subject reduces itself to this; the liasons between our ideas of each type presuppose a different order of the fibers that serve in their formation. We can therefore reasonably conjecture that the fibers of different structures are gathered together by fasciculi in the seat of the mind, approximately like colored rays are gathered together in a sunbeam, or the branches and the little twigs of a tree are gathered near the trunk. I say "somewhat" or "approximately" because these comparisons perhaps only express the intimate liason imperfectly, or the tight correspondence that exists between all the parts of the seat of the mind. (Bonnet, 1760, p. 54)

Bonnet inferred a close relationship between different brain areas. For him, there are multiple knots, "nodes," points of convergence and divergence in the brain. These hidden nodes, these intimate links form the basis of memory. Thus, all ideas come from the senses in his model. Impressions are made on a sense organ then propagated in nerve fibers, inducing movements in a specific group of fibers, leading to a change in their molecular structure. Functionally, the fibers acquire the capacity to execute new movements, that is to say, new sensations, feelings, ideas, etc. In addition, in memory there is a clear communication between fibers; each fiber may communicate with another fiber, so that an entire fasciculus moves according to induced movements; therefore, there must be physical links between fibers. The parts of each fiber responsible for inter-fiber communication Bonnet calls "pieces of a chain." Their purpose is to permit the serial recall of sensations, one after another, in other words, to coalesce at the production of a memory. These "chains," or "components of communication," have as a goal or purpose "the communication or the propagation of movement, of which the result is the diverse phenomena of memory." (Bonnet, 1760, p. 511)

Ideas connect or link up in the brain in a constant order. A word, "Geneva," for example, always recalls a precise succession of ideas, ideas like "rivers, a lake, economic welfare," etc. How is this possible? Bonnet (1760) grounded his belief on five facts that he had observed. First, the brain needs more time to acquire a series of ideas than to recall the memories of each constitutive element of the series. Second, to easily learn a series of ideas requires repetition of it in the same order. Third, in the case of a long series with many different ideas, learning is facilitated by cutting the series into equal parts. Fourth, it is more difficult to add new ideas to a series after having learned it than if the ideas were to be added at the beginning. Fifth, it is easier to learn a series when its constitutive elements bear the same characteristics, that is, have analogical features.

One may conclude that Bonnet considered memory in terms of an eighteenth-century version of information processing. He was interested not only in the mechanics but also in the processing itself. He was interested in the physical nature of memory, the organization of fibers, and the physical mechanism of ideas (Bonnet, 1760, p. 92). He imagined a special structure, an internal organization responsible for memory. There are points of contact between fibers; at these overlapping points, something analogous to a network structure, he supposed that nervous energy could pass from one fiber to another, providing a physical basis for memory structures. For Bonnet the biological basis of recognition is the same as that in recall: changes in the state of nerve fibers. In addition, he also proposed to study recognition by the functional evidence of a feeling of novelty or salience.

Memory Loss

Bonnet also offered an explanation of memory loss, particularly in brain damage: simply put, it results from a change in nerve fibers. Bonnet (1760) first postulated that during the life of an individual, nerve fibers change, modifying themselves. He argued that

the sense fibers, like all structures of the animal body, vegetate, grow, perspire and wear out. All that perfectly presupposes movements, which themselves presuppose various changes in the actual state of the fibers. (Bonnet, 1760, p. 75)

Internal or external objects lead to impressions on fibers that induce movements in them, i.e., changes in their molecular structure and arrangement. Bonnet accepted the (then current) idea that in the physical world there is a tendency for physical bodies to remain in their current states, that is, once modified by stimulation, a nerve fiber tends to keep the same molecular organization and structure. As a counterpart of that physical phenomenon, there are many other forces, troubles, or breakdowns that tend to alter this state of stability (Bonnet, 1760, p. 95). He imagined that nerve fibers were made of what he called "nourishing atoms," *atomes nourriciers*, which organize themselves in fibers according to the nature of different objects. These fundamental elements of the structure of nerve

fibers cannot modify themselves without the action of an object on the senses, or, as we will see below, an object drawn from memory or imagination. After a nerve fiber is so stimulated, these atoms organize themselves with regularity and precision "in a manner designed to accurately conserve in the fibers the impressions or determinations that they had acquired" (Bonnet, 1760, p. 75).

However, some natural events or "foreign impulses" may come along and modify the organization of these atoms. For example, a delay following a particular stimulation (action of object) may create the occasion for some other event to slowly weaken the mnestic trace. Most of the time, these "disturbances" have nothing to do with the initial action of the sensed object. Yet, the smallest events, after a lapse of time, can greatly damage a mnestic trace. In the case where events, minor or important, are strong enough to change the particular (molecular) organization of these atoms, memory loss occurs.

Other atoms, that may have come to be incorporated in the affected fibers within a particular timeframe, may not be able organize themselves in the same manner and with enough precision to permit correct recall or recognition. The more changes in the organization laid down by the original stimulation, the more noticeable is the memory deficit. Thus, Bonnet explained memory loss by a regular or irregular (it depends on the quality and frequency of the foreign impulses) change in the internal organization of nerve fibers. He assumed that, despite the fact that once modified a fiber would then tend to preserve its new structure; there are many events that may efface the recently changed state by replacing the atoms, modifying the previous change.

With time, naturally, even without these decisive events, some memory loss will still occur. Bonnet explained that

finally, in due course, the fibers no longer remain, neither the molecules nor the fibers have retained anything of these (original) impressions; the memory of the sensations will be lost to the mind; and when the (previously familiar) objects newly act on these fibers, they will move them as if they had never been transformed; the sensations that will be born in the mind thus will have the character of newness or novelty. The opposite happens if one presupposes that the objects act so frequently on the fibers as to negate the effect of these foreign impulses. Some fibers that have reached the point of losing the impressions they had received of an object are, so to speak, reactivated by this object while the new (impression) acts on them." (Bonnet, 1760, p. 76)

A summary of Bonnet's ideas on memory, as reviewed above and as found in his other publications may include these features:

1. The memory span is six objects. William James observed "By Charles Bonnet the mind is allowed to have a distinct notion of six objects at once." This very well may have been an idea original to Bonnet.
2. Recall is different from recognition.
3. It takes longer to acquire memory of a complex idea than each simpler part of the idea.
4. To learn a complex idea requires repetition (preferably, in the same order as originally presented).
5. In case of a very complex idea, learning is facilitated by dividing it up into equal parts.
6. It is more difficult to add something to an already-learned complex idea than it is to learn the new thing at the very beginning of acquiring the complex idea.
7. Novelty is more salient than that which is familiar; saliency increases the intensity (of movements or vibrations).
8. Memory loss (in brain damage or in aging) is a structural rearrangement or deterioration of the original organization of the nerve fiber bundles.
9. The biological basis of memory is found in a collection of nerve fibers. A fiber is analogous to a cell or a unit; each fiber has a primitive, molecular structure; these parts determine its type or role and they are specific to each sense and specific to each activity of each sense.
10. Memory can be described with a myriad of metaphors: branches of a tree, different colors in a sunbeam, a network of fibers for memory.
11. The points at which fibers connect with each other are called nodes; at these points, memories share features that are in common.
12. Each collection of fibers is (structurally) different for different ideas.
13. Preservation of a memory involves molecular or structural change, i.e., it is a code, not a virtual image; memory, recall, and recognition are changes-of-state of nerve fibers.
14. Movements or shaking change the molecular structure of fibers and the fibers communicate with each other via the same movements or shaking.

Visions (Now Known as The Charles Bonnet Syndrome)

Bonnet's psychophysical parallelism outlined above became a natural model for the explanation of quite curious hallucinatory phenomenona, experienced by his maternal grandfather, Mr. Charles Lullin, the first reported case of being consciously aware of visual hallucinations. We are fairly certain that Bonnet and others regarded these hallucinations with curiosity and skepticism, because the written manuscript was signed and witnessed as a legal document, on the one hand, and because it was never formally published by Bonnet, despite its being written at least 30 years before his death.

His grandfather's visual experiences are neither an illusion nor a delusion, according to Bonnet. He insists that there is no "impression from the outside," that Lullin is affected as if objects were really present and this appears to happen in a specific "part of the brain" involved in visual processing. In modern terms, a hallucination is defined as a perception of something (a visual image or sound) with no external cause usually arising from a disorder of the nervous system. Bonnet (1760) assures us of Charles Lullin's awareness of the hallucinatory perception; he is lucid and interacting with others when these hallucinations occur and does not confuse these hallucinations with reality. The following are representative summaries from *Visions de l'ancien syndic Charles Lullin de Confignion*, the original manuscript left unpublished by Bonnet and later published by Flournoy (1901).[4]

The first vision (February, 1758) of a blue handkerchief occurred 5 years after the removal of a cataract over the left eye and 9 months after the removal of a cataract over the right eye, of concern to eighteenth- and nineteenth-century physicians (Flournoy, 1901) but in our view indirectly if at all related to the visual hallucinations. The blue handkerchief appeared in Lullin's line of vision no matter the direction of his gaze. Accompanying that vision, the drapes and furniture of his apartment

---

[4] Translated from the French by both authors.

appeared to be covered with a clear brown cloth embroidered with clover leaves. Also accompanying that vision were several tall young ladies who were well groomed with nice looking coiffures, some of whom had a small container on their heads. There was also an upside-down table which moved toward and away from Mr. Lullin, while the young ladies moved to the left visual field. He also saw a small machine called a thread mill that was "artistically" composed of a number of small rotating sticks.

The second vision comprised two young men about 20-years-old, neatly dressed, one in a red suit and the other in a gray suit, each wearing hats with silver edges. In a later vision the tall young ladies reappeared, now with pink ribbons in their well-coiffed hair; they again "left" (i.e., the vision stopped) to Lullin's left side. In a subsequent vision, two very well-dressed tall ladies, almost as tall as the ceiling of the room, appeared; they left without saying anything to Lullin. Moments later, on looking out of the window, a coach appeared; the top deck on which the coachman sat appeared to be as high as the house next door, about 30 ft high.

Later visions included seeing men with various height distortions outside the window, other structures outside the window also height-distorted, and inside the apartment a continuation of visions of color and pattern variations on the tapestries seen originally, as well as new paintings on the walls, a collection of old books on wall-shelves, a fancy table setting of golden plates and dishes, an elaborate exterior wall (construction) pattern, and the ever-present aforementioned young ladies. At one point an arch appeared in Lullin's room; it opened up to a wide, vaulted, brightly illuminated office; a painting on the office wall showed trees and landscape, a river coming from a very long (sic) city in the middle of which was a white, pointed, iron bell-tower. Several smaller pictures were arrayed below this large picture, from 5 to 6 in. wide down to the size of one's hand. Also at the bottom of the wall there was a 9–10-in.[2] glass mirror with a golden frame. Another vision comprised a group of ten pigeons with tufted legs resting on a branch of wood outside of Lullin's window. One of the pigeons moved its tail; yet another vision, a group of fast-spinning atoms (sic).

Evidently, Lullin's visions only occurred when he was fully awake and alert; they typically, if not exclusively, appeared in the left visual field and disappeared to the left. In August 1758, Lullin was 90-years-old, ostensibly in good health except for residual visual acuity problems from cataracts and cataract surgery. He had evidently lost vision in his left eye before the visions started; the acuity of his right eye was, relatively speaking, all right. The visions ended in September 1758, approximately 7 months after they had begun. The document recording the visions was signed by Lullin, another witness and Charles Bonnet, in the following year, 1759; Bonnet describes Lullin, his maternal grandfather, as being of sound judgment, good memory, and perfect health at the age of 91. Lullin passed away in 1761. In Bonnet's own words:

I will limit myself therefore to simply say that I know a respectable man, full of health, disgenuousness, judgment and memory, who, in full alertness and independently of all impression from the outside, sees from time to time before him, faces of men and women, of birds, of coaches, of houses etc. He sees these things exhibiting different movements, to approach then to move away or flee, then to greatly increase or decrease in size, to suddenly appear then to disappear and reappear again. He sees buildings rising under his eyes, and present to him all of their parts that enter into their outside construction; the tapestries of his apartment seems to him to change all of a sudden, becoming tapestries of another texture and color or richer design. Once, he sees tapestries covered with pictures representing different landscapes. Another day, instead of tapestries and furniture, it is only undecorated walls. Once, it is scaffoldings. All these visual images seemed to him of a perfect clarity and to affect him with as much vivacity as if the objects themselves were present; but, it is only visual images, because the men and women do not speak and no noise affects his ear. All appears to have its seat in the part of the brain that answers to the organ of the sight . . . But, what it is very important to notice is that this old man does not take, as visionaries do, his visions for reality: he knows how to sensibly judge all these apparitions and to always separate his thoughts and judgments from what he sees. (Bonnet, 1760, pp. 426–428)

Bonnet accounted for hallucinations as intense internalized (i.e., not stimulated by external sensations) movements or impulses in nerve fibers. The origin of these movements is either in the imagination or in the memory. These movements are sufficiently intense ("shaking") that the mind is unable to control them and thus they come to overpower or dominate the rational thought processes that rely upon the external senses. His explanation strongly suggests a dissociation between two orders or types of fibers, capable of being independently put into movement, thus accounting for a hallucination without disruption of conscious thought. Bonnet's observation that these distorted images do not destroy rational thought is quite comparable to the contemporary notion of "conscious hallucinations," the hallmark of what is now called The Charles Bonnet Syndrome.

Conclusion

Bonnet has been regarded as an associationist (empiricist) philosopher, comparable to the French sensationalists. Although his cognitive neuroscience is imbued with the traditional concept that basic ideas were necessary to engender more complex ones, his interpretation of association develops a conception of brain–behavior relationships based on a psychophysiological parallelism that places emphasis on his theory of nerve fiber structure and function. Bonnet argues that each cognitive process is matched by equivalent changes in nerve fibers and vice versa. He describes the changes in the brain as "movements," not quite like the vibrations of David Hartley and others, but clearly analogous to them. Bonnet's changes in the brain's nerve fibers, unlike Hartley's, are structural changes, i.e., experiences change their "molecular" structures, they become reorganized after being stimulated. This reorganization is a change in the inner structure of the nerve fiber, resulting in an acquired "disposition" of that nerve fiber to respond to the same kind of stimulation next time, thus being the substratum for learning and memory.

Movements were thought to be provoked either by external or internal stimulation upon nerve fibers (such as in perception and in imagination). Thus, according to Bonnet, hallucinations are produced when the movements are inside, when internal impulses of the brain are intense. Such intense movements, or shaking, cannot be "contained" by the mind and subsequently come to dominate intellectual and emotional life, just as the ideas can occupy the imagination or the memory of the individual.

Bonnet's interpretation of Charles Lullin's hallucinations rests on a differential and independent pattern of movement in nerve fibers in the visual areas of the brain. Under certain circumstances, the mind can be misled by movements that, having as much vivacity or intensity as real objects, cause the individual to see imaginary objects. In the absence of any alteration of consciousness, these imaginary objects become integrated with the person's experience of the real world. In this case, in order to explain the absence of a change in consciousness, Bonnet argued that the nerve fibers responsible for thought processes or "reflection" can move independently of those accountable for vision. Obviously, Charles Lullin was not in a distorted state of consciousness. Bonnet's case of visual hallucinations in an elderly relative was, as Flournoy (1901) and Savioz (1948) mention, not only the first but also one of the classic cases of visual hallucination without change in consciousness.

Bonnet's system can be said to be cognitive for two reasons. First, it addresses how the mind internally represents the external world and the mental operations required for thinking and cognition. Second, it focuses on the mental operations associated with sensation, perception, attention, memory, language, and reasoning. Moreover, Bonnet's elucidation of the visions of Charles Lullin may be considered an early attempt to address two fundamental issues in the history of cognitive neuroscience: one being the physiological foundation of the psychological issue of association of ideas, and the other being the neurological issue of whether the whole brain is working in concert or whether parts of the brain may function independently.

According to Bonnet, in imagination, such as in dreams, a stimulus in the internal world (a mental image) imprints weaker movements in nerve fibers than do the real objects. As a consequence, the stimulation from a sensation is more effective than the one from a mental image, whether that mental image is new as from the imagination or taken from memory while awake. Bonnet assumed that any "disposition" imparted to nerve fibers fades away with time, in the absence of additional stimulation in that nerve fiber, thus accounting for memory loss. However, even the less intense mental image can be experienced vividly if attention is focused on those nerve fibers. By attention, Bonnet suggests a special power of the mind to act on nerve fibers; the mind may act on certain nerve fibers to make a formerly weak and temporary impression, a strong and durable one.

References

Bonnet, C. (1755). *Essai de Psychologie; ou Considé-rations sur les opérations de l'âme, sur l'habitude, et sur l'éducation. Auxquelles on a ajouté des principes philosophiques sur la Cause première et son Effet.* Londres: La Société Typographique.

Bonnet, C. (1760). *Essai analytique sur les facultés de l'âme.* Copenhague: C. & A. Philibert.

Bonnet, C. (1770). *La palingénésie philosophique, ou idées sur l'état passé et sur l'état futur des étres vivants: ouvrage destiné à servir de supplément aux derniers écrits de l'auteur et qui contient principalement le précis de ses recherches sur le christianisme.* Paris: Le Clerc.

Flournoy, Th. (1901). Le cas de Charles Bonnet, hallucinations visuelles chez un vieillard opéré de la cataracte. *Archives de psychologie de la Suisse Romande, 1* (1), 1–23.

Glassman, R. B., & Buckingham, H. W. (2007). David Hartley's neural vibrations and psychological associations. In H. A. Whitaker, C. U. M. Smith, & S. Finger (Eds.), *Brain, mind and medicine: Essays in 18th century neuroscience* (pp. 177–190). New York: Springer.

Nicolas, S. (2006). Introduction: La vie et l'oeuvre de Charles Bonnet (1720–1793). In Ch. Bonnet (Ed.), *Essai de psychologie* (pp. 5–32). Paris: L'Harmattan.

Savioz, R. (1948). *La philosophie de Charles Bonnet de Genève.* Paris: Vrin.

Smith, C. U. M. (1987). David Hartley's newtonian neuropsychology. *Journal of the History of the Behavioral Sciences, 23,* 123–136.

Wallace, Wes. (2003). The vibrating nerve impulse in Newton, Willis and Gassendi: First steps in a mechanical theory of communication. *Brain and Cogntion, 51,* 66–94.

14
Swedenborg and Localization Theory

Ulf Norrsell

Present-day concepts of cerebral, localized, motor areas commenced with a set of stimulation experiments with anaesthetised dogs done by two German physicians, Fritsch[1] and Hitzig[2] (1870). Applying weak electrical pulses to exposed cerebral cortex, they were able to evoke muscular contractions in the opposite body half. They varied stimulation intensity, and established its lowest effective magnitude, or threshold strength. This stimulation intensity was selected in order to avoid activation of deep structures, and they could see how muscle groups on different parts of the contralateral body half were activated from different spots on the cortical surface. Similar effects were obtained from roughly the same locations in different animals. Forelimb and facial muscles were activated from spots located lateral to those of hindlimb muscles. In their paper's introductory survey of earlier and contemporary publications, the two authors made it clear that the findings were contrary to established opinion. Neither the possibility of cortical activation, nor cortical localization of function were definitely known to exist, although sometimes surmised. The findings were not accepted immediately, but started a great number of tests internationally, and in short time engendered an atmosphere of electric excitement, to paraphrase Young (1970).

An indication that the concept of cortical localization, and even part of its details had been written down more than a hundred years earlier materialized soon afterwards. The year 1882 saw the publication of the first volume of *The Brain*, an English translation of Latin manuscripts written in the early 1740s by a polymathic, Swedish nobleman by the name of Emanuel Swedenborg, who was also a mining professional, and religious visionary (Swedenborg, 1882). The published translation did not make an immediate impact. For example, 15 years later the Austrian medical historian Max Neuburger did not mention Swedenborg at all in his monograph on the development of experimental neurophysiology prior to the nineteenth century (Neuburger, 1897). However, he remedied the omission 4 years afterwards in a speech delivered in a meeting of German naturalists and physicians in Hamburg.[3] Swedenborg's neuroscientific achievements were now acclaimed. The last sentence of Neuburger's speech reads, translated into English: "We owe indemnity to Swedenborg's memory, and fulfil a long neglected duty, when in the future we attribute more to his scientific endeavours than a spiritualist's dreams" (Neuburger, 1901, p. 2081).

---

[1] Gustav Theodor Fritsch (1838–1927).
[2] Eduard Hitzig (1838–1907).

---

[3] Swedenborg's work does not seem to have attracted much interest in the eighteenth century although he was once cited by Albrecht von Haller (Akert & Hammond, 1962). It was never entirely forgotten owing to his religious following, and some Swedish scientists. The anatomo-physiological works published during his lifetime were translated from Latin to English, and republished in the 1840s. Nevertheless, Neuburger's Hamburg lecture appears to mark a start of more general interest.

The final words were hardly chosen by mistake. In Neuburger's German original they are 'Träume eines Geistersehers' and correspond to the three first words in the title of a booklet by Immanuel Kant (1766, 1983). This work contains a withering assessment[4] of Swedenborg's mental faculties. The author's distinction implies that Swedenborg was a personage of European magnitude. He was born in Stockholm in 1688. His father Jesper Svedberg was a royal chaplain with a mining family background, who later became bishop of Skara diocese in southwest Sweden. Emanuel's mother, Sara Behm was the daughter of an affluent mining professional. The Svedberg children were ennobled in 1719, and changed their surname to Swedenborg. Emanuel was the head of this noble family, and consequently their representative for the aristocrats' chamber of the four chamber Swedish Riksdag (diet). Technological, scientific and theological activities apart, he was successfully politically active into his seventh decade. He died in 1772 in London (Åkerberg & Holmquist, 1918; Kleen, 1917; Sandel, 1772).

Swedenborg graduated from Uppsala university in 1709, although without revealing star quality. At least, in the official obituary of the Swedish Royal Academy of Science it is noted somewhat ambiguously, that his dissertation: "… may be placed among fair specimens of youthful, but often less reliable tests of scholarship" (Sandel, 1772, p. 8, translated into English). Four years of foreign travels followed, and he visited England, the Netherlands and France to study physics and mathematics. He also acquired practical skills, viz. watchmaking, carpentry, instrument-making, engravery and glass-grinding.

After returning home, his competence was recognized by King Carl XII, who made him extraordinary assessor of the board of mining[5] in 1716. At this time he published a journal called *Daedalus hyperboreus* containing, among other things, his blueprint of an aeroplane (Åkerberg & Holmquist, 1918). He constructed canal locks, and in 1718 supervised the possibly more than 25 km long, overland transportation of parts of the Swedish

royal navy. The aim was to reinforce the 1718 siege on the Norwegian city of Halden, that was subsequently raised after the Swedish king had been shot (Sandel, 1772). Swedenborg kept his position at the board of mining until retirement in 1747. It was permanent after 1724, but was interrupted repeatedly by long leaves of absence on half pay.

In the early 1720s he travelled to Germany and the Netherlands, and apart from studying mineralogy, metallurgy and mining, he published his first larger work in natural philosophy during this journey (Swedenborg, 1721). He also won the support of Prince Ludwig Rudolf of Blankenburg. Ludwig Rudolf was no run-of-the-mill prince, but eventually reigning duke of Braunschweig-Wolfenbüttel. He was father in law of the German emperor Karl VI, and therefore empress Maria Theresia's grandfather. Another daughter was married to Alexei, the disloyal, only son of Tsar Peter the Great. Ludwig Rudolf paid for the printing of Swedenborg's three folio volumes, 1,400 pages long work on mining and metals, the *Opera philosophica et mineralia*, which appeared 1734 (Åkerberg & Holmquist, 1918). Its publication perhaps triggered Swedenborg's appointment to corresponding member of the scientific academy of St Petersburg, in the same year. Its second volume, *Regnum subterraneum sive minerale de ferro* (Swedenborg, 1734b), was translated into French and published three times either in part, or in its entirety within the next 40 years.[6] Swedenborg by now must have been well recognized, and his portrait at the age of 40 in Fig. 1 shows a luminary.

The year of the *Opera* also saw the publication of an anatomical and physiological monograph (Swedenborg, 1734a) containing among other things a description of the nervous system partially based on work by Heister.[7] This was the start of serious investigations regarding the body and the nervous system. His aim was to study organic nature. His goal was to approach the soul on a scientific trail. The scientific approach would reveal the soul's materialistic face, whereas its immaterial face was turned towards God,

---

[4] The title page of Kant's booklet carried the quotation from Horace, "*velut aegri somnia, vanae Figuntur species* (delusions are created like the dreams of a sick person)".
[5] *Bergskollegium*; an agency that existed in 1630–1857, and legally supervised Swedish mining industry (Nordström, 1904).

[6] 1737 in Traité sur l'acier d'Alsace, ou l'art de convertir le fer de ponte en acier. 1761–1762 in Art des forges et forneaux à fer: avec figures en taille-douce 1774 in Descriptions des arts et métiers faites ou approuvées par messieurs de l'Académie royale des sciences de Paris.
[7] Lorenz Heister (1683–1758).

FIGURE 1. Emanuel Swedenborg at the age of 40. Copy from Kleen (1917)

and consequently might be reached through the bible (Åkerberg & Holmquist, 1918).

In the summer 1736 he left for Paris, where he stayed until March 1738. He lived at Rue de l'Observance, which was close to the Academie de Chirurgie, and convenient for studies of anatomy. His diary, translated and published by Tafel, tells little about the stay in Paris, but it cannot be excluded, and appears likely for him to have met and/or studied with Winsløw,[8] and possibly also Pourfour du Petit[9] (Jonsson, 1969; Swedenborg, 1877). After Paris he went to Italy, and visited Turin, Milan, Padua, Venice, Rome and Genoa. His activities during the Italian sojourn are also little known, but he may have

started working on a neurobiological manuscript during a five month stay in Venice (Acton, 1938). From Italy he went to Amsterdam in 1739. Here two volumes of his anatomical and physiological treatise, *De œconomia regni animalis*[10] were printed before his return to Sweden in 1741 (Swedenborg, 1740, 1741). His subsequent and final anatomical and physiological contributions were published in three volumes under the title, *Regnum animale* during his stay in the Netherlands and England 1743–1745 (Swedenborg, 1744a, 1744b, 1745).

Swedenborg's interests in natural philosophy were never devoid of religious connotations, and the year 1745 saw him turn fully to religion. The change coincided with religious crises he is described as having during 1744–1745. There are even descriptions of an acute psychotic phase during a stay in London in 1744 (Kleen, 1920).[11] His numerous publications after 1745 were with a few exceptions devoted to religious subjects, and provided basis for the founding of a Christian division after his death in the 1780s, called Church of the New Jerusalem. His religious reflections were based partially on supernatural visions, and he also acquired a reputation as medium. That reputation was the backdrop for Kant's (1766, 1983) above mentioned criticism. However, to the end Swedenborg appears to have been highly regarded by his contemporaries, despite possible psychiatric problems. Nevertheless, some eyebrows appear to have been raised. At least Sandel, lavishing praise in the obituary felt need for some justification: "If his (Swedenborg's) curiosity has gone too far, it reveals his fervent concern for enlightenment of himself and others; since with him we get no haughty disposition, conceit or intent to dupe" (Sandel, 1772, p. 23, translated into English).

---

[8] Jacob Benignus Winsløw (1669–1760). Danish-French physician, who was working in Paris for the greater part of his life. Winsløw is generally considered the best anatomist of his time (Enersen, 2005).
[9] Francois Pourfour du Petit (1664–1741). French surgeon anatomist. Attributed the contralateral paralysis caused by head wounds to lesions of the pyramidal tract in a publication 1710 (Kruger, 1963). Like his contemporary Winsløw he was a pupil of Joseph-Guichard Duvernay (1648–1730), who had performed lesion experiments, and is known for a treatise on the ear (Finger, 1994).

[10] The same title was also used for posthumously published Swedenborg manuscripts, e.g. Swedenborg (1918).
[11] Emil Anders Gabriel Kleen (1847–1923), Swedish physician, and diabetes specialist, who trained in psychiatry before writing his two-volume Swedenborg biography cited above. In its second volume he writes that he had personally discussed Swedenborg's case with the following psychiatrists: H. Maudsley (London), W. W. Ireland (Edinburgh), G. Ballet (Paris), B. Gadelius and O. Kindberg (Stockholm), F. Svenson (Uppsala). They all agreed that Swedenborg suffered from a paranoid condition with hallucinations.

In the chronological inventory of Swedenborg's texts commissioned by the Royal Swedish Academy of Science, anatomical and physiological pieces are almost entirely listed between years 1733–1745 (Stroh & Ekelöf, 1910).[12] Published books apart, there were numerous unpublished manuscripts, which were made available in the nineteenth century. Retzius (1903) and Kleen (1917) have testified about problems with reading Swedenborg's original Latin texts. They gratefully acknowledged the existence of English translations of *De œconomia regni animalis*, and *Regnum Animale*, and in addition the manuscript collection published in two volumes 1882 and 1887 and called, "The brain considered anatomically, physiologically and philosophically". Another collection of manuscripts was published in two volumes 1938 and 1940, with the title, *Three transactions on the cerebrum.*

Few persons could have been better suited to referee Swedenborg's neurophysiological contributions than Max Neuburger, who was thoroughly acquainted with the neuroscience of the eighteenth century (Neuburger, 1897). The preamble of Neuburger's above-mentioned Hamburg lecture contains a stirring, albeit verbose tribute:

I raise the great phantom from the shadows, only because his case provides a striking example of how now and then a brilliant, speculative theorist can draw conclusions from the raw empirical material, that includes the problem's heart, and penetrates deeper into its nature than the feeble conclusions of the respected representatives of scientific meticulousness. (Neuburger, 1901, p. 2079, translated into English)

Swedenborg was no experimentalist. He explained in Clissold's translation of the introduction of *Oeconomia Regni Animalis* that he had restrained his: "desire for making observations, determined rather to rely on researches of others than to trust my own" (Swedenborg, 1740, 1845, p. 8). In this manner he was the first to place the highest cognitive functions, including consciousness, exclusively in the cerebral cortex (Neuburger, 1901). Swedenborg's mode of reasoning was as follows. Like contemporaries, he considered Animal Spirits to be the basic mechanism. The glandular shape of the cells in the cortex (Malpighi) and its rich vasculature (Ruysch) made it a suitable place for animal spirit production.[13] The cerebral cortex had also been shown to be origin of the cerebral medullary fibres, from which peripheral muscle activation could be evoked (Malpighi, Swammerdam, Winsløw, and Boerhaave).[14] Hence, cortex was a logical source for decisions, and executive function.

Neuburger (1901) went on to describe how Swedenborg put voluntary motor activity under the control of the cerebral cortex. Muscle activation, on the other hand, was a second step, and depended on brain-stem and spinal structures. Swedenborg expressed the opinion, that originally cortical, voluntary motor activities, after training were taken over by deeper structures. In this way the cerebral cortex was substituted for, and could devote more power to higher spiritual functions. Swedenborg had also come to the conclusion that launching of different muscle activities issued from separate parts of cortex and, that a cortical motor map could be determined experimentally. His own way of expressing it was in Clissold's translation:

We now come to enquire what cortical tori correspond to the respective muscles in the body, and this we cannot ascertain except from experiments on living animals, by

---

[12] The inventory contains 175 entries, of which 106 are manuscripts and 68 are printed titles, in the latter case including multiple volume publications. The greatest number of entries relate to texts dealing with engineering problems, and problems of applied physics and chemistry. Such texts were mostly produced before 1740, except for two dated in the 1760s. One of the latter examined inlaying techniques for marble. The last one was a letter to the Royal Academy of Science about a method to determine longitudes, written 6 years before his death. The second largest number of items in the list of Swedenborg's writings has religious contents. These writings were almost without exception produced after

1740. However, the late start did not gainsay his lifelong religious interest. His religious assertions did not conform to Swedish canon. Therefore, he was forced to travel a great deal towards the end of life, to have books printed in the Netherlands or England. Nevertheless, many of these books could be distributed inside Sweden, thanks to influential friends. He did not lack enemies, on the other hand, whose attempts to have him indicted were unsuccessful (Åkerberg & Holmquist, 1918).

[13] Marcello Malpighi (1628–1694) and Frederik Ruysch (1638–1731).

[14] Jan Swammerdam (1637–1680) and Herman Boerhaave (1668–1738).

puncture, section, and compression in a variety of subjects, and then observing the effects produced in the muscles of the body. (Swedenborg, 1740, 1845, pp. 491–492)

Neuburger concluded that: "the thoughts and premonitions, that Swedenborg had provided triumphantly reached the verge of (present days)" (Neuburger, 1901, p. 2081, translated into English). Nevertheless, he did not mention what has become the most well known citation from Swedenborg's writings on the brain. It first became available through Tafel's translation of Swedenborg manuscripts (Swedenborg, 1882). It then probably became more generally known through a lecture by Gustaf Retzius (1903) at a meeting of the Anatomical Society at Heidelberg. A picture of the relevant manuscript page is shown in Fig. 2. The marked part reads in the translation of Ramström (1910, p. 24):

The muscles and actions which are in the ultimates of the body, or the soles of the feet, seem to depend more immediately upon the highest parts (of the anterior region of the brain)[15], the muscles which belong to the abdomen and thorax upon the middle lobe, those which belong to the face and head upon the third lobe; for they (the muscles of the body and the lobes of the brain) seem to correspond to one another in inverse order.

Ramström[16] (1910) also tried to explain how Swedenborg could have been able to make these remarkable predictions of latter day findings. Thus he found it possible to reach Swedenborg's conclusions regarding the functional supremacy of the cerebral cortex through the writings of Willis[17] (animal spirits), Malpighi (glandulæ), and Boerhaave (canals for animal spirits from glandulæ). The ideas about localization of different functions to different parts of the cerebrum could be attributed to clinical and pathological observations by, among others, Wepfer, Baglivi, and Pacchioni.[18] Swedenborg's description of an inverse representation of body parts in the cerebral motor areas may, according to Ramström, be attributed mainly to Vieussens'[19] anatomical descriptions. Vieussens had followed

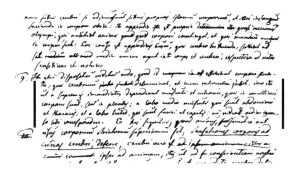

FIGURE 2. Copy from Swedenborg MS, codex 58, p. 227, with permission of Centre for history of science, Royal Swedish Academy of Sciences. The marked part of paragraph 9 reads, according to Retzius (1903, p. 11): "… imo ita ut a supremis immediatius dependeant musculi et actiones, quae in ultimis corporis sunt, seu in plantis; a lobo medio musculi qui sunt abdominis et thoracis, et a lobo tertio, qui sunt faciei et capitis: nam videntur ordine inverso sibi correspondere"

nerve tracts upwards through the corpus striatum and capsula interna, and found that they formed three regions in the centrum ovale (i.e. the great mass of white matter composing the interior of the cerebral hemisphere, and viewed in horizontal projection). These are, according to Ramström (1910, pp. 40–41): "the *regio superna*, highest up nearest the crown, the *regio media*, in the middle, and the *regio infima*, lowest down, and consequently nearest the fissure of Sylvius". Vieussens had also noted the downward projection of the same tracts, and according to Ramström (1910) had observed connections of the three regions in the spinal cord, the brainstem and the proximity of cranial nerves, which might indicate the inverse representation in the motor cortex suggested by Swedenborg.

Whatever the explanation for Swedenborg's to modern eyes apt suggestion, its detailed accuracy should not be exaggerated. The identity of the lobes mentioned in the citation of Fig. 2 remains

---

[15] The texts within parentheses had been added by Ramström, and provided with the following endnote (Ramström, 1910, p. 56): "The words enclosed in parentheses have been added by the author of this paper to make the meaning more clear, and are unmistakeably inferred from the connection."

[16] Oskar Martin Ramström (1861–1930), professor in anatomy at Uppsala University and, among other things, author of several papers on Swedenborg's anatomical and physiological observations.
[17] Thomas Willis (1622–1675).
[18] Johann Jakob Wepfer (1620–1695), Giorgio Baglivi (1668–1707), and Antonio Pacchioni (1665–1726).
[19] Raymond de Vieussens (1641–1715).

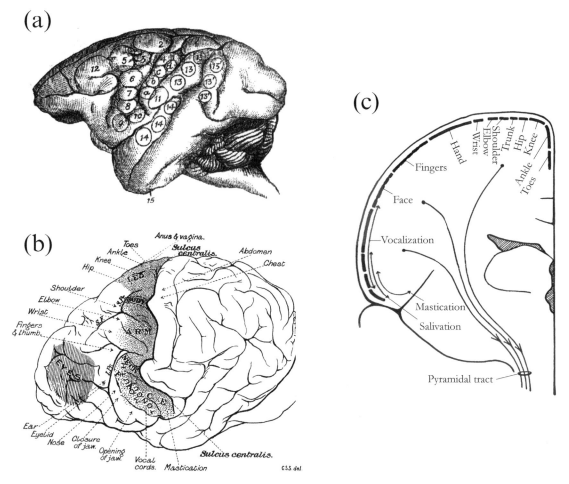

FIGURE 3. **(a)** Left hemisphere of monkey with markings of electrode positions from which different contralateral motor effects could be evoked. (Copied from Ferrier, 1876). **(b)** Left hemisphere of chimpanzee with indications of points for evoking different contralateral movements. (Copied from Grünbaum & Sherrington, 1901). **(c)** Frontal section of posterior part of frontal lobe of right, human, cerebral hemisphere with indications of different motor activities evoked from the pre-central gyrus (Redrawn after Fig. III-13 in Penfield & Jasper, 1954)

uncertain. Ramström (1910) suggested that Swedenborg divided the brain into a posterior, and an anterior part by means of the Sylvian fissure. The anterior part was divided into three lobes. Tafel, on the other hand, in his introduction of *The Brain* was of different opinion:

Swedenborg's division of the brain into lobes differs from that in use at the present day. His division is that which is presented by the median aspect of the hemispheres. His highest lobe is bounded by the marginal convolution and the quadrate lobule; his middle lobe by the lower part of the marginal convolution and the gyrus fornicatus; and his third lobe is identical with the temporo-sphenoidal lobe. (Tafel, 1882, p. XVII)

Tafel (1882) then mentions how well Swedenborg's description fitted the latest evidence described by Ferrier (1876). Ferrier[20] had followed in the footsteps of Fritsch and Hitzig and evoked contralateral muscle contractions by electrical stimulation of the cerebral cortex. His map of the motor cortex of the monkey is shown in Fig. 3(a), and conforms exactly with Tafel's interpretation of Swedenborg. Stimulation of the medial temporal cortex in the points marked 14 in Fig. 3(a) caused, among other things, 'pricking of the opposite ear'. Stimulation of medial

----

[20] David Ferrier (1843–1924).

frontal cortex caused movements of hindlimb and foot. However, techniques improved, and the map of the motor cortex had shrunk at the time of Ramström's alternative explanation of Swedenborg's lobes, and had also been tested in a higher primate. Figure 3(b) shows Grünbaum's and Sherrington's (1901) map of the motor cortex of the chimpanzee, which fits Ramström's (1910) interpretation nicely, i.e. with the third lobe, and the 'face and head' adjacent to the medial side of the Sylvian fissure.

Due to the need for functional mapping of the cerebral cortex, in connection with human neurosurgery, we now possess maps of the human motor cortex, on account of Penfield[21] and co-workers. Figure 3(c) shows critical parts of a map of the human motor cortex published by Penfield and Jasper (1954). It is displayed together with a frontal outline of one hemisphere, in order to show how ankle and toes are located on the medial surface of the hemisphere. Swedenborg (1882) was obviously referring to the human brain, and consequently it is not "the soles of the feet", but the knees, which are found at the top of the hemispheres.

In the absence of the technologies permitting Fritsch's and Hitzig's (1870) approach, there was no possibility to determine the perfect details of the cerebral motor map in the eighteenth century. Swedenborg's descriptions published and unpublished were remarkable enough. Nevertheless, his theories do not seem to have evoked any interest at all among contemporaries. Neuburger (1901) suggested several factors possibly contributing to this nothingness, Kant's censure apart; Swedenborg's secluded lifestyle, insufficiently headstrong temperament, writing technique. In addition, Swedenborg's contribution was one among several unverified hypotheses extant at the time and described by Neuburger (1897). Swedenborg's feat was to produce an amazingly accurate result by reading, and keeping track of active work of natural philosophers and men of medicine. In the introduction of his *Economy of the animal kingdom* it says:

In the experimental knowledge of anatomy our way has been pointed out by men of the greatest and most cultivated talents; such as, Eustachius, Malpighi, Ruysch, Leeuwenhoek, Harvey, Morgagni, Vieussens, Lancisi, Winslow, Ridley, Boerhaave, Wepfer, Heister, Steno, Valsalva, Duverney, Nuck, Bartholin, Bidloo, and Verheyen; whose discoveries, far from consisting of fallacious, vague and empty speculations, will for ever continue to be of practical use to posterity. (Swedenborg, 1740, 1845, p. 7, translated by Clissold)

Swedenborg's attitude invokes a citation attributed to the eminent American physicist Joseph Henry: 'The seeds of great discoveries are constantly floating around us, but they only take root in minds well prepared to receive them' (Finger, 2000, p. 201).

Acknowledgements. Supported by the Adlerbert Science Foundation. Dr. Axel Karenberg told about Kant's interest in Swedenborg. The ready cooperation of the Biomedical library of Göteborg university, the Archive of the Swedish Royal Academy of Science and the library of the Swedish military archives was of great help.

References

Acton, A. (1938). Translator's preface. In A. Acton (Ed.), *Three transactions on the cerebrum: A posthumous work* (Vol. 1, pp. XI–XXIX). Philadelphia, PA: Swedenborg scientific association.

Åkerberg, A. F., & Holmquist, H. F. (1918). Swedenborg, Emanuel. In T. Westrin, V. Leche, J. F. Nyström, & K. Warburg, (Eds.), *Nordisk Familjebok – Konversationslexikon och Realencyklopedi* (Vol. 27, pp. 906–926). Stockholm: Nordisk familjeboks förlags aktiebolag.

Akert, K., & Hammond, M. P. (1962). Emanuel Swedenborg (1688–1772) and his contributions to neurology. *Medical History, 6*, 255–266.

Ferrier, D. (1876). *The Functions of the Brain*. London: Smith/Elder & Co.

Finger, S. (1994). *Origins of neuroscience. A history of explorations into brain function*. New York: Oxford University Press.

Finger, S. (2000). Neurognostics: Question 10: And answer. *Journal of the History of the Neurosciences, 9*, 201–202.

Fritsch, G. T., & Hitzig, E. (1870). Ueber die elektrische Erregbarkeit des Grosshirns. *Archiv für Anatomie, Physiologie und wissenschaftliche Medicin – Leipzig, 37*, 300–332.

Grünbaum, A. S. F., & Sherrington, C. S. (1901). Observations on the cerebral cortex of some of the higher apes. *Proceedings of the Royal Society of London, 69*, 206–209.

Jonsson, I. (1969). Köpenhamn–Amsterdam–Paris. Swedenborgs resa 1736–1738. *Lychnos Lärdomshistoriska Samfundets Årsbok, 1967–1968*, 30–76.

---

[21] Wilder Penfield (1891–1974).

Kant, I. (1766/1983). Träume eines Geistersehers. In: W. Weischedel (Ed.), *Werke in zehn Bänden* (Vol. 2, pp. 921–989). Darmstadt: Wissenschaftliche Buchgesellschaft.

Kleen, E. A. G. (1917). *Swedenborg. En Lefnadsskildring* (Vol. 1). Stockholm: A.-B. Sandbergs Bokhandel.

Kleen, E. A. G. (1920). *Swedenborg. En Lefnadsskildring* (Vol. 2). Stockholm: A.-B. Sandbergs Bokhandel.

Kruger, L. (1963). Biographical note on Francois Pourfour du Petit. *Experimental Neurology, 7*, III–V.

Neuburger, M. (1897). *Die historische Entwicklung der experimentellen Gehirn- und Rückenmarksphysiologie vor Flourens.* Stuttgart: Ferdinand Enke.

Neuburger, M. (1901). Swedenborg's Beziehungen zur Gehirnphysiologie. *Wiener Medizinische Wochenschrift, 51*, 2077–2081.

Nordström, T. (1904). Bergskollegium. In V. Leche, B. Meijer, J. F. Nyström, K. Warburg, & T. Westrin (Eds.), *Nordisk Familjebok – Konversationslexikon och Realencyklopedi* (Vol. 2, pp. 1493–1494). Stockholm: Nordisk familjeboks förlags aktiebolag.

Penfield, W., & Jasper, H. (1954). *Epilepsy and the functional anatomy of the human brain.* Boston: Little, Brown and Company.

Ramström, M. (1910). *Emanuel Swedenborg's investigations in natural science and the basis for his statements concerning the functions of the brain.* Uppsala: Uppsala Universitet.

Retzius, G. (1903). Emanuel Swedenborg als Anatom und Physiolog auf dem Gebiete der Gehirnkunde. Eröffnungsrede des Vorsitzenden Prof. Dr. GR an dem Anatomenkongresse in Heidelberg d. 29 Mai 1903. Abdruck aus den Verhandlungen der Anatomischen Gesellschaft auf der siebzehnten Versammlung in Heidelberg von 29. Mai bis 1. Juni 1903 (pp. 1–14). Jena, Verlag von Gustav Fischer.

Sandel, S. (1772). *Åminnelse – Tal, öfver Kongl. Vetenskaps – Academiens framledne Ledamot, Assessoren i Kongl. Maj:ts och Rikses Bergs – Collegio, herr Emanuel Swedenborg, på Kongl. Vetenskaps – Academiens vägnar, hållet i stora Riddarehus – Salen den 7 octobr. 1772.* Stockholm: Direct. Lars Salvius.

Stroh, A. H., & Ekelöf, G. (1910). Kronologisk förteckning öfver Emanuel Swedenborgs skrifter 1700–1772. *K Svenska Vetenskapsakademiens Årsbok bil, 3*, 1–50.

Swedenborg, E. (1721). *Prodromus principorum rerum naturalium sive novorum tentaminum chymiam et physicam experimentalem geometrice explicandi.* Amsterdam: Johan Oosterwyk.

Swedenborg, E. (1734a). *Prodromus philosophiae ratiocinantis de infinito, et causa finali creationis: deque mechanismo operationis animae et corporis.* Dresden/Leipzig: Friedrich Hekel.

Swedenborg, E. (1734b). Regnum Subterraneum sive minerale de Ferro. *Opera Philosophica et Mineralia* (Vol. 2). Dresden/Leipzig: Friedrich Hekel.

Swedenborg, E. (1740/1845). Introduction. In A. Clissold, (Ed.), *The economy of the animal kingdom* (Vol. 1). London: W. Newbery.

Swedenborg, E. (1740). De sanguine, ejus arteriis, venis et corde agit: anatomice, physice & philosophice perlustrata cui accedit introductio ad psychologiam rationalem. *Oeconomia regni animalis in transactiones divisa* (Vol. 1). London/Amsterdam: Francois Changuion.

Swedenborg, E. (1741). Quarum haec secunda De cerebri motu et cortice et de anima humana agit: anatomice, physice & philosophice perlustrata. *Oeconomia regni animalis in transactiones divisa* (Vol. 2). London/Amsterdam: Francois Changuion.

Swedenborg, E. (1744a). De visceribus abdominis seu de organis regionis inferioris. *Regnum animale anatomice, physice et philosophice perlustratum* (Vol. 1). Haag: Adrian Blyvenburg.

Swedenborg, E. (1744b). De visceribus thoracis seu de organis regionis superioris. *Regnum animale anatomice, physice et philosophice perlustratum* (Vol 2). Haag: Adrian Blyvenburg.

Swedenborg, E. (1745). De cute, sensu tactus, et gustus; et de formis organicis in genere. *Regnum animale anatomice, physice et philosophice perlustratum.* London.

Swedenborg, E. (1877). Swedenborg's journal of travel from 1736 to 1739. In R. L. Tafel (Ed.), *Documents concerning the life and character of Emanuel Swedenborg* (Vol. 2(1), pp. 75–130). London: Swedenborg society.

Swedenborg, E. (1882). Chapter I – The cerebrum, its fabric, motion, and function in general. In R. L. Tafel (Ed.), *The brain. Considered anatomically, physiologically and philosophically* (Vol. 1, pp. 9–102). London: James Speirs.

Swedenborg, E. (1918). The medullary fibre of the brain and the nerve fibre of the body, the arachnoid tunic, diseases of the fibre. In A. Acton (Ed.), *The economy of the animal kingdom: Considered anatomically, physically and philosophically: Transaction III.* Philadelphia.

Tafel, R. L. (1882). Editor's Preface. In R. L. Tafel (Ed.), *The brain. Considered anatomically, physiologically and philosophically* (Vol. 1, pp. VII–XXX). London: James Speirs.

Young, R. M. (1970). *Mind, brain and adaptation in the nineteenth century.* London: Oxford University Press.

Section E
Medical Theories and Applications

Introduction

During the long eighteenth century there were significant developments in both theoretical and clinical (bedside) medicine. In the domain of theory, Newtonian mechanics had a great impact. With an emphasis on ethers, particles in motion, vibrations, and physical forces, neurological and other medical problems tended to be viewed more from mechanical perspectives as less in terms of medical chemistry.

Especially to many college-educated physicians, the new iatromechanical theories, with their emphases on subtle spirits traveling through a network of tubes, and with problems being caused by "blockages," had considerable appeal. Even if such spirits and blockages could only be inferred, they seemed to provide adequate explanations for a whole range of clinical phenomena. Yet to others, these new explanatory concepts seemed better when melded with earlier theories: not only iatrochemical notions, but also humoral ideas with more ancient roots. Near the close of the century, neurological theories based on an electrical fluid also entered the theoretical picture, eventually changing virtually everything.

Of course, all was not theory, especially to those bedside practitioners who did not receive formal training at the leading medical schools in the eighteenth century. Especially in the New World, where few practitioners had medical degrees and the profession was decidedly less regulated than in Britain, the orientation they adopted was often decidedly more pragmatic. Open to recommended cures for regional diseases, snake bites, and insect stings from earlier settlers and even the Indians, many practitioners in the British North American Colonies were not as concerned with the latest academic theories as their cousins, who were enrolled at Leiden (see chapter by Peter Koehler) or Edinburgh (see Julius Rocca's earlier chapter), two of the most respected medical schools at the time.

As might be inferred from these statements, bedside medicine included some new remedies and many old ones – both tending to be incorporated into new systems of medicine by physicians "of the faculty." Thus, although this was a time when the so-called "heroic therapies" – which included bleeding, sweating, and purging – were very popular, the aim of many eighteenth-century physicians was no longer to balance the four humors of antiquity, but to remove suspected obstacles, purulent matter, and blockages. For less theoretically oriented practitioners, the attraction might have been that some of these time-honored regimens seemed to be followed by some improvement (cause often mistaken), or that friends and respected sources assured them that they should work.

Medical electricity emerged during the middle of this century. To many iatromechanical physicians, electricity was the most powerful stimulant known, and therefore ideal for such things as moving sluggish fluids down tubes and tightening flaccid nerves. To others, especially those involved with "the common paralytick disorder," the palsy examined by Catherine Story in her chapter, it might have been no more than the observation that electricity could make even the paralyzed muscles of a stroke victim vigorously contract.

During the 1740s, 1750s, and 1760s, as illustrated by Hannah Locke and Stanley Finger, *Gentleman's Magazine* was filled with glowing reports on this new cure for a myriad of disorders, including various palsies, hysteria, and even deafness. John Wesley, the Methodist cleric with more than just a passing interest in medicine, as shown by James Donat, was among electricity's early supporters. Nevertheless, others quickly questioned whether it really was anything like a universal cure, including Benjamin Franklin, the century's leading electrical scientist, who began to his therapeutic "tryals" not in a European center of learning but in Colonial Philadelphia (see the chapter by Stanley Finger). Just what disorders could be cured with electricity was an issue that maintained Franklin's interest for more than four decades, and it was also a matter that continued to be studied and debated by many others. As shown by Paola Bertucci, therapeutic electricity captured just about everybody's imagination in the second half of the eighteenth century.

A second possible panacea or potential universal therapy to be examined in this section emerged a bit later and was based on a theorized, pervasive, invisible force, much like Newton's gravity. Franz Anton Mesmer claimed to be able to cure all sorts of disorders, including those related to the nerves, by directing "animal gravity," which he later termed "animal magnetism," though a patient's body. That some people seemed to be helped by his séances and the fanfare revolving around his miraculous treatments was less of an issue than his metaphysical theory, which, as shown by Douglas and Joseph Lanska, was officially put to the test in France in 1784. Franklin, Lavoisier, and the other commissioners who delved into the matter found no supporting evidence for Mesmer's theories and conjectures, confirming what Franklin thought at the outset. Instead, with the help of some very clever and systematic experiments, they showed just how powerful suggestion and imagination could be in clinical medicine, setting the stage for important nineteenth-century developments, including the use of hypnosis with patients suffering from hysteria (the disorder examined in Diana Faber's chapter), and related ailments that were increasingly being ascribed to affected or disordered minds.

With these thoughts about medical theories and the bedside practice of physic before us, and recognizing that both were varied and undergoing changes, let us now look at some of the people, teachings, issues, and complications that made eighteenth-century medicine so unsettled, but at the same time such a fascinating field of study for historians.

The Editors

15
Neuroscience in the Work of Boerhaave and Haller

Peter J. Koehler

Introduction

After his medical preparation with Professor Duverney in Tübingen (Germany), Albrecht von Haller (1708–1778) started a peregrination through Europe. The first country where he stayed was Holland.[1] His motivation to visit Leiden probably included his reading of Boerhaave's *Institutiones Medicae* (1708), which had a physiological character (Lindeboom, 1958, p. 14). Haller's impression of the first lecture by Boerhaave he attended was quite positive. "I listened to him from 1725 to 1727 for somewhat more than two years. I remember that I was filled with an unbelievable delight, when I heard him explain for the first time the true medicine with extraordinarily charming eloquence."[2]

At that time, Herman Boerhaave (1668–1738) was at the height of his career as professor of botany and medicine (since 1709) and chemistry (1718) at Leiden University. When Haller left Holland, he had finished his thesis[3] and continued his studies in London and Paris. After his return he started a detailed commentary on Boerhaave's *Institutiones* and in 1736, he was called to Göttingen as professor of anatomy, surgery, and botany. As we shall see, Boerhaave had an important influence on young Haller, who adopted a part of his teaching and in other parts disagreed and tried to prove his new insights.[4] In this chapter, I shall discuss the neuroscience in the work of these two important eighteenth-century physicians, Europe's principle medical teacher and his pupil, who was to become the most important physiologist of the period.

Herman Boerhaave and His Manuscript of the Praelectiones de Morbis Nervorum

Haller's stay in the Netherlands was a few years before the period in which Boerhaave, whom he called the Communis Europae Praeceptor, presented his *Praelectiones de Morbis Nervorum*,

---

[1] From April 1725 until July 1727.

[2] Ego quidem ab A[nno] 1725 ad 1727. biennio & quod excedit eum audivi & incredibili voluptate memini perfusum fuisse, quando primum veriorem medicinam amoenissima eloquentia ornatam proponentem audivi" Haller, *Bibliotheca medicinae-pract.*, vol. iv, p. 141, quoted by Lindeboom (1968, p. 205).

[3] *De ductu salivali Cochwiziano.*

[4] Haller dedicated several years of his life publishing a commented edition of Boerhaave's lecture notes on his own *Institutiones*, moreover he published Boerhaave's *Methodus Studii Medici* (1751) which he had provided with extensive comments (Lindeboom, 1958, pp. 14–15).

notably from 1730–1735.[5] At the time Boerhaave had resigned from his duties with respect to botany and chemistry because of his health and only held the chair of medicine. When he started the lectures series he was a 61-year-old and had much experience in practical medicine.

Before summarizing the contents of these lectures, it is important to write a few words on Boerhaave's publications. There is much confusion as to what was written by him and what has been attributed to him. He had often been disgruntled when he discovered a new fake publication under his name, full of incorrect statements and he even published a warning in the *Leydse Courant* (Schulte, 1959, pp. 31–32).[6] In fact Boerhaave was a gifted speaker rather than a talented author.

With respect to the contents of these lectures, however, we may be confident that they are Boerhaave's. The manuscripts that were found in Russia, apart from those of Boerhaave, also include Van Swieten's manuscript.[7] Moreover, the lectures had been published by Jacobus van Eems in 1761, compiled from his own lecture notes and those of Gerard van Swieten (1700–1772) and Jacobus Hovius (Eems, 1761).[8] Although Boerhaave did not recognize the publications of the lecture notes by his pupils, Schulte, comparing Boerhaave's manuscript with Van Eems publication, found that the two are remarkably similar, despite the fact that the first contains 150 pages and the latter 850, due to the fact that Van Eems wrote much more detailed, including notes on cases and therapy, apparently spoken but not written by Boerhaave. Van Eems had used both Hovius' and Van Swieten's notes and compared them with his own, when preparing the publication, in which he tried to keep to Boerhaave's words as much as possible ("ipsis summi Auctoris verbis"; Schulte, 1959, p. 42). Van Swieten's notes also seem to be reliable.[9]

Boerhaave's Doctrines

Before discussing Boerhaave's Lectures on the Nervous System, we should also look at some of his general principles in medicine (King, 1966, p. xxiii). According to the medical historian Lester King (1908–2002), Boerhaave is best characterized as a systematist. "He taught a body of doctrines whose parts integrated one with the other to form a well-organized whole." Almost every phenomenon dealing with health or disease could be explained in this system. The systems originated from observation complemented by reasoning. In fact, his system had three major sources, (a) the laws of mechanics (iatrophysics); (b) findings from microscopy; and (c) knowledge resulting from vascular injections.[10]

---

[5] The main source for Boerhaave's knowledge on the nervous system as presented in this paper, is Bento P.M. Schulte's *Hermanni Boerhaave. Praeletiones de Morbis Nervorum 1730–1735* (Schulte, 1959). Schulte held the chair of neurology in Nijmegen in the 1980s. He was well educated in Latin – it is said that he spoke Latin at home with his father – and translated the manuscripts of Boerhaave's lectures on the nervous system into Dutch, after they had been found in Russia. The present author received a signed copy of Schulte's thesis at the occasion of the completion of his thesis on Brown-Séquard in 1989, a thesis which had partly been supervised by Schulte.

[6] The article appeared a Leyden newspaper on 9 October 1726 warning against fake publications under his name.

[7] The reason for this is that Boerhaave, having only one daughter, left his books and manuscripts to his two nephews, Herman Kaau (his godchild, son of his sister Margaretha), whom he granted his name upon his baptism (Kaau Boerhaave) and the 10 years younger Abraham (who received the second name Boerhaave with consent of Boerhaave's daughter Joanna Maria, when Herman did not have sons). Herman as well as Abraham became physicians and like several Dutch physicians at the time, moved to Russia, in hoc casu St Petersburg, where Herman became court physician and Abraham medical superintendent at the navy-hospital. Schulte (1959) accurately described what happened with the manuscripts that were finally discovered in the 1920s at the Medical-surgical Academy in Leningrad. The Russian historian-chemist Menschutkin, in cooperation with the Utrecht professor Cohen worked on the material and the latter published a Catalogue in 1941. Lindeboom, the Dutch medical-historian, who visited Leningrad in 1955, and Schulte received a microphilm of the manuscripts of the lectures on nervous diseases by Boerhaave and Van Swieten's notes that formed the basis of Schulte's thesis.

[8] Jacobus van Eems (1709–?), physician in Leiden and Jacobus Hovius (1710–1786), physician in Amsterdam.

[9] He was able to write stenograph as was also known to Haller, who wrote in his diary: "Von Swieten, med. Doctor, der krafft einer Characterenschrift alle Sachen von Boerhaave von Wort zu Wort abgeschrieben [hat]." (Schulte, 1959, p. 43)

Boerhaave's system, his anatomy, physiology, pathology, and therapeutics were based on elementary principles. He believed that the body is composed of solids and fluids. The elemental particles combine to form the "smallest fiber." These elemental fibers have no independent functional existence. They constituted the real structural element, i.e., the vessel. Fibers combine into a "membrane," and the convolution of the most simple membrane constitute the simplest vessel (King, 1966, p. xxv). He recognized several orders of vessels ranging from large to small, the smallest of which he called nerve, which he thought to be hollow. "For, we call a nerve that vessel which is the ultimate and the most delicate."[11] The smaller the vessel the smaller the particles thought to be traveling through them; "the smallest vessels convey only the most subtle juices on the body."[12] The juices that go through the vessels should contain small enough particles: "... and even the nerves would be all useless without a supply of globules small enough to pervade their minutest tubuli." He believed the particles were made from food particles, transformed into chyle that is transported through the thoracic duct into the venous blood and from there to the lungs. By the movements of the lungs, the particles become smaller "adequate to the diameter of each set of globules."[13] These ideas are important as they were taught to Boerhaave's students, including Haller.

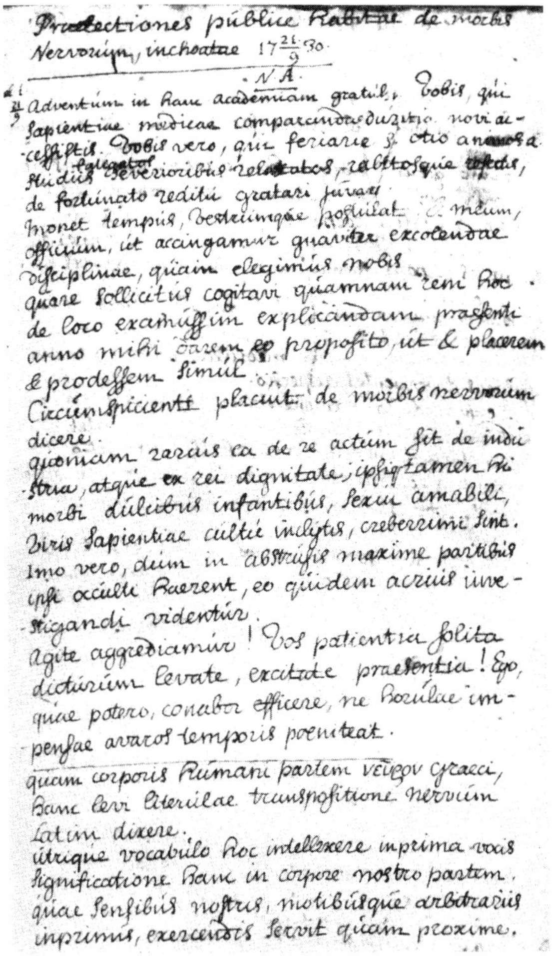
Praelectiones publice habitae de morbis Nervorum, inchoatae 17 21/9 30.
N A.
21/9 Adventum in hanc academiam gratulor vobis, qui sapientiae medicae comparandae ducitis novi accessistis. Vobis vero, qui feriarum & otio animos a studiis severioribus relaxatos, refectosque redditis, de fortunato reditu gratari juvat.
Monet tempus, vestrumque postulat & meum officium, ut accingamur graviter excolendae disciplinae, quam elegimus nobis.
quare sollicitus cogitavi quamnam rem hoc de loco examussim explicandam praesenti anno mihi darem ex proposito, ut & placerem & prodessem simul.
Circumspicienti placuit de morbis nervorum dicere.
quoniam rarius ea de re actum sit de industria, atque ex rei dignitate, ipsi tamen hi morbi dulcibus infantibus, sexui amabili, viris sapientiae cultu inclytis, creberrimi sint.
Imo vero, dum in abstrusis maxime partibus ipsi occulti haerent, eo quidem acrius investigandi videntur.
Agite aggrediamur! Vos patientia solita dicturum levate, excitate praesentia! Ego, quae potero, conabor efficere, ne horulae impensae avaros temporis poeniteat.
quam corporis humani partem νεῦρον graeci, hanc levi literulae transpositione nervum latini dixere.
utrique vocabulo hoc intellexere inprima vocis significatione hanc in corpore nostro partem, quae sensibus nostris, motibusque arbitrariis inprimis, exercendis servit quam proxime.

FIGURE 1. Boerhaave's manuscript on nervous diseases. First lecture of 21 September 1730 (from Schulte, 1959, p. 38)

Boerhaave's Lectures on the Nervous System

Boerhaave's first lecture (of 206) in this series was presented on 21 September 1730. "I congratulate you, newcomers who have the firm intention to acquire medical knowledge, with your arrival at this Academy." (Fig. 1) He also welcomed those who returned from holidays and declared that this year's subject had rarely been presented. In the

[10] With respect to the latter technique it is important to note that in his first thirty lectures on the nervous system, in which he discussed neuroanatomy and neurophysiology, Boerhaave was clearly influenced by the views of Ruysch (1638–1731) to whom he often referred. He was an admirer of Ruysch, who died in 1731 at age 92, so during the period in which Boerhaave presented his lectures on the nervous system. Although he was not the first to apply it, Ruysch was well-known because of his injection technique of blood vessels for his anatomical preparations, which in fact he had taken from Swammerdam.

[11] "Nervum enim vocamus vasorum id, quod ultimum est & tenuissimum" (Boerhaave's *Praelectiones Academicae*, II, 512 in King, 1966, p. xxvi).

[12] Boerhaave's *Praelectiones Academicae* II, 218, quoted by King (1966, p. xxviii).

[13] Boerhaave's *Praelectiones Academicae* II, 125, in King (1966, p. xxix). According to King (1966), because of this vessel/fibre theory, Boerhaave had to reject the notion of ferments, or the concept of chemistry. As King put it: "Boerhaave was the victim of overdependence on crude concepts of circulation." (1966, p. xxxi) However, as I will discuss, this is not entirely in conformity with Boerhaave's lectures.

TABLE 1. Subjects in Hermanni Boerhaave Praelectiones
Publicae Habitae de Morbis Nervorum (Schulte, 1959).

Definitio Nervi
Dura Mater
Ros Halituosus
Arachnoïdea Membrana
Pia Mater
De Spiritibus
Odontalgia
Paronychia
Calli, Verrucae, Papillae Degenerascentes
De Mente Humana
De Sensorio Común
De Sympathia
De Vertigine
Apoplexia
De Epilepsia

following description of Boerhaave's lectures several subjects are discussed, as presented in Table 1. I shall follow this sequence and comment on the Dutch translation of Boerhaave's lectures.[14]

Meninges and bloodvessels: Boerhaave defined the term "nerve" as that part that originates from the medulla of the cerebrum, the cerebellum, and the spinal cord, and from there runs to all other parts of the body, like small strings. He recognized ten pairs of cranial nerves and paid much attention to the discussion of the dura mater and pia mater (literally the hard and soft mother of the brain), to which his predecessors had attached great importance in a similar way. For centuries, it was thought that the meninges, by rhythmic contractions, caused the movement of the spirits within the ventricles. However, following a systematic discussion of all aspects concerned, Boerhaave concluded that there are no muscle fibers in the dura mater. It should be caused, he reasoned, by the movement of the blood vessels of the brain that pulsate synchronically with the heart. Previously, complaints in hysterical and hypochondric persons had been considered to be caused by movements

of the spirits due to spasms of the dura mater (Schulte, 1959, p. 370). After concluding that the dura mater has no muscular fibers, Boerhaave stated: "Based on this not a single nervous disease seems to arise from [the dura mater]."[15]

Boerhaave considered peripheral nerves to be surrounded by a continuation of the pia and dura mater. These contain blood vessels, and therefore the nerves may be subject to diseases characteristic of those of blood vessels, whereas the internal substance itself remains unaffected. "The diseases are: inflammation, erysipelas, aneurysm, varix, obstruction,.... For the same reason these diseases may affect the enclosed nerves running extracranially, whereas the brain as a whole remains unaffected."[16] In this way the function of peripheral nerves may be affected, while that of the brain and spinal cord remain intact, which, although derived from anatomical reasoning sounds rather modern. The relation between the cerebrospinal fluid (CSF) and the meninges was believed to be similar to that of the pleura, pericardium, and peritoneum, i.e., the fluid should prevent the structures to grow together.

The pia mater was considered to consist of a network of blood vessels arising from the circle of Willis. Boerhaave considered the "pia mater only an instrument of these arteries."[17] The pia would cover the wall of the ventricles and form the *plexus chorioideus*. He believed the *plexus chorioideus* to consist (a) of very thin-walled, twisted arteries within the ventricles that by the warmth of the blood and emanating of fluid nourish and relax the medulla; and (b) of veins that like absorbing sponges reabsorb the fluid and carry it back to the pia mater. "This warmth results in relaxation, and softens, and produces fluidity and preserves it, and particularly in the solid parts that are not heated otherwise, and hence become rigid easily, and thereby will contract the very narrow passages inside, impeding the functions that depend on it."[18] This is an example of his anatomy and physiology, in

---

[14] Schulte translated the Latin text into Dutch. Both the Latin text and the Dutch translation appear next to each other so that the translation may be compared with the original Latin. All quotations from the Dutch translation have been translated into English (PK), after comparison with the Latin text.

[15] Boerhaave, 1730–1735, Lecture 12, in Schulte (1959, pp. 76–77).

[16] Boerhaave, 1730–1735, Lecture 15, Schulte (1959, pp. 80–81).

[17] Boerhaave, 1730–1735, Lecture 31, Schulte (1959, pp. 125).

[18] Boerhaave, 1730–1735, Lecture 29, Schulte (1959, pp. 117).

which many of the ancient physiological ideas are still recognized.

With respect to diseases of the structure of arteries, he distinguished congenital from acquired varieties:

If the arteries of the head … deviate from the natural structure, in width, stenosis, rigidity, and flexibility, many diseases may result that are hardly curable. maybe the extraordinary size of the head? Hydrocephalus? And from this the diseases accompanied by numbness and paralysis that ultimately may be lethal. We often see decrepit, feeble-minded, and persons exposed to apoplexy in such a condition. … A delegate from France suffered from apoplexy caused by a very wide artery, Laurentius, Contr. An. (p. 803).[19] But the same may be said of too much narrowness; from these conditions, many and awful diseases may arise too.[20]

Again, this is illustrated with a case: "An ossified, stony, narrow, right carotid had caused headache of the left side of the head that was lethal (apoplexy). Only the right vertebral artery had increased. (Willis, An. Cer.c.7)."[21] In another case, taken from to Reinier de Graaf (1641–1673),[22] he described the calcified, bony, narrow arteries causing headache with lethal outcome.[23] Interestingly, he supposed a relationship between obesity and stenosis of the arteries and veins, causing the blood to flow to other arteries that are not compressed by fat, in particular to arteries within the skull and vertebral canal that are deprived of a *tuna muscularis* and hence less resistant against increased blood flow.[24]

Brain and blood: diseases caused by a deficit of blood in the pia mater are explained in a (iatro)physical way leading to veins, venous sinuses, and cavities filled with cold, thin, watery fluid, properties that for centuries had been associated with brain diseases (dysfunction of the nerves and spirits). More distally, and thus smaller blood vessels, situated far from the heart, will receive thinner fluids with less propulsion, and hence will transport it back to the heart less easily. "These vessels will be filled but subsequently the fluids will stagnate, leading to transparent, cold, soft, immovable swellings, in particular in the remote parts of the nervous system, notably the dura mater and the arachnoid … and the cavities of the brain and spinal column."

The diseases that will be caused under these circumstances, i.e., as soon as the propulsion fails, include fatigue, unconsciousness, interrupted pulse, sudden palpitations, fever, etc. These conditions occur in particular in bodies that are vulnerable by age, sex, or lifestyle. "Where the internal chyle- and blood-forming organs are unable to produce red, thick, warm blood, but only watery fluid, notably in chlorosis and in the unfavorable condition of the fluids (*cacochymia*) in dropsy of the skin."[25] Therefore, Boerhaave reasoned, one should not only pay attention to age and sex, but also to climate, region, and season, as well as other external conditions. One has to observe the symptoms including the weak and interrupted pulse, breathing, since the patient may easily become short of breath, and the heart may palpitate; furthermore cold, paleness, and weakness. Treatments should consist of means to diminish the watery fluid and increase the red and here he turned to stimulants, beer and bitter red wines, spices, and external plasters. Exercise was also a part of the regimen in particular at the end of digestion. The production of blood could also be enhanced by adding iron to other measures. With respect to iron, he referred to Thomas Sydenham (1624–1689) and Johan Baptista van Helmont (1580–1644).[26] "As soon as the production of blood is corrected, the production of pneuma will as well; and thereby all vital, natural and animal functions."[27] Iron was also considered important in

---

[19] Here Boerhaave refers to the anatomist Laurentius (1550–1609) of Montpellier and his *Contraria Anatomica*.

[20] Boerhaave, 1730–1735, Lecture 31–32, in Schulte (1959, pp. 125–126); at the end of the lectures a special chapter dedicated to apoplexy will be discussed.

[21] Boerhaave, 1730–1735, Lecture 32, Schulte (1959, p. 127); Boerhaave referred to Willis' *Cerebri anatome* Chap. 7.

[22] He referred to De Graaf's *Miscellanea Naturae Curiosorum* no. 127, 1670. De Graaf is well-known by his work on the ovaries ("De Graaf's follicle") and pancreas.

[23] Boerhaave, 1730–1735, Lecture 32, Schulte (1959, p. 127).

[24] Boerhaave, 1730–1735, Lecture 34, Schulte (1959, p. 131).

[25] Boerhaave, 1730–1735, Lecture 38, Schulte (1959, p. 137).

[26] Boerhaave, 1730–1735, Lecture 40–42, Schulte (1959, pp. 140–142).

[27] Boerhaave, 1730–1735, Lecture 1, 20/9/31, Schulte (1959, p. 145); following the first series of 45 lectures, a new series starting with no. 1 began in September 1731.

the cure of nervous diseases, in particular in hypochondric and hysterical patients.

Venous congestion causes plethora, which Boerhaave considered the opposite of what was described above. Whereas in plethora the vessels contain thick warm blood, they contain thin watery fluid if the arteries of the pia mater do not contain enough blood. Compression of the jugular vein extracranially and vomiting may lead to plethora as well as suppression of regular hemorrhages (cf. Hippocrates).

As noted above, diseases of the dura mater, the surroundings of the nerves, the arachnoid, or pia mater were believed to be caused by disorders of the blood vessels and Boerhaave tried to explain the results in an iatrophysical way. He reasoned in terms such as disorders of circulation within the skull, mainly venous congestion or signs of compression of the brain or nerves. This was thought to result in psychic disorders, headaches, blindness, deafness, loss of smell, paresis, and sensibility disorders. He demonstrated these ideas with reports from autopsies, mainly taken from Théophile Bonet (1620–1689). Diseases that cause fusing of the dura and arachnoid may result in congestion of the circulating fluid. He supposed this may cause some of the "chronic and very unusual headaches in cases, in which, at autopsy, the pia and dura mater had grown together, like the heart with the pericardium and the lung with the pleura. Maybe headache due to meningitis arises by this mechanism,"[28] referring to several cases described by Bonet, Johann Wepfer (1620–1695), and Thomas Willis (1621–1675). Most of the material described above may be considered solid pathology, but Boerhaave's humoral pathology was more influential (Schulte, 1959, pp. 376–377).[29]

Boerhaave's Teachings on *Spiritus*

Next to problems resulting from too thick and, on the contrary, thin watery blood, very fine pungent substances were believed to be able to penetrate the cortex from the vessels of the pia mater and hence disturb the functions of the nervous system. Boerhaave used the term *acrimoniae*. The supposition reminds one of the iatrochemistry of Franciscus de le Boe Sylvius (1614–1672), his predecessor as the chair of medicine in Leiden, although he did not refer directly to him. The spirit of fermented vegetable substances (alcohol) was considered the most important cause. Boerhaave described different degrees of its nature, including delirium tremens and advised gradual detoxification, dry foods, working, traveling, and exercise.[30]

Spiritus or spirit was defined as submicroscopical particles that are easily moved by air movements and may irritate nerve endings. Boerhaave recognized three groups of spirits, the first of which included the classic animal, vital, and natural spirits. The second group included toxic spirits that stop the function of the animal spirits and causes mental and physical disturbances. These *spiritus venenati* arise from the action of toxic substances originating from contagious diseases including plague, smallpox, measles, and rabies, and also from certain animals. The third group of spirits consisted of toxic spirits originating from putrefaction within living beings (Schulte, 1959, p. 379).

Boerhaave believed that these spirits, when in contact with body humors, may cause diseases of the nervous system. According to Schulte, it is not easy to find what is original and what part derived from his teachers and colleagues, with respect to his iatrochemistry. In the period when he taught the doctrines of the spirits (1732), he was quite busy with chemistry ("Chemiam dies noctesque exercuit," Schulte, 1959, p. 380). It was the year of publication of his *Elementa Chemiae* that was published in several languages. Despite the fact that in his inaugural lecture in 1718 to become professor of botany and chemistry, Boerhaave repudiated the teachings of the well-known iatrochemists (Paracelsus, Van Helmont, and Deleboe Sylvius), he later accepted some of their ideas that he tried to prove experimentally (Schulte, 1959, pp. 382–383).

---

[28] Boerhaave, 1730–1735, Lecture 21, Schulte (1959, p. 97).
[29] The ancient concepts of humoral pathology tried to explain diseases from bad mixtures of body fluids; gradually and in particular at the end of the eighteenth and early nineteenth century, solid pathology, based on disorders in solid parts of the body, gained the upper hand.
[30] Boerhaave, 1730–1735, lecture 1 18/3/32; Schulte (1959, p. 151).

Boerhaave on the *Sensorium Commune* and Mental Disorders

Boerhaave considered nerves to consist of a bundle of medullary fibers between which spirits flow in cavities.[31] It is remarkable how many diseases he discussed under the heading of diseases of the nerves, including odontalgia, paronychium, old-age blindness and deafness, verrucae, clavi, and papillae degenerascentes.[32] Starting with lecture 47, he discussed diseases of the nervous system, which he divided into four groups: mental disorders, several kinds of vertigo, apoplexies, and the epilepsies. Introducing mental disorders, he began with a philosophical discussion with respect to which we have to remember that psychopathology was still considered the territory of philosophy rather than medicine by many, including Immanuel Kant (1724–1804). The main characteristics of the mind, according to Boerhaave, were reason and will, which form the mind and are perfections, whereas emotions only confuse reason and stimulate the will. The fourth characteristic was imagination by which ideas may be imprinted without the agency of the senses.[33] These notions evoke those of Descartes, whose philosophy was very influential in Leiden and the rest of the Netherlands. Boerhaave, however, did not adopt all of Descartes' ideas. For instance, Boerhaave believed the faculty of judgment to be independent from the will.

In Boerhaave's psychopathology, the *sensorium commune*[34] played an important role. He tried to explain mental diseases from philosophical–psychological ideas in a period characterized by a large gap between philosophical psychology and practical medicine (Schulte, 1959, p. 386). He did not apply any new nosological classification and

used classic terms including insania, amentia, and dementia. According to Schulte, it is difficult to find out what exactly he understood about delirium, furor, mania, and stultitia.[35] As a result of emotions, delirium, furor, and mania may arise sometimes leading to dementia or stultitia. Insania may be caused by imagination. All were considered disorders of the *sensorium commune*.

In lectures 71–77, Boerhaave explained that the *sensorium commune* acts from the same localization as the motor force (*hormè* or *impetum faciens* in the Hippocratic writings). About the localization of the combined *sensorium commune* and *impetum faciens*, Boerhaave wrote "that this part is all those localizations where the end of the cortex of the brain and spinal cord gives off the starting points of the elements of the small nerves that all arise from there, and by which the perceptions and voluntary movements originate."[36] Boerhaave believed there are as many starting points of nerve fibers as there are endpoints in the body parts. In lecture 77 he even used the term "machina nervosa," with respect to the system of motor and sensory nerves, which reminds us of Julien Offray de la Mettrie's (1709–1751) *L'homme machine*.[37] In modern terms we could conclude that Boerhaave localized it at the transition of gray to white matter.

Disorders of the *sensorium commune* may arise from the nervous system or from other organs. With respect to mental disorders, he distinguished organic from symptomatic psychoses. Disorders of the *sensorium commune* could originate from incarcerated herniations, ileus, gall waste, black bile, toxic substances in the stomach, or tickling of certain nerves (Schulte, 1959, p. 388). The cases to which Boerhaave referred in this section were taken from Aretaeus, who was one of the first to

---

[31] The fact that he used the term nerve fibers ('fibrillae') as well as considered nerves hollow is explained in this sentence.

[32] The latter three skin disorders were considered as morbid growths of naked nerve endings.

[33] Boerhaave, 1730–1735, Lectures 47–58; Schulte (1959, p. 384).

[34] Sensorium commune: common sense in the literal classic denotation, originally derived from Aristoteles who localized its seat in the heart; 'le rendez-vous des sensations' in Lélut.

[35] Schulte noted that it is hard to say what exactly Boerhaave meant by the term stultitia. The Latin word means foolishness or madness.

[36] Boerhaave, 1730–1735, Lecture 76–77, Schulte (1959, pp. 257 and 387–388).

[37] Lamettrie attended Boerhaave's Lectures on the Nervous System in 1733 and referred to him in his anthropological materialism. *L'homme machine* was published in 1747.

mention symptoms by *consensus*, i.e., that other organs take part in a disease. In this way delirium with fever from pneumonia could be considered an affection of the brain *per consensum*, with the seat of the affection in the lungs.[38]

The potential power to which the *sensorium commune* could give rise, could be activated by external or internal (the will) causes, and once this happens, the effects in the body could take place automatically. The *sensorium commune* could incite a person's will to walk from Leiden to Amsterdam, after which the body automatically carries out the plan.[39] The *sensorium commune* was considered the central switch in the "machina nervosa" of the nervous system. The Cartesian reflex doctrine, according to Schulte, is even stronger in Boerhaave's writings than in Descartes'. Whereas Descartes localized the *sensorium commune* in the pineal gland, Boerhaave's localization, at the transition of gray and white matter resembled a switch in a more convincing way. Not surprisingly, Boerhaave, with respect to causality in his psychopathology finally referred to God. The "conditio humana" is completely dependent on the divine causality (Schulte, 1959, p. 391).

The Three Classic Nervous Diseases and Treatment

To the other three groups of nervous diseases, the classic *morbi cephalici*, vertigo, apoplexy, and epilepsy, Boerhaave devoted 61 lectures, starting with lecture 96 in March 1734. He noted that diseases that arise from the arteries, cortex, medulla, nerves, and meninges had been discussed in previous lectures.[40]

Vertigo, was considered the least malignant affliction of the three, according to Boerhaave. The description does not correspond to the present and rather resembles the current Menière syndrome (Schulte, 1959, p. 392). It was a disorder of the *sensorium commune* in which idiopathic and symptomatic forms were recognized. Ideas such as predestined and provoking causes remind one of Hippocrates.[41]

Apoplexy: in discussing apoplexy, Boerhaave applied the classic doctrine of the spirits. All types of apoplexy have in common the obstruction of animal spirits in the cortex or their transfer through the medulla or nerves. Referring to Aretaeus, he said that all external and internal perceptions, as well as all voluntary muscular movements stop.

Hence it is a disorder of the brain. And within that of the perceptive and moving part of the sensorium commune … swallowing, vomiting and bowel movements remain … Therefore these movements stand alone, independent from the sensorium commune … sometimes swallowing is disturbed by paresis of the oesophagus.[42]

The essential problem in apoplexy, irrespective of the cause, e.g., cerebral hemorrhage, is the failing of the animal spirits. Therefore it is easy to understand why he treated it with stimulating agents, almost literally resuscitating the life spirits. Disorders characterized by increased tendency to sleep, which had already been recognized by Galen, were classified within the category of apoplexies, including lethargy and cataphora (both with fever), carus and coma. Paralysis was also counted as an apoplexy. Boerhaave distinguished spastic from flaccid paresis.

Epilepsy was well described referring to Aretaeus and explained as a disorder of the principle of perception and movement, localized in the *sensorium commune*. Idiopathic and symptomatic forms were distinguished. For the first he

---

[38] Next to Aretaeus, the influences of the Hippocratic writings are abundant.

[39] Boerhaave 1730–1735, Lecture 86, Schulte (1959, pp. 269 and 390).

[40] Again Hippocrates and Aretaeus are often referred to in these lectures. With this respect Boerhaave was not the first in his time to refer to the Hippocratic writings again, as Giorgio Baglivi (1668–1707) and Thomas Sydenham (1624–1689) preceded him. It may be recognized,

among others, by the interest in the premorbid character of the patient, the differential diagnostics, and the discussions on prognosis.

[41] In fact these two factors may be recognized through the history of medicine, e.g., in Charcot's tare nerveux (constitutional factors) and agents provocateurs (short lasting triggers) up to the present etiological ideas.

[42] Boerhaave 1730–1735, Lecture 109, Schulte (1959, p. 303).

mentioned predestined and provoking causes. The predestined causes include hereditary and congenital disorders, brain abnormalities following disease or trauma, and experienced emotions. Among the provoking factors he included violent emotions, exhaustion of the *sensorium commune*, sexual intercourse, drunkenness, superfluous nourishment, full-bloodedness, and particular vapors. With respect to the predestined causes, Boerhaave was aware that these could hardly ever be removed, which is why he strongly advised avoiding provoking factors. With respect to symptomatic epilepsy, the stomach was considered an important source similar to the situation in mental diseases (Schulte, 1959, p. 395).

Treatment: Boerhaave's therapeutic methods show his eclecticism. They may be divided into advice with respect to regiment (diet, dressing, climate), medical, and physical treatment. In his medical therapies, he profited from his large knowledge of botany, applying all kinds of vegetable drugs.[43] Emetics and laxatives were among the most important categories. He also used quicksilver as a diuretic and iron for anemia. For the physical therapies he prescribed massage and balneotherapy.[44] Remarkably, he pointed to the sometimes healing power of fever therapy in paralysis. For severe mental disorders, extreme measures were sometimes applied in order to stimulate or suppress the *sensorium commune*. Among other measures he mentioned "submersio diuturna sub aqua,"[45] probably derived from Van Helmont. The patient was submersed under water to the point of coma. Sometimes toxic substances including the herb *Helleborus* and compounds of mercury, antinomy, and copper were applied.

Haller's Most Important Publications

The first edition of *Primae lineae physiologiae* (First lines of physiology), which may be considered the first modern textbook of physiology (King, 1966), dates from 1747. Haller probably published the book to supplement his lectures, realizing that Boerhaave's *Institutiones*, which he had annotated previously was no longer adequate. In a letter to his Italian colleague Giambattista Morgagni (1682–1771), Haller wrote: "I have added the proof sheets of my physiology, intended for use in my yearly lectures. Far too many things have been discovered since Boerhaave, which it would be negligence to omit."[46] (King, 1966, p. xi)

The book, in two volumes, consists of 32 chapters on subjects including animal fibers, cellular substance, arteries and veins, circulation, etc. It also contains chapters on the voice and speech, brain and nerves (XI), muscular motion (XII), and on the senses of touch, taste, smelling, and hearing. In this chapter, I shall discuss chapters X–XVIII. The second important publication that will be discussed in this paper is his *Dissertation on the sensible and irritable parts of animals* (Haller, 1755). Finally, I shall briefly discuss his third important publication, i.e., his *Elementa Physiologicae*.

Haller's *First Lines*[47]

When Haller was 39-years-old, he published the first edition of *First Lines* (1747). It was designed as a correction and improvement of Boerhaave's

---

[43] When Haller stayed in Leiden, he often witnessed his teacher striding at that early hour between the seed beds and contemplating the growth of the herbs in the *hortus bontanicus*, right behind the university building. 'Saepe vidimus ante auroram optimum senem ligneis calceis per hortum repentem, ut cominus & cultum herbarum perspiceret, & flores fructusque specularetur. Haller, *Biblioth botan*, Vol. II, p. 96, quoted by Lindeboom (1968, p. 89).

[44] Balneotherapy: treating of diseases by bathing.

[45] Daily submersion under water.

---

[46] From *Albrecht von Haller, Giambatista Morgagni, Briefwechsel. 1745–1768*. Bern, Huber, 1964; quoted by King (1966, p. xi).

[47] The first edition of *Primae lineae physiologiae* dates from 1747, the 2nd from 1751 and the 3rd from 1765. The latter was printed in Edinburgh in 1767 (Latin) and the English translation appeared in 1779 (Edinburgh). In this paper I used a facsimile reprint of the 1786 edition (Haller, 1786a), translated from the correct Latin edition that appeared in 1780 annotated by Heinrich August Wrisberg (1739–1808); printed under the inspection of William Cullen and compared with the edition published by Wrisberg.

Institutiones, by adding new discoveries including those of Morgagni, Winslow, Albinus, Douglas, and others. King considered it a link between Boerhaave's teachings and Haller's *Elementa Physiologicae*. The chapter on "Voice and Speech" (X) mainly deals with the anatomy of the speech organ with some physiology. Haller did not discuss language here.

More important of course is chapter XI: "Brain and Nerves," which starts with a description of the arterial branches from the aorta. It is remarkable that Haller was able to note that "a very great quantity of blood is in every pulsation sent to this organ, insomuch that it makes above a sixth part of the whole blood that goes throughout the body. ..." The nearby origin of the carotid stem from the heart supposes "the strongest parts of the blood go to the head, and such as are most retentive of motion." (Haller, 1786a, p. 184)

He described the cortex as follows.

... The fabric of the cortex has been a long time controverted; but it is now sufficiently evident, from anatomical injections, that much the greater part of it consists of mere vessels, which are every way inserted from the small branches of the pia mater... and conveying a juice much thinner than blood in their natural state, although in some diseases, and by strangling, they often receive even the red parts of the blood, more especially in brutes and birds. (Haller, 1786a, p. 197)

We recognize the teachings of Boerhaave and the injection method of Ruysch. "The remaining part of the cortex, which is not filled by any injection, is probably either an assemblage of veins, or of yet more tender vessels; for no other dissimilar parts are apparent in the cortex, whilst it is in an entire or natural state; from whence one may conjecture some part of it to be tubular, and the other part solid." (Haller, 1786a, p. 198)

Most of the descriptions concern the anatomy, e.g., the corpus callosum "which is streaked with transverse fibers." (Haller, 1786a, p. 198) The ideas about the ventricles are still classical: "This cavity... is naturally filled with a vapour, which is frequently condensed into water or jelly." (Haller, 1786a, p. 199) The cranial nerves are described in a particular way.

All the medulla of the brain and cerebellum goes out from the skull, through particular openings, to the parts to which it is destined. The smaller bundles of this medulla we call *nerves*; but the larger, descending through the spine, we call the *medulla spinalis*, which is a continuation of that called oblongata. (Haller, 1786a, p. 206)

In comparison with Boerhaave, there were no new insights with respect to the number of cranial nerves, i.e., ten. The brain was supposed to change "when any change happens to the body." Compression or dissection of a nerve may result in the disappearance of the peculiar sense it is taking care for, like compression or dissection of the brain causes loss of senses of the whole body. Interestingly, Haller mentioned phantom limb pain: "sensation arises from the impression of an active substance on some nerve of the human body; and that the same is then represented to the mind by means of that nerve's connection with the brain." (Haller, 1786a, p. 214) Haller was aware of the fact that irritation of the medulla of the brain "deeply in its crura" resulted in

dreadful convulsions ... whatever the part of the brain so affected ... But if the encephalon itself be compressed in any part whatever, there follows thence a loss of sense and motion in some part of the body which must be the part whose nerves are detached from the affected or compressed quarter of the brain (Haller, 1786a, p. 215).

Experiments with the spinal cord resulted in more evident effects: irritation in convulsions and compression in paralysis.

The seat of the soul: with respect to the seat of the soul, Haller wondered, "whether or not is there in the brain any principal part, in which resides the origin of all motion, the end of all the sensations, and where the soul has its seat?" (Haller, 1786a, p. 216) He had several objections against the localization of the soul in the corpus callosum (birds, for instance, have no corpus callosum). With respect to the cerebellum, he noted that it does not contribute much to the excitation of "vital motions." He believed the fifth nerve (the present seventh) is "produced by the cerebellum." Disorders of the cerebellum do not lead to "certain and speedy death. ... For certain experiments, even of our own making, show that it has borne wounds and scirrhi, without taking away life; ... not very rarely, wounds of the cerebellum cure." (Haller, 1786a, p. 217) Continuing on the seat of the soul, he writes, "we must inquire experimentally." It must be in the head, not the spinal cord. He argued from irritation experiments in which convulsions were elicited that it could not be the cerebral cortex, but "medulla" (white matter). He supposed the seat to be in the crura (mesencephali), corpora striata, thalami, pons, medulla oblongata, and cerebellum; in addition in all places, where the origin of every nerve lies, "as the first origins of all the nerves taken together make up the sensorium communae," and here we recognize again Boerhaave's teachings.

The function of nerves: Haller tried to explain how nerves become the organs for sense or motion, by reasoning that nerves stem from "medulla," i.e., white matter. This substance "is a very soft pulp." (Haller, 1786a, p. 218) Its composition is thought to be fibrous. "... that the fibers of the brain are continuous with those of the nerves, so as to form one extended and open continuation, appears, by observation, very evidently in the seventh, fourth, and fifth pair of nerves." (Haller, 1786a, p. 219) He believed the medulla to contain "a great deal of oil" (up to 1/10 part of its weight). Haller mentioned the controversy on the solid character of the fibers of the medulla and nerves, which by some scholars were supposed to work by means of vibrations transmitted to the brain. However, the comparison with elastic cords that tremble does not hold. Nerves are soft and not tensed. "Therefore, the nervous fibers cannot possibly tremulate in an elastic manner." Moreover, if nerves are cut into two pieces, they do not shorten or draw back their divided ends. He also reasoned that "the force of an irritated nerve is never propagated upward," which is also in contradiction to an elastic character. Therefore he argued that the theory of a fluid flowing from the brain through the nerves must be true. The fluid could be put "in motion by an organ of sense," transmitting sensation to the brain (Haller, 1786a, p. 220). Therefore the nerves must be hollow.

The nature of the nervous fluid: "Many of the moderns will have it to be of extremely elastic, of an ethereal or of an electrical matter. ..." Others believe it is of lymphatic or albuminous nature. "An electrical matter is, indeed, very powerful, and fit for motion ... ," but Haller believed that electricity would spread beyond the nerves (e.g., to flesh and fat), whereas in a living animal only nerves ("or such parts as have nerves running through them") are affected by irritation. "And a ligature on the

nerve takes away sense and motion, but cannot stop the motion of a torrent of electrical matter." (Haller, 1786a, p. 221)[48] He supposed a watery and albuminous nature of the juices of the body in general and the nerves in particular, and provided several supportive arguments, including the fact that a tied nerve gives rise to a tumor. "But are these properties sufficient to explain the wonderful force of convulsed nerves, observable in the dissections of living animals, and even in the lesser insects, with the great strength of mad and hysterical people?" He realized that the "nervous liquor then, which is the instrument of sense and motion, must be exceedingly moveable, so as to carry the impressions of sense, or commands of the will, to the places of their destination, without any remarkable delay: nor can it receive its motions only from the heart." (Haller, 1786a, p. 222)

Therefore, Haller reasoned, a fluid would be produced in the cortex, transmitted via the "the hollow pipies of the medulla" (white matter) to the "small tubes of the nerves," resulting in sense and motion. He also supposed two types of motion in this fluid: a slow constant flow, from the heart and a very fast one "excited either by sense or any other cause of motion arising in the brain." (Haller, 1786a, p. 223) Haller presumed that the same nerves can have a sensory as well as a motor function. If one function is defective without the other, he reasoned that "much more strength is required for the latter [motion]."

Muscular motion: in the chapter on muscular motion he admitted that "the direct manner by which the nerves excite motion in the muscles, is so obscure, that we may almost for ever despair of its discovery." (Haller, 1786a, p. 236) Like Boerhaave, Haller was known to have been a pious person, hence his appeal to God: "The motive cause which occasions the

---

[48] Despite Haller's opinion, Wrisberg, in note 105, argued that this is an essential question: "How do the nerves act in the bodies of animals?. ... The doctrine of nervous fluid, or animal spirits, is confirmed by many arguments drawn from anatomy, but refuted by as many if not more, of a similar nature'. Wrisberg further noted that "since 1766, I have been inclined to think, that it perhaps resembles the electric and magnetical fluids" (talking about the "substance, which produces the wonderful phenomena in the nerve ...").

influx of the animal spirits into the muscle so as to excite it into action, seems not to be the soul, but a law derived immediately from God." Further, he distinguished voluntary from involuntary muscles:

There seems to be this difference between the muscles obeying the will, and those which are governed by a *vis insita*; . . . as for instance, the heart and intestines. . . . Thus it happens, that, in apoplexies, the muscles which obey the will languish, and become paralytic, as being destitute of all influx from the brain; while the vital muscles, having no occasion for the operation of the brain, continue to be excited into contraction by their stimuli, the heart by the blood, and the intestines by the air and aliments. (Haller, 1786a, p. 238)

Moreover, Haller was aware of agonist and antagonist muscles: "The easy and sudden relaxations of muscles in their motion are assisted by the actions of their *antagonist* muscles." (Haller, 1786a, p. 240)

Cognitive functions: following a discussion on the senses, one of the more interesting chapters is on "the internal senses." Actually this is a chapter on cognitive functions.[49] "We now see, that it is common to them all [the organs of the senses], that the tender pulp of the nerve, being struck or impressed by external objects, conveys, by the nervous spirits, some change to that part of the brain where the impressed fibers of the nerve first arise from the arteries. . . ." He supposed new thoughts arise when changes, produced by organs of sense are "conveyed to the first origin of the nerve which receives the impression." (Haller, 1786b, p. 32)

With respect to memory he stated that

. . . if they [impressions from external senses] made a strong impression, they may for ever, and in all ages of life, be repeated to the mind; but they are weakened, and in a manner blotted out in time, by degrees, unless the representation be renewed again to the mind, either from an external object, or from the mind itself recalling the same change again into memory: so that, without this repetition, at last the change or impression will be in a manner erased and quite lost. . . . (Haller, 1786b, p. 36)

Not surprisingly, he was aware of cerebral disease that could cause memory loss. "But sometimes all of them will be suddenly destroyed by disease, in which the brain is any how compressed, either

---

[49] Imagination, memory, attention, etc. using the functions that for centuries had been mentioned in medical texts.

from the blood or other causes. . . ." (Haller, 1786b, pp. 36–37)

He also explained what happens in "mad people" with this respect. If all these mechanisms to remember are "conjoined, [they] may render the species so strong to the mind, that she will afterward receive the perception of them, as if they came from external objects, in the manner we observe in mad people." (Haller, 1786b, p. 37) He had observed that memory is almost absent in infants and gradually grows during life, to decline in old age.

Another interesting aspect of his writings on memory is *how* it is remembered. ". . . [In memory] the perceptions are commonly weaker than in the imagination, being almost only certain arbitrary signs conjoined together, with the idea that was first perceived in the mind." (Haller, 1786b, p. 36) He wrote that memory hardly represents images and pictures of things to the mind. "Only the words or signs, and certain attributes, together with the general heads of ideas." (Haller, 1786b, p. 36) With respect to diseases, e.g., by a compressing cause "acting on some part of the common sensory," he noted that it "blots out a corresponding number of species from the mind or memory, whether they be certain or all kinds of words, or even the characters by which we express words; or, lastly, the characters of our friends, and necessaries of life." (Haller, 1786b, p. 36) It is possible that Haller referred to what later became known as aphasia and therefore had some kind of functional localization in mind.

"The office of *cogitation* in the soul, is to attend to the sensations which are either brought by the sense, or recalled by the imagination; frequently also to the signs alone which recur into the mind." He also provided an interesting description of the faculty of attention, which "is said to operate when the mind observes one and the same idea alone, and for a longer time together." (Haller, 1786b, p. 37) With respect to reason, Haller stated:

The comparison of two or more ideas brought to the mind, is called *reason*; as the similitude, diversity, or relation perceived by the comparison, is called *judgment*. The principal cause of wisdom and invention lies in a slow examination of the ideas, considered in the relation of all their parts one to another in the mind, while, neglecting all other objects, she is employed with a strong attention only upon that which is under examination. (Haller, 1786b, p. 38)

Haller was aware that changes to the encephalon, by compression, irritation, or deficiency of blood are able to confound the use of reason.

Irritability Before Haller

A few years after the first edition (1747) of *First lines*, Haller published his *Dissertation on the sensible and irritable parts* (Haller, 1755). Before discussing this important publication a few words on the history of irritability are necessary. In the introduction to the English translation of the dissertation, medical historian Owsei Temkin (1902–2002) rightly stated that irritability and sensibility were not really new discoveries by Haller, as the phenomena had been known to physicians previously and the term had been coined (Temkin, 1936). Galen stated that all sensibility is bound to the nervous system, but the principle of irritability was unknown to him.[50] Thus we owe the general principle of irritability and the term itself to Francis Glisson (1597–1677),[51] whose name is attached to many anatomic structures. Next to the iatrochemical and iatrophysical models of explanation of nerve and muscle function that reigned in the seventeenth and early eighteenth century, he was the first to write about the concept of irritability. He may be considered an opponent of the materialist theories of his time. By his vitalistic doctrine he opposed the materialistic system of Descartes. His notion of "fiber" was probably that of an elementary but hypothetical unit of plant and animal tissue (Clarke & O'Malley, 1968, p. 166). From his *Tractatus de ventriculo et intestines* (1677, pp. 138–143),[52] we can conclude that he considered the fiber the basic element of the animal body. The new element in his ideas was that he attributed a vital force to the fiber.

Haller's *Dissertation on the Sensible and Irritable Parts*[53] (Fig. 2)

Despite Glisson's priority, Haller's methods may be considered new in so far that he systematically investigated a large number of body parts to observe whether these were sensible and/or irritable (Table 2). He provided an experimental definition of the two phenomena.

I call that part of the human body irritable, which becomes shorter upon being touched; very irritable if it

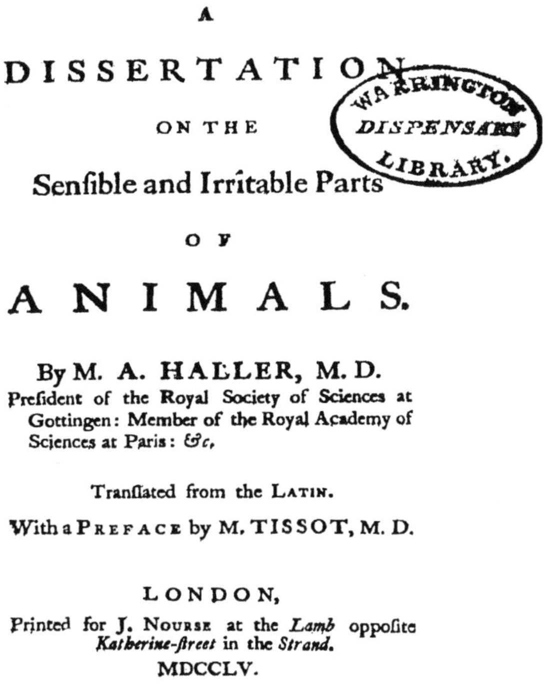

A

DISSERTATION

ON THE

Senfible and Irritable Parts

O F

ANIMALS.

By M. A. HALLER, M. D.

Prefident of the Royal Society of Sciences at Gottingen: Member of the Royal Academy of Sciences at Paris: &c,

Tranflated from the LATIN.

With a PREFACE by M. TISSOT, M. D.

LONDON,

Printed for J. NOURSE at the *Lamb* oppofite *Katherine-ftreet* in the *Strand.*
MDCCLV.

FIGURE 2. Title page of *A dissertation on the sensible and irritable parts of animals* by Albrecht von Haller (1755)

---

[50] Temkin suggested that some writings by Galen and other authors "are dimly suggestive of" the latter.

[51] Glisson, English anatomist/pathologist graduated in Cambridge (1634), where he became Regius Professor of Physic, but worked mainly in London.

[52] Francis Glisson *Tractatus de ventriculo et intestines* London, 1677, p. 138–143, quoted by (Clarke & O'Malley, 1968, p. 166).

[53] In this paper I used the English translation which is based on Simon Auguste André Tissot's French translation from Latin: *Dissertation sur les parties irritables et sensibles des animaux.* Lausanne, Bosquet, 1755. The anonymous English translation was published in the same year (1755). It was preceded by translations in German and Swedish. I shall refer to Haller's text by Haller, 1755, referring to the English translation, applying the page numbers as appeared in Temkin, 1936.

TABLE 2. Overview of sensible and irritable body parts from Haller's *Dissertation on the sensible and irritable parts of animals* (1755) as composed by PK.

Body part	Sensibility	Opposed by	Confirmed by	Irritability	Opposed by	Confirmed by
Skin	+			−		
Muscular flesh[a]	−			+		Zimmerman, Croone, Bremond
Muscle (nerve)	+			−		Oeder
Tendons	−		Van Meekren	−		
Joint capsule	−					
Joint ligaments	−		De la Motte			
Periosteum	−		Cheselden	−		
Pericranium	−					
Bone	−					
Bone marrow	−	Deventer, Paré, Du Verney				
Dura mater	−		Zinn, Mekel	−		
Pia mater	−			−		
Peritoneum	−					
Pericardium	−					
Pleura	−					
Mediastinum	−					
Arteries and Veins	−			−	Senac, Whytt	
Esophagus				+		
Stomach	+++			+		
Intestines	+			++		Wepfer, Stahl
Bladder	+			+		Wepfer
Ureters	+			−		
Vagina/uterus	+			+ (uterus)		Ruysch
Heart[b]	+			+++		
Viscera[c]	Little		Zimmerman	−		
Glands	Obtuse	De Bordeu		−		
Penis	+			+		
Tongue	++					
Eye	+			− (iris)		
Cornea	−					
Nerves	+++			−		

[a] Due to the nerves rather than the flesh.
[b] Not based on his own experiments.
[c] Lungs, liver, spleen, kidneys.

contracts upon a slight touch, and the contrary if by a violent touch it contracts but little. I call that a sensible part of the human body, which upon being touched transmits the impression of it to the soul; and in brutes, in whom the existence of a soul is not so clear, I call those parts sensible, the Irritation of which occasions evident signs of pain and disquiet in the animal. (Haller, 1755, pp. 658–659)

Methods: the first part of Haller's dissertation provides the results of his experiments on sensibility and the second part on irritability. In the introduction he informed the reader that he examined 190 animals since 1751, "a species of cruelty for which I felt such a reluctance, as could only be overcome by the desire of contributing to the benefit of mankind." (Haller, 1755, p. 657) All experiments were conducted either by himself or by his pupils, including Johann Georg Zimmermann (1728–1795). He was motivated by his observation that "the source of the great error in physic has been owing to physicians … making few or no experiments, and substituting analogy instead of them." (Haller, 1755, p. 658) He examined different kinds of animals at different ages and irritated the body parts by blowing, heat, spirit of wine, the scalpel, and chemical irritants. Some of the experiments were repeated several times.

Results: among the parts that he found not to be sensible were the tendons. He explained this by his observation that tendons do not have nerves. Previous findings to the contrary, he reasoned, were probably due to the confusion of nerves, tendons, and ligaments (Haller, 1755, p. 664). He believed that those body parts "which are exposed to a continual friction should be void of sensation." He did not find sensibility in the dura mater which many physicians "have looked upon as the seat of the most violent diseases." (Haller, 1755, p. 667) He stated that "most irritable parts are not at all sensible, and vice versa, that most sensible are not irritable." (Haller, 1755, p. 675)

He opined that we must not conclude that a part is sensible, because it is irritable. The cutting of a nerve after which sensibility in the part disappears is not associated with the disappearance of the irritability of that part. For instance, if the crural nerve is cut, the sensibility in the limb disappears, but irritation of the nerve may cause a "trembling motion" of the muscles of the leg, "wherefore at that time it is irritable, though quite insensible." (Haller, 1755, p. 676)

Another experiment which provides insight in his thoughts and methods is that in which he tied the trunk of the nerves which go the extremities, rendering the limbs insensible and *paralytic*. However, irritation of the muscles may still cause contraction, "though they were no longer subject to the command of the will." Haller continued testing parts that were separated from the body, including the intestines that preserve their peristaltic motion. The heart as well as isolated muscles, are also still irritable in that condition. Therefore, he continued, Robert Whytt (1714–1766), who had previously asserted that there is no motion of the body but by the soul, "has found himself obliged to admit the divisibility of the soul, which he believes to be separable into as many parts as the body." (Haller, 1755, p. 677)

These observations of body parts moving after disconnection from the body, made Haller think about the soul.

The soul is a being which is conscious of itself, represents to itself the body to which it belongs, and by means of that body the whole universe. I am myself, and not another, because that which is called I, is changed by every thing that happens to my body and the parts belonging to it. If there is a muscle, or an intestine, whose suffering makes impressions upon another soul, and not upon mine, the soul of that muscle or intestine is not mine, it does not belong to me. But a finger cut off from my hand, or a bit of flesh from my leg, has no connexion with me, I am not sensible of any of its changes, they can neither communicate to me idea nor sensation; wherefore it is not inhabited by my soul nor by any part of it; if it was, I should certainly be sensible of its changes. I am therefore not at all in that part that is cut off, it is entirely separated both from my soul, which remains as entire as ever, and from those of all other men. (Haller, 1755, p. 678)

He continued stating that neither the will nor the soul is changed, only the command over the amputated part is lost. The part continues to be irritable. "Irritability therefore is independent of the soul and the will." (Haller, 1755, p. 678) Reasoning in this way, the conclusion is that "the whole force of the muscles does not depend upon the nerves, because after these have been tied or cut, the muscular fibers are still capable of irritability and contraction." He expected that in the future it might well be that the nerves are found only to convey the commands of the soul to the muscle, just "to increase and excite that natural tendency which the fibers have of themselves to contract. ...." (Haller, 1755, p. 679)

Differences Between Irritable and Nonirritable Parts

Whereas irritability was associated with muscular fibers, the lack of it was with a cellular substance (e.g., in lungs, liver, kidneys, and spleen, but also ligaments and periosteum being composed of a cellular membrane). Haller considered the heart the most irritable organ: "it is constantly irritable, and much more than the intestines." (Haller, 1755, p. 687) One of the main reasons was that it "moves a great deal longer than any other part of the body, even after death, and sometime for four and twenty, or thirty hours, or longer." (Haller, 1755, pp. 687–688)

Haller concluded from these experiments that "there is nothing irritable in the animal body but the muscular fibre." Moreover he believed that the vital parts are most irritable. He also noticed that vital organs, being extremely irritable, require "only a weak stimulus to put them in motion," whereas the others "are not to be moved but by the determinations to the will, or by very strong irritations." (Haller, 1755, p. 690) Another characteristic of the most irritable parts is that they are soft, in contradistinction to the most solid parts including bones, teeth, and cartilages that "are void of Irritability." (Haller, 1755, p. 691)

Haller's findings had practical as well as philosophical implications, for instance with respect to ideas about the soul and death. The most irritable parts are not subject to the command of the soul, "which ought to be quite the reverse if the soul was the principal of Irritability." Irritability continues after death or removal of the organ from the body, which results in uncertainty about the moment of death, a problem which, according to Haller, Whytt had addressed: "the time of death is very uncertain … frequently an animal has life still remaining, after it has been looked upon as dead for some time." (Haller, 1755, p. 691) As the seat of the soul is in the head and irritability remains after the head has been removed or the nerves cut through, "it appears that this quality [irritability] … does not depend upon the soul." (Haller, 1755, p. 691) Finally he wondered whether irritability is "a property of the animal gluten, the same as we acknowledge attraction and gravity to be properties of matter in general, without being able to determine the cause of them." (Haller, 1755, p. 692)

Discussion on the Sensible and Irritable Parts of Animal

The two most important conclusions Haller drew from the series of experiments were that only those parts that are supplied with nerves possess sensibility, and irritability is a property of the muscular fibers. In the concluding part, Haller acknowledged Francis Glisson, "who discovered the active force of the elements our bodies." Haller referred to some other naturalists who wrote about the phenomena, including Baglivi whose experiments "approached nearer to a discovery of it." (Haller, 1755, p. 693) Followers of Georg Ernst Stahl (1660–1734) worked on the tone and natural contraction of fibers, but attributed it to the soul, not having made any experiments, because they had "an aversion to anatomy," according to Haller. He also acknowledged his teacher Boerhaave, who attributed an active force to the heart including a "latent principle of motion in the pieces of it which are cut; but as he attributes the cause of muscular

motion to the nerves, this proves that he did not sufficiently know, that the cause of this motion was in the muscles themselves." (Haller, 1755, p. 693)

Haller had commented on this previously.[54] "The nature of the thing obliged me to differ in opinion from my preceptor." (Haller, 1755, p. 694) He quite agreed with the opinion of "two famous [Dutch] physicians," Boerhaave's pupil Johannes de Gorter (1689–1762),[55] and Frederik Winter (1712–1760),[56] who attributed all motions of the human body to the irritability of the fibers, and the force of a stimulus.

He did not agree with Whytt, who imputed irritability to the soul, which, "feeling the impression of the irritation, occasions the contraction of the fibre."[57] Interestingly, Haller complained about "this gentleman's want of candour," adopting several of Haller's ideas, without mentioning the source. Another criticism on Whytt was that he performed only a small number of experiments (Haller, 1755, p. 695).

Criticism: soon after the publication of Haller's thesis, opponents reported their comments, e.g., on his statements that periosteum and tendons are insensible. Anton de Haen (1704–1776), one of the Dutch physicians who went to Vienna, Whytt, Claude Nicolle Le Cat (1700–1768), and Heinrich Friedrich Delius (1720–1791) were among the opponents (Temkin, 1936). Whytt, for instance, opposed the radical distinction between sensibility and irritability. He believed the distinction did not leave room for certain phenomena, such as reflex actions.[58] Le Cat and Delius, adherents of Stahl's animism, criticized Haller's separation of the soul from irritability. In *L'homme machine*, De la Mettrie believed that Haller's findings with respect to irritability were an argument against the existence of a spiritual soul (Temkin, 1936).

---

[54] Commentaries upon Boerhaave's *Institutions* of 1739, Ad. N. 187. Instit re med. Not. i. II.

[55] Upon Boerhaave's intercession he became professor of medicine in Harderwijk and when Boerhaave's nephew Herman Kaau Boerhaave died, he succeeded him as physician to Czarina Elisabeth in St. Petersburg.

[56] Phycisian in ordinary to the House of Orange, professor of medicine in Leiden. Thesis in 1746: *De certitudine in medicina practica*; and inaugural oration on *De motu vitali et irritibilitate fibrarum* (1751).

[57] This should probably be interpreted in the Whytt's well-known concept of reflex action.

[58] See Fulton, J.F. (1926). *Muscular contraction and the reflex control of movement* (pp. 32–34). Baltimore: William & Wilkins.

Many of the experimental findings by Haller were later refuted. Body parts he found to be insensible appeared to be sensible (e.g., the periosteum) and his definition of irritability appeared to have been too limited. In fact he had defined irritability as contractility and confined it to muscular fibers (Temkin, 1936). Later that century, other investigators showed that irritability should be considered a general vital principle applicable to many tissues. John Brown (1735–1788) for instance, created a medical system based on the concept of irritability, or what he called "excitability."

Neurologist and medical historian Max Neuburger (1868–1955) considered Haller's work "as far as the idea of brain localization is concerned … a retrograde step." (Neuburger, 1981, p. 118)[59] Haller's declaration that only the gray matter is insensitive, while stimulation of the white matter could elicit movements, survived for decades, according to Neuburger (1981, p. 123). However, he recognized Haller's more general importance for physiology, in particular decisively localizing the two basic properties of the organism, sensation and movement, in two different tissues.

Haller's *Elementa*

In his *Elementa physiologiae corporis humani* (1757–1766), Haller discussed all domains of physiology from antiquity to his own discoveries. In addition to physiology, he also wrote on anatomy and embryology, as well as theoretical medicine. This book is an important source for the history of medicine, as it synthesizes all previous knowledge of neuroanatomy and neurophysiology (Sigerist, 1958, pp. 173–186; McHenry, 1969, p. 109; Norman, 1991).

For the present chapter, I used Michael Foster's (1836–1907) *Lectures* (1901) in which excerpts from *Elementa* can be found (Foster, 1901). Most of the material in *Elementa* is a repetition from Haller's earlier work. Haller distinguished three types of contractile force "present not only in the animal but also in the vegetable kingdom … by

which the elements of fibers are brought nearer to each other"[60] and he mentioned: (a) elastic force; (b) a contractile force observed when tissues dead or alive shrink when treated in various ways, including heat; and (c) a special contractile force proper to muscles alone. On the latter he added that

every muscular fibre is irritable, and on the other hand you may fairly call muscular fibre everything that is irritable … [it is] a force of its own kind, different from every other power, and to be classed among the sources of the production of motion the ultimate cause of which is unknown.[61]

He did not agree with those who said it was the vital force, since "the force may for some little time survive the life of the body."[62] This *vis insita* is to be distinguished from the *vis nervosa*. The first is inherent to the muscular fibers, the second also has its seat in the muscle, but "comes from without and is carried to the muscles from the brain by the nerves, it is the power by which muscles are called into action."[63] This force may survive in cold-blooded animals as the muscles, provided they be moist and whole, are thrown into convulsions when their nerves are irritated.

Localization: Haller also returned to the question of whether particular parts of the brain function as the seat of sensation and the source of movement. He repeated that all of his predecessors believed that "the whole brain was not necessary for the full development of sensations."[64] Interpreting the phenomena of disease and the results of his experiments, however, Haller came to the conclusion that the corpora striata cannot serve as the seat of sensation and the source of movement, a view held by several of his predecessors, including Willis; nor did he consider the cerebellum to be essential in life.

Next to the interpretations from his experiments, he discussed some of his "conjectures," or one may say speculations, for instance on the nature of nervous action. As described, he rejected the vibration theory of elastic strings. He then went on to discuss the fluid/spirit doctrine, which should be "an element, too subtle to be grasped by any of the senses,

---

[59] Translated by Clarke, E. Orinally in German (1897): Die historische Entwicklung der experimentellen Gehirn – und Rückenmarksphysiologie vor Flourens, Stuttgart, Enke.
[60] From Haller's Elementa quoted by Foster (1901).

[61] From Haller's Elementa quoted by Foster (1901).
[62] From Haller's Elementa quoted by Foster (1901).
[63] From Haller's Elementa quoted by Foster (1901).
[64] From Haller's Elementa quoted by Foster (1901).

but more gross than fire, or ether or electric or magnetic matter, since it can be contained in channels and restrained by bonds and moreover is clearly produced out of and nourished by food."[65] He considered the nerves to be hollow, which is true for the peripheral as well as the nerves within the brain. In discussing the seat of the soul, Haller rejected the Stahlian opinion that it is diffused over the whole body. From his experiments he had decided that the seat of sensation and the source of muscular movement should be in the medulla of the brain. "No narrower seat can be allotted to the soul than the conjoint origin of all nerves,"[66] which in fact is quite similar to what Boerhaave taught in his Lectures on the nervous system. All localizations that were more limited were rejected by Haller, including the corpus callosum, the septum lucidum, the pineal gland, and the corpora striata. Whether different parts of the brain corresponded with different functions of the soul, for instance that the parts around the entrance of the optic nerve are especially concerned in vision, he answered by stating that "our present knowledge does not permit us to speak with any show of truth about the more complicated functions of the mind or to assign in the brain to imagination its seat, to common sensation its seat, to memory its seat. Hypotheses of this kind have in great numbers reigned in the writings of physiologists from all time. But all of them have been feeble, fleeting, and of a short life."[67]

In conclusion, it is clear that Boerhaave as well as Haller were very well informed of the contemporary medical literature that they interpreted with criticism. Haller was more clear about which knowledge derived from experiments and what should be considered "conjectures." He was critical toward physicians who did not perform experiments or drew conclusions just from analogy. He may be considered an early physiologist, whereas Boerhaave rather relied on bedside observation and the results of microscopic examination, e.g., denying the existence of muscle fibers in the dura mater. He may be considered an eclecticist and systematist and in his well-organized system, almost all aspects

of health and disease could be explained. With respect to (patho)physiology, both men applied humoral as well as solid theories. Boerhaave considered nerves to be hollow arising from the medulla of the brain and spinal cord. Spirits still formed the basis to explain the function. With this respect and many other, Haller accepted what Boerhaave had taught. Boerhaave localized the sensorium commune at the transition of white and gray matter, whereas Haller believed it should be localized in the cerebral medulla and deep structures. In localizing functions, Haller probably went further than Boerhaave, although he remained cautious and admitted that knowledge with this respect was insufficient. Haller refuted vibration and electricity to explain nerve physiology. Despite the quite mechanistic explanations in their physiology, both men referred to God with respect to ultimate question of causality. Boerhaave's as well as Haller's teachings belong to the most influential of the eighteenth century.

References

Boerhaave, H. (1730–1735). In B. P. M. Schulte (1959). *Hermanni Boerhaave Praelectiones de Morbis Nervorum 1730–1735*. Leiden: Brill.

Clarke, E., & O'Malley, C. D. (1968). *The human brain and spinal cord. A historical study illustrated by writings from antiquity to the twentieth century*. Berkeley/Los Angeles: University of California Press.

Eems, J. van. (1761). *Hermanni Boerhaave etc. Praelectiones academicae de morbis nervorum. Lugduni Batavorum*. Van der Eyk & De Pecker.

Foster, M. (1901). *Lectures on the history of physiology during the 16th, 17th and 18th centuries*. Cambridge: University Press.

von Haller, A. (1755/1936). *A dissertation on the sensible and irritable parts of animals*. Translated from Latin by Tissot. London: Nourse.

von Haller, A. (1757–1766). *Elementa physiologiae corporis humani* (Vol. 8). Lausanne, Berne.

von Haller, A. (1776–1788). *Bibliotheca medicinae practicae* (Vol. 4). Basle/Berne: Schweighauser/E. Haller.

Haller, A. (1786a). *First lines of physiology* (Vol. 1). New York: Johnson Reprint. (Reprinted in 1966)

Haller, A. (1786b). *First lines of physiology* (Vol. 2). New York: Johnson Reprint. (Reprinted in 1966)

King, L. (1966). Introduction. In A. Haller (Ed.), *First lines of physiology*. Translated from the correct Latin edition. Reprint of the 1786 edition. New York/London: Johnson Reprint Corporation.

---

[65] From Haller's Elementa quoted by Foster (1901).
[66] From Haller's Elementa quoted by Foster (1901).
[67] From Haller's Elementa quoted by Foster (1901).

Lindeboom, G. A. (1958). *Haller in Holland. Het dagboek van Albrecht von Haller van zijn verblijf in Holland (1725–1727)*. Delft: Koninklijke Gist – en Spiritusfabriek.

Lindeboom, G. A. (1968). *Herman Boerhaave, the man and his work*. London: Methuen.

McHenry, L. C. (1969). *Garrison's history of neurology*. Springfield, IL: Thomas.

Neuburger, M. (1981). *The historical development of experimental brain and spinal cord physiology before Flourens*. (original in German 1897). Baltimore/ London: Johns Hopkins University Press.

Norman, J. M. (1991). *Morton's Medical Bibliography. An annotated check-list of texts illustrating the history of medicine (Garrison and Morton)* (5th ed.). Aldershot, Hants: Scolar Press.

Schulte, B. P. M. (1959). *Hermanni Boerhaave. Praelectiones de Morbis Nervorum 1730–1735*. Leiden: Brill.

Sigerist, H. E. (1958). *The great doctors*. Translated by E and C Paul. Garden City (NY).

Temkin, O. (1936). A dissertation on the sensible and irritable parts of animals by Albrecht von Haller. *Bulletin of the History of Medicine, 4*, 651–699.

16
Apoplexy: Changing Concepts in the Eighteenth Century

Catherine E. Storey

In the eighteenth century, apoplexy was the term used to describe a clinical presentation rather than a single disease entity – a sudden catastrophic event characterised by a loss of consciousness, movement and sensation. Many of the conditions that would have been described under the term apoplexy are incorporated into what is now referred to as stroke. This chapter will explore the changes in the understanding of this spectrum of diseases included in the term as well as the shifts in the actual use of the term during the eighteenth century.

Hippocrates (c. 460–375 BCE) is credited with the introduction of the term into medical terminology (Clarke, 1963) – a term that underwent little conceptual change until the clinician-anatomists of the seventeenth century. Nominal revision was delayed well into the nineteenth. By the twentieth century, the term became synonymous with an intracranial haemorrhage, and apoplexy was finally discarded from official classifications of cerebrovascular disease in 1939. To understand the use of this term as the eighteenth century commenced there is no better description than that offered by Thomas Willis (1621–1675) when, in the *London Practice of Physick,* he explains:

The *Apoplexy*, according to the import of the Word, denotes a striking, and by reason of the stupendous Nature of the affect, as tho it contain'd somewhat Divine, it is called a *sideration*: for those that are seized with it, as tho they were *Planet-struck*, or smitten by an invisible Deity, fall on the Ground on a sudden, and being deprived of Sense and Motion, and the whole animal function (unless that they breath) ceasing, they lye dead as it were for some time, and sometimes dye out-right:

and if they revive again, they are oftentimes affected with a general Palsie or an Hemiplegia. (Willis, 1685)

How then did apoplexy fare during the period of the Enlightenment of the eighteenth century? McHenry (1969, p. 375) in his revision of Garrison's *History of Neurology* is of the opinion that "the Eighteenth Century notions on apoplexy largely reflected the experience of previous observers", and he certainly does not credit much advance to this time period. The neurologist/historian John Spillane (1981), in his comprehensive history of neurological diseases, describes the medical landscape of the eighteenth century as one that is not "universally attractive". He places the century into a period of transition – a transition from a medical practice with deep-seated roots in philosophy to one in which experimental science provided new insights into disease. This author writes that the neurological activity of this century was dominated not by a burst of experimental science that should have been anticipated as a result of William Harvey's account of the circulation in 1628, but rather "the too ambitious schemes of the classifiers and systematists, not to mention the cults of mesmerism, homeopathy, and magnetism". (Spillane, 1981, p. 111)

Superficially the understanding of apoplexy during this century looks bleak. McHenry is in part correct in that many of the theories did reflect the experiences of previous writers and there were no great discoveries or great innovations. There is evidence, however, to suggest that clinical achievements did result from a consolidation of the work of physicians of the seventeenth century that applied Harvey's ideas of the circulation to clinical problems, namely Johann Wepfer (1620–1695) and Thomas Willis (1621–1675). The vascular theories proposed in the seventeenth century were to become the underlying principles on which the history of apoplexy would later develop.

Spillane is also correct in noting that the "too ambitious schemes of the classifiers" would impact on the progress of apoplexy. For example, the Scottish physician William Cullen (1710–1790) proposed an elaborate scheme in 1769 based primarily on symptomatic presentations, which served to consolidate the accepted causes of apoplexy into a formal structure. The aim of such a classification was to provide aid to the bedside physician (Kendell, 1993). His ambitious scheme was initially popular and widely accepted, but not sustained for long after his death. The wide range of symptomatic causes was subsequently reduced into a simpler framework based on pathological findings alone. This, in turn led to the dichotomy that we now recognise in stroke: haemorrhagic and ischaemic strokes.

If the eighteenth century was not dominated by great discoveries or innovations, how best to identify any progress made during this period? As shall now be shown, progress was made at a clinical level – at the level of the medical practitioners who treated such patients, who incorporated new ideas derived from clinico-pathological correlation into their practices, who questioned long standing therapeutic options and who were guided largely by clinical experience.

The Eighteenth-Century Landscape

The works of Thomas Kirkland (1721–1798) and John Cooke (1756–1838), two busy, well respected, British physicians offer contemporary insights into the clinical perceptions of apoplexy during the eighteenth century, and some of the theories that helped set the stage.

In 1792, prompted by his long-standing dissatisfaction with the treatments used in the management of apoplexy and palsy, Kirkland published *A Commentary on Apoplectic and Paralytic Affections* (Kirkland, 1792). He aimed to provide his reader with a literature review enhanced with his own experiences, and to provide a rational approach to treatment based on symptomatic presentation.

FIGURE 1. Dr. John Cooke (1756–1838) and the frontispiece of *A Treatise on Nervous Diseases* (1820)

In contrast, the London physician Dr. John Cooke, set out with a much more ambitious agenda, hoping "to collect, to arrange and to communicate, in plain clear language, a variety of useful observations from the best authors, both ancient and modern, respecting the principal diseases of the nervous system" (Cooke, 1820, p. iii) (Fig. 1). In *A Treatise on Nervous Diseases,* he identified the principal nervous system diseases of his time as "apoplexy, epilepsy and the palsy". Cooke's "useful observations" thus provide reviews of the classical and newer literatures, intercalated with his own medical experience as physician to the London Hospital.

The Classical Era and Galen's Theories

Both Kirkland and Cooke begin with a review of the ancient origins of apoplexy in much the same manner as a "modern text" would begin with a "lay of the land" to establish the authority on which the discipline is based. Kirkland collects the scattered references to apoplexy within the Hippocratic

Corpus to construct the clinical presentation that characterised the disease for these ancient physicians, and he provides this precise summary:

He [Hippocrates] says, "A person in health is suddenly seized with a pain in the head, the voice is immediately intercepted, he snores and gapes, and if spoken to or moved, he only sighs; is insensible, and makes a great deal of water without knowing it. If he becomes mute and snores, he dies within seven days, unless a fever comes on, and then he generally recovers. This kind of disease generally happens to people advanced in years, than to young subjects; but the most usual time of attack, is betwixt forty and fifty years of age. It is more common in winter and rainy seasons. The brain suffers and becomes morbid, and if eroded, sustains a violent derangement; hence delirium, the brain is convulsed, and involves the whole man in the same perturbation; he becomes incapable of speaking, and suffocation ensues. ... If apoplectic patients are attacked with the piles, the symptoms are favourable, if with chilling and torpor, it is a bad sign. ... It is impossible to cure a vehement apoplexy; nor is a weak one very easily cured." (Kirkland, 1792, pp. 5–9)

This clinical description had served as the basis of the clinical narrative up to the time of Kirkland and Cooke.

The physiological model of this disease that was proposed by Galen (130–200) provided the basis for the understanding of the pathogenesis well into the sixteenth century and beyond. Cooke, in his discussion of Galen's physiological model of brain function, explained:

The governing principle, he thinks, is seated in the brain, and acts through the medium of the animal spirit, which is the immediate instrument of sensation and motion, being carried by the nerves to the parts, which feel and move. This animal spirit is first generated by what he calls the vital spirit, which is formed by the heart and arteries. It is further prepared by an apparatus of a curious vascular construction, which he denominates the wonderful net-like plexus [rete-mirabile]; and it is finally elaborated in the ventricles of the brain, to which it is carried from the net-like plexus. (Cooke, 1820, p. 9)

Galen reasoned that apoplexy occurred as a result of a pathological accumulation of thick phlegm within the cerebral ventricles (Karenberg, 1994). This in turn obstructed the flow of animal spirits stored within, resulting in the loss of the "instrument of sensation and motion", the key identifying clinical features of apoplexy.

Concepts of apoplexy changed very little during the period of the Middle Ages, and the ideas proposed by Galen remained unquestioned for centuries. Kirkland describes one such early challenge in this following passage:

... for what Aetius and Paulus had copied from Galen, about cold, thick phlegm being the cause of every kind of apoplexy, was handed down to about the time of Fernelius, who denied that the ventricles of the brain are in this disease either filled or obstructed with this kind of humour, and denies it as his opinion, *from reasoning upon the subject,* that it is brought on "by the arteries of the brain being compressed or obstructed. –Whence the heart does not receive the spirits usually sent to it, and sense and motion perish". (Kirkland, 1792, p. 12)

Kirkland refers to one Jean Fernel (1497–1558), an esteemed Paris philosopher, physician and according to Castiglioni (1941) a significantly influential figure in breaking down the authority of Galen in France. The case to which he alludes, originally published in 1544 (Fernel, 1544) was later included by Theophile Bonet in his *Sepulchretum sive Anatomia Practica,* published in 1679. A modern review of the *Sepulchretum,* undertaken to construct a concept of apoplexy in the seventeenth century, includes the details of this case to illustrate this nascent awareness of a vascular basis of apoplexy (Schutta & Howe, 2006, p. 255).

The case is worth examining, to explore this period of transition that was to take place from the sixteenth century, from a humoral past towards a vascular future for apoplexy. A man, after a blow to the head, "fell thunderstruck", and died some 12 h later. The post-mortem identified the brain to be intact, the ventricles empty of phlegm and undamaged, but a clot of blood lay at the base of the brain. The case illustrates several important prevailing concepts of the sixteenth century. The diagnosis of apoplexy was accepted on clinical grounds alone, there was no consideration required of the morbid anatomical findings or presumed basis of pathogenesis. Head injury was accepted as a cause of apoplexy and would remain so in Cullen's 1769 classification. Fernel reasoned from these findings, that Galen's theory of ventricular obstruction was untenable. There was no phlegm – the ventricles were empty and unobstructed. The proposed hypothesis was that the clot of blood identified obstructed the flow of "animal spirits" within the rete mirabile at the base of the brain. The existence of this vascular network was also soon to be challenged. The rete-mirabile, identified by Galen from animal experimentation and incorporated into his physiology of neural function of the human brain, was found absent from the human studies of the great Renaissance anatomists such as Andreas Vesalius (1514–1564). These long held Galenic traditions, which underpinned the conceptual framework of apoplexy to this time, were under threat as the seventeenth century approached.

Seventeenth-Century Foundations

The Swiss physician Johann Jakob Wepfer, the English physician/anatomist Thomas Willis and the Swiss physician/anatomist Theophile Bonet (1620–1689) were the main contributors to the progress of apoplexy during the seventeenth century. Wepfer and Willis were both influenced by William Harvey's (1578–1657) explanation of the circular motion of the blood published in his *De motu cordis* of 1628. Although not universally or immediately embraced as a concept, Harvey's theory of circulation was readily incorporated into the theories that would emerge from the observational studies of clinicians such as Wepfer and Willis. Theophile Bonet's contributions to the understanding arose as a result of his early attempts at the systematic review of large collections of clinico-pathological material, a discipline that would emerge during the eighteenth century under the influence of Giovanni Battista Morgagni (1682–1771).

In 1658, Wepfer published an influential work on apoplexy, his *Historiae Apoplectorum,* in which he described four cases with autopsy confirmation of an intracranial haemorrhage (Pearce, 1997; Wepfer, 1658). This finding has assured Wepfer of a place in history as the first clinician to link apoplexy with an intracranial haemorrhage, but his work had additional vascular implications. Wepfer adapted the principles proposed by Harvey to reason that apoplexy could also result from either obstruction of arterial blood to the brain or obstruction of the egress of blood from the brain. Mani (1982, p. 47) in a review of the implications of Wepfer's contribution to the scientific literature in light of Harvey's discovery, suggests that with his account, "Apoplexy

became a vascular disease affecting the cerebral circulation". Perhaps this statement is true in a retrospective historical analysis, but in clinical practice a practical link with the cerebral circulation was still a long way off.

Thomas Willis, although often cited for his eponymous description of the arterial anastomosis at the base of the brain, made other contributions to support a vascular theory for apoplexy. Others, in fact, had described parts of the arterial circle before him, including Wepfer (Meyer & Hierons, 1962; Symonds, 1955). Willis, however, extended his reasoning to propose a potential explanation for the functional significance of this arterial anastomosis:

An in the first place, though we grant that the flowing in of the blood, may be sometimes denied to the Brain; yet we do not believe, that it only happens after the aforesaid ways, nor that, for that reason, the *Apoplexy* doth arise. We have everywhere showed, that the *Cephalick* arteries, viz. the *Carotides*, and the *Vetebrals,* do so communicate one with another, and all of them in several places are so ingrassed one in another mutually, that if it happen, that many of them should be stopped or pressed together at once, yet the blood being admitted to the Head, by the passage of one artery only, either the *Carotid* or the *Vertebral,* it would presently pass thorow all those parts both exterior and interior'; which indeed we have sufficiently proved by an experiment, for that Ink being squirted in the trunk of one Vessel, quickly filled all the sanguiniferous passages, and every where stained the Brain it self. I once opened the dead carcase of one wasted away, in which the right Arteries, both the *Carotid* and the *Vertebral*, within the skull, were become bony and impervious, and did shut forth the blood from that side, notwithstanding the sick person was not troubled with the astonishing Disease; wherefore it may be doubted, whether the blood excluded from the Brain, by reason of some Arteries being obstructed or compressed, doth bring forth this Disease. Certainly there is more of danger, that the cause of the Apoplexy, should be from its too great incursion and extravasation within the Brain. (Willis, 1684, 1685; cited in Spillane, 1981, p. 69)

Willis noted the existence of arterial occlusive disease, but argued that the occlusion of a single major artery was therefore unlikely to give rise to the anticipated apoplexy ("the astonishing Disease"), as a result of the extensive anastomosis that existed between the vessels. From Willis' interpretation of these findings it is suggested that he was of the opinion that extravasation of blood into the substance of the brain was a far more likely explanation for apoplexy than from occlusive arterial disease. Notwithstanding there is evidence to suggest that Willis concurred with Wepfer's proposal that apoplexy could occur as a result of blood flow obstruction to the brain, from a rupture of intracranial vessels, or from obstruction of venous outflow. (Spillane, 1981)

Isolated case reports of apoplexy, such as that described by Fernel, had appeared in the medical literature from the sixteenth century on, but were often difficult to source and were as individual reports of little practical value for the physician. In 1679, Bonet published his *Sepulchretum sive Anatomica Practica,* a monumental work that contained over 3,000 autopsy cases with clinical notes and observations (Schutta & Howe, 2006). The work, arranged as a series of "Observations", included 70 cases of apoplexy and was the most extensive collection up to this time. It included the majority of earlier isolated reports, including those of Wepfer and Willis, together with much of his own personal material.

The *Sepulchretum* was an immediate success and, after being revised and enlarged by Mangentus in 1700, it remained an important resource for physicians until replaced in 1769 by Giovanni Battista Morgagni's (1682–1771) *De Sedibus et Causis Morborum* (Morgagni, 1761). The cases on apoplexy are mostly included in a single section *De Apoplexia* and roughly grouped into sanguineous and serous subtypes. Other cases are recorded amongst coma, headache and epilepsy. When Cooke reviewed the pathological findings associated with apoplexy in 1820, he made extensive reference to, and recommended the works of both Bonet and Morgagni. These were clearly well regarded texts amongst clinicians.

Thus, during the seventeenth century there had been significant revisions in the concepts of apoplexy. As the eighteenth century approached, although remnants of long held dogma persisted, the foundations had been laid for apoplexy to shed some ancient traditions based on symptomatic presentation and to evolve into a vascular disease based on clinico-pathological and patho-physiological correlates.

Changing Concepts During the Eighteenth Century

Kirkland published his *Commentary* in 1792 to inform his reader of the facts that would lead to more appropriate treatment options for those patients with "apoplexy and paralytic affections". When Cooke wrote his *Treatise on Nervous Disease* in 1820, he had a broader agenda. It was to acquaint his readers with the known facts, which he had sourced from expert opinions in this field, both ancient and modern. He also aimed to provide his own experiences, gained from his long service at the London Hospital in the latter decades of the eighteenth century, together with some of the salient observations from his contemporaries to provide "state of the art" statements about apoplexy. He set out to determine how one should recognise this disease and what the findings commonly associated with apoplexy might be at post-mortem. He asked what the mechanisms responsible for this disease are, how they cause their effects, and how the physician could distinguish this disorder from other diseases affecting the nervous system. In brief, he strove to tell his readers how to arrive at a correct diagnosis, provide a correct prognosis and choose the best of the various treatment options. Apoplexy and palsy, however, were considered as individual entities and dealt with in separate volumes.

Definitions Re-explored

The term Apoplexia was employed by the Greeks, and is still used, to denote a disease in which the patient falls to the ground, often suddenly, and lies without sense or voluntary motion. Persons, instantaneously thus affected, as if struck by lightning. (Cooke, 1820, p. 157)

Thus at the conclusion of the eighteenth century, Cooke begins his work with the classic definition handed down from the Greek writers. Apoplexy is a disease of sudden onset, a loss of motion and sensation. Throughout time, this has been a disease characterised and defined by this catastrophic presentation.

Both Kirkland and Cooke recognise that this is not and could not be a single entity, but a heterogeneous group of conditions with a single common presentation. Although Cooke grapples with the problem as to how his predecessors at times had debated the essential features of the definition, his own re-definition improves only on the recognition that respiratory dysfunction is an important clinical feature. In his words: "It is a disease in which the animal functions are suspended, while the vital and natural functions continue: respiration being generally laborious, and frequently attended with stertor" (Cooke, 1820, p. 166).

Definitional debates also extended to the terms "apoplexy" and "palsy". Although these designations were used somewhat interchangeably since ancient times, there remains some confusion in terminology. Aretaeus and Galen had thought of apoplexy as a general, and palsy a partial, abolition of sense and motion. Kirkland separated these conditions in his discussions and Cooke discussed these conditions in separate volumes in his *Treatise*. Important concepts had undergone revision. There is a direct relationship between apoplexy and hemiplegia, or a palsy of half of the body, and there is now a general acceptance, after a long period of debate, that a lesion of the brain can produce a contra-lateral weakness.

Cooke (1820, p. 176) quotes from a contemporary source *The Morbid Anatomy of some of the most important parts of the Human Body,* written in 1793 by physician Matthew Baillie (1761–1823) (Baillie, 1793) to illustrate these points:

When blood is effused into the substance of the brain, apoplexy is produced. When the patient is not cut off at once, but lives for some time after the attack, hemiplegia, which is almost constantly an effect of this disease, is upon the opposite side of the body from that of the brain, in which the effusion of blood has taken place. This would seem to show that the right side of body derives its nervous influence from the left side of the brain, and the left side of the body its nervous influence from the right side of the brain.

Kirkland also seeks to find a practical solution to the question of apoplexy and palsy and in doing so provides a succinct summary of his own when he concludes:

Now the moderns agree with the ancients in there being a great affinity between apoplexy and palsy, because they seize in a similar manner, and because they change vice versa from one into the other. In some instances they appear to me to arise both from the same cause, and perhaps after all, the ancients were right, and they are only different degrees of the same disease under different circumstances, though there are no diseases in which the symptoms are more unlike. (Kirkland, 1792, p. 79)

Clinico-Pathological Correlates

Bonet's *Sepulchretum*, his "burial vault" of practical anatomy, remained an important reference source for physicians into the eighteenth century until replaced by Morgagni's *De Sedibus et Causis Morborum* in 1761 and in English translation in 1769 (Morgagni, 1761). *De Sedibus* consists of five books of letters written to an unnamed colleague, in which he recorded post-mortem findings with clinical correlates. This mid-eighteenth-century work has been described as "the high water mark of clinico-pathological description" (Castiglioni, 1941, p. 602) and regarded as the foundation of modern pathological anatomy (McHenry, 1969). Thus, Morgagni, in a format similar to that begun by Bonet, provides a new means of conceptualising disease in which organ-based pathology now defines disease. The first book of the *De Sedibus, of Disorders of the Head,* is divided into various letters, and four are devoted to apoplexy, which Morgagni further divided on pathological grounds according to whether there is blood (sanguineous), serum (serous), or other pathology present. Superficially this appears to be a simple structure, the forerunner of the modern haemorrhagic or ischaemic stroke, but this would be an unfounded assumption.

After considering previous studies, both Kirkland and Cooke conclude that, although blood is by far the most common finding, it is by no means an exclusive finding. They note the vast array of pathologies observed: pus, hydatids, tumours and worms, to name but a few. They also note that these morbid appearances have been most frequently observed in the cranium, indicating that this too is not an exclusive finding. Indeed, there are still reported cases in which the suspected pathology has not been identified. Cooke (1820, p. 191) comments that "it is much to be lamented, that in the accounts of the dissection of persons supposed to have died of apoplexy, the cases have been so imperfectly described".

Clearly, despite the extensive macroscopic clinico-pathological studies that preceded his time, and to which he referred extensively, Cooke recognised the need for further clinical correlations before precise pathological relationships could be firmly established.

Causal Factors

A predisposition to apoplexy had long been recognised. The body habitus was of great import in the eighteenth century, while the patient's habits were held responsible for an increased susceptibility to this malady. The patient had borne some responsibility for his disease from the time of the ancients. Cooke describes his own research:

Persons of plethoric habit, especially those who have short thick necks, and who indulge in eating and drinking, and sleep, are predisposed to this disease. Galen, we are told, thought that apoplexy and epilepsy are sometimes occasioned by a fulness (sic) of blood; and Aretaeus remarks, that those who are of a gross habit of body; of a lax or moist constitution; slothful, and gluttonous, are predisposed to it. Dr Cullen thinks "corpulency and obesity operate much in the production of apoplexy, by occasioning a more difficult transmission of the blood through the lungs". (Cooke, 1820, p. 201)

Cooke recognises that the disease is rarely seen in the "laborious poor, unless occasioned by drinking spirits to excess" (Cooke, 1820, p. 204).

When describing the "exciting" causes, a term used to identify those conditions with a direct causal relationship, Cooke concluded that all the previously described pathologies act by pressure on the brain, dividing the pathologies into those that act "quickly and powerfully" and those that act "more gradually and feebly". He reasoned that in severe cases of sudden onset with stertorous breathing, blood as a consequence of blood vessel rupture is invariably found at post-mortem and he nominates these as "sanguineous apoplexy". In those cases that have a slower onset, often with warning symptoms, serum is the more frequent finding, a "serous apoplexy". All other pathologies such as tumours, hydatids, etc., he nominates as remote causes and most likely to present in the first instance with headache, vertigo, lethargy or blindness. Only in the later stages, and by an increase in pressure, are these individuals likely to present with apoplexy. Thus, Cooke, as a clinician, used his clinical experience to correlate presentation with the underlying pathology.

Whether Cooke provided a useful and accurate assessment was debated amongst his colleagues when he offered the opinion that:

The important conclusion to be drawn from the examination of M. Portal and others is, that in these cases it is extremely difficult to make the distinction during life. Cases seldom occur so strongly marked by circumstances and symptoms as they are described by authors in enumerating the characteristics of the two apoplexies, and even in those most distinctly marked, erroneous conclusions as to the cause have been very frequently made. (Cooke, 1820, p. 262)

Not all physicians, particularly Portal, agree that such a distinction into serous and sanguineous types can be determined in life, nor that the distinction supported by Cooke and others of sanguineous and serous types, serve the interest of the practising clinician.

Wepfer, supported by Willis, had reasoned that apoplexy might occur as a consequence of either an increase in arterial flow to the brain or of an obstruction to venous outflow. There is now sufficient reference to these mechanisms in both the works of Kirkland and Cooke to suggest that this is common and established theory. Cooke elaborates these causes from his own experiences by providing this summary:

Violent passions of the mind, violent exercise of the body, fits of intemperance, excessive straining, long continued stooping, ligatures, tumours compressing blood vessels, rarefaction of the fluids by sudden and great heat, congestions by excessive cold, and the suppression of evacuations to which the body has been accustomed. (Cooke, 1820, p. 210)

Three pages later, he also reported:

M. Portal, in his account of the plethoric apoplexy, says, this superabundance of blood in the brain may also be the consequence of strong contentions of mind, of profound study, and of violent and prolonged chagrin. Hence, we are not surprised to find that the disease often happens to literary and scientific men; in short, to such as are devoted to painful mental exertions. (Cooke, 1820, p. 213)

The question of excess arterial inflow remained unresolved and had to wait for a resolution that would come with the physiological determinants of blood pressure measurement.

Obstruction to venous outflow was an alternative explanation. Posturing of the neck (the case reported by Fothergill of a man rowing across the Thames who suffered apoplexy as he turned his neck, is commented on by both authors), a tight ligature or cravat, or the short neck are all given as examples of factors causing venous obstruction. The short neck in particular was often described in association with apoplexy and was frequently recorded as a patient characteristic in case reports during the long eighteenth century and into Cooke's early-nineteenth century. It is interesting to see, however, that there may have been some dissent to this argument, as judged by Kirkland's 1792 comment:

It is generally believe, that short-necked people are more subject to this complaint than others, owing to an unusual quantity of blood passing through the head. But has theory or facts supported this opinion? For though it does happen to persons of this structure of the body, in common with others; yet may it not be observed, that by far the greater number of short-necked people pass through life without one apoplectic symptom happening to them, and those with long necks are equally liable to this complaint. (Kirkland, 1792, p. 35)

Regardless of the underlying morbid pathology, Cooke summed up the means by which these pathologies result in the clinical presentation of apoplexy, the "proximate cause" when he writes:

The opinion, however, that apoplexy is immediately caused by an obstruction of the passage of the nervous fluid into the organs of sense and motion has been the favourite hypothesis of physiologists, and seems, most satisfactorily than any other, to explain the manner in which the exciting causes act in producing the symptoms of the disease. (Cooke, 1820, p. 251)

The concept of Galen's animal spirits stored in, and distributed from, the cerebral ventricles, first questioned in the sixteenth century has now most certainly been replaced. Cooke now reasons that it is the obstruction of a physical substance, the nervous fluid, flowing within the nerves, along predetermined paths to the parts responsible for sense and motion that ultimately is responsible for the clinical manifestations of apoplexy.

Classifications

Apoplexy had been distinguished in the early writings on the basis of the severity of symptoms. Hippocrates had distinguished the "strong" and the "weak", whereas Galen had added subcategories depending on the degree of respiratory compromise. The need to provide an accurate prognosis had driven the distinctions.

In the eighteenth century, which Spillane claims was dominated by the "too ambitious schemes of the classifiers", none might have been more ambitious than the scheme proposed in 1769 by influential Edinburgh physician William Cullen in his *Synopsis Nosologiae Methodicae* (Kendell, 1993). Motivated by the emerging importance of clinical bedside medicine and the needs of the diagnostic clinician, Cullen proposed a scheme by which all clinical presentations could be identified according to a logical taxonomy of symptoms. Cullen, inspired by the scientific approach to the classification systems proposed for botanic taxonomies, developed a similar structure to accommodate a framework for the classification of human diseases. Specifically, he divided all human diseases into four classes, 19 orders, and 132 genera according to symptomatic presentation (Kendall, 1993).

Initially Cullen's classification was widely acclaimed, and was frequently reprinted in the original Latin with subsequent English translations. The last such English translation was published in London in 1823, after which time the classification no longer suited clinical practice and lost favour. Still, a review of the means used by Cullen to classify apoplexy in 1769 provides an insight into the contemporary mid-eighteenth century concept of this disease.

Cullen allocated all diseases into one of four classes (Pyrexiae, Cachexiae, Neuroses and Locales). Apoplexy was placed in the class Neuroses and the Order of Comata (other orders within Neuroses were Adynamiae, Spasmi and Vesaniae), with subsequent divisions of Apoplexy into genera (See Table 1). This subdivision of apoplexy was wholly based on clinical presentation. Cooke provides his reader with an excellent description of Cullen's complex classification in the summary:

Dr. Cullen also places apoplexy as a genus under the order comata, and enumerates the following species of this genus, distinguishing them by their causes: the *sanguinea,* marked by signs of plethora, especially in the head; the *serosa,* chiefly occurring in the leucophlegmatic habits in the decline of life; the *hydrocephalica,* which, coming on gradually, affects infants and children, first with lassitude, slight fever and pain in the head, afterwards

TABLE 1. Cullen's classification of apoplexy – Nosologica 1769 (Kendall, 1993, pp. 226–227).

Class II	Neuroses (an injury of the sense and motion, without an idiopathic pyrexia or any local affection)
Order 1	Comata. A diminution of voluntary motion, with sleep, or a deprivation of the senses
Genus 42	Apoplexia. Almost all voluntary diminished, with sleep more or less profound; the motion of the heart and arteries remaining

The idiopathic Species:
1. Apoplexia (sanguinea) with symptoms of universal plethora, especially of the head
2. Apoplexia (serosa) with a leucophlegmasia over the whole body, especially in old people
3. Apoplexia (Hydrocephalica) coming on by degrees, affecting infants, or those below the age of puberty first with lassitude, a slight fever and pain of the head, then with slowness of the pulse, dilatation of the pupil of the Eye, and drowsiness
4. Apoplexia (Atrabiliara) taking place in those of a Melancholic constitution
5. Apoplexia (Traumatica) from some external injury mechanically applied to the head
6. Apoplexia (venerata) from powerful sedatives taken internally or applied externally
7. Apoplexia (mentalis) from a passion of the mind
8. Apoplexia (Cataplectica) in the contractile muscles, with immobility of the limbs by external force
9. Apoplexia (suffocata) from some external suffocating power

The apoplexy is frequently symptomatic
1. Of an intermittent fever
2. Continued fever
3. Phlegmasiae
4. Exanthema
5. Hysteria
6. Epilepsy
7. Podagra
8. Worms
9. Ischuria
10. Scurvys

with a slow pulse, dilated pupil, and somnolency; the *atrabiliaria,* which takes place in persons of the melancholic temperament; the *traumatica,* arising from external mechanical injuries of the head; *venerata,* occasioned by sedative powers internally taken or externally applied; the *mentalis* produced by passions of the mind; the *cataleptica,* in which the limbs remain in whatever situation they are placed by external power; and the *suffocata,* arising from an external suffocating cause. (Cooke, 1820, p. 255)

Only 8 years separates Cullen's classification from Morgagni's determination of the main pathologies into sanguineous and serous, yet this had failed to make a significant impact on his classification. When Cooke reviewed the classification it is clear that he was influenced by Morgagni and hoped to reduce the distinction into a more manageable, clinically relevant distinction. He opined that from his position in the early years of the nineteenth century how "many eminent modern physicians" now accept the distinction of sanguineous and serous types, explaining:

The *sanguineous,* they say, is marked by plethora, especially in the head, a strong, robust constitution, suddenness

of the attack, a swelled, red or purple countenance, a full, strong pulse, laborious stertorous breathing, prominence and inflammation of the eyes, and increased heat of the body, the disease happening not infrequently in the vigour of life, The *serous* apoplexy, takes place, they say, more especially in advanced, or old age, in debilitated or leucophlegmatic habits, and makes its attack more gradually than the sanguinous. In this species the countenance is said to be pale; the pulse weak and often irregular. (Cooke, 1820, pp. 257–258)

From Cullen's very complex classification of apoplexy, a dichotomy of the term had emerged, a schism that would broaden, as the pathogenetic mechanisms of causation were determined.

Treatment Options

Throughout the ages the prognosis of apoplexy had been regarded as very poor and this pessimistic outlook has had a detrimental effect on treatment trials. Many physicians had reiterated the original Hippocratic aphorism "It is impossible to cure the strong apoplexy, and not easy to cure the weak"

(Clarke, 1963). Cooke discouraged this nihilistic view explaining that the dogma of Hippocrates had had "a most mischievous influence, both on ancient and modern practice" (Cooke, 1820, p. 290).

To support his stance, he also provides many examples of cases with good prognoses and recovery. Spontaneous recovery is possible, he concludes, and the condition is not necessarily fatal. How then should one approach the treatment of apoplexy? Cooke explains the principles of management:

Under this head I shall point out, first, the means to be employed when symptoms appear threatening an apoplectic attack; secondly, the mode of proceeding in the paroxysm of the disease; and lastly, the remedies to be used on recovery from the fit, with a view of preventing its return. (Cooke, 1820, p. 285)

The aforementioned approach may be familiar to stroke physicians today but, the remedies Cooke discussed are far from recognisable. In this chapter, the largest of all of Cooke's chapters, there is a seemingly endless array of treatment options, which contrasts greatly with the impression of therapeutic nihilism expressed in the Hippocratic writings. The acute catastrophic onset calls for acute arresting therapies. The use of cathartics, revellents, cataplasms, clysters, sialogues, sinapisms, rubefacients, cupping and leeches are amongst the long list of available choices. There appears to be no shortage of options for active intervention.

There is one remedy, however, that is discussed at great length – that of blood letting. Bloodletting was a long held remedy based on the original humoral theories of disease. The public's expectations rather than established clinical guidelines often determined therapeutic interventions. Cooke informs his readers that, in spite of continuing positive support from well-respected physicians, there have been recent objections voiced to this form of therapy.

I am desirous of removing every objection which can be opposed to blood-letting, which I am convinced is not only the most effectual remedy in apoplexy, but is much more effectual than all others in use. (Cooke, 1820, p. 293)

It is clear that ancient remedies were still in use, though now being challenged.

There is little evidence of any recent advances or novel treatments in Cooke's chapters on apoplexy. However, as has been alluded to, both apoplexy and hemiplegia were considered as separate clinical entities or as varying degrees of severity of a spectrum of disease. A review of the management options for hemiplegia offers a different perspective on treatment choices. In this case, stimulating applications are recommended and a novel therapy includes the more recent experience of medical electricity.

Both Cooke and Kirkland credit Jean Louis Jallabert (1712–1768), professor of philosophy and mathematics at Geneva, as the first to attempt a cure of a palsied limb – a limb paralysed for a period of 14 years, by the use of electrical shocks in a series of treatments in 1747–1748 (Cooke, 1820, p. 154; Kirkland, 1792, p. 163). Although many centres and individual practitioners have reported success with the technique of medical electricity, there are, according to our nineteenth century physicians, reputable operators who have not. Both authors report a lack of success by Dr. Benjamin Franklin, one of the most prominent practitioners, when it comes to the treatment of palsy. In spite of large volume of reports in leading publications, a great number of favourable testimonials, notwithstanding Cooke is led to the conclusion that "it seems of late years to have fallen into disrepute" (Cooke, 1821, p. 155).

In spite of all these varied options for both apoplexy and palsy, Cooke believes that prevention is the most advantageous approach to management and concludes with the advice from a respected colleague:

We ought not, in these cases, to wait for the enemy, but to be beforehand with him by cupping, and a short course of purging every spring and autumn, whether threatenings of the disease occur or not. (Cooke, 1820, p. 358)

Conclusions

Superficially, it may appear that the disorder known as apoplexy had changed little over many centuries. The definition introduced in the Hippocratic Corpus was still in use when Cooke published his *Treatise* in 1820, and acute clinical presentation still dominated definitions. Treatment options were many, but in many instances derived from the humoral theories of antiquity. And Hippocratic nihilism with regard to prognosis continued to dominate medical thinking.

Nevertheless, there had been significant modifications; changes brought about by the great efforts of those who correlated the clinical features with which the patient presented and the findings at post-mortem study of their brains. Physicians still used the term apoplexy to imply a catastrophic event. But those physicians who linked pathological findings with clinical experience were able to advance the understanding of this disorder. The progress of apoplexy was not marked by great scientific innovations or landmark discoveries during the eighteenth century. Rather, changes in understanding arose from careful clinical observations and time-consuming pathological studies. Through these clinical efforts apoplexy had indeed become a "vascular disease".

References

Baillie, M. (1793). *The morbid anatomy of some of the most important parts of the human body*. London: J. Johnson and G. Nicol.

Castiglioni, A. (1941). *A history of medicine*. New York: Alfred A Knopf.

Clarke, E. (1963). Apoplexy in the hippocratic writings. *Bulletin of the History of Medicine, 35*, 301–314.

Cooke, J. (1820). *A treatise on nervous diseases: Volume 1: On apoplexy*. London: Longman, Hurst, Rees, Orme and Brown.

Cooke, J. (1821). *A treatise on nervous diseases: Volume 2: History and method of cure of the various species of palsy*. London: Longman, Hurst, Rees, Orme, and Brown.

Fernel, J. (1544). *Universa Medicina, Notis Observationibus et Remedis Secretis, cum Casibus et Observationis Rarioribus*. Geneva: De Tornes.

Karenberg, A. (1994). Reconstructing a doctrine: Galen on apoplexy. *Journal of the History of the Neurosciences, 3*, 85–101.

Kendell, R. E. (1993). William Cullen's Synopsis Nosologiae Methodicae. In A. Doig, J. P. S. Ferguson, I. A. Milne & R. Passmore (Eds.), *William Cullen and the eighteenth century medical world* (pp. 216–233). Edinburgh: Edinburgh University Press.

Kirkland, T. (1792). *A commentary on apoplectic and paralytic affections and on diseases connected with the subject*. London: William Dawson.

Mani, N. (1982). Biomedical thought in Glisson's Hepatology and in Wepfer's work on Apoplexy. In L. G. Stevenson (Ed.), *A celebration of medical history* (pp. 37–63). Baltimore/London: The Johns Hopkins University Press.

McHenry, L. C., Jr. (1969). *Garrison's history of neurology*. Springfield: Charles C Thomas.

Meyer, A., & Hierons, R. (1962). Observations on the history of the `Circle of Willis'. *Medical History, 6*, 119–130.

Morgagni, G. B. (1761). *De Sedibus, et Causis Morborum per Anatomen Indagatis Libri Quinque*. Vienna: ex typographica Remondiana.

Pearce, J. M. (1997). Johann Jakob Wepfer (1620–1695) and cerebral haemorrhage. *Journal of Neurology, Neurosurgery, and Psychiatry, 62*, 387.

Schutta, H. S., & Howe, H. M. (2006). Seventeenth century concepts of "Apoplexy" as reflected in Bonet's "Sepulchretum". *Journal of the History of the Neurosciences, 15*, 250–268.

Spillane, J. D. (1981). *The doctrine of the nerves. Chapters in the history of neurology*. Oxford: Oxford University Press.

Symonds, C. (1955). The Circle of Willis. *British Medical Journal, 1*, 119–124.

Wepfer, J. J. (1658). *Observationes anatomicae, ex cadaveribus eorum, quos sustulit apoplexia, cum exercitatione de ejus loco affecto*. English translation from Baglivi's Practice of Physick, London 1704; cited in Major RH. Classic descriptions of disease, 3rd ed. Illinois: Charles C Thomas, 1945:474–477: J. C. Suteri.

Willis, T. (1684). *Practice of Physick. Being the whole works of that renowned and famous physician*. London.

Willis, T. (1685). *The London practice of physick: Or the whole practical part of physick contained in the works of Dr Willis*. London: Thomas Bassett; William Crooke.

17
Benjamin Franklin and the Electrical Cure for Disorders of the Nervous System

Stanley Finger

Introduction

The eighteenth century was the century of the Enlightenment, a movement that permeated all aspects of life in Europe and her colonies, yet one with deep roots in the previous century (Outram, 1995; Porter, 1990, 2000). In natural philosophy, the drive to understand the mysteries of nature, including human nature, was strongly influenced by Francis Bacon's (1561–1626) call for better instruments, more detailed observations, careful experiments, and reasoning with clear eyes and unbiased minds (Bacon, 1620, 2000; Pérez-Ramos, 1988). In medicine, Thomas Sydenham (1624–1689) recognized the merit of Bacon's philosophy and, although he did not live into the eighteenth-century, his more data-driven approach to bedside medicine attracted many younger followers (Dewhurst, 1966; Sydenham, 1848, 1850).

Making a new discovery in and of itself was not the ultimate goal of most individuals who gravitated to experimental natural philosophy in the eighteenth century. Rather, what these natural philosophers and physicians hoped to discover more than anything else was something that would help suffering humanity. Foremost on the minds of many such individuals was discovering cures for dreaded, life-threatening diseases and a myriad of disabling disorders.

The call for useful knowledge was not just heeded by academicians and learned members of prestigious societies, such as the Royal Society of London that had been established in 1660. In many places, the quest was more of a democratic undertaking. It was open to rural healers without university degrees, well-positioned gentlemen, and even tradesmen with calloused hands and a desire to participate.

Often cited for his diplomacy, democratic ideals, exceptional wit, best-selling almanac, and moral philosophy, to name but a few of his interests, Benjamin Franklin (1706–1790) was one such tradesman (Fig. 1). Franklin came from a large, working-class family in Boston (Isaacson, 2003; Van Doren, 1938). Strapped for cash, his father Josiah took him out of school after just 2 years, hoping he could find a trade that would interest him. After trying soap and candle making in his father's shop, he began an apprenticeship in printing under his brother James. In 1723, annoyed by how he was being treated by James, who was now in trouble with the law, he made his way to Philadelphia, where he first worked for other printers. He then went into business for himself, and soon garnished a reputation as an exceptionally skilled printer, a talented writer, and a good businessman.

In part because he was always interested in the natural world, and also because it was good for his chosen trade as a writer and printer of useful information, Franklin followed new developments in natural philosophy and medicine. He included many pieces on these subjects in his highly successful *Poor Richard's Almanack* and in his newspaper, the *Pennsylvania Gazette* (Finger, 2006, pp. 19–36). From the start, he seemed to understand the power of experiments and the need for data, which he favored far more than grandiose theories that seemed almost impossible to test. And with his fertile and practical mind, he firmly believed that even seemingly mundane

FIGURE 1. Benjamin Franklin (1706–1790) (engraving by John Thomson (1805) after an original oil by Joseph-Siffred duplessis)

discoveries could have unexpected consequences that might ultimately benefit humankind.

Franklin was an exceptional man, but he was also a man of the times. Dabbling with electricity, whether to set a glass of brandy ablaze with a spark before unsuspecting onlookers or to determine how charged objects might attract and repel one another, was in vogue at mid-century (Bertucci & Pancaldi, 2001; Heilbron, 1979; Schaffer, 1983, 1993). Experimenting with glass tubes and disks that could produce electricity by friction was, in fact, one of the most exciting things that anyone with an interest in natural philosophy could wish to do (for a history of electrical machinery, see Hackmann, 1978). It was in this Zeitgeist, and after personally witnessing an electrical demonstration while on a trip to Boston and then receiving a gift of electrical equipment from Peter Collinson (1694–1768) in England, that Franklin began his own forays into electricity, in the 1740s (Cohen, 1941). These experiments on the nature of electricity would in turn lead him into therapeutics and what we could now call the neurosciences.

Mastering Electricity

Franklin knew that the demands on his time were such that he could not pursue his electrical experiments by himself, and that he would benefit by

collaborating with others who shared his interest. The Philadelphians who joined him on the experiments were also amateurs in experimental natural philosophy: Philip Syng Jr. (1703–1789) was a silversmith, Thomas Hopkinson (1709–1751) was a merchant-lawyer, and Ebenezer Kinnersley (1711–1778) was an unemployed Baptist minister. Further, he handed over the day-to-day operations of his printing business to his Scottish assistant, David Hall (1714–1772), in order to devote more time to his "philosophical amusements," which he aspired to approach in the manner befitting a proper gentleman.

Franklin and his associates began by replicating some published findings. They then made a significant discovery: that points are better than "blunts" for throwing off sparks and attracting them. The first member of the group to realize this was Hopkinson, and it led Franklin to the pointed lighting rod:

From what I have observed on experiments, I am of the opinion that houses, ships, and even towns and churches may be effectually secured from the stroke of lightning by their means: for if . . . there should be a rod of iron 8 or 10 feet in length, sharpen'd gradually to a point like a needle . . . the electrical fire would, I think, be drawn out of a cloud silently, before it could come near enough to strike. (Labaree, 1960, pp. 472–473)

The importance of the lightning rod stemmed from more than the fact that it could render a powerful force harmless, so as to save property and lives. It also made lightning a more comprehensible natural event at a time when many people were still associating the destructive powers of lightning storms with the devil and the supernatural. In addition, it showed that basic science could have immense practical ramifications. Franklin realized these things when he gifted the lightning rod to humankind, something he would do with all of his successful inventions (see Mitchell, 1998, for more on the reception of the lightning rod).

The Philadelphia experiments led to many other discoveries and insights. One was that there probably were not two different electrical fluids (vitreous and resinous), as was believed by Charles du Cisternay DuFay (1698–1739) and Abbé Jean-Antoine Nollet (1700–1770) in France. Instead, Franklin suggested that objects will be "plus" or "electricized positively" if they have an "over quantity" of electrical matter, whereas they will be "minus" or "electricized negatively" if they have an "under quantity." He further hypothesized that Nature always strives to create an equilibrium condition, which is why lightning occurs and why a charged body might attract certain objects (of opposite sign) while repelling others (of the same sign).

Franklin also designed an experiment at about mid-century to determine whether lightning and man-made electricity might be qualitatively identical. He was not, however, the first person to perform the "critical" experiment with an iron rod that would rise high into the air above a sentry box. Thomas François d'Alibard (1703–1779), a Frenchman who read what Franklin (1751) had written about the needed experiment, captured lightning in this way and graciously gave Franklin his credit. Not knowing what had transpired in France, Franklin captured lightning with a modified kite shortly after d'Alibard conducted his experiment, and further showed that it had the same properties as the electrical fluid produced with glass instruments and stored in then newly invented Leyden jars (for the history of Van Musschenbroek's Leyden jar, see Dorsman & Grommelin, 1957).

Franklin's experiments and letters about electricity, which appeared in his 1751 pamphlet, *Experiments and Observations on Electricity*, made him famous among philosophers interested in the natural world. Immanuel Kant (1724–1804) called him "the new Prometheus" and Joseph Priestley (1733–1804) referred to him as "the father of modern electricity." His achievements also led to honorary degrees. In 1753, he received his first two: an honorary masters from Harvard and another from Yale (his first honorary doctorate would come from St. Andrews in 1759). He was also awarded a Copley Medal from the Royal Society in 1753, and 3 years later was given full membership in the Royal Society.

Overlooked by Franklin biographers and even those scholars who described Franklin's work on the nature of electricity in some detail, is just how much he hoped to find a beneficial use for the force he was soon understanding better than anyone else at the time. Franklin was obviously pleased that he was able to protect people from lightning, but this was not a practical use for the electrical fluid. Stimulated by reports from Europe that were appearing in *Gentleman's Magazine* (Locke & Finger, this volume), Franklin found himself thinking about the potential of medical electricity, especially for curing disorders affecting the nervous system.

The Palsies

Franklin was particularly interested in the palsies, and particularly the "common paralytic disorder," an older term for the loss of movement caused by stroke. In 1744, when Johann Krüger (1715–1759) of Halle first suggested in writing that electricity might have some usefulness in medicine, the German professor specifically predicted, "The best effect would be found in paralyzed limbs" (Krüger, 1744; Licht, 1967, p. 5). After all, even small electrical shocks had a way of making muscles come to life.

Krüger's student Christian Gottlieb Kratzenstein (1723–1795) was the first person to test this hypothesis, and he reported that electricity did help some people with chronic movement problems (Kratzenstein, 1745; Snorrason, 1974). Although his first cases seemed to be individuals suffering from arthritic conditions, and not stroke victims, his findings stimulated people to think that electricity might be able to cure other and perhaps all movement disorders.

In the mid-1740s, Nollet reported to the Académie des Sciences in Paris that various experiments had shown that electricity could decrease the viscosity and increase the motions of sluggish fluids in glass tubes (Nollet, 1749a,b, 1752; the *Mémoires de l'Académie* were always published a few years after the papers were given). In the context of the iatromechanical physiological and medical theories of the day, this finding and additional related observations led him to think that electricity might help brain-damaged paralytics. Hence, he conducted some clinical trials with surgeons Sauveur François Morand (1697–1773) and Joseph Marie François de la Sône (1717–1788), to see if voluntary movements could be restored to a small sample of men with paralyzed limbs. These trials utilized the latest technologies (e.g., Leyden jars) and produced some tingling sensations and involuntary movements, but they did not produce miraculous recovery of voluntary motor functions – at best, they were inconclusive (Nollet, 1751).

Among the first claims of success with brain-damaged patients was a case report from Switzerland in 1748. Jean Louis Jallabert (1712–1768), professor of experimental philosophy and mathematics at Geneva was enthusiastic (Jallabert, 1748). His patient was a master locksmith, who fell while forging an iron bar and landed on the back of his head. The accident knocked him unconsciousness and caused a loss of speech and a paralysis of his right side. Before being treated electrically some 14 years after his accident, his speech returned, he was able to walk with a limp, and he had regained at least some movement in his right arm. After extensive daily treatments with electricity, in which the muscles were "convulsed" and "agitated," coupled with heat and massage, Jallabert claimed that he recovered muscle mass, and with it all voluntary movement.

Franklin's protégée, Joseph Priestley, would write in the section "On Medical Electricity" of his *History and Present State of Electricity* that "Mr. Jallabert's own account of this cure is very circumstantial" and that, according to the Abbé Nollet's later account of the case, "this person relapsed to the condition in which Mr. Jallabert found him" (Priestley, 1775, 1966, p. 473). Nevertheless, when Jallabert's findings first became public, optimism swelled and Nollet took it upon himself to conduct some further clinical trials with Morand on patients at the Hôtel des Invalides in Paris.

These men selected four patients from a sample of 20. One was an older man, who they had treated 2 years earlier, but he died from a fever during the trials. Two of the three remaining patients treated in 1748 had partial paralyses caused by head injuries, and the third seemed to have a slowly growing paralytic condition. Shocks were administered to their paralyzed limbs and their bodies were electrically charged, after which sparks were drawn from the limbs. Nevertheless, voluntary movements were not restored before their subjects grew tired of the occasionally painful daily experiments that were not providing cures (Morand & Nollet, 1753).

With negative findings to his credit, and increasingly wary of what others were claiming with questionable methodologies and perhaps vivid imaginations, Nollet became even more skeptical about claims of curing hemiplegia with electricity (Bertucci, 2005). And he was not alone (e.g., Louis, 1747). Still, others remained enthusiastic.

The motto selected by the Royal Society of London was *Nullius in Verba*, meaning "On the Words of No One." These words signified the philosophical organization's commitment to experiments and careful observations, rather than to unsubstantiated and untestable opinions, old or new. Well before he received the news that he had been elected into this august body, Franklin had embraced this philosophy. And with vastly different opinions being aired about medical electricity, he felt it necessary to find out for himself whether it worked with paralytics.

Franklin's patients included some of the most important people in the British North American Colonies, in addition to numerous other people whose names may never be known (Finger, 2006, pp. 80–101). The gaps stem from the fact that he left no known casebook and did not communicate in writing with most of his patients. Hence, names, details of the disorders, and precisely what he did are limited to people who exchanged letters with him or were important enough to have others write about them.

One such person was James Logan (1674–1751), a scholarly Quaker with a library of some 3,000 books (Myers, 1904; Tolles, 1957). Logan had come to America in 1699 as William Penn's secretary and stayed in Pennsylvania after Penn returned to England. He made his fortune in the fur trade and also held public offices, including acting Governor of Pennsylvania. Logan had known Franklin for some time and had been following his electrical experiments since 1747. He was also knowledgeable about early electrical machines and had been following claims being made for medical electricity.

Franklin tried medical electricity on Logan in 1750, after Logan suffered a stroke that severely affected his right side and impaired his speech. Notably, this was not the first time he had suffered a stroke, but his previous stroke(s) preceded the advent of medical electricity. To his dismay, Franklin found that applying electrical shocks to Logan's "disordered side" did not help. Logan died a paralytic in 1751, at age 77.

A second famous person to ask Franklin for help was Jonathan Belcher (1682–1757), a deeply religious man who was born in Massachusetts, had received two degrees from Harvard College, and was a leading figure in colonial politics (Batinski, 1966; Burr, 1757). Belcher had earlier served as Governor of the Massachusetts Bay Colony and neighboring New Hampshire, and he had become Royal Governor of New Jersey in 1746. Franklin also knew him personally, with his presses printing high-quality paper money and other documents for New Jersey.

Belcher was "stricken with the palsy" in 1750, at age 68, while attending a graduation at the school he helped found, the College of New Jersey (later Princeton University). Although pressing political matters preempted Franklin from treating the governor as requested, he sent him electrical equipment and instructions for its use. Through an exchange of letters, Franklin was informed that some of the equipment arrived broken, but that Dr. Aaron Burr (1716–1757), Belcher's close friend, president of the college, and father of the future vice president, was able to find suitable replacements and was able to administer the shocks. Nevertheless, as was true for Logan, the treatment did not help. Quoting Belcher: "I have . . . been electrifyd several times but at present without any alteration in my Nervous disorder" (Labaree, 1961, pp. 255–256). These words were penned in 1752. Belcher grew progressively weaker and died 5 years later.

Franklin sailed to England the year Belcher died, and once settled in London he signed the register to become an official member of the Royal Society, which had elected him into the organization in 1756. At the time, the merits of medical electricity were being assessed by the membership, and Franklin was asked to share his thoughts on treating the palsies. His communication was dated December 21, 1757, was read on January 12, 1758, and appeared in the *Philosophical Transactions* later that year (Franklin, 1758). Franklin deemed his carefully worded document so important that he included it in future editions of his *Experiments and Observations on Electricity* (e.g., Franklin, 1774).

Franklin explained that after "the News papers made Mention of great Cures perform'd in Italy or Germany by means of Electricity, a Number of Paralytics were brought to me from different Parts of Pensilvania [sic.] and the neighboring Provinces, to be electricis'd, which I did for them, at their Request." He mentioned how he charged their bodies and drew sparks from the "affected Limb or Side," and also how he shocked the palsied part with "two 6 Gallon Glass [Leyden] Jarrs . . . repeating the Stroke commonly three Times each day."

Franklin did not indicate the number of palsied patients he dealt with in his report. But he did write that, while some improved a little while being treated, these effects were only temporary, "so that I never knew any Advantage from Electricity in Palsies that was permanent." He then added: "And how far the apparent temporary Advantage might arise from the Exercise in the Patients Journey and coming daily to my house, or from the Spirits given by the Hope of Success, enabling them to exert more Strength in moving their Limbs, I will not pretend to say."

Franklin never altered his negative assessment of medical electricity as a cure for the palsies of long duration, and specifically for the "common paralytic disorder." He did not even recommend it for his wife when she suffered a paralytic stroke in 1768. Even more telling is a communication to Mather Byles, who had been instrumental in persuading Harvard College to award Franklin his first honorary degree in 1753. In 1787, Byles asked Franklin whether he thought the electrical cure would be worth trying for his palsy. The note he received back from the aged and sickly Founding Father read, "I wish for your sake that Electricity had really prov'd what at first it was suppos'd to be, a Cure for the Palsy" (Smyth, 1906, p. 656).

Smallpox

Franklin saw why people feared contracting smallpox after a deadly epidemic broke out in Boston in 1721 (for details on smallpox in the colonies, see Fenn, 2001; for Franklin and smallpox, see Finger, 2006, pp. 49–65). At the time, Cotton Mather (1663–1728) and Zabdiel Boylston (1676–1766) were promoting variation (inoculation with human scabs) as a new means of protection from this dreaded disease, while James Franklin and the staff at his newspaper were lambasting the new procedure as dangerous and unproven.

As a teenage apprentice, and perhaps not knowing who was right, Franklin remained quiet. But during the 1730s, after settling in Philadelphia, he launched a public campaign based on statistics to promote variation. In 1759, he even teamed up with physician William Heberden to produce a free pamphlet for the poor on the benefits of inoculation and how to do it at minimal cost (Heberden, 1759).

As a result of the Boston epidemic and others that followed, Franklin knew that about 15% of the survivors of the "speckled monster" were rendered blind or experienced severe hearing losses. And one of his letters reveals that he tried to cure the deafness caused by this disease with medical electricity. The letter was in response to a request he had received from Humphrey Stenhouse in 1765 (Labaree, 1968, pp. 27–28).

Stenhouse was a Cambridge graduate and a fellow of Pembroke College with many interests, including archeology, mining, and politics. Then in his thirties, he had been deaf half his life because of smallpox. By his own admission, his disability had left him "in a great Measure deprived of the Pleasures of Conversation." It also prevented him from entering the ministry or pursuing a career in politics. Hence, "although an utter Stranger" to Franklin, whom he knew only as an authority on electricity, he asked him "whether you think any Benefit might arise to me from the Application of the Electrical Machine."

"I wish I could give you any Encouragement to hope Relief in your Case by means of Electricity," Franklin replied. He then explained: "No Instance of the kind has fallen within my Knowledge. On the contrary, I have try'd it on some Patients, but without the least Success" (Labaree, 1968, p. 29).

An Organic Disorder with Convulsions

Two years after responding to Stenhouse's letter about deafness, and while still in England, Franklin was asked to treat a girl "with spasms and convulsions." Precisely what caused these ominous clinical signs is unknown. Three of the more likely possibilities are a brain tumor, an abscess, and epilepsy without mass. What can be stated with more certainty is that her disorder was organic, progressing aggressively, and now life threatening.

Lady Mary Catherine (1754–1767) was the 12-year-old daughter of the Duke of Ancaster. Dr. John Pringle (1707–1782), an expert on military medicine and hospital care, one of Franklin's closest friends, and a future president of the Royal Society, personally requested Franklin's help with this case, if for no other reason than to show the family that he was doing everything in his power to

save the girl (Pringle, 1750, 1752; Selwyn, 1982). Pringle's urgent note to Franklin was written in March 1767, and it read:

I take the liberty to beg that You would come as soon as You can to the Duke of Ancaster's in Berkeley Square, as His Grace and the Duchess are in the greatest distress about their daughter, who has been long in a most Miserable condition with spasms and convulsions. After all that we have done the distemper remains obstinate, and therefore the Parents have thought of electrifying Her. I have recommended the Operation to be performed by Spence and the rather as the present spasm has shut the Young Lady's jaw and deprived Her both of speech and swallowing. I ventured to name You as the person the most proper for directing the operation, trusting to your friendship to me and humanity towards the distressed. (Labaree, 1970, p. 95)

Franklin's surviving letters provide no more information about what he was thinking on the trip, what he observed when he arrived, or what then transpired. Nevertheless, he might have been fairly skeptical about the efficacy of the electrical cure for this patient. This is suggested by a very brief note published a year earlier in *Gentleman's Magazine*. It stated that Franklin had observed the electrical treatment of a woman who "had for above 6 weeks lost her speech by convulsive fits." The anonymous author did not state whether the woman might have had a case of hysteria, but he did write that "Mr. *Franklin* of *Philadelphia* . . . expressed his astonishment" that the cure seemed to work (*Gentleman's Magazine*, 1766, p. 147).

If Franklin had been skeptical about helping Lady Mary Catherine, and did so only because Pringle asked for his assistance in dealing with the young daughter of a duke and duchess, his doubts and concerns would have been justified. After being seen and presumably treated, the dying girl was moved from Chelsea to the spa at Clifton, a location near Bristol. She died shortly afterward, on Palm Sunday, April 12, 1767, just days before her 13th birthday.

Hysteria

Despite his inability to cure paralyses, deafness, or a possible brain tumor, Franklin never claimed that medical electricity was just another in a long line of quack remedies. One reason for this is that he only experimented on a few conditions that might have been helped by the new cure. But even more importantly, he had a notable, well-publicized success early on. It involved a young woman who was suffering from hysteria.

At the time, it was believed that just as a sick body could affect a person's mind, a sick mind could affect the physical machinery of the body, including the status of the nerves. Hence, most (but not all) physicians tended to look upon hysteria as a physical disorder triggered by mental problems, as opposed to a purely mental disorder, which is how "conversion reactions" are viewed at present.

Franklin joined Cadwallader Evans (1716–1773), a younger man who had learned his medicine through the apprenticeship system, to help "C.B.," who might have been Evans' sister (Klein, 1967; Thacher, 1828). The treatment took place in Philadelphia in 1752, a full report with Evans as sole author appeared in *Medical Observations and Inquiries* in 1757, and a summary of the case was published in *Gentleman's Magazine* that same year (Evans, 1757a,b).

Evans gave this background:

C.B, in the summer of 1742, and about the fourteenth year of her age, was seiz'd with convulsion fits, which succeeded each other so fast, she had near 40 in 24 hours after the first attack. . . . [Thereafter] her disorder continued in one shape or other, or, return'd after an intermission of a month or two, at farthest. Sometimes she was tortur'd almost to madness with a cramp in different parts of the body; then with more general convulsions of the extremities, and a choaking deliquium; and, at times with almost the whole train [of] hysteric symptoms . . . [which] continued and harrass'd her alternatively for 10 years, tho' she had the best advice the place afforded, and took a great number of medicines. (Evans, 1757a, pp. 83–84)

A letter written by C.B. now provided more information:

About this time there was a great talk of the wonderful power of electricity . . . Accordingly I went to Philadelphia, the beginning of September 1752, and apply'd to B. Franklin, who I thought understood it best of any person here. I receiv'd four strokes morning and evening . . . and indeed they were very severe. . . .
   The symptoms gradually decreased, till at length they entirely left me. I staid in town but two weeks, and when I went home, B. Franklin was so good as to supply me with a globe and bottle, to electrify myself everyday for

three months. The fits were soon carried off, but the cramp continued somewhat longer, tho' it was scarcely troublesome, and very seldom return'd. I now enjoy such a state of health, as I wou'd have given all the world for. . . . (Evans, 1757a, pp. 84–86)

At the conclusion of the article, the reader is informed that C.B. was still in good health 5 years later, in 1757. Subsequent records show that she exhibited "uncommon powers of reasoning" and no further signs of hysteria before dying at age 79 (Thatcher, 1828).

Interestingly, neither Evans nor Franklin mentioned why electricity was used to treat C.B. or why they thought it worked. In this context, three possibilities have to be considered. First, following popular medical beliefs, they might have thought that hysteria could be treated effectively with a stimulant, in this case the most powerful stimulant yet discovered and one capable of affecting sluggish nerve juices. Second, they might only have wanted C.B. to believe that a great new cure administered by none other than Franklin, the authority on electricity, would work – hence, playing on her gullibility and the power of suggestion, and really administering just a placebo. As Franklin's creation Poor Richard put it in 1734, "As Charms are nonsense, Nonsense is a Charm" (Labaree, 1959, p. 14). A third possibility, of course, is that electricity might have been tried without any hypothesis behind its use. Indeed, Franklin, like most self-educated individuals in America, tended to approach most medical problems empirically, relying more on trial-and-error than theory, so long as the treatment to be tested did not harm the patient.

Regardless of precisely what might have been going through Franklin's mind while treating C.B., three things are clear. First, he knew that he could apply electricity safely. Second, experimental results, meaning facts, meant far more to him than opinions. And third, pretty much the same protocol would have been followed no matter what his motivation had been.

Melancholic Madness

Although melancholia was considered a close cousin of hysteria in the eighteenth century, Franklin's hope that melancholics could be cured with electrical shocks to the cranium did not stem from his success with C.B., from medical theories, or from trials that might have been done by others. Instead, it can be traced to some electrical accidents that involved Franklin and then his close friend in the Austrian court, Dutch-born physician Jan Ingenhousz (Beaudreau & Finger, 2006; for more on Ingenhousz see Conley & Brewer-Anderson, 1997; Reed, 1947–1948).

Franklin described his own first serious electrical accident in a letter to his brother John in 1750 and one to Peter Collinson in 1751 (Labaree 1961, pp. 82–83, 112–113). He wrote that a jolt equal to that of 40 common Leyden jars produced a "universal Blow that affected him from head to foot," producing "a violent quick Trembling in the Trunk," and some temporary confusion (Labaree 1961, p. 113). This accident, in which the electricity entered his body via his arm, showed that a person could survive a fairly intense shock to the body, in which the electricity spreads to the brain, without serious consequences (he knew, of course, that a much greater jolt could kill a person).

Franklin's second accident occurred a few years later and was briefly mentioned in a letter dated 1755 to physician and experimental natural philosopher John Lining, who lived in South Carolina (for more on Lining, see Mendelsohn 1960). Franklin provided the missing details of this accident in 1785, after having been asked by Ingenhousz for more information about it (Smyth 1906, pp. 308–309). He explained that he was shocked through the head and knocked to the floor, while attempting to treat a paralytic patient.

As a result of these two personal experiences, an experiment he conducted on six men in which electricity entered their heads, plus an accident in which a female patient was inadvertently shocked through the head, Franklin learned that a skilled "operator" could apply a fairly large amount of electricity directly to the head without putting a person in danger. He also learned that cranial shocks strong enough to buckle a person would probably produce a distinct amnesia for the event (Finger & Zaromb, 2006).

Franklin described the amnesia after his second accident as follows:

I had a Paralytick Patient in my Chamber, whose Friends brought him to receive some Electric Shocks. . . . I was oblig'd to quit my usual Standing, and plac'd myself inadvertently under an Iron Hook which hung from the Ceiling down to within two inches of my Head, and communicated by a Wire with the outside of the [Leyden] Jars. I attempted to discharge them, and in fact did so; but I did not perceive it, tho' the charge went thro' me, and not through the Persons I entended it for. I neither saw the Flash, heard the Report, nor felt the Stroke. When my senses returned, I found myself on the Floor. I got up, not knowing how that had happened. I then again attempted to discharge the Jars; but one of the Company told me they were already discharg'd, which I could not at first believe, but on Trial found it true. . . . On recollecting myself, and examining the Situation, I found the Case clear . . . but I do not remember any other Effect good or bad. (Smyth, 1906, pp. 308–309)

From a clinical perspective, the amnesia was not important to Franklin. This was because the memory lacuna was more or less confined to when the shock occurred, and because it did not prevent a man from immediately returning to his job or a woman from still running a household. Yet the shock-induced amnesia was interesting enough for him to make his new discovery public in later editions of his pamphlet on electricity (e.g., Franklin, 1774).

Franklin's information about surviving shocks to the head without serious consequences became more important to Ingenhousz after he experienced his own serious accident, which sent a large jolt of electricity through his brain. The event took place in 1783 and it initially left him confused and extremely fearful that his previously exceptional mind would never recover. To his surprise, however, he felt elated the very next morning. As he explained to Franklin:

My mental faculties were at that time not only returned, but I felt the most lively joye in finding, as I thought at the time, my judgment infinitely more acute. It did seem to me I saw much clearer the difficulties of every thing, and what did formerly seem to me difficult to comprehend, was now become of an easy Solution. I found moreover a liveliness in my whole frame, which I never had observed before. (Packard Humanities Institute 2003, Vol. 40, u. 209, p. 2)

The idea that some good might come out of this frightful experience was not lost on Ingenhousz. Hence, he further informed Franklin that the sense of wellbeing he felt the next morning

has induced me to advise som[e] of the London mad-Doctors, as Dr. Brook, to try a similar experiment o[n] mad men, thinking that, as I found in my self, my mental faculties impro[ved] and as the world well knows, that your mental faculties, if not improved [by] the two strooks you received, were certainly not hurt, by them, it might perhaps be[?] a remedie to restore the mental faculties when lost. (Packard Humanities Institute, 2003, Vol. 40, u. 209, p. 2)

Equally anxious to explore a possible application for a new finding, even one stemming from embarrassing accidents, Franklin informed Ingenhousz that he "communicated that Part of your Letter to an Operator, encourag'd by Government here [in France] to electrify epileptic and other poor Patients, and advis'd his trying the Practice on Mad People according to your opinion" (Smyth 1906, pp. 308–309).

Prior to these communications, mild electrical shocks to the head had been tried with some patient groups. John Wesley (1703–1791), for example, wrote in his *Desideratum* that cranial shocks had been used to treat some cases of headache, "fits" (seizures), and hysteria (Wesley, 1760). Other than indicating that the shocks were light, he provided minimal details about these cases and did not cite melancholia as disorder to be treated by applying electricity directly to the head. In his popular *Primitive Physic,* he praised electricity (not specifying to the head) along with valerian root for "Nervous Disorders," a category he never defined, and even in editions updated in 1780, under the heading "Lunacy," he only commented, "electricity: tried," without adding another word about whether sparks were drawn or applied, and if applied, where (see Wesley, 1791). Also notable is Giuseppe Veratti (1703–1793), who in 1748 reported how he successfully treated a man with a nervous affliction that might today be classified as an anxiety disorder. But in his case, sparks were only "drawn out from the head," as well as various other body parts, and not applied from Leyden jars (Veratti, 1748).

From these writings, it seems reasonable to conclude that cranial electrical shock treatments were not widely used for melancholia, if used at all in places like London, Paris, and Bologna, when Ingenhousz and Franklin issued their calls for clinical trials. This possibility also stems from examining *Gentleman's Magazine* and some other widely read publications from the third quartile of the 1700s (see Locke & Finger, this volume).

Moreover, it is consistent with the fact that Ingenhousz and Franklin presented the idea of treating melancholia with cranial shocks as something new, or at least now being worthy of a proper test. Because the idea was generated and presented by two of the greatest minds of the Enlightenment, this proposal was probably further circulated by word of mouth.

Although Ingenhousz informed Franklin in his 1783 letter that he "could never persuade any one" to attempt the new treatment, and Franklin replied back in his lengthy note that he had not yet heard from the French "Operator," fairly detailed case studies of cranial electricity to treat melancholia now began to appear in books, some of which were written by authoritative figures. Most notably, good results were reported by John Birch (1745–1815), a medical operator at a major hospital in London, and by Giovanni Aldini (1762–1834), the nephew of Luigi Galvani (1737–1790) and a prominent physician-scientist in Bologna (Aldini, 1803; Birch, 1792). A considerably lesser-known practitioner, but one who also reported favorable results, was T. Gale (c. 1800) from New York State (Gale, 1802).

The cranial shocks used by these three pioneers did not appear to be of sufficient intensity to cause a notable loss of consciousness, full-body convulsions, or falling down. More accurately, the practitioners, who did not want to harm their patients, did not report such effects. Nevertheless, the shocks were made fairly strong for a few of the patients, and some trembling was noted (see Beaudreau and Finger, 2006).

Interestingly, when Ugo Cerletti (1877–1963) and his assistant Lucio Bini (1908–1964) brought cranial shock treatments fashionably to the fore in 1938, their treatments called for full-body convulsions, and it was soon believed that the accompanying amnesia was important for the cure (Alexander, 1953; Cerletti 1954; Cerletti & Bini, 1938; Kalinowsky & Hoch, 1952; Steinfeld, 1951). These more modern practitioners showed no awareness of what Ingenhousz and Franklin had experienced and proposed, nor did they cite the positive reports with cranial electricity for melancholics (and even possibly schizophrenics) that were beginning to appear in the literature as the long eighteenth century came to a close.

Commentary

The advent of medical electricity in the eighteenth century attracted two types of natural philosophers. One group was made up of well-educated "men of the faculty," such Christian Gottlieb Kratzenstein and the highly regarded Abbé Nollet, who conducted some of the first clinical trials on patients with movement problems, albeit from different causes. The second group comprised individuals who were not formally trained in experimental natural philosophy or in medicine. John Wesley (1703–1791), the Methodist cleric who hoped to treat bodies economically while also saving souls, was one of these people (Wesley, 1760). Benjamin Franklin, who was admired and cited by Wesley and almost everyone else for his work on electricity, was another.

Unlike most university-trained physicians, but in accord with many individuals who gravitated to medical electricity to help others, to achieve fame, or to make a living, Franklin's approach was highly empirical. With his background as a tradesman, especially in the New World with its emphasis on survival and immediate needs, he was much more interested in whether something worked than why it worked. Hence, his letters on medical electricity do not mention physiological theories, such as those dealing with imagined or "secret" obstructions, although he knew far more than most empiricists about the various, largely iatromechanical ideas being promoted by the leading physicians and academicians of the eighteenth century (see King, 1971, for more on iatromechanical medicine).

To his credit, Franklin was very successful in showing some people, including John Wesley, the dangers in generalizing too freely about medical electricity. As he saw it, electricity was neither a panacea that could cure all disorders of the nervous system, nor a quack remedy without any utility at all. For Franklin, the truth could be found somewhere between these two extremes. The new cure was clearly ineffective for palsies of long duration or deafness caused by smallpox. Although none of his surviving letters discuss the case of the Lady Mary Catherine, he probably felt the same way about severe seizures and contractures indicative of serious brain diseases, such as tumors and abscesses. Nevertheless, Franklin observed that medical electricity could be used to cure hysteria. Further, he agreed with Jan Ingenhousz that electrical shocks applied directly to the head might help lethargic patients classified as melancholic, another disorder thought by many academic physicians to involve sluggish or flaccid nerves.

Benjamin Franklin was a highly pragmatic man who, by his own admission, did things with specific goals in mind. With his explorations into medical electricity he bettered himself, served people in need, contributed to the community, and tangibly honored his God. These were highly motivating forces to Franklin, one of the most enlightened figures of the eighteenth century – and arguably the first major contributor from the New World to what today would be called neurology or the neurosciences.

Acknowledgments. The author would like to thank Sherry Beaudreau for the research she did on the clinical experiments that were performed by Birch, Gale, and Aldini, and Franklin Zaromb for his assistance in investigating the history of electrical shock-induced memory problems.

References

Aldini, J. (1803). *An account of the late improvements in Galvanism.* London: Wilks and Taylor.

Alexander, L. (1953). *Treatment of mental disorders.* Philadelphia: W. B. Saunders.

Bacon, F. (2000). *The new organon.* Cambridge: Cambridge University Press. (Originally published 1620).

Batinski, M. C. (1966). *Jonathan Belcher.* Lexington: University of Kentucky.

Beaudreau, S., & Finger, S. (2006). Medical electricity and madness in the eighteenth century: The legacies of Benjamin Franklin and Jan Ingenhousz. *Perspectives in Biology and Medicine, 49*, 330–345.

Bertucci, P., & Pancaldi, G. (Eds.). (2001). *Electric bodies.* Bologna: Università di Bologna.

Bertucci, P. (2005). Sparking controversy: Jean Antoine Nollet and medical electricity south of the Alps. *Nuncius, 20*, 153–187.

Birch, J. (1792). A letter to Mr. George Adams, on the subject of medical electricity. In G. Adams (Ed.), *An essay on electricity* (4th ed., pp. 519–573). London: Hindmarsh.

Burr, A. (1757). *A servant of god: Jonathan Belcher.* New York: Hugh Gaine.

Cerletti, U. (1954). Electroshock therapy. *Journal of Clinical and Experimental Psychopathology, 15*, 191–217.

Cerletti, U., & Bini, L. (1938). Un nuevo metodo di shockterapie "L'electro-shock." *Bollettino Accademia Medicina Roma, 64*, 136–138.

Cohen, I. B. (1941). *Benjamin Franklin's experiments*. Cambridge: Harvard University Press.

Conley, T. K., & Brewer-Anderson, M. (1997). Franklin and Ingenhousz: A correspondence of interests. *Proceedings of the American Philosophical Society, 141*, 276–296.

Dewhurst, K. (1966). *Dr. Thomas Sydenham*. Berkeley: University of California Press.

Dorsman, C., & Grommelin, C. A. (1957). The invention of the Leyden Jar. *Janus; revue internationale de l'histoire des sciences, de la médecine, de la pharmacie, et de la technique, 46*, 275–280.

Evans, C. (1757a). A relation of a cure performed by electricity. *Medical Observations and Inquiries, 1*, 83–86.

Evans, C. (1757b). Account of a cure by electricity. *Gentleman's Magazine, 27*, 260.

Fenn, E. A. (2001). *Pox Americana*. New York: Hill and Wang.

Finger, S. (2006). *Doctor Franklin's medicine*. Philadelphia: University of Pennsylvania Press.

Finger, S. (2006). Benjamin Franklin, electricity, and the palsies: On the 300th anniversary of his birth. *Neurology, 66*, 1559–1563.

Finger, S., & Zaromb, F. (2006). Benjamin Franklin and shock-induced amnesia. *American Psychologist, 61*, 240–248.

Franklin, B. (1751). *Experiments and observations on electricity*. London: Cave.

Franklin, B. (1758). An account of the effects of electricity in paralytic cases. *Philosophical Transactions of the Royal Society, 50* (Part 2), 481–483.

Franklin, B. (1774). *Experiments and observations on electricity* (5th ed.). London: Newberry.

Gale, T. (1802). *Electricity, or ethereal fire, considered*. Troy: Moffitt & Lyon.

Hackmann, W. D. (1978). *Electricity from glass*. Alphen aan den Rijn: Sijthoff & Noordhoff.

Heberden, W. (1759). *Some account of the success of inoculation for the small-pox in England and America*. . . . London: Strahan. (Preface by Franklin).

Heilbron, J. L. (1979). *Electricity in the seventeenth and eighteenth centuries*. Berkeley: University of California Press.

Historical chronicle. (March 1, 1766). *Gentleman's Magazine, 36*, 147.

Isaacson, W. (2003). *Benjamin Franklin*. New York: Simon & Schuster.

Jallabert, J. L. (1748). *Experiences sur l'Electricité*. . . . Geneve: Barrillot et Fils.

Kalinowsky, L. B., & Hoch, P. H. (1952). *Shock treatments, psychosurgery, and other somatic treatments in psychiatry*. New York: Grune & Stratton.

King, L. S. (1971). *The medical world of the eighteenth century*. Huntington: Krieger.

Klein, R. S. (1967). Dr. Cadwalader Evans (1716–1773). *Transactions and Studies of the College of Physicians of Philadelphia, 35*, 30–36.

Kratzenstein, C. G. (1745). *Schreiben von dem Nutzen der Electricität in der Arzneywissenschaft*. Halle: Hemmerde.

Krüger, J. G. (1744). *Zuschrift an seine Zuhörer, worinnen er Gedancken von der Electricität mittheilt und Ihnen zugleich seine künftigen Lectionen bekannt macht*. Halle: Hemmerde.

Labaree, L. W. (1959). *The papers of Benjamin Franklin* (Vol. 1). New Haven: Yale University Press.

Labaree, L. W. (1960). *The papers of Benjamin Franklin* (Vol. 3). New Haven: Yale University Press.

Labaree, L. W. (1961) *The papers of Benjamin Franklin* (Vol. 4). New Haven: Yale University Press.

Labaree, L. W. (1968). *The papers of Benjamin Franklin* (Vol. 12). New Haven: Yale University Press.

Labaree, L. W. (1970). *The papers of Benjamin Franklin* (Vol. 14). New Haven: Yale University Press.

Licht, S. (1967). *Therapeutic electricity and ultraviolet radiation* (2nd ed.). Baltimore: Waverly Press.

Locke, H. S., & Finger, S. (this volume). Gentleman's Magazine, the Advent of Medical Electricity, and Disorders of the Nervous System.

Louis, A. (1747). *Observations sur l'Électricité*. . . . Paris: Delaguette.

Mitchell, T. A. (1998). The politics of experiment in the eighteenth century: The pursuit of audience and the manipulation of consensus in the debate over lightning rods. *Eighteenth-Century Studies, 31*, 307–331.

Mendelsohn, E. (1960). John Lining and his contribution to early American science. *ISIS, 51*, 278–292.

Morand, S.-F., & Nollet, J.-A. (1753). Expériences de l'électricité appliquée à des paralytiques. *Mémoires de l'Académie Royale des Sciences de Paris (pour l' année 1749)*, 29–38.

Myers, A. C. (1904). *Hannah logan's courtship*. Philadelphia: Ferris & Leach.

Nollet, J.-A. (1749a). Sur les causes de l'électricité des corps. *Mémoires de l'Académie Royale des Sciences (por l'année 1745)*, 107–151.

Nollet, J.-A. (1749b). *Reserches sur les Causes Particulières des Phénoménes Électriques, et sur les Effects Nuisibles ou Advantageux qu'on Peut Attendre*. Paris: Chez les Frères Guérin. (Read to l'Académie Royale des Sciences April 20, 1746).

Nollet, J.-A. (1751). Observations sur quelques nouveaux phénoménes d'électricité. *Mémoires de l'Académie Royale des Sciences de Paris (pour l'année 1746)*, 1–23.

Nollet, J.-A. (1752). Des effets de la vertu électrique sur les corps organises. *Mémoires de l'Académie Royale des Sciences de Paris (pour l'année 1748)*, 164–199.

Outram, D. (1995). *The enlightenment*. Cambridge: Cambridge University Press.

Packard Humanities Institute. (unpublished). *Papers of Benjamin Franklin*. (Prepublication CD ROM #102)

Pérez-Ramos, A. (1988). *Francis Bacon's idea of science and the Maker's knowledge tradition*. Oxford: Clarendon Press.

Porter, R. (1990). *The enlightenment*. Atlantic Highlands: Humanities Press International.

Porter, R. (2000). *Enlightenment*. London: Allen Lane.

Priestley, J. (1775). *History and present state of electricity* (3rd ed., Vol. 1). London: Bathurst and Lowndes. (Reprinted by Johnson Reprint Co., New York, 1966).

Pringle, J. (1750). *Observations on the nature and cure of hospital and Jayl-Fevers*. London: Millar and Wilson.

Pringle, J. (1752). *Observations on the diseases of the army in Camp and Garrison*. London: Millar, et al.

Reed, H. S. (1947–1948). Jan Ingenhousz. *Chronica Botanica, 11*, 288–391.

Schaffer, S. (1983). Natural philosophy and public spectacle in the eighteenth century. *History of Science, 21*, 1–43.

Schaffer, S. (1993). The consuming flame: electrical showmen and Tory mystics in the world of goods. In J. Brewer, & R. Porter (Eds.), *Consumption and the world of goods* (pp. 489–526). London: Routledge.

Selwyn, S. (1982). Sir John Pringle. *Medical History, 10*, 266–274.

Smyth, A. H. (1906). *The writings of Benjamin Franklin* (Vol. 9). New Haven: Yale University Press.

Snorrason, E. S. (1974). *C. G. Kratzenstein and his Studies on electricity during the eighteenth century*. Odense: Odense University Press.

Steinfeld, J. I. (1951). *Therapeutic studies on psychotics*. Des Plaines: Forest Press.

Sydenham, T. (1848, 1850). *The works of Thomas Sydenham* (2 Vols.). London: Sydenham Society.

Thacher, J. (1828). Cadwallader Evans. In *American medical biography* (p. 265). Boston: Richardson, et al.

Tolles, F. B. (1957). *James Logan and the culture of provincial America*. Boston: Little Brown & Co.

Van Doren, C. (1938). *Benjamin Franklin*. Cleveland: World Publishing Company.

Veratti G. (1748). *Osservazioni Fisico-Mediche Intorno alla Ellectrità. . . .* Bologna: Lelio dalla Volpe.

Wesley, J. (1760). *The desideratum*. London: Flexney.

Wesley, J. (1791). *Primitive physic*. Philadelphia: Parry Hall.

18

Gentleman's Magazine, the Advent of Medical Electricity, and Disorders of the Nervous System

Hannah Sypher Locke and Stanley Finger

Introduction

The idea that electricity might have a place in mainstream medicine represented one of the most significant therapeutic developments in the eighteenth century. By looking at articles written or sponsored by Fellows of the Royal Society in the *Philosophical Transactions* and by examining what was written by the leadership of other elite organizations, one can begin to appreciate what the academy physicians were encountering, discovering, and believing. Still, this more traditional approach is limited in scope.

In 1731, businessman Edward Cave began publication of a magazine that would include news and potentially useful information for the broader British public. He targeted inquisitive people in cities and rural areas who wanted to keep abreast of new developments, including medical breakthroughs. Moreover, his innovative *Gentleman's Magazine* welcomed submissions and commentary from his readers, even if they were not college educated or members of prestigious societies.

Cave's publication is often regarded as the first modern magazine. But of even greater importance to medical historians, it also provides a unique glimpse of how new developments in medicine were perceived and reported by "nonacademy" practitioners, observers, and even patients. In this chapter, we shall present what people read about medical electricity in *Gentleman's Magazine* from 1745, when the new treatment began to be made public, to 1760, by which time it had become one of the most popular fads in the history of medicine.

Our emphasis will be on disorders of the nervous system, which then included not just strokes and other neurological disorders, but also hysteria.

By appreciating what the public was reading, it will become easier to understand how medical electricity reflected the humanitarian goals of the Enlightenment, how it stemmed from natural philosophy, and why it quickly rose to become a panacea for almost every human ailment (for more on the scope, culture, fascination, and promise of Enlightenment science, see Bertucci, 2003, 2005a; Golinski, 1992; Schaffer, 1983, 1993; Stewart, 1992; Sutton, 1995). Yet, as will also be pointed out, *Gentleman's Magazine* was not without its faults. Given the need of its editors to publish intriguing and even sensational material, and the biases and aspirations of the people submitting reports, it tended to present an overly optimistic portrayal of how well the new cure worked. With these thoughts in mind, we must begin with Edward Cave and how his widely disseminated magazine came into existence.

Edward Cave

Edward Cave was born in 1692 to parents of modest means in Newton, a town in Warwickshire, England (Carlson, 1938, pp. 3–28). His parents sent him to the Rugby school, where he was considered gifted. But young Cave fell into disgrace after participating in a prank on the headmaster's wife. This prank resulted in more than just a slap on the wrist; Cave's life was now made miserable. The headmaster accused him of accepting money for tutoring other students, and he gave him more work than he could really do. Not willing to put up with the abusive treatment, Cave opted to leave the Rugby school.

Cave eventually made his way to London with a serious interest in literature, a penchant for winning arguments, and little regard for authority. Between 1712 and 1714, he apprenticed to Freeman Collins, an established printer, and quickly grasped the rudiments of journalism and running a newspaper. Collins was so impressed with his apprentice that he asked Cave, then only 22, to head to East Anglia to establish a new rural newspaper, the *Norwich Courant*. Unfortunately, Collins died soon after the newspaper started. Unable to get along with Collins' widow, and with his own apprenticeship now completed, Cave left Norwich and headed back to London to look for another newspaper job.

Cave now began to work for John Barber, a Tory printer with extensive political connections in the city. He quickly realized that news from the countryside was hard for city papers to get, and that country newspapers were badly in need of quick, reliable, and comprehensive information from the city. The solution to the two-sided problem emerged when Cave's wife helped him obtain a job at the London post office, where he began as a sorter in 1723. His position made it easy for him to see what was being printed in the country newspapers before those publications reached the city market, and he quickly became a supplier of this valuable information. Similarly, he also began supplying information from the city to select country newspapers – also for fees.

Cave's methods for getting information were not always in accordance with the law. Parliamentary news was in great demand and he seemed to get more than his share of privileged information before it was officially released. In 1728, he was tried and convicted of selling information about the activities of Parliament, while both houses were still in session. He was kept in jail for 10 days, but was pardoned after he gave a formal apology while on his knees. Despite his conviction, fine, and reprimand, he somehow managed to keep his position at the post office.

Cave took his next big step when he bought his own printing press in 1731. He installed it at St. John's Gate, a landmark in Clerkwell, his central London location. The early-sixteenth-century structure (once a priory for the Knights Hospitallers) had two towers connected by a wide arch. It became the symbol on all Cave publications, but none became better known than the *Gentleman's Magazine or Trader's Monthly Intelligencer*, which first appeared in February 1731.

Gentleman's Magazine

The basic idea behind Cave's new magazine was that it would present rural and urban news, as well as other types of articles, under a single cover, so that the gentlemanly reader could get everything in one place, rather than having to buy several city and country newspapers (Carlson, 1938, pp. 29–30). As its chief editor, Cave took the *nom de plume* of Sylvanus Urban. Sylvanus was the Roman god of the woods and Urban referred to *metropolitan*. This was just one way for Cave to show that he wanted his new publication to be relevant to people in both the city and the countryside.

Gentleman's Magazine has been hailed as the first modern magazine, and it might well have been the most important periodical for readers of English during the volatile eighteenth century. Yet because its innovations were so successful, it soon engendered competition.

From every account, "Ned" Cave remained totally dedicated to the management of his magazine, always looking for ways to stay ahead of his competitors. The great Samuel Johnson once commented that Cave "never looked out the window, except with a thought to its improvement" (Carlson, 1938, p. 14). Johnson had become a writer on Cave's staff in 1737 and he quickly rose to become an active editor. He contributed poems, essays, and political pieces for the periodical through the 1740s, and, along with Alexander Pope, judged literary submissions for cash prizes. Johnson also wrote biographical sketches of leading physicians for *Gentleman's Magazine*, including one on Hermann Boerhaave and another on Thomas Sydenham (McHenry, 1959).

What started off as a collection of news items first published elsewhere, to which were added political articles, sketches, and essays by the staff, soon emerged as a depository for a much wider range of submissions. Importantly, *Gentleman's Magazine* increasingly covered new discoveries in natural philosophy, with special emphasis on those with practical utility, including applications to bedside medicine. To keep readers on top of the latest scientific and medical developments, the editors summarized reports to prestigious professional societies and took material from their associated journals, as shall be shown below.

Above and beyond this, the editors also began to include original reports from the readership. Some of these reports came from amateur natural philosophers and physicians without affiliations to colleges or prestigious societies; frequently men who hoped to make names and establish more rewarding careers for themselves. Others came from eyewitnesses and concerned members of the laity, who were excited by the wonders they were observing and the opportunity to participate in the "Republic of Letters."

In effect, citizens from all facets of life were given the opportunity to publish an observation, experiment, or an opinion in *Gentleman's Magazine* – unlike the less democratic academy publications, including the *Philosophical Transactions*. Partly owing to the fact that William Watson, a leading scientist at the Royal Society with a strong interest in electricity, was skeptical and afraid of deceptions, the *Philosophical Transactions* avoided publishing reports on electrical cures prior to 1748. Only then did it slowly begin to include selective pieces, meaning items written by, or supported by, people considered trustworthy authorities on the subject.

With the door open to them, less well-known practitioners of the healing arts and lay people sent communications to Cave. The nonacademic contributors also replied to one another on the pages of the magazine, sometimes agreeing among themselves and with the summaries taken from the authorities, and sometimes not. Although there were exceptions, most medical pieces tended to be brief, ranging from just a few lines to some short paragraphs. In addition, and as if they were following an unwritten rule, the direct contributors to Cave's periodical tended to be courteous to each other – and their reports made good reading.

Reflecting the spirit of the Enlightenment, even the most established practitioners strove to present themselves as sympathetic servants of the public, and not as jealous guardians of their trade. The public, in turn, was pleased to have access to their good advice. During the eighteenth century, the head of the household was still the major decision maker when it came to medical decisions. It was ultimately up to this person to decide whether to treat himself, his family, and his staff on his own, or whether to negotiate a course of action with a more knowledgeable physician (Jewson, 1976; Porter, 1985a, 1985b). Thus, *Gentleman's Magazine* helped laymen make informed diagnostic and therapeutic decisions and, through the further exchange of information, public-spirited laymen also helped to "educate" open-minded medical professionals.

Looking back, Cave (and his brother-in-law David Henry and nephew Richard Cave after the founder's death in 1754) can be thought of as contributing to the medical Enlightenment in two important ways. First, they supported and encouraged the dissemination of new information, ideas, and practices to a wider audience than just the medical elite. And second, they fostered dialogues and exchanges between common and not-so-common people, promoting a more democratic approach to experimental natural philosophy, which during the Enlightenment called for using the methods of good science to improve the human condition.

The Success of the Publication

By any measure, *Gentleman's Magazine* was innovative, entertaining, informative, and a great success. In 1746 it was being read by more than 3,000 paying subscribers and was growing. Not only did it rapidly enlarge its subscription base to about 10,000, but it was also being circulated among a far larger number of inquisitive readers who were congregating at pubs and coffeehouses, not only in London and throughout the British Isles, but also in the North American Colonies.

The varied lay and professional readership of *Gentleman's Magazine* awaited each new issue, hoping to learn more about governmental politics, news from distant lands, natural philosophy, and increasingly, its extension into practical medicine. Indeed, by mid-century, *Gentleman's Magazine* was awash in pieces on health regimens, nostrums, first aid, and new medical developments. In fact, every issue had at least a few articles related to health and medicine. In addition, Cave included material from the London *Bills of Mortality*, important obituaries, and meteorological pieces that attempted to get at the source of epidemics.

The tone of the medical material was almost always empirical. That is, the emphasis was on observations and experiments, or what did and did not work, although theory sometimes crept into the picture when articles summarized news from the academies.

Not surprisingly, opposing or contradictory findings occasionally followed one another. Much more often than not, the editors tended to stand back when there was a conflict, a debate, or insufficient evidence. In effect, they left it up to the readers to draw their own conclusions.

The medical contributions tended to be widely praised by the readership and played a major role in the periodical's success. In 1747, N. Elles sent a note to Cave stating, "Amongst the agreeable variety which your Magazine contains, I shall always esteem that part the most useful, which can contribute towards relieving those whom it has pleased providence to afflict with sickness" (Elles, 1747, p. 77). In response to Elles' note, the editors wrote: "To find that we have in any degree contributed to the benefit of the public, or of any individual, gives us great pleasure" (*Gentleman's Magazine*, 1747, p. 362).

Seven years later, R. B. praised the steadily increasing medical coverage in Cave's periodical with the words: "I am pleased to observe the increase in your medical correspondents, by whom the public must be great gainers . . . none ought to be more esteemed than those that contribute to the knowledge and removal of those causes that obstruct the health of mankind" (R. B., 1754, p. 362).

As put by the late medical historian Roy Porter, "Cave and his successors were astute entrepreneurs, with their fingers securely on the pulse of public taste, and we may assume that this weighty attention to medical matters reflected, or created, real consumer demand" (Porter, 1985b, p. 292). Writing well before Porter, C. Lennart Carlson, one of the magazine's historians, opined:

In these pages where dull prose and duller verse alternate with accounts of brilliant discovery and extracts from poetical masterpieces, where politics jostle with theology and the sciences with art, here we have a kaleidoscopic view of the culture of a century. We sense the development of popular taste, we feel the surge of popular opinion. . . . A chronological survey of the various medical items published in the magazine would amount to a cross-section of the history of medicine in the eighteenth century, so varied is the material and so accurately does it show the public interest in various of the medical problems which were being discussed at the time. (Carlson, 1938, pp. 58, 158)

For determining what many people were first learning about medical electricity, and particularly for understanding how the literate middle class and bedside practitioners were responding to this promising new development, much can be gleaned from *Gentleman's Magazine*. Of particular significance is the era in which patients began to be treated with shocks from electrical machines – from 1745 when the technology was first applied, to 1760, when it already had a large following.

Needless to say, the periodical was filled with numerous articles on other facets of electricity before and within this time frame – electricity, lightning rods, new branches of natural philosophy, and the "love of the marvelous" being so much on the public's mind (Heilbron, 1979; Mitchell, 1998; Schaffer, 1983, 1993). As put by Simon Schaffer, who examined the lay and professional appeal of electricity, as well as its politics, economics, theological implications, and utilitarian promises:

Electrical fire dominated London natural philosophy. Public interest in the new electrical demonstrations was intense . . . It was used by natural philosophy lecturers to swell their audiences . . . Instrument-makers catered for an expanding market with ranges of electrical devices . . . accompanied by handbooks . . . Electricity also intersected the concerns of the learned professions . . . Electrical phenomena could capture an audience, satisfy customers, cure the body and save the soul. (Schaffer, 1993, pp. 490–491)

Medical Electricity

The idea of employing electricity to treat paralyses, seizure disorders, and other neurological problems is sometimes traced to the use of electric fish by the Romans (Kellaway, 1946; Schechter, 1971). Yet it would not be until the 1770s that John Walsh, a fellow of the Royal Society who was encouraged by Benjamin Franklin, would firmly associate the shocks of these strange creatures with electricity (Cavallo, 1795; Piccolino, 2003; Piccolino & Bresadola, 2002; Walsh, 1773, 1774). Moreover, these fish were never a part of mainstream Western medicine. This was partly because they were difficult to catch and keep healthy in captivity, and also because it was virtually impossible to control their shocks.

Hence, the history of medical electricity really begins with the advent of frictional machines for generating electrical sparks on demand. In this context, Otto von Guericke is often cited for building the first static electricity machine, although his intention had been to model the earth to study its gravitation. In 1672, he described how he rotated a large sulfur globe on a spindle and how, after being rubbed, it attracted and repelled small objects, sometimes with crackling sounds and sparks (von Guericke, 1672).

Francis Hawksbee, curator of experiments for the Royal Society, took the next step. He introduced better electrical machines that used glass early in the eighteenth century (Hawksbee, 1709). A third important pioneer was Stephen Gray, who showed that electricity from static machines could be transmitted over threads and wires. Gray further demonstrated that the human body could be electrified safely (Gray, 1731).

Johann Gottlob Krüger, a professor in Halle, is often cited as the first person to suggest that the electrical machines might have utility in medicine (Krüger, 1744; also see Bertucci, 2001a; Licht, 1967). The idea was presented in one of his lectures in 1743, and he even hypothesized that "the best effect would be found in paralyzed limbs" (Trans. in Licht, 1967, p. 5). These lectures were published in 1744, and that year his student Christian Gottlieb Kratzenstein set out to test his ideas (for more on Kratzenstein, see Snorrason, 1974).

Kratzenstein began with two patients who could not move their fingers, most likely because of arthritic conditions. He found that gentle electrification allowed the woman to overcome a contracture and the man to play his harpsichord again. He described these cases in some personal letters and in a monograph published in 1745 (Kratzenstein, 1745). Joseph Priestley, who consulted with Benjamin Franklin when writing his landmark *History and Present State of Electricity*, called Kratzenstein's writings "the first account I have met with of the application of electricity to medical purposes" (Priestley, 1767, 1966, p. 472).

Cromwell Mortimer, physician and secretary of the Royal Society, promoted the idea that electricity may have physiological and perhaps even medical significance in the same year (Mortimer, 1745). Moreover, working with electricity became decidedly easier with Pieter van Muschenbroek's invention of the Leyden jar in 1745 (Brazier, 1984, pp. 178–180; Dorsman & Grommelin, 1957; Hackmann, 1978, pp. 90–103). Made of glass, covered with metal foil, and filled with water or lead shot, the jar could store an electrical charge that could be released on demand.

An extract of a letter describing van Muschenbroek's discovery of the power of the Leyden jar appeared in *Gentleman's Magazine* in 1746 (*Gentleman's Magazine*, 1746). We read that "the famous professor . . . was struck with such a violent blow [from his charged jar] that he thought his life at an end" (163). Readers are also informed that his technological breakthrough had just been confirmed in France by the Abbé Nollet and Monsieur de Menniers "of the academy."

The advent of the Leyden jar led to the development of a new line of portable electrical machines that could be used at the bedside (Bertucci, 2003). New, inexpensive machines, in turn, stimulated more case studies, many of which were reported on the pages of *Gentleman's Magazine* by all sorts of people, on all sorts of patients, and with all sorts of outcomes. Nevertheless, most high ranking physicians preferred to disseminate their material elsewhere. For example, the Abbé Nollet treated paralytics in Paris with electricity in the 1740s (e.g., Morand and Nollet, 1753; Nollet, 1746). But with his aristocratic leanings and royal appointment, he was not one to submit his communications to *Gentleman's Magazine*.

This is not to say that the Abbé's experiments and conclusions never appeared in *Gentleman's Magazine*, because they did. His followers and Cave and his editors took some of the Abbé's medical observations from other sources and summarized them in English for readers of the periodical.

Late in the 1740s, for example, the Abbé learned about some incredible findings with medical electricity. Giovannifrancesco Pivati along with several other Italians had been claiming that camphor, Peruvian bark, opium, and other drugs could be put into sealed glass tubes, from which the medicines could be transported electrically to patients to produce wonderful cures (see Bertucci, 2001a). The readers of *Gentleman's Magazine* were first told about the cures in 1749. The information came from Johann Heinrich Winkler, a professor of experimental natural philosophy at Leipzig, who had been communicating with the Italians and had recently replicated some of their findings (Winkler, 1749). A year earlier, Winkler (1748) penned his findings and positive outlook to the Royal Society. At the conclusion of the more broadly disseminated *Gentleman's Magazine* article bearing his name, we find the statement: "These are such remarkable instances as, I think, leave no room to doubt of the usefulness and assistance of electricity in medicine" (453).

But not everyone was able to confirm these effects, including William Watson who had translated Winker's letter to the Royal Society. To help set the record straight, the Abbé Nollet set out for Italy in 1749 to learn more about these experiments, which did not require the usual ingestion of medicines. He presented himself as an impartial but informative arbiter, although he had been unable to confirm the effects in his own studies (Bertucci, 2001a, 2005b).

During the summer of 1751, *Gentleman's Magazine* presented the "Extract of the Abbé Nollet's Examination of certain Phenomena in Electricity published in Italy," which was based on material first presented to the Académie des Sciences in Paris and the Royal Society (Nollet, 1751a, 1751b, 1753). The article began by telling how, after hearing of the Italian successes, "Every one was accordingly desirous of repeating what M. Pivati said had been done at Venice, M. Verati at Bologna, and M. Bianchi at Turin" (Nollet, 1751b, p. 261). Winkler's subsequent work was also mentioned, as were some failures to replicate, including Nollet's own experiments in Paris.

The Abbé first stop, readers were informed, was in Turin, to visit Professor Bianchi. "I begged him that all his experiments, which had not succeeded either with me or many others, might be repeated between us" (Nollet, 1751b, p. 261). Bianchi obliged and tried to replicate some experiments using two different types of purgatives, scammony, and gamboges, which the Abbé held in his hands, while receiving electric shocks. Others then underwent the procedure, but only a servant experienced an effect.

The young man, who said he had two stools, said also, that having . . . visited with his wife, an hour after his being electrized . . . he had communicated this electricity to her, and that she had been purged as well as himself: This, with his idle manner of boasting to everyone he met, of the operations he had undergone, rendered his testimony more than suspicious. (Nollet, 1751a, 1751b, p. 262)

What transpired reflected a serious design flaw in the experiments, which was recognized by Nollet. Many participants were in the service of the Italian experimenter and were led to believe there would be a positive effect and clearly wanted to please him. Hence, the Abbé decided to use only "reasonable people, that might not be suspected of reporting any nothing [sic] but the truth" (Nollet, 1751a, 1751b, p. 262). Additional testing with even stronger shocks, fresher medicines, and on a higher class of subjects led the Abbé to conclude that Bianchi's claims could not be substantiated.

Nollet then visited Pivati in Venice, and this only added to his skepticism (Bertucci, 2005b). Among Pivati's excuses for not demonstrating any effects with his medicated tubes were that "there was too much company, the weather too hot, and consequently, the electricity too weak" (Nollet, 1751b, p. 349). Nollet then asked him about follow-up examinations with some of his most celebrated cases, including the bishop of Sebenico. "He owned, that the prelate was not cured, and that since his electrification he had been as he was before" (349). Nollet concluded that Pivati "may be suspected of too much credulity" – *Qui Vult Decipi Decipiatur* [whoever wishes to be deceived (351), let them be deceived].

The medicated tube controversy did not diminish future interest in medical electricity. As noted by Bertucci (2001a, 2005b), it drew new attention to it from practitioners and philosophers, and others who had been more concerned with the nature of electricity and performing popular demonstrations, such as showing how an electrically charged person could set a glass of brandy afire. In effect, the medicated tube controversy helped bring the highly respected methods of experimental natural philosophy even more into the practical healing arts. As Nollet himself recognized: "The possibility of medical applications of electricity immediately upgraded the status of electrical experiments from salon entertainment to frontline research" (Bertucci, 2005b, p. 66).

Some Theory

Although theory was secondary to outcomes in *Gentleman's Magazine*, a few of the early articles on medical electricity put the new cure in the context of the iatromechanical theories that were so popular among academicians at the time. The basic assumption of iatromechanical medicine was that illnesses could be accounted for by such things as blockages of the nerves, constrictions of the blood vessels, or internal fluids that had become too turgid or dense to allow particulate matter to flow properly through them. This sort of thinking reflected the application of Newtonian mechanics to the internal physiology of the body. That these changes could not be seen or measured directly was of little concern to many erudite physicians, who had been trained in theory at the leading medical institutions of the day.

The 1745 *Gentleman's Magazine* article bearing the title "An historical account of the wonderful discoveries made in Germany, &c concerning Electricity" has been associated with Albrecht von Haller, although it does not provide the name of the author (Heilbron, 1977). Although the thrust of the article is basic science, with no case studies or cures being mentioned, electricity is referred to as "the subject in vogue" and "as surprising as a miracle." Moreover:

It has already been discover'd, or believ'd to be so, that electricity accelerates the movement of water in a pipe, and that it quickens the pulse. There are hopes of finding in it a remedy for the sciatica or palsy. (von Haller, 1745, p. 197)

Four years later, *Gentleman's Magazine* reprinted in English the contents of a letter from Abbé Nollet to Martin Folkes, President of the Royal Society. In it, the Abbé alluded to the widely accepted idea that obstructed pores can trigger poor health, because these blockages prevent the body from discharging noxious humors or waste products. He then argued that electricity could remove obstructions from the pores, scour them of noxious humors, and increase the "insensible perspiration" (Nollet, 1749).

Thus, even during the 1740s, the readers of *Gentleman's Magazine* were given theoretical reasons to believe that electricity administered to sick and debilitated patients might make them better. Notably, they were informed that it could make sluggish fluids less turgid, that it could overcome blockages, and that it could aid in the expulsion of noxious products from the body. The possibility that the nerves might function by electricity was also beginning to circulate at this time. It fuelled the related idea that applying electricity might be able to stimulate a machine that is functioning improperly. With the potential to be beneficial from several theoretical perspectives, providers and patients could be optimistic.

Nevertheless, for lay readers and noncollege-educated practitioners, the more numerous case reports that were devoid of theory had to be much more influential. And of those reports, the ones that dealt with curing disorders of the nervous system that seemed resistant to traditional medicines probably captured the imagination most of all.

Palsies of the Tongue and Body

The palsies drew considerable attention from early medical electricians, because they tended to be highly visible, disabling, and resistant to other treatments. Included under the umbrella term "palsies" were some cases in which speech was affected – the belief being that some losses of speech might be due to paralyses of the tongue.

One such case was published in *Gentleman's Magazine* by W. Watts in April 1751. Watts was a London physician and electrical scientist. His report does not give specifics about the condition of the rest of the man's body, although it too was affected. Nor does he address the palsy's likely cause or type, other than to state that it came on rapidly. This would suggest that his patient might have had a small stroke or what was then called the "common paralytic disorder."

To quote:

Robert Moubray, who in the beginning of *January* was struck with a complete palsy of the tongue, and since that time entirely lost the use of his speech, was taken into the royal infirmary some weeks thereafter, where, by the use of remedies, he was, in great measure, relieved of other symptoms that attended him; but the palsy of the tongue remaining obstinate, he was at last order'd by the physicians to lay aside the use of all medicines, that he might fairly try what electrifying would do in such a case: Accordingly last week it was begun, and by *Saturday* he was able to extend and put out his tongue, which till then had remained dead and motionless: On *Monday* he could plainly articulate a few words; and, after repeating the experiment on *Tuesday*, he spoke distinctly, to his own great joy, and the surprise of all who were present. (Watts, 1751, p. 152)

A year later, surgeon Patrick Dickson described the treatment of his physician-friend Philip Brown, who was "seized with a nervous fever and asthma . . . and . . . lost entirely the use of his tongue. Many were the attempts to remove his unhappy disorder, but all in vain" (Dickson, 1752, p. 363). Following a myriad of unsuccessful treatments, Brown visited Edinburgh professor of medicine John Rutherford, who agreed to try electricity. Early on, Brown's condition worsened as the electricity was applied. But after showing some recovery a few months later, he was willing to try it again.

Dickson now purchased an electrical machine to conduct in-home treatments with Brown (for more on inexpensive and even portable machines and their patents, see Bertucci, 2001b, 2003). Quoting Dickson: "I desired he would receive the shock upon his tongue, which he did, tho' very gently on the first day; on the second he received about 80 pretty strong shocks; this made him salivate a full pint in a few minutes." He then increased his shocks, noting that "every shock gave him more freedom in the motion of his tongue." Nevertheless, the restored tongue motion was by itself insufficient to restore Brown's speech. He improved significantly only after he was given specific instructions for repeating the alphabet with specific positions for his tongue, some practice with syllables, and then mastering short words. "The more he spoke, he perceived his tongue grow stronger; and in a few hours could speak as well and distinctly as ever," wrote Dickson (1752, p. 364).

Dickson was mentioned again in 1753, in "An Extract of a Letter from Bernard Castle, in Yorkshire."

Two gentlemen and a young girl in this neighbourhood having lost their speech for several years, were lately restored to the use of it again, by electricity, and duly observing the direction of Mr. *Dickson*, the surgeon there. A gentleman [presumably Brown] who was restored to his speech about a year ago, by the like experiment, continues to speak very well. (Castle, 1753, p. 432)

Yet another case from 1753 involved a 35-year-old woman diagnosed with a "paralytic stroke" that

also rendered her speechless. But in contrast to most patients who lose the ability to speak, the reader is told that she was paralyzed on the left side, not the right. Further, although she seemed to show some improvement even before being electrified, her left side continued to remain functionally useless, "so that she was obliged to be dress'd and undress'd, and to have a person on each side to support her." After being treated with electricity by Mr. Thomas Howell, however, "she was able to dress herself, make her bed, and to do many things about the house." A day later, "she was most surprisingly mended, and assisted in some very laborious housework, and the next day the use of her limbs was perfectly restored to her" (Watson, 1753, pp. 268–269).

The aforementioned report was submitted to *Gentleman's Magazine* by I. Watson, who added:

I have mentioned nothing in this wonderful cure but what I was eye-witness to, having attended the electrical operations every night . . . She made use of no medicine, of any kind, during the course of the electrical operations, or for a considerable time before. (Watson, 1753, p. 268)

Watson's words were followed by some poetry from "Cynthio," which was dedicated "To Mr. Thomas Howell, on his late surprising cure by Electricity." The second stanza reads:

From deadly maladies that grieve us
Thy art can presently relieve us,
Without the help of pill or bolus.
When *Palsie*, with benumbing power,
Spreads half the human fabrick o'er.
And makes the body, while we breathe,
A living spectacle of death!
By touch electric thou, with ease,
Canst rescue from the dire disease;
Bid us receive the friendly shock,
And triumph o'er the death-like stroke!
(Cynthio, 1753, p. 269)

Although electricity is praised here, Dr. Cheney Hart, who worked at a hospital in Shrewsbury, questioned the use of electricity to help paralytics. In 1754, after his own electrical treatments failed to cure a woman with a "wasted" paralytic arm, Hart sent a letter to the Royal Society (Hart, 1754). In a 1755 summary of some of the material appearing in the *Philosophical Transactions*, an unnamed *Gentleman's Magazine* editor summarized what Hart had reported in one sentence: "The patients on whom it was tried were paralytic; on one it produced no effect, and on the other it twice rendered a partial palsy universal" (*Gentleman's Magazine*, 1755, p. 489). Negative statements such as this one, however, were in the minority.

Seizure Disorders

Some cases of palsy were also associated with convulsive fits. For example, M. Allaman wrote in 1758 that "A girl of thirteen or fourteen years old, being in a house alone, heard a violent knocking at the door; the surprise threw her into a fit of strong convulsions, which was scarce over before it was succeeded by a palsy." Most of the resultant paralysis dissipated, but "the tongue was rowled up in the lower part of the mouth . . . [and] remained stubbornly inactive." The electrician drew sparks from her tongue. "The first day of the operation he fancied that he observed some motion . . . Upon the fourth experiment the tongue was unrowled. Seven or eight more electrifications, and constant exercise of her tongue brought her perfectly to her speech again" (Allaman, 1758, pp. 467–468).

Two years later, readers learned about Mary Beaumont of High Twyford, who was seized with a feverish disorder that led to convulsions. According to Joseph Hornblower,

These fits left her in a little time . . . but [she] had totally lost the use of her limbs, so that she was not able to stand with the help of a pair of crutches. In this condition she continued about two years . . . every method was taken and every medicine tried . . . but without effect. At last she was brought to me, and was electrified and shocked every day for about a month before she perceived any amendment. She then began to gather strength, and having the operation continued every day 10 weeks more . . . her weakness and lameness were entirely removed; her strength and vigour returned; she was able to go about her business, and has continued well ever since. (Hornblower, 1760, p. 477)

In a number of other reported cases, there were convulsions without paralyses. For instance, in May 1753, I. G., a Devonshire practitioner, recounted the story of a sailor, who

applied to me, to try the effects of electricity, in an epileptic disorder, with which he had long been troubled, and which increased so fast upon him, that he was apprehensive it would quite incapacitate him from going any

more to sea. The first operation producing great alterations in him, I was encouraged to proceed; and by repeating it twice more, at about a week's interval each, he found such benefit as to have no return of his disorder since. (I. G., 1753, pp. 227–228)

Just a few months later, William Morris of Kenton described how he drew sparks from the neck and chest of a man who was "seized with convulsions in his breast and throat," which led to "falling down, deprivation of the senses, and . . . stammering." Electricity was subsequently applied to his breast and throat.

The first night, having received several shocks, by means of a large coated phial, he went home and sweat plentifully, slept better than usual, and next day found himself somewhat eased from that weight in his breast, he used to complain about. He came to me about a week afterwards, and received shocks in the same manner; but the force of the shocks was so much increased . . . that he could scarce bear it . . . The week following I gave him several more shocks from the two coated phials [Leyden jars], as before, and since that time he has had no more symptoms of his disorder. (Morris, 1753, p. 379)

Morris concluded that there should be a wider use of electricity in medicine:

For is it not reasonable to think that the electrical machine would greatly add to the efficacy of the materia medica? And, if the use of it was brought more into practice, might we not expect great improvements in healing, by removing many obstinate nervous disorders, which now baffle the skill of physicians? (Morris, 1753, p. 379)

Neither I. G. nor Morris referred to their patients as hysterical, yet some of the cases described on the pages of *Gentleman's Magazine* probably suffered from hysteria. One of the hallmarks of hysterical symptoms is the way in which they can migrate from one limb to the other, or from one side of the body to the other. Another is that they tend to wax and wane. During the mid-eighteenth century, it was widely believed that disorders of the mind could physically affect the body (Foucault, 1965; Porter, 1987; Trillat, 1995; Veith, 1970). Hence, stimulating those body parts that seemed prone to hysterical palsies or hysterical convulsions, in order to remove suspected blockages and to strengthen them, made sense, just as did treating a paralysis caused by an actual stroke.

Unlike I. G. and Morris, Cadwallader Evans, a "student in Physick at Philadelphia," specifically mentioned hysteria in his report on a young girl, identified only as C. B., whom he and Benjamin Franklin treated successfully with electricity (for more on this case, see Finger, 2006, also previous chapter). She was seen in 1752, and the case was published in its entirety in *Medical Observations and Inquiries* in 1757 (Evans, 1757a). In a one-paragraph summary of the case, the readers of *Gentleman's Magazine* were told that

the patient was a girl about 14, who was seized with convulsion fits, which succeeded each other so fast, she had near 40 in twenty-four hours . . . she was sometimes . . . tortured by the cramp in different parts of her body, and sometimes with more general convulsions of the extremities, attended with a choking deliquium, and almost the whole train of hysteric symptoms. (Evans, 1757b, p. 260)

C. B., who might have been Evans' sister, was anxious to go to Philadelphia to be treated by Franklin, who was already famous for his experiments and understanding of the nature of electricity. After applying electricity, Franklin even gave her "a globe and bottle" to generate and store the electricity, so she could electrify herself at home according to his instructions. The patient dutifully followed Franklin's instructions and showed an excellent recovery from her hysterical seizures – there being no relapses 5 years later, when her case report was sent to press.

It is worth noting that Franklin was skeptical about curing seizure disorders that seemed to be organic. Less than a decade after the Evans report was published, he joined a number of gentlemen to witness an "experiment in electricity" on a convulsive woman in a British workhouse. The treatment on the woman who had lost her speech 6 weeks earlier apparently worked, and in *Gentleman's Magazine* it was dutifully reported that "Mr. Franklin of Philadelphia . . . expressed his astonishment" (*Gentleman's Magazine*, 1766).

Pain and Other Sensory Disorders

Pain and sensory disorders also received mention in the context of medical electricity, although not as frequently as palsies and convulsive disorders. In some cases, the pain was brought on by non-neurological disorders. For example, in 1747, J. W. described a gentleman who had "a torment-

ing pain, which he apprehends to have been rheumatic, in the two smallest joints of the fore finger of his right hand." Upon witnessing a public demonstration of

the wonders of electricity, [he] had the curiosity to try its effect upon the pained finger. . . . In consequence of this, he observed the pain to be somewhat abated before he went to bed; that on the morrow it grew much easier; and on the third day his finger was perfectly well, and so continues. – Who knows what further experiments and discoveries such an incident may lead to? (J. W., 1747, p. 238)

Although the majority of the early medical electricity pieces presented single cases, several cases were put together in a few articles, some of which dealt with pain and sensory problems.

In 1755, for example, physician M. Lindult presented "A succinct Account of Disorders lately cured at Stockholm by Electrification." Lindult, "By order of his Swedish Majesty," explained that electricity could be an excellent cure for rheumatic pains, muscle pains, and toothaches. The reader finds out that one Swedish man, who had a "violent toothache" and had been deaf for years as a result of a head injury, experienced improvements in both sensory domains. Lindult then went on to present several other cases of electrically restored hearing, writing:

A young man of 22, who had almost lost all his hearing for six months, by violent vomitings . . . was cured . . . and continues to hear perfectly well. A girl of seven, born deaf, who consequently could not speak, began to hear words spoken very loud in her ear, and could repeat some of them in a few days. In the year 1744, a young fellow of 19 fell into a well . . . since which time his hearing had been very weak; he has found much benefit from being electrified, and it is hoped will be perfectly cured. (Lindult, 1755, pp. 111–112)

Cave's periodical ran a second piece on medical electricity from Sweden later in the same year, and it too summarized multiple case studies (*Gentleman's Magazine*, 1755). Originally written in Latin, it was passed on to *Gentleman's Magazine* by a member of the Royal Society and published in English. It was based on the work of a medical practitioner, Dr. Rosen, and a respected academic, Professor Stromer. For headache, the conclusion was that electricity by itself does not produce long-term cures. The reader is told, however, "that where proper internal remedies were given at the same time, the desired effect was speedily produced" (314). For sciatica, "In some the pains have been greatly appeased by electrification, but returned again afterwards" (314).

Confusions and Conclusions

The pages of *Gentleman's Magazine* reveal a growing interest in medical electricity in its formative years, defined here as the period from 1745 to 1760. More specifically, the articles reveal that practitioners and philosopher-electricians, men with and without medical degrees and hospital appointments, quickly applied the new cure to patients with a myriad of different conditions. Strokes, seizure disorders, sensory problems, and hysteria were among the neurological conditions treated with electrical machines, and medical electricity is presented as the most promising new therapeutic development in this period. To many readers, it showed how enlightened experimental science could be used for the betterment of society.

Nevertheless, Cave's periodical also fostered its share of misconceptions and confusions. Particularly frustrating to the eighteenth-century subscriber untrained in medicine or experimental natural philosophy was that Cave and his editors almost always left it up to his readers to decide which reports should be believed or weighted most heavily. His open door policy led to some healthy dialogues and exchanges, and it also gave everyday people the opportunity to achieve some worldly fame. But it left people vulnerable to exaggerated claims from quacks with financial motivation, as well as from easily seduced observers and participants with overly active imaginations and their own agendas.

As might be expected, amateurs with little training in medicine or the methods of science tended to write less critically and more enthusiastically about medical electricity than their better-trained cousins. For example, contributors unfamiliar with the demands of good experimentation tended to be more likely to attribute effects of medical electricity to this intervention alone, although several factors, including drugs, bed rest, and time, might also have been operative.

Of course, not all trained practitioners had a strong "scientific" mindset or were as skeptical as the Abbé Nollet, and not all amateurs were naive. Benjamin Franklin, for one, lacked formal schooling and a medical apprenticeship, yet he was always extremely careful about what he was willing to conclude in the realm of practical medicine (Finger, 2006). Still, the fact remains that the staff of *Gentleman's Magazine* did little to guide the reader through the reports, which might have led to uncritical acceptance of some questionable material by those who did not know one contributor from another.

A second problem facing Cave's readers had to do with how little information was actually presented in most of the case studies. During the mid-eighteenth century, many disorders that are now classified separately were called by the same name, because they shared a major sign or symptom. One need only examine the umbrella term "palsy." Under this term we might find patients with the "common paralytic disorder" (stroke), what would later be called "the shaking palsy" (Parkinson's disease), unilateral facial palsies (soon to be called *Bell's palsy*), hysterical paralyses, and so forth. But without adequate case histories, even the periodical's more sophisticated readers would have been left wondering just what the individual really looked like and whether another case had the exact same problem. This lack of detail is significant because some palsies may remit fairly quickly, while others might come and go, and still others could progress or remain debilitating for the rest of the patient's life, with or without medical electricity.

A third problem faced by Cave's subscribers, which also could have led them to erroneous conclusions, has to do with selection biases. Because practitioners, observers, and reviewers were all anxious to send in positive reports, the editors of *Gentleman's Magazine* were considerably more likely to receive articles praising the new cure than questioning it. Especially among nonacademic physicians and amateurs, the feeling could well have been that their own skills would be questioned if they failed to obtain or confirm a positive finding. And, more likely than not, only those patients that were helped would have returned to inform the therapist of the outcome.

Consider a report from 1752, in which the contributor described an unexpected encounter with a young man in the audience during a public demonstration. The man had been suffering from ague (a term often signifying *malaria*), and he wanted to be cured. The author recounted:

I gave him the common shock, which was repeated pretty strongly. He went away and returned the next day and told me that he was in better spirits than usual; and had not the least symptom of the disorder; whereas he expected to have a return of the fit that morning: neither has he had the least return of it since. (D. H., 1752, p. 442)

Would this patient have bothered to come back if he had not been cured? And would any demonstrator or even practicing physician have bothered to submit a negative finding under such circumstances?

In addition, there were selection biases inside the magazine itself. Cave wanted to project an image of fairness, but at the same time he knew that sensational positive findings would attract more readers than negative results. Thus, positive findings not only stood a better chance than negative findings of being submitted, but also of being accepted for publication.

Hence, the ratio strongly favoring medical electricity as an effective cure for palsies, seizure disorders, pain, deafness, and other neurological problems cannot be taken as an accurate measure of what was actually being seen at the bedside. Given the biases of the submitters and the editors, the lack of critical thinking among some of the reporters, the inadequate terminology, and the absence of clinical details on the pages of *Gentleman's Magazine*, it is easy to understand why practitioners and lay people saw so much promise in medical electricity in the 1700s. It is also easy to understand the mass illusion that followed, particularly among those laymen and healers not affiliated with the leading academies.

Although the impact and power of Cave's enlightened press cannot be scientifically quantified, it clearly played an important role in generating widespread interest, whether justified or not, in using electricity to treat disorders of the nervous system.

References

Allaman, M. (1758). Account of the cure of an extraordinary kind of palsy, both as to the cause and the part affected, by means of electricity. *Gentleman's Magazine*, 28, 467–468.

B., R. (1754). Uncommon tumour. *Gentleman's Magazine*, 24, 362.

Bertucci, P. (2001a). The electrical body of knowledge. In P. Bertucci, & G. Pancaldi (Eds.), *Electric bodies* (pp. 43–68). Bologna: Università di Bologna.

Bertucci, P. (2001b). A philosophical business: Edward Nairne and the patent electrical machine (1782): Medical electricity in mid-18th-century London. *History of Technology*, 23, 41–58.

Bertucci, P. (2003). The shocking bag: Medical electricity in mid-18th-century London. *Nuova Voltiana*, 5, 31–42.

Bertucci, P. (2005a). Promethian sparks: Electricity and the order of nature in the eighteenth century. In S. Zielinski, & M. Wagnermaier (Eds.), *Variantology 1: On deep time relations of arts, sciences and technologies* (pp. 41–56). Koln: Walther König.

Bertucci, P. (2005b). Sparking controversy: Jean Antoine Nollet and medical electricity south of the Alps. *Nuncius*, 20, 153–187.

Brazier, M. (1984). *A history of neurophysiology in the 17th and 18th centuries*. New York: Raven Press.

Carlson, C. L. (1938). *The first magazine: A history of the Gentleman's Magazine*. Providence: Brown University Press.

Castle, B. (1753). Extract of a letter from Bernard Castle, in Yorkshire. *Gentleman's Magazine*, 23, 432.

Cavallo, T. A. (1795). *Complete treatise of electricity . . .* (Vol. 2, 4th ed.). London: Dilly.

"Cynthio." (1753). To Mr. Thomas Howell, on his late surprising cure by electricity. *Gentleman's Magazine*, 23, 269.

Dickson, P. (1752). A curious case of a gentleman, who has entirely recovered his speech by the use of the electrical machine alone, after having lost it for above 20 months. *Gentleman's Magazine*, 22, 363–364.

Dorsman, C., & Grommelin, C. A. (1957). The invention of the Leyden jar. *Janus*, 46, 275–280.

Elles, N. (1747). Of methods of exercise within doors. *Gentleman's Magazine*, 17, 77.

Evans, C. (1757a). A relation of a cure performed by electricity. *Medical Observations and Inquiries*, 1, 83–86.

Evans, C. (1757b). Account of a cure by electricity. *Gentleman's Magazine*, 27, 260.

Finger, S. (2006). *Doctor Franklin's medicine*. Philadelphia: University of Pennsylvania Press.

Foucault, M. (1965). *Madness and civilization*. New York: Vintage Books.

G., I. (1753). *Gentleman's Magazine*, 23, 227–228.

Gentleman's Magazine. (1746). Extract of a letter from Paris, March 25. *Gentleman's Magazine*, 16, 163.

Gentleman's Magazine. (1755). The efficacy of electricity in the cure of diseases, determined by a great variety of experiments made by appointment of the Royal College of Physicians in Stockholm. *Gentleman's Magazine, 25,* 313–315.

Gentleman's Magazine. (1747). *Gentleman's Magazine, 17,* 77.

Gentleman's Magazine. (1755). An account of the effects of electricity in the county hospital at Shrewsbury; by Dr. Hart. *Gentleman's Magazine, 25,* 489.

Gentleman's Magazine. (1766). *Gentleman's Magazine, 36,* 147.

Golinski, J. (1992). *Science as public culture: Chemistry and Enlightenment in Britain.* Cambridge: Cambridge University Press.

Gray, S. (1731). A letter to Cromwell Mortimer. *Philosophical Transactions of the Royal Society, 37,* 18–44.

von Guericke, O. (1672). *Experimenta Nova (ut Vocantur) Magdeburgica.* Amsterdam: J. Jansson-Waesberg.

H., D. (1752). An ague cured by electricity. *Gentleman's Magazine, 22,* 442.

Hackmann, W. D. (1978). *Electricity from glass* (pp. 90–103). Alphen aan den Rijn: Sijthoff & Noordhoff.

von Haller, A. (1745). An historical account of the wonderful discoveries made in Germany, &c concerning electricity. *Gentleman's Magazine, 15,* 193–197.

Hart, C. (1754). Part of a letter from Cheney Hart, M.D. to William Watson. *Philosophical Transactions of the Royal Society, 49,* 558–563.

Hawksbee, F. (1709). *Physico-mechanical experiments on various subjects.* London: Printed for the author.

Heilbron, J. L. (1977). Franklin, Haller and Franklinist history. *Isis, 68,* 539–549.

Heilbron, J. L. (1979). *Electricity in the seventeenth and eighteenth centuries.* Berkeley: University of California Press.

Hornblower, J. (1760). [Editor's heading: "Lameness cured by electricity."] *Gentleman's Magazine, 30,* 477.

Jewson, N. D. (1976). The disappearance of the sick-man from medical cosmology, 1770–1870. *Sociology, 10,* 225–244.

Kellaway, P. (1946). The part played by electric fish in the early history of bioelectricity and electrotherapy. *Bulletin of the History of Medicine, 20,* 112–137.

Kratzenstein, C. G. (1745). *Schreiben von dem Nutzen der Electricität in der Arzneywissenschaft.* Halle: Hemmerde.

Krüger, J. G. (1744). *Zuschrift an seine Zuhörer, worinnen er Gedancken von der Electricität mittheilt und Ihnen zugleich seine künftigen Lectionen bekannt macht.* Halle: Hemmerde.

Licht, S. (1967). History of electrotherapy. In S. Licht (Ed.), *Therapeutic electricity and ultraviolet radiation* (2nd ed.) (pp. 1–70). Baltimore: Waverly Press.

Lindult, M. A. (1755). Succinct account of disorders lately cured at Stockholm. *Gentleman's Magazine, 25,* 111–112.

McHenry, L. C. (1959). Dr. Samuel Johnson's medical biographies. *Journal of the History of Medicine and Allied Sciences, 14,* 298–310.

Mitchell, T. A. (1998). The politics of experiment in the eighteenth century: The pursuit of audience and the manipulation of consensus in the debate over lightning rods. *Eighteenth-Century Studies, 31,* 307–331.

Morand, S. F., & Nollet, J. A. (1753). Expériences de l'electricité appliqué a des paralitiques. *Mémoires de l'Académie des Sciences pour l'année, 1749,* 29–38.

Morris, W. (1753). Editor's heading: "Electrical cure." *Gentleman's Magazine, 23,* 379.

Mortimer, C. A. (1745). Letter concerning the natural heat of animals. *Philosophical Transactions of the Royal Society, 43,* 473–480.

Nollet, J. A. (1746). *Essai Sur l'électricité Des Corps.* Paris: Frères Guéri.

Nollet, J. A. (1749). New remarks on electricity by the Abbé Nollet. *Gentleman's Magazine, 19,* 449–450.

Nollet, J. A. (1751a). Extract of a letter from the Abbé Nollet, F.R.S. etc. to Charles Duke of Richmond, F.R.S. accompanying an examination of certain phenomena in electricity. *Philosophical Transactions of the Royal Society, 46,* 261–263, 368–397.

Nollet, J. A. (1751b). Extract of the Abbé Nollet's examination of certain phenomena in electricity published in Italy. *Gentleman's Magazine, 21,* 261–263, 349–351.

Nollet, J. A. (1753). Expériences et observations en différens endroits d'Italie. *Mémoires de l'Académie des Sciences pour l'année, 1749,* 444–488.

Piccolino, M. (2003). *The taming of the ray.* Florence: Olschki.

Piccolino, M., & Bresadola, M. (2002). Drawing a spark from darkness: John Walsh and electric fish. *Trends in Neuroscience, 25,* 51–57.

Porter, R. (1985a). Lay medical knowledge in the eighteenth century: The evidence of the Gentleman's Magazine. *Medical History, 29,* 138–168, 154–156.

Porter, R. (1985b). Laymen, doctors and medical knowledge in the eighteenth century: The evidence of the Gentleman's Magazine. In R. Porter (Ed.), *Patients and practioners: Lay perceptions of medicine in pre-industrial society* (pp. 283–314). Cambridge: Cambridge University Press.

Porter, R. (1987). *Mind-forg'd manacles.* Cambridge: Harvard University Press.

Priestley, J. (1767/1966). *History and present state of electricity* (Vol. 1, 3rd ed.). New York: Johnson Reprint Corporation.

Schaffer, S. (1983). Natural philosophy and public spectacle in the eighteenth century. *History of Science, 21,* 1–43.

Schaffer, S. (1993). The consuming flame: Electrical showmen and Tory mystics in the world of goods. In J. Brewer, & R. Porter (Eds.), *Consumption and the world of goods* (pp. 489–526). London: Routledge.

Schechter, D. C. (1971). Origins of electricity. *New York State Journal of Medicine,* 997–1008.

Snorrason, E. S. (1974). *C. G. Kratzenstein and his studies on electricity during the eighteenth century.* Odense: Odense University Press.

Stewart, L. (1992). *The rise of public science: Rhetoric, technology, and natural philosophy in Newtonian Britain, 1670–1750.* Cambridge: Cambridge University Press.

Sutton, G. (1995). *Science for a polite society: Gender, culture, and the demonstration of Enlightenment.* Boulder: Westview Press.

Trillat, E. (1995). Conversion disorder and hysteria. In G. E. Berrios, & R. Porter (Eds.), *A history of clinical psychiatry* (pp. 433–450). New York: New York University Press.

Veith, I. (1970). *Hysteria.* Chicago: University of Chicago Press.

W., J. (1747). *Gentleman's Magazine, 17,* 238.

Walsh, J. (1773). On the electric property of the torpedo. *Philosophical Transactions of the Royal Society, 63,* 461–477.

Walsh, J. (1774). Of torpedos found on the coast of England. *Philosophical Transactions of the Royal Society, 64,* 464–473.

Watson, I. (1753). *Gentleman's Magazine, 23,* 268–269.

Watts, W. (1751). Editor's heading: "Cure by electricity." *Gentleman's Magazine, 21,* 152.

Winkler, J. H. (1748). Novum reique medicae utile electricitatis inventum. *Philosophical Transactions of the Royal Society, 45,* 262–271.

Winkler, J. H. (1749). A new discovery in electricity, of use in medicine. *Gentleman's Magazine, 19,* 450–453.

19
Therapeutic Attractions: Early Applications of Electricity to the Art of Healing

Paola Bertucci

In the past few decades a number of studies dealing with eighteenth-century natural philosophy in England have pointed out its inextricable links with spectacle and public display. The commodification of cultural products, which was one of the main features of the Enlightenment, extended to science and scientific instruments, textbooks, and demonstrations, as well as to medicine. Pivotal works by Roy Porter have indelibly portrayed the vibrant marketplace in which medical practitioners operated. Even when they had a formal degree, "regular" healers had to compete both with "irregulars" and with a widespread culture of self-treatment (Porter, 1985, 1990, 1995; Porter & Porter, 1989; Schaffer, 1983; Stewart, 1992). In such competitive arena recently invented therapies attracted the attention of both patients and practitioners. From the 1740s onward, "medical electricity" was among the most attractive ones. The term indicated the applications of electric shocks and sparks to the treatment of various diseases, in particular palsies and "nerve disorders."

Electrical healing was first presented to the eighteenth-century public as a branch of experimental philosophy (Bertucci, 2001a). This essay analyzes the early diffusion of medical electricity, setting it in the context of the experimental culture from which it emerged. I deal with a relatively short span of time – the few decades during which almost instantaneously medical electricity came to be practiced in different European states – and I highlight the role played by itinerant demonstrators and instrument-makers in spreading what would soon become a fashionable, though controversial, healing practice.

Spectacular Treatments

It is commonly held that it was the invention of the Leyden jar, in the mid-1740s, that triggered wide interest in the new science of electricity (Hackmann, 1978; Heilbron, 1979). However, it is difficult to make causal connections among the events that in the span of a few years made electrical science the craze of the century. The Leyden jar became widely known in 1746, the same year in which most electricians began to investigate systematically the healing properties of the electric matter. The possibility that electrical machines, generally employed for entertaining paying crowds with amusing demonstrations, could also provide new ways of curing long standing diseases began to be discussed around 1744, mostly in Germany. The country was homeland to a group of electricians who envisaged the potential of the new science for attracting powerful patrons. The Wittenberg professor Georg Matthias Bose was a main actor in the transformation of electricity from an academic branch of natural investigation into a fashionable amusement. If, during the 1730s, both Stephen Gray at the Royal Society of London and Charles Du Chisternay Dufay at the Académie des Sciences in Paris devoted themselves to the investigation of the electric matter, it was Bose and other "Saxon virtuosi" that, as the *Gentleman's Magazine* acknowledged in 1745, made electricity a subject à-la-mode, with "princes willing to see this new fire that man produced from himself, and which did not descend from heaven." (*Gentleman's Magazine*, 1745)

In his several writings Bose highlighted the spectacularity of electrical experiments, providing also a theory of "male" and "female" fire that would explain attractions, repulsions, and luminous appearances. He also published a two-book poem dedicated to the most entertaining electrical experiments, such as the "beatification" (which he invented), the "flying boy," the circuit experiment, and the most exciting of all, the "electric kiss" (or *Venus electrificata*) (Figs. 1–3):

Once only, what temerity!
I kissed Venus standing on pitch.
It pained me to the quick. My lips trembled
My mouth quivered, my teeth almost broke
(Bose, 1754; Heilbron, 1979, p. 267n)

All these experiments required that spectators would directly experience the passage of electricity through their bodies. If their active involvement

was one of the most thrilling aspects of electrical demonstrations, it was because of the inclusion of the human body in such amusing experiments that the physiological responses to the passage of electricity began to be noticed.

FIGURE 2. The "flying boy" experiment. From Nollet, *Essai sur l'électricité des corps*, Paris, 1746

FIGURE 1. Bose's "beatification." From Benjamin Rackstraw, *Miscellaneous observations, together with a collection of experiments on electricity*, London, 1748

FIGURE 3. "The electric circuit" experiment. From Windler, *Tentamina de causa electricitatis*, Neapoli, 1747

Around 1744, Johann Gottlob Krüger, professor of medicine and philosophy at Halle, reported that all the people who subjected themselves repeatedly to the "electric kiss" presented small red spots on their hands, a result that would disappear after several hours (Krüger, 1744, p. 544). Krüger noticed that electrification produced involuntary muscular motion and anticipated that electricity could be therefore applied to the treatment of palsies. Together with his pupil Christian Gottlieb Kratzenstein, he was the first to claim that electricity might be applied to medical purposes. Kratzenstein, in particular, measured the pulse before and after electrification and reported a constant increase, amounting to a third of the normal rate.

Aware of his colleagues' results, Bose conceived of an experimental demonstration that seemed to provide a mechanical explanation of the increase of the pulse and of other physiological responses to the passage of electricity. He prepared a metallic siphon and put some water in it. In normal condition, the liquid would intermittently drop off, but when the metal was electrified, the water flew in full stream (Bose, 1744–1745, p. 420), [Fig. 4, bottom right]. Bose also noticed that blood issuing from an opened human vein "streams off more quickly when the man is electrified [. . .] blood drops appear luminous like fire" (Nollet, 1750, p. 118).

As a consequence, he concluded that the electrification of the human body increased blood circulation, insensible perspiration, and the pulse. All diseases that may benefit from such effects could be treated by electricity. Kratzenstein, on his part, reported that electricity produced tiredness and concluded that it would be helpful to all whose

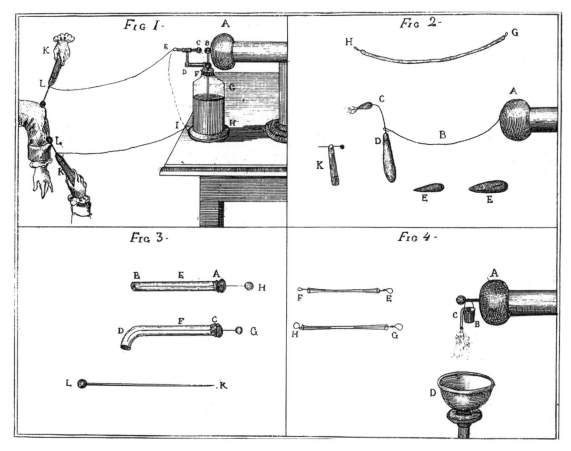

FIGURE 4. Table from Tiberius Cavallo's *Essay on medical electricity*, London, 1781. Fig. 1 (in the plate) shows the medical bottle, Fig. 4 represents the "electrified siphon"

"riches, sorrows, and worries prevent them from closing their eyes at night." (Kratzenstein 1745 in Snorrason, 1974, p. 37).

Kratzenstein's electric treatments soon became known in the rest of Europe and began to be replicated in Copenhagen, Vienna, Uppsala, Stoccolma, Rouen, Lipsia, where electricians engaged in applying medical electricity and reported positive results. News of prodigious cures was published in popular magazines more often than in academic publications. The possibility that electricity might have healing virtues added to the attractiveness of the spectacular demonstrations; Bose's "experiment of the pulse" became one of the attractions offered during electrical soirées. But the medical applications of electricity demonstrated also that the new science was useful as well as delightful; indeed, "electricity made useful" became synonym with medical electricity in a series of booklets on the healing properties of the electric matter (Cavallo, 1780; Lovett, 1756; Wesley, 1760).

In 1745, during the War of the Austrian Succession (1740s), the Saxon army physician Christian Xavier Wabst – later to become physician to the Empress Maria Teresa – settled in Venice, where he began to astonish local audiences with the spectacular effects of his electrical machine. It was the first time that the instrument reached Italian audiences and his demonstrations initiated a new fashion south of the Alps. As an army physician, Wabst was obliged by his Court not to leave Venice, but his terrific shows attracted spectators also from other towns. A number of them were looking for entertainment, others wanted to learn how to replicate the most recent experimental discoveries.

At the University of Padua, the professor of experimental philosophy, Giovanni Poleni, was particularly keen on public demonstrations. His "theatre of experimental philosophy," which was inaugurated in 1738, was the place where he offered his academic lectures, as well as the venue where he carried out experimental demonstrations for prestigious visitors. In an unpublished diary, he listed all the experiments he performed in front of cardinals, administrators, military captains, princes, and princesses.[1] When he

got to know of Wabst's electrical performances, he became anxious to learn how to reproduce them in his theatre. Since Wabst could not go to Padua, Poleni sent his assistant Vitaliano Donà to Venice. The University refunded the fee of 24 *soldi* required by Wabst and, in 1747, Poleni began to perform electrical experiments during the public demonstrations he offered to his audiences (Salandin & Pancino, 1987, p. 654).

Whereas Wabst could not leave Venice, several itinerant demonstrators, mostly coming from abroad, traveled through the Italian peninsula showing the marvels of electricity to local aristocrats, academics, and the simply curious. As a Florentine literary magazine reported, in 1747, there were numerous "practical experimenters, or circumforaneous philosophers, who have found their way of living on the electric virtue by traveling the world with electrical machines coarsely made by themselves, as we have seen all over Italy." (*Novelle letterarie pubblicate in Firenze*, 1747, p. 654)

The "circumforaneous" (itinerant) philosophers announced their arrival by means of printed leaflets that listed their experiments. Locals would offer their own houses to host electrical soirées in exchange for a certain percentage on the entrance fee that each guest would pay. The shows would be replicated for one or two weeks in the same place. It was thanks to the activities of such itinerant demonstrators that electrical experiments became quickly known in the Italian states. Together with entertaining demonstrations, they spread the idea that sparks and shocks could be employed in the art of healing.

When the news of Bose's "experiment of the pulse" arrived to Padua, Poleni hastened to repeat it. In his theatre of experimental philosophy he involved his noble spectators, measuring their pulses before and after electrification, and included himself in the measurements. His results confirmed the increase predicted by Kratzenstein.

Italian aristocrats delighted themselves with electric shocks and attractions and, when news of the healing virtue of electricity spread in the peninsula, they engaged in putting it to trial. In Vicenza, a town in the Venetian mainland, the marquis Luigi Sale was one of the first to apply electricity as a healing therapy. He owned a large electrical machine in his villa and he corresponded with local electricians, such as the above mentioned Giovanni Poleni in Padua and Scipione Maffei in Verona, an

---

[1] Biblioteca Nazionale Marciana, Venice. Mss Lat, VIII, 158.

antiquary who in 1747 published accounts of his electrical experiments. After reading about the newest application of the electric matter as a medical remedy, Sale decided to test medical electricity on his servant Giambattista Negretti, whose unusual problem was quite well known in the area. Since the age of eleven, Negretti had suffered from a "most extravagant disease," which presented itself each year at springtime. In the middle of the night, he would get up from his bed and, still sleeping, he would look for food, drinks, money . . . he would even go to the local tavern without ever waking up (Maffei, 1747, p. 144).

The sleep-walker from Vicenza was a living marvel. Several booklets had been published on his case; they described his symptoms and the vain (sometimes cruel) attempts to wake him up. In February 1747, Sale began to electrify Negretti. The result was immediate. A few painful sparks drawn from his body – without making use of the Leyden jar – sufficed to wake up the irreducible sleep-walker, who declared himself extremely happy, thanked the marquis with "tears of consolation" and blessed the electrical machine.[2]

The cure of the sleep-walker from Vicenza circulated widely, but mainly in private circles. The medical virtues of electricity were still discussed in the Republic of letters, yet Sale and his entourage were not interested in taking part in an international debate. The cure of the sleep-walker can be regarded as an instance of the inclusion of electricity in the aristocratic culture of curiosity, with its love for prodigies, marvels, and strange facts.

Not far from Vicenza, there were other electricians who were pursuing different goals. A few months after Wabst's departure, an ambitious gentleman who lived in Venice, Gianfrancesco Pivati, claimed to have discovered a new method of applying electricity to cure instantaneously inveterate diseases. Having obtained the support of the prestigious Institute of Sciences of Bologna, Pivati published a short booklet in which he presented his invention, the "medicated tubes," and described the prodigious

cures he obtained by this means. Among the people he healed, there was the bishop of Sebenico, a 75-year-old man who had been suffering of podagra and chiragra (gout) for decades (Pivati, 1747). The prodigious cure of the bishop, toured the learned world.

Pivati's method was quite different from others. His "medicated tubes" were sealed glass cylinders filled with healing substances; when the glass was electrified, according to Pivati, the medicines evaporated through the pores of the glass and diffused themselves in the surrounding air. Patients would only have to breathe in the healing atmosphere to be cured.

Such news in our electrical experiments! How perfect! Such a success! [. . .] We substantiated them in such a way that made us think we were not guessing, and as to myself, I believe we surpassed everyone in matters electrical.[3]

With such words, the Secretary of the Bologna Institute of Sciences communicated Pivati's prodigious cures to his colleague Jean Jallabert in Geneva. In 1748, the Bologna Institute sponsored the publication of Giuseppe Veratti's *Physico-medical observations on electricity*, which confirmed Pivati's results. Only one year later, however, the physician Gianfortunato Bianchini discredited Pivati's tubes in his *Essay on medical electricity* published in Venice. In a span of few months, all over Europe several other electricians published their opinions about the medicated tubes. By making public the results of their trials, both supporters and opponents of Pivati's method spread news about the possibility that electricity might have healing virtues. I have elsewhere analyzed the controversy on the medicated tubes that climaxed with Jean Antoine Nollet's journey to Italy in 1749; it may be enough to note here that its great impact in the Republic of Letters triggered further work on the healing virtues of electricity and increased the public's interest toward electrical treatments (Bertucci, 2005).

The proliferation of presumed cures by means of electricity engendered caution among electricians working in academic contexts, especially the Abbé

---

[2] Bibliothèque Municipale, Nîmes. Vol. 9352, "Copia della lettera del Sig. Marchese Luigi Sale di Vicenza al Sig. Gian Ludovico Bianconi Medico di S.A. il Principe e Vescovo di Augusta" (12 March 1747). I am grateful to Ivano Dal Prete for attracting my attention to this document.

[3] Bibliothèque Publique Universitaire, Genève. Collection Jallabert, Correspondance, SH 242, ff. 186 (Zanotti to Jallabert, 1 July 1748).

Nollet at the Académie Royale des Sciences in Paris (the leading authority in electrical matters at the time) and William Watson at the Royal Society. Although the Royal Society refrained from publishing about medical electricity, the British public came to know of the healing virtues of electricity by means of itinerant lecturers. By the mid-eighteenth century traveling demonstrators were numerous in the country; it was by their means that Newtonian experimental philosophy had become so widely popular (Morton & Wess, 1993; Schaffer, 1983; Stewart, 1992).

Electricity was soon incorporated in the repertoires of popular demonstrators such as Benjamin Rackstrow, Benjamin Martin, and James Ferguson and, exactly like in other European countries, the spectacular display of shocks, sparks, and attractions became feverishly requested by the upper classes (Millburn, 1976, 1983, 1988). In 1770, James Ferguson published *An introduction to electricity*, a booklet clearly tailored to his audience. Although he was mainly interested in mechanical instruments, the popularity of electrical experiments induced him to add electricity to the other subjects of his lectures. His *Introduction* described simple experiments and included medical electricity as one of the wondrous effects of the electric fire. Hence, the electrical machines that he sold were equipped with accessories for the performance of both electrical experiments and electrical treatments.

Both Richard Lovett and John Wesley, the first authors who wrote texts on medical electricity in English, declared that they got to know the properties of electricity in the course of public lectures they were attending. Lovett was a lay-clerk at Worcester Cathedral, whereas Wesley was an Anglican preacher. Their advocacy of electrical healing, informed by their theological views on Christian piety, highlighted its cheapness and universality: electrical machines were easy to make, their cost could be shared among neighbors, anyone could learn how to administer electrical treatments (Bertucci, 2006).

Instruments and Therapies

In England, "electricity made useful" – as medical electricity was called by Wesley and Lovett – began to be advertised as a therapy that could meet the needs of the lower classes of society. In 1746, John Reddall, who corresponded with Lovett, organized a course of lectures on medical electricity in London that attracted more than a hundred people a day, even from the countryside. Reddall was convinced that "electricity in a little time will be generally practis'd" and that the several successful cures would bring it "into universal Practice" (Lovett, 1760, p. 35). Wesley included medical electricity among the free treatments offered by his Dispensaries in London and published lists of people who had been healed with the help of electrical machines. John Read, a cabinet-maker who had been healed by electrical treatments in Wesley's Dispensaries, devoted himself to making electrical machines especially designed for medical purposes (Wesley, 1760, p. 60). He put his manual skills to work and designed low-price portable electrical machines that could be bought by people living in the same neighborhood:

[Read] has just invented a smaller One, that will take to Pieces, and pack up in a Box of about a Foot Square, and is endeavouring to reduce them to a very low Price, in order to make them as public as possible. (Lovett, 1760, p. 40)

Read's machines were acknowledged in Joseph Priestley's *History and Present State of Electricity* (London, 1767) as especially practical for medical purposes (Fig. 5).

Instrument-makers in England played a crucial role in spreading medical electricity. In a context in which experimental philosophy was firmly

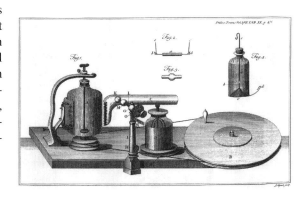

FIGURE 5. John Read's electrical machine with Timothy Lane's discharging electrometer. From Timothy Lane, "Description of an electrometer invented by Mr Lane," *Philosophical Transactions of the Royal Society* (1767), 57, 451–460, table XX, p. 431

grounded in a vibrant marketplace, medical electricity was for them business opportunity. This was specific to the English case, whose instrument-makers were world famous and their products were requested from customers all over the world. They soon realized that the medical applications of electricity would represent an effective means of advertising their goods. John Neal, a London instrument-maker whose shop was in Leaden Hill, specialized in the making of electrical machines. In 1747, he published *Directions for Gentlemen, who have electrical machines, how to proceed in their experiments*, in which he invited readers to submit cases of electrical cures, which would be included in a forthcoming booklet on medical electricity (Neale, 1747, p. 76).

The fact that medical electricity was profitable can be inferred from the analysis of a manuscript notebook that belonged to John Fell, a surgeon at Ulverstone, who, in the last quarter of the eighteenth century, decided to specialize in medical electricity. At the time there was no formal training to become medical electrician, so he got practical information from the several textbooks that had been published in the course of the years. He also got in touch with apothecaries, instrument-makers, and surgeons who practiced medical electricity in London. His expense-book shows that within a few months he completely covered all he spent to set up his laboratory, which he equipped with the best available instruments.[4]

The medical applications of electricity generally worried the academic establishment. In Italy, the physician Giovanni Bianchi was outraged by the craze for electricity that, in his opinion, congested the minds of almost all Italian philosophers. William Watson, on his part, regarded electricity an idle activity, with no useful application:

If it should be asked to what useful purposes the effects of electricity can be applied, it may be answered, that we are not as yet advanced in these discoveries as to render them conducive to the service of mankind. (Watson, 1746, p. iv)

In France, the leading authority in electrical matters, Jean Antoine Nollet, spent a lot of his time testing the healing virtues of electricity upon a group of paralytic soldiers at the Hôpital des Invalides. He was the first to engage in systematic

trials carried out in collaboration with the physician Morand and the head surgeon De La Sone, but he regarded the results they obtained "too uncertain to be worth mentioning." (Nollet, 1751, p. 19)

The first report that was taken seriously by the academic world came from Geneva in 1748. The physics professor Jean Jallabert, a respected electrician, published the account of his successful cure of a paralytic man called Nogues. On the 26th of December 1747, he began the electrical therapy in front of several witnesses, among whom there were a number of members of the Faculty of Medicine and Surgery (Jallabert, 1749, p. 173). Treatment consisted in drawing sparks from the paralyzed arm for 2 hours in the morning and 2–3 hours in the afternoon, for a month and a half. Nogues was also subjected to electric shocks at least four or five times a day. Jallabert notified his success to academies all over Europe, and above all to his correspondent and friend Nollet in Paris.[5]

Prompted by Jallabert's report, Nollet resumed his trials at the Hôpital des Invalides, but again, after months of failed attempts, he had to conclude that the movements restored by electrification were always involuntary (Morand & Nollet, 1753). Notwithstanding his own results, Nollet was aware that, however controversial, medical electricity increased public interest in the new science of electricity, as well as demand for booklets, demonstrations, and instruments. As an author, public demonstrator, and instrument-maker, he was unsurprisingly pleased. On returning thanks to Jallabert for his report on Nogues' cure, he acknowledged that it was "thanks to you that electricity has begun to sell really quickly." (Benguigui, 1984, p. 161)

Similar tones were used by another French academic, Boissier de Sauvages. He was Royal professor of Medicine at Montpellier and corresponded with Jallabert, who gave him advice on how to apply electricity to medical purposes. In a few months, De Sauvages cured a number of paralyzed patients by means of electricity. Among them, a 70-year-old beggar, whose left arm and right leg were completely motionless, received such benefit and in such a short time, that his wife believed he was granted a miracle (Jallabert, 1749, p. 367).

---

[4] Wellcome Library, London. MS 1175.

[5] Archives de l'Académie des Sciences, Paris. *Procès Verbaux*, Tome 67 (1748), ff. 43–44 (3 February 1748).

On the wave of enthusiasm, De Sauvages wrote to Jallabert that "thanks to your instructions electricity has become fashionable [;] in this town everybody wants to be electrified."[6] In Montpellier, electrical healers began to have portable machines built so as to be able to offer electrical treatments in the patients' homes, a practice that would become customary in England.

The rumor that such reports made in the learned world prompted even the skeptical William Watson to test the effects of electricity. In 1756, he published in the *Philosophical Transactions* the report of his successful cure of a case of tetanus by means of electricity. In the second half of the eighteenth century, in Russia, Sweden, the Netherlands, as well as in France and Britain, reports of electrical cures began to appear in academic publications. Yet the growing attention of the academic world toward this new healing practice did not entail approval or consensus. On the contrary, electrical healing retained its controversial character even in the following centuries (Bertucci & Pancaldi, 2001). But one result of the increased interest, both at the academic and popular level, was that treatments and instruments began to be more standardized.

In 1756, the first English textbook on medical electricity explained how to apply three methods of electrical therapy: simple electrification, drawing of sparks from the patient, and electric shock. The three methods relied on the conception of the healthy body as a container of a natural amount of electricity, whose accidental variation provoked disorders. According to the author, Richard Lovett, electricians should begin any treatment with simple electrification: the patient, standing on an insulated chair, was connected to the metallic conductor of the electrical machine; this way, lost quantities of electricity would be replaced. By drawing sparks from the patient's body, on the contrary, excesses of electricity would be taken away; placed on an insulated chair, the patient was connected to the electrical machine, while the operator drew sparks from him or her by means of a pointed metallic rod which he kept from its insulated end. The electric shock, used to remove obstructions of bodily fluid, was produced by discharging a Leyden jar through a selected part of the patient's body; the inner and

the outer coating of the jar would be connected to the extremities of the limb, or the head, or other parts of the body, by means of insulated conductors (Fig. 6). In spite of this simple presentation, in their practice, medical electricians chose therapies without referring to theoretical conceptions. The real efficacy of simple electrification was controversial and most often practitioners applied only sparks and shocks.

In the 1750s, a kit of medico-electrical instruments was made of a small number of essential objects: a portable machine, similar to those made by Read, Leyden jars of different sizes, metallic chains to convey the electric fluid to the part of the body affected by the disorder, and an insulating stool for the patient to stand on while receiving treatment. With these instruments and with contemporary booklets on medical electricity, anyone could try the healing properties of the electric fire.

A new generation of medico-electrical instruments would appear in the following decade. In 1765, the London apothecary Timothy Lane, who worked also as a medical electrician, invented a discharging electrometer, meant to regulate the intensity of the electric shock (Fig. 5). Lane's electrometer was incorporated in Leyden jars, that came to be called "medical bottles," or directly in the prime conductors of the electrical machine (Fig. 4, top left).

If we look at the catalogues of instrument-makers, we realize that, by the 1780s, the range of instruments especially designed for electrical treatments had consistently increased. The treatments of specific disorders such as toothaches, deafness, and problems of the eye, led to the design of new tools for conveying the electric fluid from the electrical machine to the body. They consisted of a metallic wire fixed inside a glass tube, normally 15 cm in length, and two and a half in diameter, terminating with a brass knob outside the tube (Fig. 4, bottom left). During the treatment, the patient would be placed on an insulated stool and connected to the prime conductor of the machine. Subsequently, the instrument would be directed toward the affected part of the patient's body, while the operator would bring the knuckles of his fingers at a small distance from the brass knob. As a result of these operations, a series of sparks issued simultaneously between the patient and the tube and between the knob and the operator's knuckle.

---

[6] Bibliothèque Publique Universitaire, Geneva, MS 82: Collection Jallabert f. 41 (15 August 1746).

to face the title Page

J. Lodge sculp

FIGURE 6. Frontispiece of George Adams's *An Essay on Electricity explaining the Theory and Practice of that useful Science; and the mode of applying it to Medical Purposes*, London, 1785

This type of instrument proved versatile. Not only could it be used for cases of deafness or eye-disorders, but it could also be employed to produce a new electrical treatment, called the "electric stream." This was literally a stream of little electric sparks emitted by a wooden or metallic point connected to the machine. The operator could select from wooden or metallic points of various sizes, according to the intensity of the stream he wished to produce (Fig. 4, top right). The wire could be bent according to the therapy, and the electric stream could then be applied to any part of the body. The method could be employed together with "drawing sparks through a piece of flannel," another new therapy that was particularly recommended for cases of rheumatism. The flannel was placed over the part of the body under treatment, and the insulated patient was connected to the conductor. The operator would then use a wire, terminating at the extremities in a sharp point and a brass knob. He would place the point into contact with

the flannel, while he quickly shifted the knob from place to place, producing a great number of very small sparks (Cavallo, 1780, p. 46).

Medico-electrical practitioners introduced several variations on this theme. The instrument-maker to the King, George Adams, designed a new instrument that could be employed for a therapy that he called "gentle and refreshing stream." The instrument consisted of a glass tube, terminating in a capillary filled with rose water, or any other perfumed fluid. The connection of the tube with the machine produced a light shower of the scented fluid, which passed from the capillary to the patient.

This constellation of new instruments suggests that, in spite of academic skepticism, medical electricity was quite popular in the medical marketplace. A range of disorders were treated by electricity. Apart from palsies, rheumatisms, and eye–ear–teeth problems, tumors, inflammations, intermittent fevers, nervous disorders, headaches, ulcers, the gout, and St Anthony's fire, were also treated by means of electricity. In the late 1770s, the electric shock became common treatment for the "removal of female obstructions" (or interruption of the menses), as indicated in a booklet by the London surgeon Thomas Birch (Birch, 1780).

From the point of view of instrument-makers, the success of medical electricity climaxed with the patent awarded in 1782 to Edward Nairne for his medico-electrical machine (Fig. 7, Bertucci, 2001a). By then medical electricity had become very popular. In the booklet in which he gave directions on how to operate the patent machine, Nairne had to remind his customers that he was not a medical electrician.

It is also significant that, in 1785, George Adams chose as the frontispiece for his *An essay on electricity* explaining the theory and practice of that useful science; and as the mode of applying it for medical purposes, an image representing the application of electricity to the forearm of a young girl. The unfinished print showed the big electrical apparatus in full detail, while the human figures (the young girl, the operator, and the woman who accompanies the girl) were only sketched (Fig. 6). The image, a visual introduction to the contents of the book, highlighted the usefulness of electricity for medical purposes and drew the reader's attention to the electrical apparatus. Adams, like other instrument-makers in the same period, saw medical electricity as an activity that could stimulate new demand of electrical instruments.

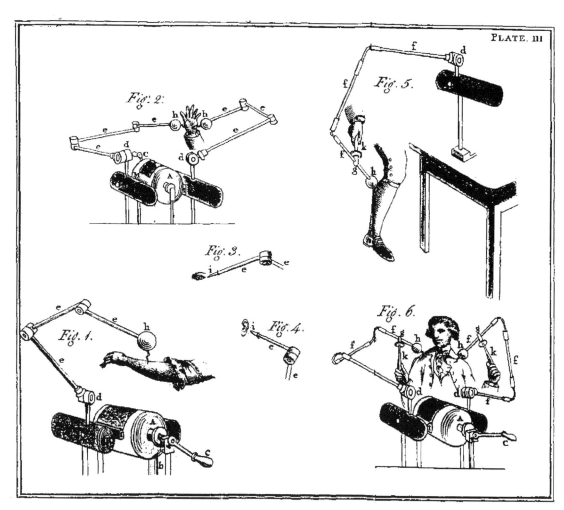

FIGURE 7. Edward Nairne's patent medical electrical machine. From Edward Nairne, *The Description and Use of Nairne's Patent Electrical Machine*, London 1782

Conclusion

If medical professionals were skeptical about a therapy whose success relied mostly on patients' assessments, medical electricity throve thanks to the variegated eighteenth-century market for health matters. In 1780 medical electricity became one of the most flamboyant attractions offered in James Graham's Temple of Health. Graham was an imaginative quack and the "Temple" that he set up in the fashionable quarter of Adelphi in London consisted of ten large rooms, in which several alternative remedies were offered to the rich as well as the poor. Relying on contemporary society's dissatisfaction with medical professionals, Graham made the entrance in the Temple a "total" experience for his patients; all the senses were excited by sounds, smells, colors; healthy goddesses of health recited odes to Apollo, while music played in the background. In his theatrical scenery for the cult of health, with related rituals, he displayed "the largest and most elegant medico-electrical apparatus in the world." (Graham, 1778; Porter, 1982)

Graham's use of electricity in his eccentric enterprise was multifaceted. Rather than actually using electrical machines to administer electricity to his patients, he exploited the fashion enjoyed by electricity as a further extravaganza for his healing centre; the "largest electrical apparatus in the world" was on display, rather than in use, in the Temple, where electrical vapors wrapped up the patients,

gently pervading the whole system with a copious tide of that celestial fire, fully impregnated with the purest, most subtle, and balmiest parts of medicines, which are extracted by, and flow softly into the blood and nervous system, with the electric fluid, or restorative aetherial essences. (Graham, 1780, p. 29)

The actual practice of medical electricity was looked upon by Graham as an "ignorant and improper application of this awful element" in the "hands of ignorant and rash people," such as barbers, surgeons, tooth-drawers, apothecaries, or common mechanics "turned into electrical operators," who were sprouting "in almost every street in this great metropolis." (Graham, 1780, p. 29)

Graham's enterprise was short lived; he went bankrupt two years later. However, the ephemeral story of the Temple of Health highlights the wide variety of eighteenth-century electrical healers and the heterogeneity of their views on the role of electricity in the animal economy. Whereas Graham was not interested in the nature of the "electric fluid" and its action on the human frame, other eighteenth-century electrical practitioners asked important questions on the relationship between electricity and life. Some believed that the electric fluid was produced by the brain and that, being one and the same thing as the "nervous fluid," it circulated in the nerves. According to this theory, palsies were caused by obstructions of the nervous fluid that could be removed by electric shocks or, in other words, by forcing the electric fluid through the blocked vessels. Others maintained that electricity exerted only a mechanical action on the human frame, accelerating the natural motion of the vital fluids (Rowbottom & Susskind, 1984). In the last quarter of the eighteenth century experimental enquiries into the nature of muscular motion and on the electric organs of fish such as the torpedo or the gymnotus electricus opened up new fields of electrical researches that proved crucial to shaping the neurosciences in the following centuries (Piccolino, 2003; Piccolino & Bresadola, 2003). However, the success of medical electricity in the public domain is to be understood by taking into account the intersection of experimental philosophy, spectacle, and business that it represented. Eighteenth-century electricity was sensational and its therapeutic applications, independently of the theoretical convictions of electrical healers, highly attractive – and financially profitable.

References

An historical account of the wonderful discoveries made in Germany &c. concerning electricity. (1745). *Gentleman's Magazine, 15*, 193–197.

Benguigui, I. (1984). *Théories électriques du XVIIIe siècle. Correspondence entre l'abbé Nollet (1700–1770) et le physicien genevois Jean Jallabert (1712–1768)*. Genève: Georg.

Bianchini, G. (1749). *Saggio d'esperienze intorno alla Medicina Elettrica*. Venezia: Pasquali.

Bertucci, P. (2001a). A philosophical business: Edward Nairne and the patent medical electrical machine (1782). *History of Technology, 23*, 41–58.

Bertucci, P. (2001b). The electrical body of knowledge: Medical electricity and experimental philosophy in the mid-eighteenth century. In P. Bertucci & G. Pancaldi 2001 (pp. 43–68).

Bertucci, P. (2005). Sparking controversy: Jean Antoine Nollet and medical electricity south of the Alps. *Nuncius, 20*, 153–187.

Bertucci, P. (2006). Revealing sparks. John Wesley and the religious utility of electrical healing. *British Journal for the History of Science, 39*, 341–362.

Bertucci, P., & Pancaldi, G. (Eds.). (2001). *Electric bodies. Episodes in the history of medical electricity. Bologna Studies in the History of Science*. Bologna: CIS, University of Bologna.

Birch, J. (1780). *Considerations on the efficacy of electricity in removing female obstructions*. London: Cadell.

Bose, G. M. (1744). *Tentamina Electrica*. Wittenberg: Ahlfeldium.

Bose, G. M. (1744–1745). Abstract of a letter from Monsieur De Bozes. *Philosophical Transactions of the Royal Society, 43*, 419–421.

Bose, G. M. (1754). *L'électricité, son origine et ses progrès, poème en deux livres*. Leipsic: Lankisch.

Cavallo, T. (1780). *An essay on the theory and practice of medical electricity*. London.

Della formazione dei fulmini trattato del marchese Scipione Maffei. (1747). *Novelle letterarie pubblicate in Firenze, 8*, 649–656.

Ferguson, J. (1770). *An introduction to electricity*. London.

Graham, J. (1778). *The general state of medical and chirurgical practice exhibited*. Bath.

Graham, J. (1780). *Medical transactions of the temple of health*. London.

Hackmann, W. (1978). *Electricity from glass*. The Netherlands: Alphen aan den Rijn.

Heilbron, J. (1979). *Electricity in the 17th and 18th century. A study of early modern physics*. Berkeley: University of California Press.

Jallabert, J. (1749). *Expériences sur l'électricité avec quelques conjectures sur la cause de ses effects*. Genève: Barillon & Fils.

Kratzenstein, C. G. (1745). *Abhandlung von dem electricitat in der Urznenwissenschaft*. Halle: Hemmerde.

Krüger, J. G. (1744). *Der Weltweisheit und Artzneygelahrheit Doctors und Professors auf der Friedrichs Universität Naturlehre nebst Kupfern und vollständigem Register*. Halle: Hemmerde.

Lane, T. (1767). Description of an electrometer invented by Mr Lane. *Philosophical Transactions of the Royal Society, 57*, 451–460.

Lovett, R. (1756). *Subtil medium prov'd*. London: Hinton.

Lovett, R. (1760). *The Reviewers review'd*. London: Lewis.

Maffei, S. (1747). *Della Formazione de' Fulmini*. Verona: Tumermani.

Millburn, J. (1976). *Benjamin Martin: Author, instrument-maker and country showman*. Leyden: Noordhoff International.

Millburn, J. (1983). The London evening courses of Benjamin Martin and James Ferguson, eighteenth-century lectures on experimental philosophy. *Annals of Science, 40*, 437–455.

Millburn, J. (1988). *Wheelwright of the heavens: The life & work of James Ferguson, FRS*. London: Vademecum.

Morand, & Nollet. (1753). Expériences de l'Electricité appliqué a des Paralitiques. *Mémoires de l'Académie des Sciences de Paris pour l'année, 1749*, 29–38.

Morton, A. Q., & Wess, J. A. (1993). *Public and private science. The King George III collection*. Oxford: Oxford University Press in association with the Science Museum.

Neale, J. (1747). *Directions for gentlemen, who have electrical machines, how to proceed in their experiments*. London: printed for the author.

Nollet, J. A. (1750). Conjectures sur les causes de l'électricité des corps. *Mémoires de l'Académie des Sciences de Paris pour l'année, 1745*, 109–151.

Nollet, J. A. (1751). Observations sur quelques nouveaux phénomènes d'électricité. *Mémoires de l'Académie des Sciences de Paris pour l'année, 1746*, 1–23.

Piccolino, M. (2003). *The taming of the ray. Electric fish research in the enlightenment from John Walsh to Alessandro Volta*. Firenze: Olschki.

Piccolino, M., & Bresadola, M. (2003), *Rane, Torpedini e Scintille. Galvani, Volta, e l'elettricità animale*. Turin: Bollati Boringhieri.

Pivati, G. (1747). *Della Elettricità Medica Lettera al Celebre Signore Francesco Maria Zanotti*. Lucca.

Porter, R. (1982). The sexual politics of James Graham. *British journal for eighteenth-century studies, 5*, 201–206.

Porter, R. (Ed.) (1985). *Patients and practitioners: Lay perceptions of medicine in pre-industrial society*. Cambridge: Cambridge University Press.

Porter, R. (1990). *Health for sale: Quackery in England* (pp.1660–1850). Manchester : Manchester University Press.

Porter, R. (Ed.) (1995). *Medicine in the enlightenment*. Amsterdam: Rodopi.

Porter, D., & Porter, R. (1989). *Patient's progress. Doctors and doctoring in eighteenth-century England*. Oxford: Polity.

Rowbottom M., & Susskind, C. (1984). *Electricity and medicine. History of their interaction*. San Francisco: San Francisco University Press.

Salandin, G. A., & Pancino, A. (1987). *Il "teatro" di filosofia sperimentale di Giovanni Poleni*. Trieste: Lint.

Schaffer, S. (1983). Natural philosophy and public spectacle in the eighteenth century. *History of Science, 21*, 1–43.

Snorrason, E. (1974). *C.G. Kratzenstein and his studies on electricity during the eighteenth century*. Odense: Odense University Press.

Stewart, L. (1992). *The rise of public science. Rhetoric, technology and natural philosophy in Newtonian Britain* (pp. 1660–1750). Cambridge: Cambridge University.

Veratti, G. (1748). *Osservazioni Fisico-Mediche intorno all'Elettricità*. Bologna: Della Volpe.

Watson, W. (1746). *Experiments and observations tending to illustrate the nature and properties of electricity*. London.

Wesley, J. (1760). *The Desideratum, or electricity made plain and useful*. London: Flexney.

20
John Wesley on the Estimation and Cure of Nervous Disorders

James G. Donat

The Reverend John Wesley (1703–1791) is historically remembered as the founder of the Methodist Church. Largely forgotten is his long-standing involvement in medical matters, as an advisor of healthy regimens and collector of helpful remedies. Still active at 79 years of age he publishes the anonymous pamphlet, *An Estimate of the Manners of the Present Times* (1782), and at 81 he writes "Thoughts on Nervous Disorders" (1784), sending it to press in the *Arminian Magazine* for January–February 1786. In both of these compositions, the aged Wesley is aware of the growing problem of nervous complaints in English society, with an eye to its effect on Methodist people, who are part of that society: *An Estimate* especially focuses on the numbers; and "Thoughts" on the cause and treatment of said disorders.

These complaints are called nervous because at the time Wesley was writing, they were commonly associated with the nervous system. Many medical writers in the eighteenth century presumed this connection (Cheyne, 1733; Robinson, 1729; Smith, 1768; Whyte [Whytt], 1765), while others, especially in the first half of the century, viewed them as the disorders of other bodily systems, with different functional explanations (Blackmore, 1725; James, 1745, "Hypochondriacal Morbus"; Mandeville, 1730; Midriff, 1721; Purcell, 1707). Wesley, like other non-medical authors of essays, novels, plays and poetry, mirrors the contemporary fashions of medical thinking. Although the theoretical mechanics envisioned for these disorders vary from writer to writer, the terminology for these distempers is more or less the same throughout the century: spleen, vapours, hypochondriasis, hysteria, lowness of spirits, melancholy; with spleen and hypochondriasis usually being restricted to men; vapours and hysteria to women. In fact, these terms can also be found in the sixteenth to seventeenth centuries, though rarely associated with the nervous system.

These complaints are taken for granted as an unfortunate part of life. But when their frequency begins to rise, the reaction is one of alarm.

Back in 1675, Dr. Thomas Sydenham (1624–1689) observes, "Many die by violent deaths; but, with the exception of these, two thirds of our race die of fevers."[1] He is speaking about fevers as an acute form of disease. Dr. George Cheyne (1671–1743), however, alters this medical horizon with his publication of *The English Malady* in 1733, which focuses on the surge of another kind of medical phenomena, not acute like fevers, rather chronic in nature. He describes it as

… a Class and Set of Distempers, with atrocious and frightful Symptoms, scarce known to our Ancestors, and never rising to such fatal Heights, nor afflicting such Numbers in any other known Nation. These nervous Disorders being computed to make almost one third of the Complaints of the People of Condition in England. (1733, p. ii)[2]

---

[1] Sydenham, "Medical Observations Concerning the History and Cure of Acute Diseases [dated Dec 30, 1675]," Sect VI, Chap VII [Quinsy], #10, in *The Works of Thomas Sydenham*, I: 268; this is also affirmed in "Epistle I – Epidemic Diseases [dated Feb 12, 1679]," II, p. 22, para 52.

[2] "Spleen, Vapours, Lowness of Spirits, Hypochondriacal, and Hysterical Distempers" are disorders known to Sydenham: "Of all chronic diseases hysteria – unless I err – is the commonest; since just as fevers – taken with

And in 1807, Dr. Thomas Trotter (1760–1832) revises Cheyne's computation on the dramatic growth of this new category of complaint.

Sydenham at the conclusion of the seventeenth century, computed fevers to constitute two thirds of the diseases of mankind. But, at the beginning of the nineteenth century, we do not hesitate to affirm, that *nervous disorders* have now taken the place of fevers, and may be justly reckoned two thirds of the whole, with which civilized society is afflicted: from which we are led to believe, they were then, little known among the inferior orders. But from causes, to hereafter investigated, we shall find, that nervous ailments are no longer confined to the better ranks in life, but rapidly extending to the poorer classes. (1807, p. xvii)[3]

Wesley's essays are written a century after Sydenham's computation on fevers, half a century after Cheyne[4] puts nervous disorders on the medical map under the rubric of the "English Malady", and nearly a quarter of a century before Trotter envisions these nervous disorders "rapidly extending to the poorer classes".

Background to an Estimate of the Manners of the Present Times

Cheyne's assumption in *The English Malady* is that these nervous diseases, which he alleges to compose one third of medical complaints of the English nation, are a by-product of changes in the civilization, chiefly caused by the increased luxury and sloth enjoyed by the upper classes. In *An Estimate of the Manners of the Present Times*, Wesley admits that luxury and sloth[5] are on the increase in England, but no more than in France, Spain, Portugal, Italy, Germany, Holland, or the Northern Kingdoms. Moreover he denies their widespread and lasting effect, thus challenging the impression that the disorders brought about by these changes are so numerous, or at all a special characteristic of the English nation as a whole.

---

their accompaniments – equal two thirds of the number of all chronic diseases taken together, so do hysterical complaints (or complaints so called) make one half of the remaining third. As to females, if we exclude those who lead a hard and hardy life, there is rarely one who is wholly free from them – and females, be it remembered, form one half of the adults of the world. Then, again, such male subjects as lead a sedentary or studious life, and grow pale over their books and papers, are similarly afflicted; since however, much, antiquity may have laid the blame of hysteria upon the uterus, hypochondriasis (which we impute to some obstruction of the spleen or viscera) is as like it, as one egg is to another. True, indeed, it is that women are more subject than males. This, however, is not on account of the uterus, but for reasons which will be seen in sequel." Sydenham, *Works*, Vol. 2, Epistolary Dissertation to Dr. Cole [Jan 20, 1682]," II, p. 85, para 59. What is new in Cheyne's book is the reconfiguration of all the above-mentioned disorders under the category of nervous diseases, and, of course, observing their growing numbers.

[3] Historians of nervous disorders in England are prone to accept at face value, for lack of more accurate figure, the Cheyne–Trotter estimates of one-third/two-third. See Bynum (1985, pp. 91–93); Porter (1993, p. 590); and Logan (1997, p. 19). It may only be an odd historical coincidence, but this one-third/two-third tandem also appears in seventeenth-century fever literature, with Sydenham's two thirds being a revision of Willis' estimate of one third. See Bates (1981, p. 45, and fns 3–4).
[4] In Wesley's *Journal* entry for June 28, 1770, he says, "When I grew up, consequence of reading Dr. Cheyne, I

chose to eat sparingly, and drink water. This was another great means of continuing my health, till I was about seven-and-twenty." *JWJD* (22, p. 237). It is possible that he had a personal conversation with Cheyne on December 15, 1741. On March 12, 1742 he read Cheyne's last book, *Natural Method of Curing Diseases* (1742), commenting, "But what epicure will ever regard it? for 'the man talks against good eating and drinking'!" And in the last pages of the 1747 preface to *Primitive Physick*, #16, Wesley quotes extensively from the rules in Cheyne's *An Essay on Health and Long Life* (1724, pp. 17–18, 72–78, 87–88, 104–108, 143, 170–172).
[5] Wesley chooses the word sloth, which has a long history with roots back into the middle ages and the "seven deadly sins", which is a kind secularized version the word "acedia", or spiritual sloth. In the *Minutes of Several Conversations Between the Reverend Messieurs John And Charles Wesley. And Others* [better known as "The Large" Minutes]. Wesley uses the word "indolence" instead of "sloth", in his answer to Q. 28, "What reason can be assigned why so many of our Preachers contract nervous disorders? A. "…Indolence. Several of them used too little exercise, for less than they wrought at their trade. And this will naturally pave the way for many, especially nervous disorders." (1780, pp. 507–508). For a parallel confirmation of this point, see the anonymous author of *A Dissertation upon the Nervous System to show its Influence upon the Soul*, "But the most general cause of Nervous disorders is Indolence. The active and laborious are seldom troubled with them. They are reserved for the Children of ease and affluence." (1780, p. 42). For a literary analysis of the hazards of "indolence" see Thomson's poem, *The Castle of Indolence* (1748).

1. Some years ago an ingenious man published a treatise with this title. According to Him, the Characteristics of the English at present are *Sloth* and *Luxury*. And thus much we may allow, That neither the one nor the other ever abounded in England as they do at this day. With regard to Sloth, it was the constant custom of our Ancestors, to rise at four in the Morning. This was the stated hour, summer and winter, for all that were in health. The two Houses of Parliament met at five; *Hora quinta antemeridiana*, says their Journal. But how is it with people of fashion now? They can hardly huddle on their clothes, before Eight or Nine o'clock in the morning: perhaps some not before twelve. And when they are risen, what do they do?

   "They waste away
   In gentle inactivity the day!"

   How many are so far from working with their hands, that they can scarce set a foot to the ground? How many, even young, healthy men, are too lazy either to walk or ride? They must loll in their Carriages day by day: and these can scarce be made easy enough! And must not the *minor Gentry* have their coaches too? Yea, if they only ride on the outside. See here the grand cause (together with Intemperance) of our innumerable nervous complaints? For how imperfectly do either Medicines or the Cold Bath supply the place of Exercise? Without which the human body can no more continue in health than without Sleep or Food.

2. We allow likewise the abundant increase in *Luxury*, both in meat, drink, dress, and furniture. What an amazing profusion do we see, not only at the Nobleman's table, but at an ordinary City Entertainment? Suppose the Shoemakers' or Taylors' Company? What variety of Wines, instead of the good, home-brewed Ale, used by our Fore-fathers? What Luxury of Apparel (changing like the moon,) in the City and Country, as well as the Court? What superfluity of expensive furniture glitters in all our great men's Houses? And Luxury naturally increases Sloth, unfitting us for Exercise either of body or mind. Sloth, on the other hand, by destroying appetite, leads to farther Luxury. And how many does a regular kind of Luxury betray at last into Gluttony and Drunkenness? Yea, and lewdness too of every kind; which is hardly separable from them?

3. But allowing all these things, still this is not a true Estimate of the present Manners of the English Nation. ... But neither Luxury nor Sloth is either Universal or Constant in England.

4. Whatever may be the case of many of the Nobility and Gentry, (the whole body of whom are not a twentieth part of the nation,) it is by no means true, that the English in general, much less *universally*, are a slothful people. There are not only some Gentlemen, yea, and Noblemen, who are of the ancient stamp, who are patterns of Industry in their calling to all that are round about them, but it is undeniable that a vast majority of the middle and lower ranks of people are diligently employed from morning to night, and from the beginning to the end of the year. And indeed those who are best acquainted with other nations, will not scruple to testify, that the bulk of the English are at this day as diligent as any people in the Universe.

5. Neither is Sloth the Constant, any more than the Universal character of the English Nation. Upon many occasions even those that are most infected with it, arise and shake themselves from the dust. Witness the behaviour of those of the highest rank, when they were engaged in war. Did any one charge Sloth on the late Duke of *Marlborough*, or the Marquis of *Granby*? Witness the behaviour of many eminent men in the Militia, setting an example to all their troops! Yea, some of them were neither afraid nor ashamed to march on foot at the head of their men! ...

6. Neither is *Luxury*. For it is not Universal, no, nor General. The food which is used by nine tenths of our Nation is (as it ever was) plain and simple. A vast majority of the Nation, if we take in all living souls, are not only strangers to Gluttony and Drunkenness, but to Delicacy of either meat or drink. Neither do they err in quantity any more than quality, but take what nature requires, and no more.[6]

To understand Wesley's perspective in *An Estimate* it is necessary to identify the original "ingenious man" who "published a treatise with this title". He is clergyman-essayist John Brown (1715–1766). The full title of his book is *An Estimate of the Manners and Principles of the Times*. Although it is anonymously published in 1757, the author is well known, and the book's sudden popularity encourages seven editions in the years 1757–1758.[7] Wesley, however, omits the words "and Principles" from the title of his own essay. The principles to which Brown is referring are religion, honour and public spirit. Of these principles, the only one Wesley discusses is religion.[8] For Brown "... the Character of the

---

[6] Wesley, *Estimate* (1782, #1–7, pp. 3–8); *WJW* (1872, 11, pp. 154–158).

[7] The latter editions of 1758 grew into two volumes, and a third volume with the title, *An Explanatory Defence of the Estimate of Manners and Principles of the Times*. For a discussion of the reaction to Brown's book see Sekora (1977, pp. 93–95).

[8] Wesley, *Estimate* (1782, #11–24, pp. 10–22); *WJW* (1872, 11, pp. 159–63).

Manners of our Times: which, on a fair Examination, will probably appear to be that of a 'vain, luxurious, and selfish EFFEMINACY.'"[9] Of these three items, Wesley only discusses "luxury".[10] Brown also believes that the effeminate character of manners is based on habits formed in childhood.[11] In his view the reformation of manners and principles begins with the early habit training of those at the top of the social hierarchy, with its influence descending down the social ladder. Wesley accepts this notion of habit training, with respect to effeminacy and indolence, but applies it to the discipline of children in Methodist schools,[12] regardless of their rank on the social ladder, in effect, leveling traditional class distinctions. In other words, Wesley uses the popular format of Brown as a springboard for the shaping of his own view of the English character.

In 1745, or 12 years before the appearance of Brown's book, Wesley renders his own vision of the upper class, "Who in Europe can compare with the sloth, laziness, luxury, and effeminacy of the English gentry…"[13] Yet he is careful to note that the Nobility and Gentry only amount to 5% of the population, whose character does not necessarily mirror that of the other 95%. Among the 95% of the population are the "vast majority of the middle and lower ranks of people … diligently employed from morning to night, and from the beginning to the end of the year". Included in this "vast majority"[14] are the tradesmen who are attracted to the Methodist movement. In R. Campbell's statistical survey of *The London Trades* (1747) nearly 80% of tradesmen work 12 hours a day, 6 days a week, many 14–15 hours. Such demanding work schedules are hardly indicative of a luxurious standard of living, or conducive to sloth and the opportunity to "lying too long in bed"[15] beyond early rising.[16] Add to these the Gentlemen and Noblemen, who are military heroes, "who are of the ancient stamp, who are patterns of Industry in their calling to all that are round about them …"[17] They are all poor candidates for nervous disorders. Thus the impression of nervous distempers rising to epidemic proportions may be true for a large portion of the upper classes, or 5% of the population, but distorted if applied to the remaining 95%.

Wesley's Other Writings on Nervous Disorders

In Wesley's more famous medical work, *Primitive Physick*, that went through 23 editions from 1747 until his death in 1791, the subject of nervous disorders is of minor importance. Of the cumulative 333 categories of medical complaints that appear in the various editions of *Primitive Physick*, only three are notably associated with nerves: A Nervous Head-Ach, which appears in all editions; Nervous Disorders, which doesn't make its appearance until the 8th edition of 1759, where it remains through the 23rd edition of 1791; and A Nervous Fever, which appears briefly in editions 15–16, 1772–1774. From the point of view of *Primitive Physick*, then, fevers are numerically far more commonplace than nervous disorders: For an Ague [1–23], A Tertian Ague [1–23], A Double Tertian [12–23], A Quartan Ague [1–23], A Fever [1–23], An Acute Fever [15–16], A Burning Fever [1–16], A Continual Fever [1–16], A Hectick Fever [1–17], A Strong Hectick Fever [1–2], A High Fever [1–23], An Intermitting Fever [1–23], A Fever with Pains in the Limbs [1–23], A Rash Fever [1–23], A Slow Fever [1–23], A Worm Fever [23]. Although several of the above fevers are deleted (among them "A Nervous Fever" and its remedy) after the 16th edition, owing to a

---

[9] Brown, *Estimate* (1758, 1, p. 29).

[10] Wesley, *Estimate* (1782, #1–10, pp. 3–10); *WJW* (1872, 11, pp. 154–59). And the "sloth" that Wesley emphasizes is not found in Brown.

[11] Brown, *Estimate* (1758, 1, pp. 29, 99–100).

[12] See Wesley's, *A Short Account of the School in Kingswood, Near Bristol* (1768, #4), *WJW* (1872, 13, p. 284); *A Plain Account of Kingswood School* (1781, #11), *WJW* (1872, 13, p. 293).

[13] See Wesley, *A Word in Season: Or, Advice to an Englishman* (1745, #4, p. 3); *WJW* (1872, 11, p. 183).

[14] Wesley, *Estimate* (1782, #4, pp. 6–7); *WJW* (1872, 11, p. 157).

[15] Wesley, *Thoughts* (1786, #7, p. 95); *WJW* (1872, 11, p. 519). "And if we sleep longer than is sufficient, and of course grow weaker and weaker. And if we lie longer in bed, though without sleep, the very posture relaxes the whole body. … I cannot therefore but account this, the lying too long in bed, the grand cause of Nervous Disorders."

[16] Campbell, (1757, pp. 331–340). Also see Lane, "Work and leisure" (1996, pp. 95–116).

[17] Wesley, *Estimate* (1782, #4, p. 6); *WJW* (1872, 11, p. 157).

public criticism by William Hawes (1736–1808) in 1776,[18] the majority are retained,[19] indicating that fevers continue to be problematic for Wesley,[20] even in an age of ascending nervous complaints.[21] Alas *Primitive Physick* is composed with the intention of providing readers with a broad source book for the treatment of the common complaints in alphabetical order. It is not his purpose to organize and classify all known diseases, like William Cullen does in *Synopsis nosologiæ methodicæ* (1769). For Wesley, nervous disorders exist along side a greater multitude of other disorders that a common man might experience.

Wesley first mentions the word "nerves" in the 1747 Preface to *Primitive Physick*, "Coffee and Tea are extremely hurtful to persons who have weak nerves."[22] It is a reference to an advice found in Cheyne's, *An Essay on Health and Long Life* (1724).[23] The term "nervous disorder" follows shortly after, in 1748, in *A Letter to a Friend Concerning Tea*, where he describes his own

---

[18] Hawes (1776, pp. 44–52). For Wesley's response see his letter to Hawes, dated July 29, 1776, that appears in The London Evening Post of July 31–August 2 (1776, p. 113). Hawes charges Wesley with quackery in Primitive Physic, "2. ...'An injudicious collection of pretended remedies: a publication calculated to do essential injury to mankind. He gives, at his first setting out, a satisfactory evidence of his total want of medical knowledge. He says, neither the knowledge of astronomy, natural philosophy, nor even anatomy itself, is absolutely necessary to the cure of most diseases.'" Which, under the same number, Wesley rebuts by saying, "I do say so still: and yet I may possibly know a little anatomy and natural philosophy. Neither is the "evidence of my total want of medical knowledge quite satisfactory." I have read a few medical books, (more I perceive than Mr. Hawes ever saw) I have conversed with many physicians; and I have attended sick beds for more than fifty years. By this time, therefore, I may be fairly supposed to have gained a little even of medical knowledge."

[19] Hawes (1776). With respect to fevers, deleted or reaffirmed remedies, Wesley answers, "4. ... Strange is Mr. Hawes as little acquainted with Dr. Wills's [Willis's] works, as with those of Dr. Sydenham? ... He seems not to know that an ague and an intermitting fever are the same disease. I know the contrary: Every ague is an intermitting fever; but every intermitting fever is not an ague. P. 48. 'Mr. Wesley classes fevers under the following heads.' No: it never was my design to *class fevers*; but merely to speak of a few, one after another. ..." 6. However, I again thank Mr. Hawes for his publication; I am glad to learn of him, or of any man. I have accordingly read it over with care: in consequence whereof, in the edition of the *Primitive Physic*, which I am now preparing for press, I have omitted forty or fifty remedies. I have endeavoured to guard others: I have adopted several of his practical observations; so that I have reason to hope, the present edition will be more useful than any of the former."

[20] As aptly described by Lawrence (1994, p. 7), "Epidemic diseases, such as small pox and measles, and endemic ones, such as the ubiquitous 'fever', were part of everyone's experience."

[21] In a paper read before the American Philosophical Association in Philadelphia, February 4, 1774, Benjamin Rush (1745–1813) cites William Cullen (1710–1790) as a contemporary authority on the classification of diseases, "... that the number of diseases which belong to civilized nations, according to Doctor Cullen's nosology [*Synopsis nosologiæ methodicæ*, 1769], amounts to 1387; the single class of nervous diseases form 612 of this number." In Rush's reading of Cullen, 612 of 1387 diseases are nervous in origin, or 45%, which is significantly higher percentage than that found in Wesley's *Primitive* Physick, where less than 1% is clearly demarcated as nerve related. The implication of this comparison is that Wesley's *Primitive* Physick is written for a bygone age of medical thought. With respect to changes in medical climate Rush says, "... formerly that fevers constituted the chief diseases of the Indians. According to Doctor Sydenham's computation, about 66,000 out of 100,000 died of fevers in London, about 100 years ago; but fevers now constitute but a little more than one-tenth part of the diseases of that city. Out of 21,780 persons who died in London, between December 1770 and December 1771, only 2273 died of simple fevers. I have more than once heard Doctor Huck complain, that he could find no marks of epidemic fevers in London, as described by Dr. Sydenham. London has undergone a revolution in its manners and customs since Doctor Sydenham's time. New diseases, the offspring of luxury, have supplanted fevers; and the few that are left are so complicated with other diseases, that their connection can no longer be discovered with an epidemic constitution of the year. The pleurisy and peripneumony, those inflammatory fevers of strong constitutions, are now lost in catarrhs, or colds, which, instead of challenging the powers of nature or art to a fair combat, insensibly undermine the constitution, and bring on an incurable consumption. Out of 22,434 who died in London between December 1769 and the same month in 1770, 4,594 perished with that British disease. Our countryman, Doctor Maclurg, has ventured to foretel that the gout will be lost in a few years, in a train of hypochondriac, hysteric, and bilious diseases. In like manner, may we not look for a season when fevers, the natural diseases of the human body, will be lost in an inundation of artificial diseases, brought on by the modish practices of civilization?" *Medical Inquiries and Observations* (1805, 1, pp. 35, 40–41).

[22] *Primitive Physick* (1747, Preface, 16/II/10).

[23] Cheyne (1724, pp. 62–63).

nervous complaint, which he attributes to excessive tea drinking,[24] and cures it through abstinence.[25] Indeed, in his "Thoughts", Wesley envisions tea drinking as one of the chief culprits in the rise of nervous disorders,[26] where he again testifies to his own case in point, and advises that if it is to be consumed at all, "Drink but little, and none with out eating, or without sugar or cream".[27]

One would assume from this opinion that Wesley, in future, would personally refrain from tea drinking. In fact, this is not true. In his *Journal* entries for June 10 and July 31, 1764 (published in 1768), and October 28, 1767 (published in 1771), he reveals to his readers that he is again drinking tea. Critic Richard Hill (1733–1809) made a point of this hypocrisy in his *Review of all the Doctrines Taught By The Rev. Mr. John Wesley.... To Which is Added a Farrago* (1772).[28] After which, Wesley defends his renewed tea consumption in *Some Remarks on Mr. Hill's Review* (1772),

"I did set them an example for twelve years. Then at the close of a consumption, by Dr. *Fothergill's* direction, I used it again."[29]

Ironically Wesley uses medical advice to continue a habit that he considered medically harmful to health. The upshot of the controversy obliged Wesley to remove his *Letter to a Friend Concerning Tea* from circulation.

Wesley's next mention of "nervous disorders" comes in 1759, in two places: in the 8th edition of *Primitive Physick*, in a new section labelled Nervous Disorders; and in *The Desideratum*, or his work on the new practice of electrotherapy, which actually appears in 1760. The bulk of the new sec-

tion in *Primitive Physick* was extracted in bits and pieces from John Hill's (1714?–1775) treatise on, *The Construction of the Nerves. And Causes of Nervous Disorders: With A Regimen and Medicines which have proven successful*, newly published in 1758. However, Wesley amended this extraction from Hill with an electrical remedy that harmonizes with a conclusion in *The Desideratum*.

Primitive Physick: "But I am firmly persuaded that, there is no Remedy in Nature, for Nervous Disorders of every Kind, comparable to the proper and constant Use of the *Electrical Machine*."[30]

The Desideratum: "7. But still one may upon the whole pronounce it the *Desideratum*, the general and rarely failing Remedy, in nervous Cases of every Kind (Palsies excepted); as well as in many others. Perhaps if the Nerves are really perforated, (as now generally supposed) the Electric Ether is the only Fluid in the Universe, which is fine enough to remove through them. And what if the *Nervous Juice* itself, be a Fluid of this Kind? If so, it is no Wonder that it has always eluded the Search of the most accurate Naturalists...." 9. I doubt not, but more nervous Disorders would be cured in one Year, by this single Remedy, than the whole *English Materia Medica* will cure, by the end of the Century.

Wesley again affirms this belief in the efficacy of the electrical cure for nervous disorders in his *Journal* entry for January 4, 1768, in spite of having just read Joseph Priestley's recently published work on *The History And Present State or Electricity, With Original Experiments* (1767), who is skeptical about its medicinal value.[31] Further, Wesley's affirmative opinion is reprinted in all of the succeeding editions of *Primitive Physick* and *The Desideratum* published in his lifetime.[32]

---

[24] Wesley, *A Letter to a Friend Concerning Tea* (1748, #13/2); *WJW* (1872, 11, p. 507).

[25] *Ibid.* (1748: #26/4); *WJW* (1872, 11, p. 513).

[26] Wesley, *Thoughts* (1786, #4, pp. 53–54); *WJW* (1872, 11, p. 520).

[27] *Ibid.* (1786,10/2, p. 96); *WJW* (1872, 11, p. 520).

[28] Hill, *Review* (1772, p. 138).

[29] Wesley, *Some Remarks* (1772, #XX/69, p. 27); *WJW* (1872, 10, #XX/69, p. 393). Wesley also consulted Dr. John Fothergill (1712–1780) during his near fatal bout with consumption in 1753. See the entries in his *Journal* for Friday, October 19–November 28, 1753, *JWJD* (20, pp. 479–483).

---

[30] The reference here is to a gentle electrification from machines like that on exhibit at the Wesley Museum on City Road in London, not to a severe shock or electroshock in the twentieth-century psychiatric sense. See Hackmann (2003).

[31] See Priestley, "Of Medical Electricity" (1767, pp. 408–422; Wesley in particular, 417–418).

[32] In Risse's study of the case books relating to "Hysteria at the Edinburgh Infirmary" (1988, pp. 1–22), he discovers two important facts: first, that some poor women were diagnosed with "hysteria or hysteric complaints", albeit 1%, a category by then subsumed under Cullen's "Class II – Neuroses"; and second, that electrical treatments were administered to some of those patients in the form of "sparks" and "shocks", which supports Wesley's contention of their use for nervous disorders.

What Wesley means by a nervous disorder, in his *Letter to a Friend Concerning Tea*, is literally, a disorder of the nerves. Here he is following the solidest assumption of George Cheyne, without admitting any hereditary taint.[33] With respect to his own experience with tea as an effecting agent that weakens the nerves, he says,

13. ...That it gives Rise to numberless Disorders, particularly those of the nervous Kind: And that, if frequently used by those of weak Nerves, it is no other than a slow Poison.

26. ...You need not go far to see many good Effects of leaving it off: You may see them in me. I have recovered hereby that healthy State of the whole nervous System, which I had in a great Degree, and I almost thought irrecoverably, lost, for considerably more than twenty Years.[34]

But not all disorders of the nerves are the result of personal self-indulgence. Some are accidental side effects of other diseases. For example, in his *Journal* entry for March 17, 1783, Wesley records,

My fever was exactly the same kind with that I had in the north of Ireland. On *Monday, Tuesday, Wednesday, and Thursday*, I was just the same: The whole nervous system was violently agitated. Hence arose the cramp, with little intermission, from the time I lay down in bed till morning: Also a furious, tearing cough, usually recurring before each fit to the cramp. And yet I had no pain in my back or head, or limbs, the cramp only excepted. But I had no strength at all, being scarce able to move, and much less to think. In this state I lay till *Friday* morning, when a violent fit of the cramp carried the fever away.[35]

After 1759, Wesley continues to reflect on the effects of nerves on mental life, and religious life. For example, in a 1762 sermon entitled "Wandering Thoughts", Wesley says,[36]

Let but the blood move irregularly in the brain, and all regular thinking is at an end. Raging madness ensues, and then farewell to all evenness of thought. Yea, let only the spirits be hurried or agitated to a certain degree, and a temporary madness, a delirium, prevents all settled thought. And is not the same irregularity of thought in a measure occasioned by every nervous disorder? So does the 'corruptible body press down on the soul, and cause it to muse about many things'.[37]

And in a letter to Lady Maxwell, dated February 17, 1770,

Indeed, nervous disorders are, of all others, as one observes, enemies to the joy of faith.[38]

And in a letter to Miss Loxdale, dated October 8, 1780,

I believe Mr. W—'s nervous disorder gave rise to many, if not most, of those temptations to which many persons of equal grace, but firmer nerves, are utter strangers.[39]

Background to Thoughts on Nervous Disorders

An Estimate of the Manners and "Thoughts on Nervous Disorders" both reflect Wesley's interest in the central nervous system as a source of pathology. But he is not satisfied with the existing explanations for these complaints,

1. WHEN Physicians meet with disorders which they do not understand, they commonly term them *Nervous*: a word that conveys to us no determinate idea, but is a good cover for learned ignorance.[40]

He stands not alone in this insecurity. Robert Whyte's authoritative book, *Observations on the Nature, Causes, and Cure of those Disorders which have been Commonly called Nervous,*

---

[33] See Cheyne (1733, pp. 14–15): "...Diseases are chiefly and properly called Nervous, whose Symptoms imply that the System of the Nerves and their Fibres, are evidently relax'd and broken.... In treating of Nervous Distempers, the Disorders of the Solids are chiefly what are to be had regard to; yet they rarely or never happen alone (except perhaps in those Nervous Disorders that proceed from acute Diseases, preternatural Evacuations, external Injuries, or a wrong and unnatural Make and Frame) but even in original nervous Distempers there is always some Viscidity or Sharpness attending them from the bad Constitutions of the Parents, from whom they have derived their material Organs."

[34] Wesley, *op. cit.* (1748, pp. 7, 14); *WJW* (1872, 11, pp. 508, 513).
[35] Wesley, *JWJD* (23, p. 265).
[36] Wesley, Sermon 41, "Wandering Thoughts (1762)," *JWS* (2, pp. 129–130).
[37] *Cf* Wisdom of Solomon 9.15, KJV: "For the corruptible body presseth down the soul, and the earthly tabernacle weigheth down the mind that museth upon many things."
[38] Letter 341, *WJW* (1872, 12, p. 247).
[39] Letter 876, *WJW* (1872, 13, p. 132).
[40] Wesley, *Thoughts* (1786, #1, p. 52); *WJW* (1872, 11, p. 515).

Hypochondriac, or Hysteric (2nd ed., 1765), expresses a similar skepticism,

The disorders which are the subject of the following Observation, have been treated of by authors, under the names of Flatulent, Spasmodic, Hypochondriac, or Hysteric. Of late, they have also got the name of Nervous; of which appellation having been commonly given to many symptoms seemingly different, and very obscure in their nature, has often made it to be said, that Physicians have bestowed the character of *nervous*, on all those disorders whose nature and causes they were ignorant of.[41]

For lack of a better expression, both Wesley and Whytt employ the term "nervous disorder" in order to participate in a contemporary thought world that is actively reclassifying certain traditional diseases under the rubric of "nerves". Whytt defends this move with the following qualified statement,

Since, in almost every disease, the nerves suffer more or less, and there are few disorders which many not, in a large sense, be called *nervous*, it might be thought that a treatise on nervous diseases should comprehend almost all the complaints to which the body is liable. The design, however, of the following Observations is far different. In them, it is only proposed to treat of those disorders, which in a *peculiar* sense deserve the name *nervous*, in so far as they are, in a great measure, owing to an uncommon delicacy or unnatural sensibility of the nerves, and are therefore observed chiefly to affect persons of such a constitution.[42]

However this change in orientation is not based on any discoveries of anatomical certainty about the brain and nervous system, rather more on variations in speculations about their function. This proclivity toward speculation not only leaves Wesley and Whytt insecure in their knowledge, but other notable writers from the past are dissatisfied as well. Already in 1664 Thomas Willis (1621–1675) criticized the faulty diagnosis of hysteria.[43] And in 1669 Nicolaus Steno (1638–1686) publicly lamented about the sad state of skills available for the dissection of the brain,[44] resulting is "fantastic interpretations" as to the details of its form and function.[45]

As the full title of Wesley's paper "Thoughts on NERVOUS DISORDERS; particularly that which is usually termed LOWNESS of SPIRITS"

indicates, depression is for him the chief feature of the mental side of nervous disorders, although it can arise from other distempers as well.

2. But undoubtedly there are Nervous Disorders, which are purely natural. Many of these are connected with other Diseases, whether Acute or Chronical. Many are the forerunners of various Distempers, and many the consequences of them. But there are those, which are not connected with others, being themselves a distinct, original Distemper. And this frequently arises to such a height, that it seems to be one species of madness. So one man imagines himself to be made of glass: another thinks he is too tall to go at the door. This is often termed the *Spleen* or *Vapours*: often *Lowness of Spirits*: a phrase that having scarce any meaning, is so much the fitter to be given to this unintelligible Disorder. It seems to have taken its rise from hence. We sometimes say, A man is in *high spirits*. And the proper opposite to this is, *He is low-spirited*. Does this not imply, that a kind of faintness, weariness, and listlessness affects the

---

[41] Whytt (1765, Preface, p. iii).
[42] Whytt (1765, p. iv).

[43] Willis, "The hysterical passion is of so ill fame, among the Diseases belonging to women, that like one half damn'd, it bears the faults of many other Distempers: For when at any time, a sickness happens in a woman's body, of an unusual manner, or more occult original, so that its Cause lyes hid, and the Curatory Indication is altogether uncertain, presently we accuse the evil influence of the womb, (which for the most part is innocent) and in every unusual Symptom, we declare it to be something hysterical, and so to this Scope, which oftentimes is only the subterfuge of Ignorance, the medical Intentions, and use of Remedies are directed." (1681, p. 76).
[44] Steno, "Thus, since anatomical research has not yet reached the degree of perfection that allows for correct dissection of the brain, we should deceive ourselves no further, we should rather acknowledge our ignorance, so that we do not first delude ourselves and then others by promising to show the correct structure." (1965 [1669], p. 125).
[45] Steno, "2. Fantastic interpretations are made in describing the Ventricles. (Animal Spirits and Excrements). The ventricles, or cavities of the brain, are no less unknown than its substance. Those who place the animal spirits there believe that they are as much in the right as those who mark them out as receiving the excrements: but both parties find that they are equally trammeled when it becomes necessary to ascertain the origin of these excrements or these spirits. These might come as readily from the vessels that are seen in these cavities as from the substance of the brain itself, and it is not any easier to mark where they make their exits. ... In short our standard dissections cannot clarify any of these difficulties concerning animal spirit." (1965, pp. 23–24). For a more thorough discussion of Willis and Steno, see Spillane (1981), pp. 53–107.

whole body, so that he is disinclined to any motion, and hardly cares to move hand or foot? But the mind seems chiefly to be affected, having lost its relish of every thing, and being no longer capable of enjoying the things it once delighted in most. Nay, every thing round about, is not only flat and insipid, but dreary and uncomfortable. It is not strange if to one in this state, life itself is become a burden: yea so insupportable a burden, that many who have all this world can give, desperately rush into an unknown world, rather than bear it any longer.[46]

By going directly to "lowness of spirits",[47] Wesley skips over the vagueness he associates with "spleen" and "vapours", to a condition that almost any reader can understand. He further refines the idea to a mental state where the mind has "lost its relish for everything", incapable of "enjoying the things it once delighted in most", "flat", "insipid", "dreary", "uncomfortable", "life itself has become a burden", a desperate state that many who are unable to bear it "rush into an unknown world", in other words, in search for relief they commit suicide. Other synonyms of the time could be added to the list: despondent, gloomy, sullen, irascible, morbid, etc.

Wesley's list, in particular, avoids any mention of the extreme condition of deep melancholy or madness, or conditions that would require bodily medication. He is aware that in a nervous disorder, the body does frequently press down a person's spirit.[48] But he prefers to mock medical advice in favour of a religious hypothesis of the problem,

If he is not fully employed, will he not frequently complain of lowness of spirits? – an unmeaning expression, which the miserable physician usually no more understands than his miserable patient. We know there are such things as nervous disorders. But we know likewise, that what is commonly called "nervous lowness" is a secret reproof from God; a kind of consciousness that we are not in our place; that we are not as God would have us to be: We are unhinged from our proper centre.[49]

Contrary to the notion that an author writes about a condition he knows best from personal experience, Wesley claims to be a stranger to "lowness of spirits (which I do not remember to have felt one quarter of and hour since I was born)".[50] Nevertheless, he encounters it in others, recording the instances in his *Journal*[51] and *Letters*.[52] Viewing it as a kind of litmus test for one's spiritual life.[53] To suffer from "lowness of spirits" is a sign of being "unhinged from our proper centre". It follows, then, with so little down time due to "lowness", that Wesley considered himself a model of both natural and spiritual health.

The Causes of Nervous Disorders

In the *Minutes Of Several Conversations Between The Reverend Messieurs John And Charles Wesley, And Others,* (5th ed. 1780), better known as "The Large Minutes" Wesley is asked,

Q. 28. What reasons can be assigned why so many of our Preachers contract nervous disorders?

A. The chief reason, on Dr. Cadogan's principles, is either indolence or intemperance.

1. Indolence. Several of them use too little exercise, far less than when they wrought at their trade: and this will naturally pave the way for many, especially nervous disorders.

2. Intemperance (though not in the vulgar sense). They take more food than they did when they laboured more: and let any man of reflection judge, how long this will consist with health. Or they use more sleep than when they laboured more: and this alone will destroy the firmness of the nerves. If then our Preachers would avoid nervous disorders, let them: 1. Take as little meat, drink, and sleep, as nature will bear: and, 2. Use full as much exercise daily as they did before they were Preachers.[54]

---

[46] Wesley, *Thoughts* (1786, #2, pp. 52–53); *WJW* (1872, 11, p. 516).

[47] Wesley does not go into any detail about what theory of spirits he is using, such as, whether they are animal spirits or vital spirits.

[48] Wesley, Letter 765, to Mrs. Knapp, Mar 25, 1781, *WJW* (1872, 13, p. 69).

[49] Wesley, Sermon 84, "The Important Question," September 11, 1775, III, 7, *JWS* (3, p. 193).

[50] Wesley, Sermon 77, "Spiritual worship" December 22, 1780, III, 2. *JWS* (3, p. 98).

[51] See the *Journal* entries for July 28, 1739, *JWJD* (19, p. 79; July 26, 1752, *JWJD* (20, p. 435).

[52] Wesley, Letter 765, to Mrs. Knapp, *WJW* (1872, 13, p. 69).

[53] Wesley, Letter 365, to Miss Pywell, January 22, 1772, *WJW* (1872, 11, p. 364); Letter 776, to Hester Anne Roe, afterwards Mrs. Rogers, September 16, 1776, *WJW* (1872, 13, p. 78).

[54] Wesley, *Minutes*, London: John Mason (1862, p. 507–09); *WJW* (1872, 8, pp. 313–14).

Wesley's answers to the question put a social perspective on nervous disorders within this one aspect of Methodist society. Tradesmen can become preachers. When they were tradesmen, they worked 12–14 hours a week, or 70–80 hours a week; whereas the new occupation of preacher is probably less time consuming, and clearly less physically demanding. Unfortunately, the extra time and lack of physical strain made them fitting candidates for nervous distempers.

Although increased luxury is depicted by Cheyne as the vital factor in the rise of nervous pathology, it is not part of Wesley's description of Methodist life. Excesses of the poor and working class Methodists are limited, when compared to the wealthy. Yes it is easier for the wealthy to indulge in indolence,[55] intemperate gluttony, drunkenness, and excessive sleep, than the poor. But the availability of cheap gin and other drams in the eighteenth century increased the possibility of its exposure to the poor. The habit of tea drinking had also become popular, as Wesley himself complains, which he later deemed, in a guarded way, as an acceptable beverage for Methodists, a habit certainly preferable to dram drinking.

The Cure of Nervous Disorders

10. But is there no cure for this sore evil? Is there no remedy for Lowness of Spirits? Undoubtedly there is, a most certain cure, if you are willing to pay the price of it. But this price, is not silver, or gold; nor any thing purchaseable thereby. If you would give all the substance of your house for it, it would be utterly despised. And all the medicines under the sun avail nothing in this distemper. The whole *Materia Medica* put together, will do you no lasting service: they do not strike at the root of the disease: but you must remove the cause, if you wish to remove the effect. But this cannot be done by your own strength: it can only be done by the mighty

power of God. If you are convinced of this, set about it trusting in him, and you will surely conquer.[56]

In Wesley's "cause and effect" view, there is no cure without first removing the cause. And the cause is found in some disturbance among the "six non-naturals" that were commonly considered at the time to be necessary for health: air; meat and drink; sleep and waking; motion and rest; excretion and retention; affections of the mind. Any of these, either through chronic abuse or accident, can become the cause of a malady. Indolence, intemperance, and indulgence in "irregular passions", all fit under the rubric of the "six non-naturals".[57] If the resultant disorder is "nervous", the cure is a correction of these abuses; added exercise for indolence; restriction and alternation of diet for intemperance; and avoidance excessive passions. But Wesley is well aware of the hardship that these corrective self-disciplines impose, such that the sufferer needs help from the power of God to overcome the "lowness of spirits" that is characteristic of nervous disorders, as he sees it.

By "lowness of spirits" Wesley implicitly means nervous depression. However the word "spirit" has other connotations in the "physick" of his time that are not necessarily related to the nerves. Spirits in the context of the human body were characteristic of something that is alive: animal spirits are generally associated with the brain; vital spirits with the heart and natural spirits with the blood producing liver. "Lowness" in any of the three would be a condition calling for medical treatment, in the natural rather than the religious domain.

But Wesley does his best to solve the problem of "lowness" without the intervention of "physick" or physicians. In his view, indolence and intemperance, or the conditions causing "lowness of spirits", can be overcome through self-discipline, in the sense of bringing the body back to natural health.

Of these conditions, indolence is the most difficult to define because Wesley uses it in connection with a variety of images. On the one hand, he

---

[55] Wesley uses the word "indolence" instead of "sloth". For a parallel confirmation of this point, see the anonymous author of *A Dissertation upon the Nervous System to show its Influence upon the Soul*, "But the most general cause of Nervous disorders is Indolence. The active and laborious are seldom troubled with them. They are reserved for the Children of ease and affluence." (1780, p. 42) For a literary analysis of the hazards of "indolence" see the Thomson poem, *The Castle of Indolence* (1748).

[56] Wesley, *Thoughts* (1786, #10, pp. 96–97); *WJW* (1872, 11, pp. 519–20).
[57] See Wainewright (1737); Rather (1968, p. 337–347).
[58] Wesley, *Advice to an Englishman* (1745, #5, pp. 3–4); *WJW* (1872, 11, p. 183–184); *A Farther Appeal Pt II* (1746, II, 19), *WJW* (1872, 8, p. 163–164): *WJW* (1975, 11, p. 231).

associates it with the life-styles of the rich.[58] On the other hand, he employs synonyms that can cross social boundaries. As Samuel Johnson (1709–1784) says, "Words are seldom exactly synonymous," making it "necessary to use a proximate word".[59] In the case of the word indolence, Wesley associates it, in a proximate sense, with "laziness",[60] "idleness",[61] "inactivity",[62] "directly opposite to the spirit of industry",[63] "love of ease",[64] "softness",[65] and "effeminacy".[66]

Indolence is his preferred word in the "Thoughts"; as sloth is in the *Estimate*. Elsewhere in his writings the two words appear together in tandem, making it difficult to comprehend precisely what he means by either, be they synonymous or with a subtle difference.[67] Moreover, indolence and sloth, as he uses them in the *Estimate* and the "Thoughts" are burdened with non-physical implications of a religious nature. Does this mean he is really talking about both the secular and religious meaning of the words? The word sloth has a long history of both meanings going back into the Middle Ages and the "seven deadly sins", into the term "acedia", or "spiritual sloth".[68] Likewise Wesley uses the term "spiritual sloth" outside of the context of nervous disorders.

But on a practical advice level Wesley's opinion carries weight within the Methodist organization. And at the Leeds meeting of the annual conference on August 4, 1778, he is asked,

Q. 24. What advice would you give to those that are *Nervous?*

A. Advice is made for them that take it. But who are they? One in ten, or twenty?[69]

Then I advise: –

1. Touch no dram, tea, tobacco, or snuff.
2. Eat very light, if any supper.
3. Breakfast on nettle or orange-peel tea.
4. Lie down before ten; – Rise before six.
5. Every day use as much exercise as you can bear: – Or,
6. Murder yourself by inches.[70]

As hinted in his answer, advice is easy to give and difficult for the afflicted who "take it" on. A cure is possible but not necessarily probable. Wesley is aware, along with Cadogan and others that "chronic maladies are of our creating"[71]; that sufferers from any social class are probably not in good physical condition, if their debilitation is traceable to the long-standing indolence or intemperance. Sydenham describes this condition clearly a century before, "For when the *stamina vitae* are much debilitated, and in a manner work not, either by age, or by remarkable and continued irregularities in the use of

---

[59] Johnson, (1755, Preface, #48).

[60] Wesley, *An Address to the Clergy,* February 6, 1756, II, 1, (5), *WJW* (1872, 10, pp. 491–92); Letter 211, to Mr. S., April 24, 1769, *WJW* (1872, 12, pp. 247–248).

[61] Wesley, *A Farther Appeal* II (1746), II, 19, *WJW* (1872, 8, p. 163), *WJW* (1975, 11, p. 163); Letter 71, to Charles Wesley, December 20, 1764, *WJW* (1872, 12, p. 129); *A Short Account of the School at Kingswood,* 1768, #4, *WJW* (1872, 13, p. 284).

[62] Wesley, Letter 136, to Mr. Ebenezer Blackwell, July 20, 1752, *WJW* (1872, 12, p. 178); *Thoughts on a Single Life* (1770), #13, *WJW* (1872, 11, p. 461), *WJW*; Letter 84, to Charles Wesley, March 25, 1772, *WJW* (1872, 12, p. 138).

[63] Wesley, Letter 71, to Charles Wesley, December 20, 1764, *WJW* (1872, 12, p. 129).

[64] Wesley, Sermon 68, "The Wisdom of God's Counsel (1784)," #18, *JWS* (2, pp. 562–63).

[65] Wesley, *A Farther Appeal Pt II* (1746), II, 19, *WJW* (1872, 8, p. 163), *WJW* (1975, 11, p. 231); *Thoughts on a Single Life* (1770), #13, *WJW* (1872, 11, p. 461; Sermon 68, "The Wisdom of God's Counsels (1784)," #18, *JWS* (2, pp. 562–63); Sermon 107, "On God's Vineyard," October 17, 1787, V, 4, *JWS* (3, p. 516).

[66] Wesley, *Advice to an Englishman* (1745), #4, p. 3, *WJW* (1872, 11, p. 183); *A Farther Appeal Pt II* (1746), II, 19, *WJW* (1872, 8, p. 163), *WJW* (1975, 11, p. 231); *A Short Account of the Kingswood School* (1768), #4, *WJW* (1872, 13, p. 284); *A Plain Account of the Kingswood School* (1781), #11, *WJW* (1872, 13, p. 293); *Thoughts on a Single Life* (1770), #13, *WJW* (1872, 11, p. 561); Sermon 107, "On God's Vineyard," 17 Oct 1784, V, 3, *JWS* (3, p. 515).

[67] Wesley, Sermon 4, "Spiritual Christianity," August 21, 1744, 1, IV, 7, *JWS* (1, p. 176); *A Farther Appeal Pt II* (1746), II, 19, *WJW* (1872, 8, p. 163), *WJW* (1975, p. 231); Letter to Rev. Dr. Conyers Middleton, 4 Jan 1748–49, VI, 4, *WJW* (1872, 10); Letter 211, to Mr. S., April 24, 1769, *WJW* (1872, 12, 5, (1), p. 248); *A Plain Account of Christian Perfection* (1777), Q/A 35, *WJW* (1872, 11, p. 432).

[68] See Wenzel (1967) and Bloomfield (1952) for the transition of these sins to the English-speaking world.

[69] Wesley is here suggesting that 5–10% of the Methodist clergy might be candidates for these nervous disorders, or at least those who have ceased to perform in the amount of healthy labour required by their former trades.

[70] *Minutes Of The Methodist Conferences, From the First, Held in London, By the Late Rev. John Wesley, A.M. In the Year 1744* (1812, 1, 2, p. 136).

[71] Beddoes, *Hygëia*, "Essay Seventh. Essays on the Means of Avoiding Habitual Sickliness, and Premature Mortality," (1803, 2, p. 98).

the *six naturals,* especially in relation to food and drink … occasioning different diseases, according as they are variously vitiated and depraved."[72] What kind of exercise is he willing to take on with discipline? What kind of food and drink is the sufferer willing to give up? For without these changes, the prognosis is grim – "Murder yourself by Inches"!

Abbreviations

Estimate = John Wesley, *An Estimate of the Manners Of the Present Times* (1782).

JWJD = John Wesley, *The Works of John Wesley, Journals and Diaries* (1991–2003).

JWS = John Wesley, *The Works of John Wesley, Sermons* (1984–1987).

Thoughts = John Wesley, "Thoughts on Nervous Disorders." (1786).

WJW (1872) = John Wesley, *The Works of the Rev. John Wesley, A.M.,* Thomas Jackson ed.

WJW (1975) = John Wesley, *The Appeals to Men of Reason and Religion* (1746).

References

Anonymous. *A dissertation upon the nervous system to show its influence upon the soul.* (1780). London: Privately printed.

Bates, D. G. (1981). Thomas Willis and the fevers literature of the seventeenth century. In W. F. Bynum & V. Nutten (eds.), *Theories of fever from antiquity to the enlightenment* (pp. 45–70). London: Wellcome Institute for the History of Medicine.

Beddoes, T. (1760–1808). *Hygëia: Or essays moral and medical, on the causes affecting the personal state of our middling and affluent classes* (3 Vols.). Bristol: Printed by Mills J. (1802) *Essay seventh. Essays on the means of avoiding habitual sickliness, and premature mortality* (Vol. 2).

Blackmore, R. (d 1729). *A treatise of the spleen and vapours: Or, hypochondriacal and hysterical affections. With three discourses on the nature and cure of the cholick, melancholy, and palsies.* London: Printed for Pemberton, J. (1725).

Bloomfield, M. W. (1952). *The seven deadly sins; An introduction to the history of a religious concept, with special reference to medieval English literature.* East Lasing: Michigan State University Press.

Brown, J. (1715–1766). *An estimate of the manners and principles of the times* (7th ed., 2 Vols.). London: Printed for Davis, L., & Reymers, C. (1758).

Brown, J. (1715–1766). *An explanatory defence of the estimate of the manners and principles of the times. Being an appendix to that work, occasioned by the Clamours lately raised against it among certain ranks of men.* London: Printed for Davis, L. & Reymers, C. (1758).

Bynum, W. F. (1981). Cullen and the study of fevers in Britain, 1760–1820. In W. F. Bynum & V. Nutten (eds.), *Theories of fever from antiquity to the enlightenment* (pp. 135–147). London: Wellcome Institute of the History of Medicine.

Bynum, W. F. (1985). The nervous patient in eighteenth- and nineteenth-century Britain: The psychiatric origins of British neurology. In W. F. Bynum, R. Porter, & M. Shepherd, (eds.), *The anatomy of* madness (Vol. 1, pp. 89–101). London: Tavistock.

Cadogan, W. (1711–1797). *A dissertation on the gout, and all chronic diseases jointly considered, as proceeding from the same causes; What those causes are; And a rational and natural method of cure proposed. Addressed to all invalids.* Quod petis in te est. London: Printed for Dodsley, J. (1771).

Cadogan, W. (1711–1797). 10th ed., 1772, reprint appearing in John Ruhräh (1925) William Cadogan and His Essay on Gout. *Annals of Medical History, 7,* 67–90; also in Ruhräh's, *William Cadogan. His Essay on Gout.* New York: 1925, preface to 10th ed. dated "Nov. 20, 1771".

Cadogan, W. (1711–1797). 11th ed., London: J. Dodsley, 1772.

Campbell, R. (1747). *The London tradesman. Being a compendious view of all the trades, professions, arts, both liberal and mechanic, now practised in the cities of London and Westminster* (3rd ed.). London: Printed by T. Gardner. 1757, is identical.

Carlson, E. T., & Meribeth M. S. (1969). Models of the nervous system in eighteenth century psychiatry. *Bulletin of the History of Medicine, 43/2,* 101–115.

Cheyne, G. (1671–1743). *The English malady: Or, a treatise of nervous diseases of all kinds, as spleen, vapours, lowness of spirits, hypochondriacal, and hysterical distempers, &c.* London: Printed for George Strahan, 1733. Reprinted in *George Cheyne: The English Malady (1733),* edited with an introduction by Roy Porter. London: Tavistock/Routledge, 1991.

Cheyne, G. (1671–1743). *An essay of health and long life.* London: Printed for George Strahan (1724) (pp. 60–70, 76), discuss tea.

---

[72] Sydenham, (1763, p. 484).

Cheyne, G. (1671–1743). *The natural method of cureing the diseases of the body, and the mind depending on the body* (3rd ed.). London: Printed for Geo. Strahan (1742).

Cullen, W. (1710–1790). *First lines of the practice of physic, for the use of students in the University of Edinburgh* (Vol. 3). Edinburgh: Printed for William Creech (1782); see Chap. 3, *Of hypochondriasis; or the hypochondriac affection, commonly called vapours or low spirits* (pp. 120–142).

Cullen, W. (1710–1790). *Synopsis nosologiæ methodicæ*. Edinburgi, 1769.

Hackmann, W. (2003). *John Wesley and his electrical machine*. London: John Wesley's House and The Museum of Methodism.

Hawes, W. (1736–1808). *An examination of the Rev. Mr. John Wesley's primitive physic: Shewing that a great number of the prescriptions therein contained, are founded on ignorance of the medical art, and of the power and operation of medicines; and that it is a publication calculated to do essential injury to the health of those persons who may place confidence in it*. London: Printed for the Author, 1776.

Hill, J. (1714?–1775). *The Construction of the nerves, and causes of nervous disorders: With a regimen and medicines which have proven successful* (2nd ed.). London: 1758.

Hill, R. (1733–1809). *Review of all the doctrines taught by the Rev. Mr. John Wesley … to which is added a farrago*. London: Printed for E. and C. Dilly (1772).

Hill, R. (1733–1809). *Logica Wesleiensis: Or The farrago double distilled. With an heroic poem in praise of Mr. John Wesley*. London: Printed for E. and C. Dilly, 1773.

James, R. (1705–1776). *A medicinal dictionary; including physic, surgery, anatomy, chemistry, and botany. in all their branches relative to medicine* (2 Vols.). London: Printed for T. Osborne (1743–1745).

Johnson, S. (1709–1784). *A dictionary of the English language* (2 Vols.). London: Printed by Strahan W. (1755).

Lane, J. (1996). *Apprenticeship in England, 1600–1914*. London: UCL Press.

Lawrence, C. (1994). *Medicine in the making of modern Britain*. London: Routledge.

Logan, P. M. (1997). *Nerves & narratives; A cultural history of hysteria in 19th century British prose*. Berkeley: University of California Press. (Forward by Roy Porter.)

Mandeville, B. (1670–1733). *A treatise of the hypochondriack and hysterick diseases. In three dialogues* (2nd ed.). Corrected and Enlarged by the Author. London: Printed for Tonson, J. (1730).

Midriff, J. (pseudonym). *Observations on the spleen and vapours: Containing remarkable cases of persons of both sexes, and all ranks … who have been miserably afflicted with those melancholy disorders since the fall of the South-Sea, and other publick Stocks*. London: Printed for Roberts, J. (1721).

Porter, R. (1985). Lay medical knowledge in the eighteenth century: The evidence of the Gentleman's Magazine. *Medical History*, 29, pp. 138–168.

Porter, R. (1993). Diseases of civilization. In F. B. William & P. Roy (Eds.) *Companion encyclopedia of the history of medicine* (Vol. 1, pp. 585–600). London: Routledge.

Porter, R. (1993). Religion and medicine. In F. B. William & P. Roy (Eds.) *Companion encyclopedia of the history of medicine* (Vol. 2, pp. 1449–1468). London: Routledge.

Priestley, J. (1733–1804). *The history and present state of electricity, with original experiments*, Causa latet, vis est notissima. *Ovid*. London: Dodsley, J., Johnson, J., Davenport, B., & Cadell, T. (1767).; Sect 14, *Of medical electricity* (pp. 408–422).

Priestley, J. (1733–1804). 3rd ed., 2 Vols. London: Printed for Bathurst, C., & Lowndes, T. (1775). Vol. 1, pp. xxxvi, 503, Vol. 2, pp. viii, 376, Corrected And Enlarged; Vol. 1, Sect 14, Of Medical Electricity, (pp. 472–489).

Purcell, J. (1674–1730). *A treatise of vapours, or, hysterick fits. Containing an analytical proof of its causes, mechanical explanations of all its symptoms and accidents, according to the newest and most rational principle. Together with its cure at large* (2nd ed.). Revised and Augmented. London: Printed for Edward Place (1707).

Rather, L. J. (1968). The 'Six Things Non-Natural': A note on the origins and fate of doctrine and a phrase. *Clio Medica*, 3, pp. 337–347.

Riese, W. (1945). History and principles of classification of nervous diseases. *Bulletin of the History of Medicine*, 18, pp. 465–512.

Risse, G. B. (1988). Hysteria at the Edinburgh infirmary: The construction and treatment of a disease, 1770–1800. *Medical History*, 32, pp. 1–22.

Robinson, N. (1697?–1775). *A new system of the spleen, vapours, and hypochondriak melancholy: Wherein all the decays of the nerves, and lownesses of the spirits, are mechanically accounted for. To which is subjoin'd, A discourse upon the nature, cause, and cure, of melancholy, madness, and lunacy*. London: Printed for Bettesworth, A., Innys, W., & Rivington, C. (1729).

Rush, B. (1745–1813). (1805). An inquiry into the natural history of medicine among the Indians of North-America; and a comparative view of the their diseases and remedies with those of civilized nations [1774]. In *Medical inquiries and observations* (Vol. 1, 2nd ed., pp. 3–68). Philadelphia: J. Conrad & Co.

Sekora, J. (1977). *Luxury; the concept in Western thought, Eden to Smollett.* Baltimore: The Johns Hopkins Press.

Smith, W. (1768). *A dissertation on the nerves; containing an account, 1. Of the nature, of man. 2. Of the nature of brutes. 3. Of the nature and connection of soul and body. 4. Of the threefold of life of man. 5. Of the symptoms, causes and cure of nervous diseases.* London: Printed for the Author.

Spillane, J. D. (1981). *The doctrine of the nerves; chapters in the history of neurology.* Oxford: Oxford University Press.

Steno, N. [Niels]. (1638–1686). (1965). *Lecture on the anatomy of the brain* [1669]. Copenhagen: Nyt Nordisk Forlag. (Introduction and translation by Gustav Scherz.)

Sydenham, T. (1624–1689). In John Swan (Ed.), *The entire works of Dr Thomas Sydenham, newly made English from the originals: Wherein the history of acute and chronic diseases, and the safest and most effectual methods of treating them, are faithfully, clearly, and accurately delivered* (4th ed.). London: Printed by Cave, R. (1763). Wherein *A Treatise of the Gout and Dropsy* [1683] is found on pp. 461–510; and "Of the Gout," on pp. 659–663.

Sydenham, T. (1624–1689). *The works of Thomas Sydenham, M.D.* Translated from the Latin Edition of Dr. Greenhill with *A Life of the Author* by R. G. Latham, M.D., 2 vols. London: Printed for the Sydenham Society. Vol. 1, 1848; Vol. 2. Reprinted, Birmingham AL: The Classics of Medicine Library, 1979, 2 Vols. in 1.

Thomson, J. (1700–1748). *The castle of indolence. An allegorical poem. Written in imitation of Spenser* (2nd ed.). London: Printed for A. Millar, 1748.

Trotter, T. (1760–1832). *A view of the nervous temperament; Being a practical enquiry into the increasing prevalence, prevention, and treatment of those diseases commonly called nervous, bilious, stomach and liver complaints; indigestion; low spirits; gout, &c.* London: Printed by Edw. Walker, 1807.

Wainewright, J. (b 1673). *A mechanical account of the non-naturals: Being a brief explication of the changes made in human bodies, by air, diet, &c.* (5th ed.). London: Printed for John Clarke, 1737.

Wenzel, S. (1967). *The sin of sloth: Acedia in medieval thought and literature.* Chapel Hill: The University of North Carolina Press.

Wesley, J. (1756). *An address to the clergy,* in *WJW* (1872, 10, pp. 480–500).

Wesley, J. (1760). *The desideratum: Or, electricity made plain and useful. By a lover of mankind, and of common sense.* London: Printed by W. Flexney.

—(1771), (2nd ed.), Bristol: Printed by William Pine.

—(1773), (*W* ed.), *The works of the Rev. John Wesley, M.A.* (24, pp. 284–368). Bristol: Printed by William Pine.

—(1778), (4th ed.), London: Printed by R. Hawes.

—(1790), (3rd ed.), London: Printed and sold at the New-Chapel, City Road.

Wesley, J. (1782). *An estimate of the manners of the present times.* London: np.

Wesley, J. (1746). *A farther appeal to men of reason and religion part II,* in *The works of John Wesley* (1975, 11: 203–71).

Wesley, J. (1748). *A letter to a friend concerning tea.* London: Printed by W. Strahan.

—(1749), (2nd ed.), Bristol: Printed by Felix Farley.

—(1773), (*W.* ed.), *The works of the Rev. John Wesley, M.A.* (24, pp. 264–283). Bristol: Printed by William Pine.

Wesley, J. (1776). Letter to Mr. Hawes, apothecary and critic. *The London Evening Post* (July 31–August 2: p. 113).

Wesley, J. (1812). *Minutes of the Methodist conferences, from the first held in London, in the year 1744.* London: Printed at the Conference Office by Thomas Cordeaux, 2 vols.

Wesley, J. (1862). *Minutes of several conservations between the Reverend Messieurs John and Charles Wesley, and others.* London: John Mason.

Wesley, J. (1777). *A plain account of Christian perfection, as believed and taught by the Reverend Mr. John Wesley, from the year 1725–1777,* in *WJW* (1872, 11: pp. 356–446).

Wesley, J. (1747). *Primitive physick: Or, an easy and natural method of curing most diseases* [1st ed.], London: Printed by Thomas Trye.

—[nd], [2nd ed.], Printed [by Felix Farley, "Inlarged".]

—[3rd ed., not found.]

—[4th ed., not found.]

—(1755), (5th ed.), Printed by J. Palmer, "corrected and enlarged".

—[6th ed., not found.]

—[7th ed., not found.]

—(1759), (8th ed.), Bristol: Printed by John Grabham, "corrected and enlarged".

—(1761), (9th ed.), London: Printed by W. Strahan, "corrected and enlarged".

—(1762), (10th ed.), Bristol: Printed by William Pine, "corrected and enlarged".

—[11th ed., not found.]

—(1765), (12th ed.), Bristol: Printed by William Pine, "corrected and much enlarged".

—(1768), (13th ed.), Bristol: Printed by William Pine, "corrected and much enlarged".

—(1770), (14th ed.), Bristol: Printed by William Pine, "corrected and much enlarged".

—(1772), (15th ed.), London: Robert Hawes, "Corrected and much Enlarged".

—(1773), (*Works* ed.), Primitive physic: Or, an easy and natural method of curing most diseases, in *The Works of the Rev. John Wesley, M.A,* Bristol: Printed by William Pine, vol 25, pp. 3–149, "Errata".

—(1774), (16th ed.), *Primitive physick: Or, an easy and natural method of curing most diseases*, London: Printed by R[obert] Hawes.

—(1776), (17th ed.), London: Printed by R[obert] Hawes.

—[18th ed., not found.]

—[19th ed., not found.]

—(1781), (20th ed.), *Primitive physic: Or, an easy and natural method of curing most diseases.* London: Printed by J. Paramore.

—(1785), (21st ed.), London: Printed by J. Paramore.

—(1788), (22nd ed.), London: Printed at New Chapel.

—(1791), (23rd ed.), London: Printed at New Chapel. For a modern version of this edition see *Primitive physic*, London: The Epworth Press, 1961, with an Introduction by A. Wesley Hill.

—(1792), (24th ed.), London: Printed by G. Paramore.

Wesley, J. (1768), *A short account of the school in Kingswood, near Bristol,* in *WJW* (1872, 13, pp. 283–89).

Wesley, J. (1772, 2nd ed.). *Some remarks on Mr. Hill's review of all the doctrines taught by Mr. John Wesley.* Bristol: Printed by W. Pine.

Wesley, J. (1786). Thoughts on nervous disorders; particularly that which is usually termed lowness of spirits, *The Arminian Magazine* (9, pp. 52–54, 94–97).

Wesley, J. (1770). *Thoughts on a single life,* in *WJW* (1872, 11: 456–63).

Wesley, J. (1745, 8th ed.). *A word in season: Or, advice to an Englishman.* Printed for W. Strahan.

Wesley, J. (1771–1774). *The works of the Rev. John Wesley, M.A., late Fellow of Lincoln College, Oxford.* Bristol: William Pine, 32 vols.

Wesley, J. (1872). *The works of the Rev. John Wesley,* ed. Thomas Jackson, London: Wesleyan Conference Office, 14 vols.

Wesley, J. (1975–2003). *The works of John Wesley,* Oxford edition. Oxford: Clarendon Press; after 1983–2003, Bicentennial edition, Nashville: Abingdon Press. *Sermon* series, 1–4, edited by Albert C. Outler, *Journal and diaries* series, 18–24, W. Reginald Ward & Richard P. Heitzenrater (eds.). Vol. 1, "Sermons I, 1–33," 1984; Vol. 2, "Sermons II, 34–70," 1985; Vol. 3, "Sermons III, 71–114," 1986; Vol. 4, "Sermons IV, 115–151," 1987; Vol. 11, The appeals to men of reason and religion and certain related open letters. Gerald R. Cragg (ed.), 1975; Vol. 18, "Journal and diaries, I, (1735–1738)," 1988; Vol. 19, "Journal and diaries, II, (1738–1743)," 1990; Vol. 20, "Journal and diaries, III, (1743–1754)," 1991; Vol. 21, "Journal and diaries, IV, (1755–1765)," 1992; Vol. 22, "Journal and diaries, V, (1765–1775)," 1993; Vol. 23, "Journal and diaries, VI, (1776–1786)," 1995; Vol. 24, "Journal and diaries, VII, (1787–1791)," 2003.

Whyte [Whytt], R. (1714–1766). *Observations on the nature, causes, and cure of those disorders which have been commonly called nervous, hypochondriac, or hysteric: to which are prefixed some remarks on the sympathy of the nerves.* The Second Edition, Corrected. Edinburgh: Printed for T. Becket, and P.A. De Hondt, London; and J. Balfour, Edinburgh, 1765.

Willis, T. (1621–1675). *An essay of the pathology of the brain and nervous stock: in which convulsive diseases are treated of.* Translated out of Latine [*Ceribri anatome*, 1664] into English, by S.P. London: Printed by J.B., 1681.

21

Franz Anton Mesmer and the Rise and Fall of Animal Magnetism: Dramatic Cures, Controversy, and Ultimately a Triumph for the Scientific Method

Douglas J. Lanska and Joseph T. Lanska

In the late eighteenth century, Franz Anton Mesmer (1734–1815) promulgated "animal magnetism" as a pervasive property of nature that could be channeled as an effective therapy for a wide variety of conditions (Fig. 1). His claims of dramatic therapeutic success were supported by glowing testimonials, in some cases from socially prominent individuals. However, mainstream medical practitioners, professional societies, and political bodies rejected Mesmer and his treatment, and ultimately moved to eliminate Mesmer's practice and that of his disciples. In retrospect it is clear that traditional physicians in the late eighteenth century had little to offer their patients therapeutically that had any real possibility of benefit,[1] and instead, often harmed their patients with their treatments, whereas Mesmer could demonstrate cases "cured" by his treatment that had previously failed all conventional approaches. While one might be tempted to dismiss his therapeutic successes as only applicable to hysterical or imagined illness, some of his patients went on to lead quite functional lives when before they were deemed hopeless invalids, a point that even his detractors acknowledged.

Mesmer and the Introduction of Animal Magnetism

Mesmer, a German by birth, studied medicine for 6 years in Vienna before presenting his dissertation for the degree of doctor of medicine in 1766. His dissertation, De Planetarum Influxu ("On the influence of the planets") (Mesmer, 1766, 1980), attempted to relate the motion of the planets with effects in humans, but was largely plagiarized from a book published in 1704 by the acclaimed English physician Richard Mead (Pattie, 1956; Pattie, 1994). Mesmer's dissertation is of consequence

---

[1] As noted by Golub (1994, p. 55–57): "[Therapeutics] had not changed significantly during almost two millennia prior to 1800 . . . For all practical purposes, Galen and the gentleman physician of eighteenth century London or Paris treated patients virtually the same way." The few effective preventatives or treatments available to eighteenth-century physicians included variolation for the prevention of smallpox (e.g., Boyslton, 1726; Franklin, 1759; Jurin, 1723; Massey, 1723; Montagu, 1717, 1861, 1970; Nettleton, 1722, 1723; Woodward, 1714) (and later vaccination with cowpox as introduced by Edward Jenner at the very end of the eighteenth century), fresh fruit or fruit juice for the prevention and treatment of scurvy as recommended by James Lind (though largely ignored at that time) (Lind, 1753, 1762), cinchona for treatment of fevers and malaria (introduced from Peru by the 1600s), willow bark (containing salicin) for fever or pain (Stone, 1764), narcotics such as opium and laudanum (a tincture of opium) for pain (known to Hippocrates), marginally effective mercurials for syphilis (introduced perhaps in the late fifteenth century), and foxglove (containing digitalis) for "dropsy" (Withering, 1785).

FIGURE 1. Franz Anton Mesmer (1734–1815) promulgated "animal magnetism" as a pervasive property of nature that could be channeled as a universal cure for disease. He achieved his height of fame and fortune in Paris before his magnetic doctrine was demolished by the scientific evaluation of the Royal Commission in 1784. Courtesy of the National Library of Medicine

only in retrospect, because Mesmer subsequently cited it in an attempt to claim priority for his conceptualization of animal magnetism. Despite the later course of his career, Mesmer's approach to medicine was basically orthodox during his first eight years of practice.

After his marriage in 1768 to a wealthy widow, Anna Maria von Posch, Mesmer was prosperous and socially well-positioned in Vienna, even to the point of entertaining the family of the young Wolfgang Amadeus Mozart (1756–1791) and perhaps staging the first performance of Mozart's opera *Bastien et Bastienne* in his garden theater in 1768 (when Mozart was 12).

The Hysterical Miss Österlin and a Treatment from Hell

A defining case for Mesmer's career was that of Franziska ("Franzl") Österlin, a 28-year-old woman with hysteria (she would now meet diagnostic criteria for somatization disorder) (American Psychiatric Association, 2000), who "since her

childhood, seemed to have a very weak nervous manner, had undergone terrible convulsive attacks since the age of two ... [and] had an hysterical fever to which was joined periodically, persistent vomiting, inflammation of various visceral organs, retention of urine, excessive toothaches, earaches, melancholic deliriums, opisthotonos ... blindness, suffocation, and several days of paralysis and other irregularities" (Mesmer, 1775, 1980, p. 26). Mesmer initially tried to treat the young woman in his home using "the most accredited remedies to counteract these different ailments ... without, obtaining, however, a lasting cure, for the irregularities always returned after some time" (Mesmer, 1775, 1980, p. 26). Despite Mesmer's efforts using orthodox medical treatments, including blistering, bleeding, and various medicines, no progress was made over a period of 2 years.

In late 1774, Mesmer was introduced to a new form of treatment with magnets by the Reverend Father Maximillian Hell (1720–1792), a Jesuit priest and the Austrian Astronomer Royal. Several months earlier, in June 1774, Hell had lent a heart-shaped steel magnet (magnetized by repetitive stroking with a lodestone) to a baroness afflicted with intractable abdominal pain. Four days later, the baroness was restored to health, and Hell ultimately concluded that the magnet had produced curative effects by acting on the nervous system. Hell suggested Mesmer try his magnets on Miss Österlin, who had suffered a relapse of hemiplegia in July 1774.

Taking Father Hell's advice, Mesmer attached Hell's magnets to Miss Österlin's feet and another heart-shaped magnet to her chest with dramatic results.

She soon underwent a burning and piercing pain which climbed from her feet to the crest of the hip bone, where it was united with a similar pain that descended from one side – from the locality of the magnet attached on the chest – and climbed again on the other side to the head, where it ended in the crown. This pain, in passing away, left a burning heat like fire in all the joints. (Mesmer, 1775, 1980, pp. 26–27)

Despite pleas from the patient and Mesmer's assistants that the treatment be terminated, Mesmer not only persisted, but added further magnets, continuing the treatment through the night. Gradually after the symptoms waned and ultimately disappeared, Mesmer pronounced her cured. Several subsequent relapses were easily addressed with further magnetic applications, so Mesmer advised her to wear several magnets as a prophylactic.

A controversy over the distribution of credit for this apparent therapeutic success followed with a series of alternating public "letters" by Hell and Mesmer (Pattie, 1994). Father Hell published the first letter on January 6, 1775, reporting Mesmer's successful application of the magnetic therapy to Miss Österlin, but claiming for himself the idea of treating such patients with magnets. Affronted by Hell's attempt to take credit for the magnetic cure, Mesmer immediately published his account in the newspapers and as a pamphlet.

In his public rebuttal to Father Hell, Mesmer claimed priority for the concept of using magnets therapeutically, stating that he had written in his doctoral thesis in 1766 on a property of the animal body that makes it sensitive to universal gravitation, a property he said he had labeled "gravity … or animal magnetism" (Mesmer, 1775, 1980, p. 25). However, the term "animal magnetism" was not, in fact, used in the dissertation, and the property that was described ("gravitus animalis") subsequently shifted in Mesmer's usage from a force that acts upon the body to a property of the body itself (Pattie, 1994).

In any case, Mesmer claimed that Hell's magnets were superfluous for the magnetic therapy, because virtually any object could be magnetized and used therapeutically.

I observed that magnetic material is almost the same thing as electrical fluid, and that it is propagated by intermediary bodies in the same way as is electrical fluid. Steel is not the only substance that attracts the magnet; I have magnetized paper, bread, wool, silk, leather, stones, glass, water, different metals, wood, men, dogs – in one word all that I touched – to the point that these substances produced the same effects upon the patient as does the magnet. (Mesmer, 1775, 1980, pp. 27–28)

Mesmer claimed to be able to fill bottles with this previously unrecognized magnetic material, and to direct it from a distance of 8–10 ft, even through other people or walls, so as to produce "jolts in any part of the patient that I wanted to, and with a pain as ardent as if one had hit her with a bar of iron" (Mesmer, 1775, 1980, p. 28). Despite the apparent brutality of the treatment, Mesmer was able to produce seemingly miraculous cures for a wide range of conditions.

By means of magnetism I restored menstrual periods and hemorrhoids to their normal condition … I cured hemoptysis, a paralysis following an apoplexy, an unexpected trembling after a fit of passion, and all kinds of hypochondriac, convulsive, and hysterical irregularities in the same way. (Mesmer, 1775, 1980, p. 28)

Mesmer proposed that "magnetic matter, by virtue of its extreme subtlety and its similarity to nervous fluid, disturbs the movement of the fluid in such a way that it causes all to return to the natural order, which I call the harmony of the nerves" (Mesmer, 1775, 1980, p. 29). But how could such a powerful force have escaped previous notice? Mesmer explained (conveniently so as to preclude refutation of his thesis) that such magnetic effects could not be perceived by healthy persons, but only by persons in whom "the harmony is disturbed" (Mesmer, 1775, 1980, p. 9).

Failed Solicitations in Vienna

Around 1775, Mesmer sent statements of his ideas on animal magnetism to a majority of the academies of science in Europe and to a few selected scientists, inviting their comments (Mesmer, 1779, 1980). The only reply he received, from the Berlin Academy in March, 1775, was dismissive, arguing reasonably that: (1) Mesmer's statements that magnetic effects could be communicated to materials other than iron and concentrated in bottles contradicted all previous experiments; (2) Mesmer's evidence – based on "the sensations of a person afflicted with convulsions" (Berlin Academy quoted in Pattie, 1994, p. 45) – was not adequate or even appropriate for proving the existence of the postulated animal magnetism; (3) the absence of detectable effects in healthy persons made the report of "animal magnetism" highly suspect; and (4) other explanations could account for the results obtained in patients (and indeed the Academy suspected Mesmer had "fallen into the fallacy of considering certain things as causes which are not causes") (Berlin Academy quoted in Pattie, 1994, p. 46).

Mesmer's attempts around this time to demonstrate the effects of animal magnetism to physician-scientist Jan Ingenhousz (1730–1799) were even more negative and publicly humiliating (Mesmer, 1779, 1948; Mesmer, 1779, 1980; Pattie, 1994). While Mesmer demonstrated the magnetism of a single teacup in a group and elicited convulsions by pointing a magnet toward a relapsed Miss Österlin,

Ingenhousz surreptitiously tested the effects of strong magnets which he had concealed. Ingenhousz found that the patient reacted only to objects which she believed were magnets or that were connected with Mesmer. As a result, Ingenhousz publicly denounced Mesmer as a fraud. In response, an incensed Mesmer publicly attacked Ingenhousz's scientific ability and demanded a court-ordered commission to establish the facts concerning his treatment of Miss Österlin. Mesmer's treatment was ultimately observed for 8 days by a local hospital physician, but the physician became cold and indifferent, a response Mesmer attributed to the machinations of Ingenhousz who "succeeded in having those who suspended judgment or who did not share his opinion classed as feeble-minded" (Mesmer, 1779, 1980, p. 55). Mesmer then temporarily abandoned efforts both to obtain a court-appointed commission and to disseminate his treatment into hospitals.

Controversy over Mesmer's Treatment of the Blind Miss Paradis

Through 1775 and 1776, Mesmer accumulated testimonials from several prominent individuals who reported being successfully treated by Mesmer, including Professor Bauer of the Vienna Normal School, Baron Hareczky de Horka, and Peter von Osterwald, Director of the Munich Academy (Pattie, 1994). However, controversies stemming from Ingenhousz's denouncement as well as Mesmer's failure to obtain public recognition from physicians, scientists, or scientific academies, caused Mesmer to attempt a dramatic cure of a difficult case, which he hoped would redeem his reputation and demonstrate to all observers the effectiveness of his discovery. Therefore, in 1777, Mesmer began treatment of the blind pianist, Maria Theresa Paradis (1759–1824), but the outcome of this therapeutic gamble was far worse for Mesmer than he anticipated.

Miss Paradis, the only child of a secretary to the Holy Roman Emperor Francis I (1708–1765) and Queen-Empress Maria Theresa (1717–1780), reportedly awoke with acute blindness at the age of 3 years and 7 months. She was treated by the most prominent Viennese physicians – with blistering plasters for two months, cauterization, leeches, purgatives, diuretics, and thousands of electric shocks through the eyes from discharging Leyden jars – but without the least success (Mesmer, 1779, 1980). She was ultimately

deemed incurable. Her parents tried to enrich the poor girl's life with music lessons, and she eventually became a talented singer and player of the clavichord and organ. The Empress attended one of her performances and became her patron when she was just eleven, providing her with a pension so she could continue her musical education.

Mesmer began treating Miss Paradis when she was 18 – at that time, totally blind with bulging eyes "so much out of place that as a rule only the whites could be seen" (Mesmer, 1779, 1980, p. 72), depressed, and with "deliriums which awakened fears that she had gone out of her mind" (Mesmer, 1779, 1980, p. 72). Under Mesmer's treatment, as attested by her father, she experienced trembling in her limbs, hyperextension of the neck, increased "spasmodic agitation in her eyes" (Mesmer, 1779, 1980, p. 72), severe head pains radiating to the eyes, dizziness, and other symptoms. Suddenly light bothered her eyes, and she was kept with her eyes bandaged in a dark room as "the slightest sensation of light on any part of her body affected her to the extent of causing her to fall" (Mesmer, 1779, 1980, p. 74). Only very gradually was she exposed to light and then was reportedly able to distinguish light and dark, as well as various colors, shapes, and faces, although with some reported distortion and limited understanding of what she saw.

She was frightened on beholding the human face: the nose seemed absurd to her and for several days she was unable to look upon it without bursting into laughter . . . Not knowing the name of the features, she drew the shape of each with her finger. One of the most difficult parts of the instruction was teaching her to touch what she saw and to combine the two faculties. Having no idea of distance, everything seemed to her to be within reach, however far away, and objects appeared to grow larger as she drew near to them . . . Nothing escaped her, even the faces painted on miniatures, whose expressions and attitudes she imitated. (Mesmer, 1779, 1980, p. 75)

Unfortunately, partial restoration of Miss Paradis' sight did not make her happy and threatened her financial support from the Empress. She became increasingly irritable, annoyed with the constant questions and testing, and prone to attacks of crying and syncope. Light bothered her, yet when her eyes were covered she became unable to take a step without guidance, whereas before, she was able to walk about her house in complete confidence. Her musical performances also suffered dramatically, and her father fretted that her royal pension might be terminated. In addition, a prominent Professor of Diseases of the Eye, Dr. Joseph Barth (1745–1818), became convinced that Miss Paradis could not really see, undermining Mesmer's claims of therapeutic success. A fracas ensued between the parents, patient, and Mesmer, with the absurd chain of events reportedly including a convulsion by the patient, an angry mother throwing her head-first against a wall, a sword-wielding father loudly demanding Mesmer release his daughter, the mother fainting, the servants disarming the father, the father swearing oaths and curses, and a relapse into blindness, vomiting, and rages by the patient.

Still Mesmer kept the patient under treatment, even in opposition to the pleading of the chief court physician, saying that Miss Paradis could not be released without danger of death. Within a month, Miss Paradis' vision had again been restored and her health was improved, her father was apologetic, and the public was invited to witness her recovery. When the patient was ultimately released after nearly six months of care, though, her family soon reported that she was still blind and prone to convulsions. Mesmer bitterly responded (possibly correctly) that the parents had a conflict of interest and "compelled her to imitate fits and blindness" (Mesmer, 1779, 1980, p. 63) so as to retain her pension. In any case, Miss Paradis forever after lived the life of a blind person.

Following his ignominious public failure in the treatment of Miss Paradis, Mesmer found himself thoroughly discredited and derided in Vienna – with absolutely no supporters among the medical profession – and he ultimately left for Paris in January, 1778.

Dissemination of Animal Magnetism: Lay Versus Professional Channels

A Lucrative Practice in Paris

Mesmer arrived in Paris in February 1778 and, despite his previous humiliation in Vienna, quickly established an extremely lucrative practice, fostered by his charismatic personality and his unshakeable belief in the importance of his discovery of animal magnetism. Prerevolutionary Paris society was much more open than Vienna, and Parisians were periodically "carried away by sensational reports of novelties, inventions, and scientific and medical marvels … [making] Paris a fertile ground for dissemination of the magnetic doctrine" (Pattie, 1994: 69). Patients, many of them from the nobility and upper classes, flocked to Mesmer for treatment, even while others labeled him a charlatan who had been forced to flee Vienna (Pattie, 1994). Mesmer was soon operating at the top of the Parisian social pyramid, actively seeking patients and admirers of high prestige and ultimately collecting among his adherents Queen Consort Marie Antoinette (1755–1793), a fellow Austrian; Charles-Phillip, Count d'Artois (1757–1836), one of the two younger brothers of King Louis XVI (and later, himself, King Charles X); and Marquis de Lafayette (1757–1834), a young aristocrat who would later become an American Revolutionary War hero and proselytizer for mesmerism in America.

Mesmer was in fact so inundated with patients in Paris that he devised a method of mass treatment using various rituals and paraphernalia, including most notably a device called a *baquet*, a large wooden vat of "magnetized" water with 20 or so protruding bent metal rods (Fig. 2). The *baquet* was placed in the center of the magnetization room so that numerous patients could simultaneously stand or sit around it while applying the metal rods to their afflicted areas. Simultaneously, Mesmer and his assistants moved about the room directing magnetic energy at the afflicted, either with metal wands or manually: "Patients are magnetized by the laying of hands & the pressure of fingers on the hypochondria & lower abdominal areas; the contact often maintained for a considerable time, sometimes a few hours" (Franklin et al., 1784, 1997, p. 69). The flow of animal magnetism was facilitated further by having patients hold hands, by careful placement of mirrors (purportedly to reflect the magnetic energy toward the patients), by looping a knotless rope around them (as knots supposedly would impede the flow of the magnetic fluid), and by certain sounds (which also would communicate the postulated fluid). Ethereal sounds were provided either by a glass harmonica[2] (Finger, 2006), a piano or singing.

---

[2] A glass harmonica (or armonica) was a musical instrument invented by Benjamin Franklin that incorporated a series of graduated revolving glass bowls made to vibrate like water glasses by contact with the fingertips.

FIGURE 2. *The Magnetism*, drawn by Sergent, engraved by Toyuca, ca. 1785. Fashionable Parisians are shown participating in a group treatment or séance around a *baquet* (French for tub or vat), which is filled with mesmerized water. The therapeutic magnetism was purported to be transferred through the moveable iron rods protruding through the *baquet* to the ailing body parts, thereby resolving obstructions to the free flow of animal magnetism within the body. Some patients experienced convulsive crises (as in the woman on the right) and had to be carried off to a padded crisis room (background). Courtesy of the Bakken Library and Museum, Minneapolis

Responses to the magnetic treatment varied widely but were sometimes quite dramatic, in both Mesmer's practice and that of his followers. As noted later in the practice of one of Mesmer's disciples, "Some are calm, quiet, & feel nothing; others cough, spit, feel slight pain, a warmth either localized or all over, & perspire; others are agitated & tormented by convulsions" (Franklin et al., 1784, 1987, p. 69). Some patients experienced violent convulsions during the treatments, sometimes requiring further management in an adjoining padded room.

These convulsions are extraordinary in their number, duration, & strength. As soon as a convulsion begins, many others follow ... some lasting for more than three hours ... These convulsions are characterized by quick, involuntary movements of limbs & the entire body, by a tightening of the throat, by the twitching of the hypochondria & epigastric area, by blurred & unfocused vision, by piercing shrieks, tears, hiccups & excessive laughter. They are preceded or followed by a state of languor & dreaminess, of a kind of prostration & even sleepiness. (Franklin et al., 1784, 1987, p. 69)

Thwarted Dissemination Through Academic and Professional Channels

During his time in Paris, Mesmer sought testimonials attesting to the value of his discovery from the Royal Academy of Sciences (*Académie des Sciences*), the Royal Society of Medicine (*Société Royale de Médecine*), and the Faculty of Medicine (*Faculté de Médecine*), believing that these societies would confirm what his many patients and the general public already acknowledged de facto. However, Mesmer was repeatedly rebuffed or ignored.

An attempt to demonstrate animal magnetism before a meeting of the Academy of Sciences in early 1778 was received poorly and failed to convince any of the attendees. Later Mesmer was asked by two members of the Academy to demonstrate the utility of his supposed discovery by curing patients. Mesmer embarked on several months of treatment of a group of patients in a village near Paris, but Mesmer's subsequent entreaties for a review of the success of his treatment by the Academy were discussed and dismissed without a reply to Mesmer. There was indeed no way of validating any treatment response by interviewing or examining these patients at the end of their treatment – no clear baseline had been established, and other potential factors impacting on outcome (e.g., natural history of the conditions, placebo effects, etc.) had not been addressed.

Mesmer's subsequent attempt to solicit members of the newly founded Royal Society of Medicine fared no better. The Royal Society was responsible for oversight and regulation of new remedies, and on this basis its representatives suggested the appointment of a commission to investigate Mesmer's animal magnetism. However, Mesmer refused the Society's proposal on the grounds that he had no medicine to patent or license, that he did not wish to trust the fate of his doctrine to commissioners unknown to him, and further that he did not wish his therapy lumped among the licensed drugs (that he undiplomatically alleged were nothing more than poisons). Instead, Mesmer entreated the society to simply accept the testimonials of his patients and "be witnesses of the salutary effects of my discovery, to assert its truth while rendering homage to it, and by this simple means to merit the gratitude of the nations" (Pattie, 1994, p. 82). After further haggling, the Royal Society and Mesmer at least temporarily agreed that he would treat patients previously certified by physicians of the Faculty of Medicine so that the success of his treatment could be judged; however, when the physicians charged with this certification had difficulty establishing the presence of disease in Mesmer's patients, Mesmer doubted that they would be any less hesitant to certify the cures he anticipated *after* his treatments. When a commission was nevertheless appointed, Mesmer adamantly stated he would not even receive the commissioners, whereupon the society discharged the commission and terminated any further consideration of Mesmer and his treatment: the official response stated, "The commissioners whom the Society has appointed at your request to follow your experiments, cannot and should not render any opinion without having previously certified the condition of the patients by mean of a careful examination" (Pattie, 1994, p. 83).

Never very diplomatic, typically grandiose, and frequently somewhat paranoid, Mesmer in frustration charged that if his techniques were disseminated among even a small number of physicians, the rest of the medical profession would be forced to see him and his disciples as dangerous enemies who threatened their profits, and in their greed would attempt to undermine and destroy his doctrine (Pattie, 1994).

By 1780, Mesmer was able to recruit only one disciple of high professional and social standing – Dr. Charles d'Eslon or Deslon (1750–1786), who held the highest rank (*docteur-régent*) in the Faculty of Medicine and who was the personal physician (*premier médecin*) of Count d'Artois. d'Eslon observed Mesmer's practice and became a true believer in Mesmer's ability to cure patients using animal magnetism, although d'Eslon admitted he did not understand fully how Mesmer

accomplished this. d'Eslon tried to raise interest among members of the Faculty of Medicine and selected three physicians to observe Mesmer's work every two weeks over a period of seven months. However, the doctors remained unconvinced and could not decide how many of the apparent cures could be attributed to treatment and how many resulted from spontaneous recoveries. When d'Eslon defended Mesmer to the Faculty of Medicine and wrote a book supporting Mesmer's therapy (d'Eslon, 1780), the Faculty became openly hostile and unanimously censured d'Eslon. Mesmer nevertheless refused to acknowledge d'Eslon as a qualified disciple. Later, when Mesmer learned that d'Eslon had established a clinic of 60 patients where he produced cures using animal magnetism, Mesmer became enraged and charged d'Eslon with betrayal, breach of promises, and theft of his ideas and techniques.

The Society of Harmony: Dissemination Through Lay Disciples

Nicolas Bergasse (c. 1750–?), an unhappy young lawyer, began seeing Mesmer as a patient in 1781 and believed that Mesmer significantly improved his health. So, with growing ambivalence and eventual resentment, Bergasse began serving Mesmer as an unpaid secretary, writer, and tutor of French. Bergasse wrote public defenses of Mesmer's ideas (in much better French than Mesmer could muster), and became among the clearest expositors and disseminators of mesmerism, trying to establish a coherent doctrine from among Mesmer's vague and inconsistent statements and writings. In 1783, Bergasse proposed and was the primary architect and developer of the Society of Harmony (*Société de l'Harmonie*), a secret society of wealthy patrons who paid handsomely to ensure Mesmer's fortune and signed nondisclosure covenants with severe penalties for any breech, with the understanding that when sufficient subscriptions had been sold Mesmer would reveal his system to them for their own use. However, although Mesmer collected an incredible sum – some 400,000 *livres* – he continued to manipulate the members, while never fulfilling his verbal agreement. Still, such mesmeric societies proliferated across France and eventually spread to other countries.

Evaluation of Animal Magnetism: The Royal Commissions (1784)

Appointment of the Royal Commissions

The popularity of mesmerism alarmed the physicians and the government. The orthodox practitioners saw Mesmer – with his lucrative practice, his aristocratic patronage, and his recruitment of one of their most prominent members – as an economic threat to their own practices. The monarchy, nobility, and police also began to see mesmerism and its secret societies as a threat, especially as Bergasse and other revolutionary agitators in the Society for Harmony opposed the established order of the *ancien regime* and helped propagate subversive ideas (Darnton, 1968). The controversy over animal magnetism escalated with open dissention among Mesmer's disciples and increasing hostility from various academic and professional opponents.

Eventually, King Louis XVI (1754–1793), being less enthralled than his wife with Mesmer and his treatments, and concerned with the intensifying controversy, established a Royal Commission of the Royal Academy of Sciences and the Faculty of Medicine to evaluate Mesmer's claims (Franklin et al., 1784, 1997; Franklin et al., 1784, 2002). The distinguished Commission included four members from the Faculty of Medicine and five members from the Royal Academy of Sciences, including diplomat-scientist Benjamin Franklin (1706–1790), America's Minister Plenipotentiary to France, as well as chemist Antoine-Laurent Lavoisier (1743–1794), astronomer Jean-Sylvian Bailly (1736–1793), physician Joseph-Ignace Guillotine (1738–1814), Jean François Borie, professor Charles Louis Sallin, physician and chemist Jean Darcet (1725–1801), geographer and cartographer Gabriel de Bory (1720–1801), and physician Michel Joseph Majault (Duveen & Klickstein, 1955). A second commission was also established, drawn from the Royal Society of Medicine, but their report was largely redundant and will not be further discussed.

Justification of the Commissioners' Investigative Approach

The Commission was charged "to examine & report on animal magnetism practiced by *Monsieur* Deslon" (Franklin et al., 1784, 1987, p. 68) and not the practice of Mesmer himself. The rationale for this choice was not disclosed (Pattie, 1994), but Mesmer naturally objected: "I do not want him [d'Eslon] to determine the destiny of a doctrine which belongs to me, and whose importance and extent I alone know, I am bold enough to say …" (Pattie, 1994, p. 144). The Commissioners disagreed that there was any significant difference in the practices of d'Eslon and Mesmer, and in any case believed that their evaluation applied to the practice of animal magnetism in general and not to the specific practice of an individual practitioner:

These principles of M. Deslon are the same as those in the twenty-seven propositions that M. Mesmer made public through publication in 1779 … Now it is easy to prove that the essential practices of magnetism are known to M. Deslon. M. Deslon was for several years the disciple of M. Mesmer. During that time, he constantly saw the employment of the practices of Animal magnetism & the means of exciting it & directing it. M. Deslon himself has treated patients in front of M. Mesmer; elsewhere, he has brought about the same effects as at M. Mesmer's. Then, united, the one & the other combined their patients & treated them without distinction, & consequently following the same procedures. The effects correspond as well. There are crises as violent, as multiplied & as pronounced by similar symptoms at M. Deslon's as at M. Mesmer's; these effects therefore do not belong to a particular practice, but to the practice of magnetism in general. (Franklin et al., 1784, 1987, p. 83)

The Commissioners understood their purpose was:

to unravel the causes & to search for proofs of the existence & the utility of magnetism. The question of existence is primary; the question of utility is not to be addressed until the first has been fully resolved. Animal magnetism may well exist without being useful but it cannot be useful if it does not exist. (Franklin et al., 1784, 1987, p. 70)

The Commissioners judged that Mesmer's theory supporting the practice of animal magnetism was irrelevant to the question of whether the phenomenon actually existed.

If M. Mesmer announces today a more encompassing theory, there is no need whatsoever for the Commissioners to know this theory to decide on the existence and utility of magnetism. They had only to consider the effects. It is by the effects that the existence of a cause manifests itself; it is by the same effects that its utility may be demonstrated. Phenomena are known through observation a long time before one can reach the theory that links them & which explains them … The theory of M. Mesmer is immaterial & superfluous here; the practice, the effects, it has been a question of examining these. (Franklin et al., 1784, 1987, p. 83)

They recognized that, "The most reliable way to ascertain the existence of animal-magnetism fluid would be to make its presence tangible" (Franklin et al., 1784, 1987, p. 70). However, the existence of the animal magnetism could not be proven by its physical properties, because the magnetic fluid was claimed to be an intangible agent.

[This] fluid escapes detection by all the senses. Unlike electricity, it is neither luminescent nor visible [as is lightning]. Its action does not manifest itself visibly as does the attraction of a magnet; it is without taste or smell; it spreads noiselessly & envelops or penetrates you without your sense of touch warning you of its presence. (Franklin et al., 1784, 1987, p. 70)

Thus, the existence of animal magnetism could only be determined by any effects it might have on human behavior or disease.

In 1780, on behalf of Mesmer, d'Eslon had proposed a comparative trial of animal magnetism versus conventional medical therapy to the Faculty of Medicine (Donaldson, 2005; Mesmer 1781, 2005), and in 1784, d'Eslon similarly advised the Commissioners to study principally the therapeutic effects of animal magnetism, but the Commissioners rejected an assessment of the effects of animal magnetism in the treatment of diseases. They acknowledged the existence of cases where seriously ill patients had not responded to "all means of ordinary medicine" (Franklin et al., 1784, 1987, p. 71) and yet had fully recovered after treatment with magnetism. However, it was impossible, the Commissioners reasoned, to separate the effects of spontaneous recovery from the effects of treatment.

Observations over the centuries proves [sic] & Physicians themselves recognize, that Nature alone & without the help of medical treatment cures a great number of patients. If magnetism were inefficacious, using it to treat patients would be to leave them in the hands of Nature. In trying to ascertain the existence of this agent, it would be absurd to choose a method that, in attributing to the agent all of Nature's cures, would tend to prove that it has a useful & curative action, even though it would have none. (Franklin et al., 1784, 1987, p. 71)

To defuse potential arguments that the Commissioners had ignored the evidence that animal magnetism

cured disease, the Commissioners cited Mesmer's own statement in this regard.

The Commissioners are in agreement on this with M. Mesmer. He rejected the cure of diseases when this way of proving magnetism was proposed to him by a Member of the Académie des Sciences: *it is*, said he, *a mistake to believe that this kind of proof is irrefutable; nothing conclusively proves that the Physician or Medicine heals the sick*. The treatment of diseases, therefore can only furnish results that are always uncertain & often misleading. (Franklin et al., 1784, 1987, p. 71)

Therefore, the Commissioners chose to restrict their investigations "to the temporary effects of the fluid on the animal body, by stripping these effects of all illusions possibly mixed up with them, & making sure that they cannot be due to any cause other than animal magnetism" (Franklin et al., 1784, 1987, p. 71).

Observational Studies and Hypothesis Generation

The Commissioners verified the absence of an electrical charge or magnetic field associated with the *baquet* used during the group treatments:

The Commissioners used an electrometer[3] & a non-magnetic, metal needle to check that the vat did not contain any electrical or charged matter; and upon the declaration of M. Deslon [d'Elson] regarding the composition of the inside of the vat, they agreed that no physical agent

---

[3] The precise instrument used by the Commissioners is unknown. Various electroscopes and electrometers were in use at the time and the terminology employed was not consistent. Eventually "electroscope" was used for instruments that could detect the presence of an electrostatic charge, whereas "electrometer" was used for instruments that could quantify such charges. John Canton made one of the first portable electroscopes in 1754 (Canton, 1754; Herbert, 1998). This instrument utilized a pair of pith balls hung on linen threads, while later electroscopes utilized a pair of thin gold leaves attached to a conducting rod and held in an insulated frame. When a charge was applied to the instrument, the balls or leaves moved apart, due to mutual repulsion of like charges. In 1772, William Henley described a quadrant electrometer which utilized a single cork ball hung by a thread from a stem; when the electrometer was charged, the ball was repelled from the stem and the divergence of the ball from the stem was measured on a quadrant scale.

capable of contributing to the reported effect of magnetism was present. (Franklin et al., 1784, 1987, p. 69)

The Commissioners also observed group treatment sessions to familiarize themselves with the practice of animal magnetism, witness the range of apparent effects, and formulate their own initial hypotheses for the observed phenomena. They were absolutely astounded by the magnitude of the responses of patients during the séances:

Nothing is more astonishing than the spectacle of these convulsions; without seeing it, it cannot be imagined: & in watching it, one is equally surprised by the profound response of some of these patients & the agitation that animates others . . . All submit to the magnetizer; even though they may appear to be asleep, his voice, a look, a signal pulls them out of it. Because of these constant effects, one cannot help but acknowledge the presence of a great power which moves & controls patients, & which resides in the magnetizer. (Franklin et al., 1784, 1987, p. 69)

The Commission realized that the group séances were too complex to sort out the factors responsible for the observed effects. A simpler setting was needed in order to isolate and control the underlying factors: "The freedom to isolate the effects was necessary in order to distinguish the causes; one must like them have seen the imagination work, partially in some way, to produce its effects separately & in detail, so as to conceive of the accumulation of these effects, to get an idea of its total power & take account of its wonders" (Franklin et al., 1784, 1987, p. 82). Therefore, the Commission chose to observe the responses to the treatment of individual subjects separated from the communal psychological influences of the group treatment.

The Commissioners themselves were magnetized in a private setting so they could experience the effects, if any, firsthand. They were magnetized once a week by d'Eslon or a disciple in a separate room.

[They] stayed for two to two & a half hours at a time, the iron rod resting on the left hypochondrium, & themselves surrounded by the rope of communication, & from time to time making the chain of thumbs . . . they were magnetized, sometimes with the finger & iron rod held & moved over various parts of the body, sometimes by applying hands & finger pressure to either the hypochondria or on the pit of the stomach. None of them felt a thing, or at least, nothing that could be attributed to the action of magnetism. (Franklin et al., 1784, 1987, p. 72)

Nor did they experience any effects when they were magnetized for 3 days in a row. The contrast could not have been greater between the dramatic effects they observed among patients during the group treatments and the absence of effects they experienced during their own private treatments. They concluded that "magnetism has little or no effect on a state of health, & even on a state of slight infirmity" (Franklin et al., 1784, 1987, p. 72).

The Commissioners next observed the effects of private application of the magnetic treatment to sick patients. Of the first seven patients, all commoners, three felt some effects (e.g., local pain, headache, or shortness of breath), and four felt nothing. The next seven patients were "chosen from high society who could not be suspected of ulterior motives & whose intelligence would permit them to discuss their own sensations & report on them" (Franklin et al., 1784, 1987, p. 72) and *none* of these felt anything that could be attributed to magnetism. The difference, the Commissioners reasoned, stemmed from the commoners' expectations and desire to please.

Let us take the standpoint of a commoner, for that reason ignorant, struck by disease & desiring to get well, brought with great show before a large assembly composed in part of physicians, where a new treatment is administered which the patient is persuaded will produce amazing results. Let us add that the patient's cooperation is paid for, & that he believes that it pleases us more when he says he feels effects, & we will have a natural explanation for these effects; at the least, we will have legitimate reasons to doubt that the real cause of these effects is magnetism. (Franklin et al., 1784, 1987, p. 74)

The Commission observed that magnetism "seemed to be worthless for those patients who submitted to it with a measure of incredulity [and] that the Commissioners . . . did in no way feel the impressions felt by the three lower-class patients" (Franklin et al., 1784, 1987, p. 74). Therefore, the Commissioners hypothesized that the effects observed in the lower-class patients "even supposing them all to be real, followed from an anticipated conviction, & could have been an effect of the imagination" (Franklin et al., 1784, 1987, p. 74). The generation of this rival hypothesis to Mesmer's animal magnetism focused all subsequent investigations: "From now on, their research is going to be directed toward a new object; it is a question of disproving or confirming this suspicion, of determining up to what point the imagination can be the cause of all or part of the effects attributed to magnetism" (Franklin et al., 1784, 1987, p. 74).

Experiments to Decide Between Rival Hypotheses: Animal Magnetism and Imagination

To decide between the rival hypotheses, the Commissioners conducted a series of experiments, actively intervening to systematically isolate and independently vary each possible explanatory factor (e.g., magnetization, expectation, knowledge of the body part magnetized), while holding all other factors constant. By this experimental approach, the Commissioners demonstrated that magnetization had no effect: subjects developed the characteristic mesmeric crises if and only if they *expected* to be magnetized, regardless of whether they were actually magnetized.

By misleading subjects to believe they were being magnetized when they were not, the Commissioners were able to demonstrate the full range of mesmeric effects, including the characteristic crises. In one experiment the Commissioners seated a woman by a door and told her that d'Eslon was magnetizing her from the other side when in fact she was not being magnetized at all.

It was barely a minute of sitting there in front of that door before she began to feel shivers. A minute after that she started to chatter even though she felt generally warm; finally, after the third minute she fell into a complete crisis. Her breathing was racing, she stretched both arms behind her back, twisting them strongly & bending her body forward; her whole body shook. The chatter of teeth was so loud that it could be heard from outside; she bit her hand hard enough to leave teeth marks. (Franklin et al., 1784, 1987, p. 77)

Such demonstrations showed that the effects attributed to animal magnetism could be produced solely by suggestion in the absence of magnetization.

Demonstrations confirming that suggestion could produce apparently similar consequences to that achieved by practitioners of animal magnetism were not sufficient to falsify the rival animal magnetism hypothesis. To provide convincing evidence, the Commissioners conducted simple controlled experiments that would unambiguously support one

hypothesis while refuting the other (experiments probably designed mostly by Lavoisier with input from Franklin and the other Commissioners) (Duveen & Klickstein, 1955; Pattie, 1994). Predictions based on the logical consequences of each provisional explanation could be objectively tested by assessing the observed consequences of the experiments – assuming that the experimental methods were sound, correspondence between the predictions and observed consequences of experiment provided some support for the hypothesis, while lack of correspondence meant that the hypothesis should be rejected (Harré, 1981).

A woman – the door-keeper of Commissioner le Roy – felt heat or moving flames on whatever area of her body was magnetized, but the Commissioners found by blindfolding her that this correspondence was present only if she knew where the magnetization was applied: "when the woman could see, she placed her sensations precisely on the magnetized area; whereas when she could not see, she placed them haphazardly & in areas far from those being magnetized" (Franklin et al., 1784, 1987, p. 74). In further experiments, she experienced similar mesmeric effects even if *nothing* was done to her if she believed that she was being magnetized: "The results were the same, even though nothing was done to her from near or afar; she felt the same heat, the same pain in her eyes & ears; she also felt heat in her back & loins" (Franklin et al., 1784, 1987, p. 74).

Another young woman (previously established as magnetically sensitive) was invited to an apartment on the pretext that she was being considered for a job as a seamstress. There she conversed cheerfully with a female confederate of the experimenters while without her knowledge one of the Commissioners magnetized her through a concealed doorway for a half hour to no effect.

In Passy she had fallen into a crisis after three minutes; here she endured magnetism for thirty minutes without any effect. It is just that here she did not know she was magnetized, & in Passy she believed that she was. (Franklin et al., 1784, 1987, p. 78)

When the same Physician-Commissioner moved so that the patient was aware of his magnetization efforts, she was easily magnetized, even to a characteristic crisis.

[After] three minutes, [she] felt ill at ease & short of breath; then followed interspersed hiccups, chattering of the teeth, a tightening of the throat & a bad headache; she anxiously stirred in her chair; she complained about lower back pain; she occasionally tapped her feet rapidly on the floor; she then stretched her arms behind her back, twisting them strongly.... She suffered all this in twelve minutes whereas the same treatment employed for thirty minutes found her insensitive. (Franklin et al., 1784, 1987, p. 79)

Suggestion was also enough to terminate the effects. Even when continuing the magnetization efforts, the Commissioner said it was time to finish.

[Nothing] therefore had changed, the same treatment should have continued the same impressions. But the intention was enough to calm the crisis; the heat & headache dissipated. The areas that hurt were attended to one after the other, while announcing that the pain would disappear. In this way, the [Commissioner's] voice, by directing the [subject's] imagination, caused the pain in the neck to stop, then in succession the irregularities in the chest, stomach & arms. It took only three minutes; after which [she] declared that she no longer felt anything & was absolutely back in her natural state. (Franklin et al., 1784, 1987, p. 79)

For another experiment, the Commissioners had d'Eslon magnetize an apricot tree in Franklin's garden in Passy, while four other trees were left nonmagnetized. According to the magnetic doctrine, "When a tree has been touched following principles & methods of magnetism, anyone who stops beside it ought to feel the effect of this agent to some degree; there are some who even lose consciousness or feel convulsions" (Franklin et al., 1784, 1987, p. 76). A young man, deemed by d'Eslon to be magnetically sensitive, was blindfolded, led to each nonmagnetized tree, and asked to hug the tree for 2 min. At the first nonmagnetized tree he experienced diaphoresis, coughing, and mild headache. At each successive nonmagnetized tree, he experienced progressively more severe effects with increasing dizziness and headache until he collapsed unconscious with limbs stiffened under the fourth nonmagnetized tree, 24 ft from the magnetized apricot tree. d'Eslon of course objected when the observed results conflicted with his predictions, but the Commissioners simply discounted d'Eslon's objections.

M. Deslon [d'Eslon] tried to explain what happened by saying that all trees are naturally magnetized & that their own magnetism was strengthened by his presence. But in that case, anyone sensitive to magnetism could not

chance going into a garden without incurring the risk of convulsions, an assertion contradicted by everyday experience. (Franklin et al., 1784, 1987, p. 76)

In a similar experiment, a magnetically sensitive woman was seen in Lavoisier's Arsenal and offered several cups of water, one after the other, only one of which was magnetized. With each successive nonmagnetized cup, she too experienced progressive effects until she developed a crisis with the fourth cup. When she then asked for some water to drink, a shrewd Commissioner passed her the *magnetized* cup from which "she drank quietly & said she felt relieved" (Franklin et al., 1784, 1987, p. 77). Later, while her attention was focused elsewhere, the same magnetized cup was held at the back of her head for several minutes, yielding no effect.

The experimental subjects were deliberately misled about the purpose and conduct of these experiments, and one subject was experimented upon without her knowledge or consent with a treatment that was reputed to produce painful crises. The Commissioners justified such actions by recourse to a higher authority (the King) and by weighing the anticipated benefits to the common good above the rights of individuals. As the Commissioners commented,

such examination requires a sacrifice of time, & much follow-up research which one does not always have the leisure to pursue for the purpose of instruction or satisfying one's own curiosity, or which one does not have even the right to undertake unless one is like the Commissioners charged by the King's orders & honored with the group trust. (Franklin et al., 1784, 1987, p. 82)

The Commission's Conclusions

The Commission's evidence supported their hypothesis that the effects attributed to animal magnetism were due to the subjects' own expectations of magnetization ("imagination"), and clearly refuted any effect of animal magnetism. The Commissioners had successfully induced and terminated crises by manipulating only the subjects' imaginations, demonstrating that suggestion was sufficient to produce the effects attributed to animal magnetism. Magnetization itself produced no effects without suggestion. Thus the Commissioners concluded, "The experiments just reported are consistent & also decisive; they authorize the conclusion that the imagination is the real cause of the effects attributed to magnetism" (Franklin et al., 1784, 1987, p. 78).

The Commission also criticized the genesis of the magnetic theory.

New causes are not to be postulated unless absolutely necessary. When the effects observed can have been produced by an existing cause, already manifested in other phenomena, sound Physics teaches that the effect observed must be attributed to it; & when one announces the discovery of a cause hitherto unknown, sound Physics also demands that it be established, demonstrated by effects that cannot be attributed to any known cause, & that can only be explained by the new cause. It would thus be up to the followers of magnetism to present other proofs & to look for effects that were entirely stripped of the illusion of the imagination. (Franklin et al., 1784, 1987, p. 78)

The Commissioners here effectively allude to Occam's razor, the principle of philosophy that states that explanatory assumptions must not be invented or multiplied unnecessarily, and therefore the simplest hypothesis based on existing knowledge is best.

Beyond their devastating scientific critique, the commissioners had further concerns about the potential moral dangers of animal magnetism that they communicated in a separate secret report to the King (Franklin et al., 1784, 2002): the prolonged close physical proximity between the magnetizers (all men) and their patients (predominantly women), and the sensitive condition of the patients, made the Commissioners fear that the practitioners of animal magnetism could take improper advantage of their patients. This document had little impact though on the practice of animal magnetism as it was not published until long after animal magnetism was already abandoned in France.

Abandonment of Animal Magnetism

More than 20,000 copies of the Commission's report were rapidly and widely distributed. Publication of the report eroded much of Mesmer's support base; greatly decreased his clientele; led to a series of satirical pamphlets, books, and stage plays; and helped shift popular opinion from support to scorn and ridicule (Pattie, 1994). Furthermore, the Faculty of Medicine soon acted to suppress professional practice or support of animal magnetism by expelling any partisan members. Despite such favorable public and professional response to the report, Franklin was not confident that it was sufficient to cause the abandonment of mesmerism, as he confided to his grandson William Temple Franklin on August 25, 1784.

The Report makes a great deal of talk. Everybody agrees that it is well written, but many wonder at the force of imagination described in it as occasioning convulsions, etc., and some feel that consequences may be drawn from it by infidels to weaken our faith in some of the miracles of the New Testament. Some think it will put an end to Mesmerism, but there is a wonderful deal of credulity in the world and deceptions as absurd have supported themselves for ages. (Duveen & Klickstein, 1955, p. 299)

Proponents of animal magnetism mounted a campaign to counteract the Commission report, using a barrage of hundreds of lay articles and pamphlets, including critiques of the Commission report and compilations of testimonials, but this had little effect and interest in animal magnetism dissipated. Mesmer threatened to leave France to avoid the spreading conflicts but was persuaded by members of the Society for Harmony to stay at least temporarily so his departure would not imply his acquiescence to the Commission's findings. Mesmer continued to practice animal magnetism for a short time in a greatly diminished capacity while trying to arrange an alternative evaluation of his own patient outcomes as opposed to those of d'Eslon. On April 29, 1785, eight months after the Commission report was published, Franklin wrote in a letter to Ingenhousz:

Mesmer continues here and has still some Adherents and some Practice. It is surprising how much credulity still subsists in the World. I suppose all the Physicians in France put together have not made so much money during the Time he has been here, as he has done. (Duveen & Klickstein, 1955, p. 301; Hirschmann, 2005, p. 832; Parish, 1990, p. 110; Pattie, 1994, p. 229)

However, Mesmer soon left Paris and lived the rest of his life in relative obscurity, ultimately dying in 1815 in Switzerland.

Discussion

Animal magnetism was a failed or aborted therapeutic technology that gained temporary popular
support but was never accepted by orthodox medicine. Certainly, during the period from 1778 until
the Commission reports in 1784, animal magnetism was in vogue and accepted by a wide spectrum
of Parisian society: patients flocked to Mesmer's
clinic for treatment and willingly paid the high
fees, in part because of Mesmer's self-confident,
charismatic personality; the novelty and relative
innocuousness of the treatment (e.g., compared
with bleeding, blistering, and purging); and various
public communication channels claiming dramatic
efficacy in the face of treatment failures with orthodox medicine. However, although Mesmer was
himself a physician, he failed to gain professional
support or endorsement from colleagues or any
medical or scientific societies. Mesmer did obtain
the support of a single initially influential colleague
(d'Eslon), but that colleague was then censured and
ostracized by the medical establishment and subsequently denounced by Mesmer. Mesmer himself
limited the dissemination of animal magnetism by
seeking to maintain sole control of the practice: he
never sanctioned anyone other than himself as
adequately qualified to use animal magnetism
therapeutically, but instead sought to acquire assistants and disciples with indefinite (and seemingly
perpetual) periods of apprenticeship. Ultimately,
animal magnetism was abandoned when its erroneous theoretical foundations were exposed.

Mesmerism had a limited resurgence in Britain in
the 1840s and 1850s (Winter, 1998), in the United
States in the early nineteenth century (Gravitz,
1994; McCandless, 1992; Roth, 1977; Tomlinson &
Perret, 1974; Wester, 1976), and in Germany
(Frankau, 1948). To this day mesmerism continues
to resonate in numerous cultural echoes, in the form
of carnival hypnotists, fringe healers, spiritualists,
Christian Science,[4] continued belief in the therapeutic value of magnets (Shermer, 2002), mainstream

advertising,[5] movies (Spottiswoode, 1993), and
indeed in the very fabric of language (e.g., with
continued, albeit altered, usage of the terms "animal
magnetism" and "mesmerize"). Webster's dictionary defines animal magnetism as "the power to
attract others through physical presence, bearing,
energy, etc. [or the] power enabling one to induce
hypnosis" (*Webster's Universal College Dictionary*,
1997, p. 32), while mesmerize is defined as "to hyp-
notize ... to spellbind, fascinate ... [or to] compel
by fascination" (*Webster's Universal College
Dictionary*, 1997, p. 504); none of these definitions
quite capture the eighteenth-century realities of
Mesmer's treatment (e.g., the word "hypnosis" was
introduced in the nineteenth century after Mesmer's
death) (Braid, 1843; Kihlstrom, 2002).

Was Mesmer a Quack?

Every era has had their "quacks" – fraudulent pre-
tenders to medical skill, knowledge, or qualifications who operate outside of mainstream medicine
and who are deemed by orthodox providers to be
unqualified charlatans. Many in his era and subsequently have labeled Mesmer a quack or charlatan
(MacKay, 1852, 1932; Pattie, 1994). Indeed,
although Mesmer had the credentials of an orthodox physician, he certainly adopted many of the
features of a quack (Mermann, 1990; Smith, 1985;
Wolf, 1980): focusing on a single treatment as a
panacea – claiming the treatment dramatically
cures or alleviates suffering for a wide range of
dissimilar conditions, including especially chronic,
disabling, or stigmatizing conditions felt to be
beyond the abilities of orthodox therapies; promoting the treatment outside of the conceptual framework of contemporary orthodoxy and unsupported
by accepted medical doctrine; incorporating complex rituals or paraphernalia into the administration
of the treatment; applying the treatment to individual patients without first establishing a clear diagnosis through accepted procedures; announcing
the new treatment in the lay press before it is presented in the traditional medical literature; ignoring or actively avoiding formal investigation of the
efficacy of the treatment, and producing instead

---

[4] Christian Science is a religious body founded in the
1870s by Mary Baker Eddy – a woman plagued with
emotional and physical illnesses, who initially claimed
she was cured by mesmerist Phineas Parkhurst Quimby
in 1862.

[5] For example, the term "animal magnetism" has been
used in advertising copy for animal-print lingerie by J. C.
Penny and other companies (J. C. Penny Co, 2004).

testimonials of patients – particularly celebrities "cured" by the treatment – whose diagnoses were not appropriately established in the first place; advertising the treatment directly to the public and to nonprofessional disciples using publicity in the lay media to increase public demand; and employing unseemly self-promotion with apparent avarice.

Nevertheless, even with the distance of two centuries, Mesmer is not so easily categorized as a simple quack or charlatan. Indeed, some have questioned whether he was possibly a "sincere believer, deluded no less than his patients in mistaking the power of suggestion for the physical effects of an actual substance" (Gould, 1989, p. 16), and others have considered him a "thoughtful student of medicine" (Waterson, 1909), a "student of human nature" (Walsh, 1923, p. 88), a "scientific pioneer" (Eden, 1957, p. v), a "brilliant innovator" (Schneck, 1959, p. 463), a "blind prophet" (McGrew & McGrew, 1985, p. 200), and the "father of modern psychotherapy" (Frankau, 1948, p. 9).

Mesmer was indeed a complex figure with a number of faults, but he played an important role in understanding the effects of suggestion on the imagination, and was a pivotal figure in the history of psychosomatic illness, psychotherapy, and therapeutic hypnosis (Kihlstrom, 2002). Arguably even more important, though, was the role that he and his therapy played in shifting therapeutic evaluation from anecdotes and testimonials to a critical scientific methodology. Without being sufficiently threatening to established medical and political order, a Commission would not have been necessary.

The Franklin Commission provided a devastating attack on the theory of "animal magnetism." The Commission focused not on the changes in health or quality of life of the treated patients, but instead focused on whether the supposed effects of animal magnetism could be consistently demonstrated, and on whether simpler explanations (e.g., suggestion and imagination of the subjects) could suffice to explain the observations. Unlike "mineral magnetism," whose effects could be repeatedly and consistently demonstrated (e.g., by attraction of ferromagnetic materials, by lines of force shown with scattered iron filings, etc.), the Commission found that animal magnetism varied most obviously with the expectations of the subjects. Without evidence to support the very existence of animal magnetism, there was deemed little need to study treatment outcomes.

The "Tomato Effect": Was a Therapeutic "Baby" Thrown Out with the Magnetic Water?

Tomatoes are of South American origin and were introduced to Europe in the sixteenth century, but even through the eighteenth century tomatoes were not cultivated in North America, because, belonging to the nightshade family, they were presumed to be poisonous (regardless of obvious evidence to the contrary) (Goodwin & Goodwin, 1984). This historical curiosity explains the derivation of the so-called "tomato effect," where an efficacious treatment is rejected because it does not conform to prevailing concepts of disease pathogenesis (Goodwin & Goodwin, 1984). In many historical cases, efficacious therapies were initially rejected if they did not make sense at the time, while physicians instead employed various placebos that were presumed to be efficacious based on contemporary concepts of disease pathogenesis and therapeutic action (Lanska, 2002). In this sense, the abandonment of animal magnetism under the impetus of the negative findings of the Commission can be considered as an example of the tomato effect to the extent that the therapy incorporated a therapeutically efficacious component, even if the theoretical basis was faulty.

Although Mesmer's "theory" of animal magnetism was vague, mystical, largely incomprehensible, and scientifically unsupportable, one cannot discount that he held tremendous influence over his patients and disciples. His empirically developed psychotherapeutic techniques – even if lacking a supportable theoretical foundation – were certainly believed to be extremely beneficial by numerous patients, while orthodox medicine was not (Kihlstrom, 2002; Parish, 1990; Pattie, 1994; Perry & McConkey, 2002). Although one should not accept either such beliefs or the numerous collective anecdotes of (even sometimes dramatic) therapeutic benefit as being adequate evidence of efficacy of some aspect of the global treatment, it is fair to say that treatment outcomes per se were not actually scientifically assessed (Parish, 1990; Pattie, 1994; Perry & McConkey, 2002) as both Mesmer and d'Eslon bitterly complained.

Before the Commission report, d'Eslon (1780) acknowledged that he did not know how animal magnetism produced its effects, but

[If] Mr. Mesmer had no other secret than that of making the imagination act to produce health, would not that be a marvelous benefit? If the medicine of imagination is the best, why shouldn't we practice it? (d'Eslon, 1780, p. 46–47; Pattie, 1994, p. 105)

d'Eslon's concession was quoted by the commissioners themselves (Franklin et al., 1784, 1997, p. 82) as was d'Eslon's similar testimony during the investigation.

[d'Eslon] declared ... that he believed he could in fact lay down the principle that the imagination had the greatest part in the effects of animal magnetism; he said that this new agent may be only the imagination itself, the power of which is so great that it is little understood: at the same time he certifies that he has constantly been cognizant of this power in the treatment of his patients, & he certifies also that several have been healed or remarkably relieved. He has remarked to the Commissioners that the imagination directed in this way toward the relief of human suffering would be a great blessing in the practice of Medicine. (Franklin et al., 1784, 1997, p. 82)

Benjamin Franklin, the titular head of the Commission, acknowledged (in a letter to La Sablière de la Condamine, on March 8, 1784, just prior to his appointment to the Royal Commission) that the imagination might be directed in a positive therapeutic sense and at the very least this approach was bound to be less toxic than the questionable therapies of the orthodox physicians.

As to the animal magnetism, so much talk'd of ... there being so many disorders which cure themselves and such a disposition in mankind to deceive themselves and one another on these occasions; and living long have given me frequent opportunities of seeing certain remedies cry'd up as curing everything, and yet so soon after totally laid aside as useless, I cannot but fear that the expectation of great advantage from the new method of treating diseases will prove a delusion. That delusion may however and in some cases be of use while it lasts. There are in every great rich city a number of persons who are never in health, because they are fond of medicines and always taking them, whereby they derange the natural functions, and hurt their constitutions. If these people can be persuaded to forbear their drugs in expectation of being cured by only the physician's finger or an iron rod pointing at them, they may possibly find good effects tho' they mistake the cause. (Lopez, 1993, p. 327; McConkey & Perry, 2002, p. 324; Pattie, 1994, pp. 143–144)

Benjamin Rush (1745–1813) – the most famous American physician of the time and with

Benjamin Franklin a signer of the Declaration of Independence – while denouncing Mesmer's theory, acknowledged in 1789 in his "Duties of a Physician" (Rush, 1818) that Mesmer's global approach had therapeutic value even if his theory of its effects did not.

I reject the futile pretensions of Mr. Mesmer to the cure of diseases, by what he has absurdly called animal magnetism. But I am willing to derive the same advantages from his deceptions.... The facts which he has established clearly prove the influence of the imagination, and will, upon diseases. Let us avail ourselves of the handle which those faculties of the mind present to us, in the strife between life and death. I have frequently prescribed remedies of doubtful efficacy in the critical stage of acute diseases, but never till I had worked up my patients into a confidence, bordering upon certainty, of their probable good effects. The success of this measure has much oftener answered, than disappointed my expectations; and while my patients have commended the vomit, the purge, or the blister, which was prescribed, I have been disposed to attribute their recovery to the vigorous concurrence of the will in the action of the medicine. (Schneck, 1978, p. 10)

The commissioners also accepted that imagination or suggestion may have therapeutic value, but strongly disagreed with the way in which the imagination was directed toward violent crises by Mesmer and other practitioners of animal magnetism.

No doubt the imagination of patients often has an influence upon the cure of their maladies. ... It is a well-known adage that in medicine faith saves; this faith is the product of the imagination ...: the imagination therefore acts only through gentle means; through spreading calm through the senses, through reestablishing order in functions, in reanimating everything through hope. ... But when the imagination produces convulsions, it acts through violent means; these means are almost always destructive. (Franklin et al., 1784, 1997, p. 82; McConkey & Perry, 2002, p. 322)

The commissioners felt that potentially harmful treatments should be applied only out of necessity and then judiciously so as to move the patient toward health, rather than indiscriminately, lest the treatment cause more harm than good.

[There] are some desperate cases where all must be disturbed in order to be put in order anew. These dangerous upsets may only be used in Medicine the way poisons are. It must be necessity that dictates their use & economy that controls it. This need is momentary, the upset must be unique. Far from repeating it, the wise physician busies himself with repairing the damage it has necessarily produced; but at the group treatment of magnetism, crises repeat themselves everyday, they are long, violent; the situation of these crises being harmful, making a habit of them can only be disastrous. ... How can one imagine that a man, whatever his disease, in order to cure it must fall into crises where sight appears to be lost, where limbs stiffen, where with furious & involuntary movements he batters his own chest; crises that end with an abundant spitting up of mucous & blood!.... These effects therefore are real afflictions & not curative ones; they are maladies added to the disease whatever it may be. (Franklin et al., 1784, 1997, p. 82)

Conclusion

The process by which animal magnetism was introduced, disseminated, evaluated, discredited, and abandoned remains instructive for the evaluation of therapies today.[6] Mesmer's animal magnetism was introduced as a panacea based upon a vague and poorly supported theory, supported by glowing testimonials, disseminated primarily through lay channels when support could not be obtained through professional channels, and ultimately formally tested – long after initial dissemination – when the therapy was already accepted by a significant segment of the populace. The Commission charged with investigating animal magnetism ignored Mesmer's poorly formulated theory and focused instead on the observable effects of the treatment. The methodology utilized by the Commission was truly groundbreaking; whereas previous therapies were judged based on experience and authority, animal magnetism was evaluated using carefully designed controlled experiments. By actively intervening to systematically isolate and independently vary each possible explanatory factor, while holding all other factors constant, the Commissioners demonstrated that magnetization had no effect. Instead they provided strong support for their rival hypothesis that the observed effects were due to suggestion and the imagination of the subjects: subjects developed the characteristic mesmeric crises if and only if they *expected* to be magnetized, regardless of whether they were actually magnetized. The application of a scientific approach to the evaluation of therapies had rarely been applied and never before with such sophistication. The devastating arguments of the Commissioners unleashed a flood of satire and ridicule that eroded support for Mesmer and led to abandonment of animal magnetism as a treatment in France. Nevertheless, animal magnetism was subsequently briefly revived in other countries by disciples of Mesmer in the early nineteenth century, and distorted cultural echoes of this therapy persist today.

References

American Psychiatric Association. (2000). 300.81 Somatization disorder. In *Diagnostic and statistical manual of mental disorders* (4th ed., pp. 485–490). Washington, DC: American Psychiatric Association (text revision).

---

[6] Throughout the ensuing nineteenth century, the effectiveness of medical treatments continued to be assessed primarily by the results of uncontrolled case series (Lanska & Edmonson, 1990), a process "fraught with difficulty, uncertainty, and error" (Moses, 1984, p. 709). Moreover, investigators often failed to identify a clearly defined group of individuals with a specific condition for study, disregarded the natural history of the conditions under study (particularly for conditions which may remit), failed to adequately consider placebo effects, did not establish objective measures of baseline status or degree of change in clinical condition with treatment, generally did not establish an otherwise similar untreated comparison group, and had no clear way of determining whether the results could reasonably have resulted from chance or bias (Lanska & Edmonson, 1990). Some increase in sophistication came with Pierre Louis' analytical method in the 1830s (Louis, 1836) in which he demonstrated increased mortality associated with early bloodletting in a retrospective case–control study of patients with pneumonia. However, with rare (and methodologically limited) exceptions, comparative prospective clinical trials were not employed until the twentieth century, thus allowing many ineffective and harmful traditional therapies to remain in routine use (e.g., bleeding, blistering, purging, and administration of highly toxic heavy metals) (Gehan & Lemak, 1994). Sadly, even today, ineffective therapies continue to be disseminated based on marginally supported theoretical rationales and such limited empiric evidence as a favorable case series before any formal experimental evaluation.

Boylston, Z. (1726). *Historical account of the small-pox inoculated in New-England, upon all sorts of persons, whites, blacks, and of all ages and constitutions: With some account of the nature of the infection in the natural and inoculated way, and their different effects on human bodies. With some short directions to the unexperienced in this method of practice*. London: Printed for S. Chandler, at the Cross-Keys in the poultry. (Humbly dedicated to her Royal Highness the Princess of Wales. [2nd ed.]).

Braid, J. (1843). *Neurypnology: On the rationale of nerveous sleep, considered in relation with animal magnetism*. London: John Churchill.

Canton, J. (1754). A letter to the Right Honourable the Earl of Macclesfield, President of the Royal Society, concerning some new electrical experiments. *Philosophical Transactions of the Royal Society of London, 48*, 780–785.

Darnton, R. (1968). *Mesmerism and the end of the enlightenment in France*. Cambridge: Harvard University Press.

Donaldson, I. M. L. (2005). Mesmer's 1780 proposal for a controlled trial to test his method of treatment using 'animal magnetism'. *Journal of the Royal Society of Medicine, 98*, 572–575.

Duveen, D. I., & Klickstein, H. S. (1955). Benjamin Franklin (1706–1790) and Antoine Laurent Lavoisier (1743–1794). Part II. Joint investigations. *Annals of Science, 11*, 271–302.

Eden, J. (1957). Translator's preface. In F. A. Mesmer (1799/1957) (Ed.). *Memoir of F. A. Mesmer on his Discoveries: 1799* (pp. 5–7). New York: Eden Press.

d'Eslon, C. (1780). *Observations sur le magnétisme animal*. London: Didot.

Finger, S. (2006). *Doctor Franklin's Medicine*. Philadelphia: University of Pennsylvania Press.

Frankau, G. (1948). Introductory monograph. In *Mesmerism by Doctor Mesmer (1779): Being the First Translation of Mesmer's Historic "Mémoire sur la découverte du Magnétisme Animal" to appear in English* (pp. 7–26). London: MacDonald.

Franklin, B. (1759). *Some account on the success of inoculation for the small-pox in England and America*. London: W. Strahan.

Franklin, de Borey, Lavoisier, Bailley, Majault, Sallin, et al. (1784/2002). Secret report on mesmerism, or animal magnetism. *International Journal of Clinical and Experimental Hypnosis, 50*, 364–368.

Franklin, B., Majault, le Roy, Salin, Bailly, J.-S., d'Arcet, et al. (1784/1997). The first scientific investigation of the paranormal ever conducted: Testing the claims of mesmerism: commissioned by King Louis XVI: designed, conducted, & written by Benjamin Franklin, Antoine Lavoisier, & others. [Translated by Salas, C., Salas, D.] *Skeptic, 4*, 66–83.

Franklin, B., Majault, le Roy, Salin, Bailly, J.-S., d'Arcet, et al. (1784/2002). Report of the commissioners charged by the king with the examination of animal magnetism. *International Journal of Clinical and Experimental Hypnosis, 50*, 332–363.

Gehan, E. A., & Lemak, N. A. (1994). *Statistics in medical research: Developments in clinical trials*. New York: Plenum Medical Book Company.

Golub, E. S. (1994). *The limits of medicine: How science shapes our hope for the cure*. New York: Times Books division of Random House.

Goodwin, J. S., & Goodwin, J. M. (1984). The tomato effect: rejection of highly efficacious therapies. *Journal of the American Medical Association, 251*, 2387–2390.

Gould, S. J. (1989). The chain of reason vs. the chain of thumbs: why did several eminent eighteenth-century scientists – including Benjamin Franklin – sit around a table playing thumbsies? *Natural History, 7*, 12–21.

Gravitz, M. A. (1994). Early American mesmeric societies: a historical study. *American Journal of Clinical Hypnosis, 37*, 41–48.

Harré, R. (1981). *Great scientific experiments: 20 experiments that changed our view of the world*. Oxford: Phaidon.

Herbert, K. B. H. (1998). John Canton FRS (1718–72). *Physics Education, 33*, 126–131.

Hirschman, J. V. (2005). Benjamin Franklin and medicine. *Annals of Internal Medicine, 143*, 830–834.

J. C. Penny Co. (2004). *Fall & Winter '04* (pp. 203) [Catalog].

Jurin, J. (1723). A letter to the learned Dr. Caleb Cotesworth, F. R. S. of the college of physicians, London, and physician to St. Thomas's Hospital; containing a comparison between the danger of the natural small pox, and that given by inoculation. *Philosophical Transactions of the Royal Society of London, 32*, 213–227

Kihlstrom, J. F. (2002). Mesmer, the Franklin Commission, and hypnosis: A counterfactual essay. *International Journal of Clinical and Experimental Hypnosis, 50*, 407–419.

Lanska, D. J., & Edmonson, J. (1990). The suspension therapy for tabes dorsalis: a case history of a therapeutic fad. *Archives of Neurology, 47*, 701–704.

Lanska, D. J. (2002). James Leonard Corning, and vagal nerve stimulation for seizures in the 1880s. *Neurology, 58*, 452–459.

Lind, J. (1753). *A treatise of the scurvy. In three parts. Containing an inquiry into the nature, causes and cure, of that disease. Together with a critical and chronological view of what has been published on the subject.* Edinburgh: Printed by Sands, Murray and Cochran for A Kincaid and A Donaldson.

Lind, J. (1762). *An essay on the most effectual means of preserving the health of seamen, in the Royal Navy.* London: D Wilson.

Lopez, C.-A. (1993) Franklin and Mesmer: an encounter. *The Yale Journal of Biology and Medicine, 66*, 325–331.

Louis, P. C. A. (1836). *Researches on the effects of bloodletting in some inflammatory disease, and on the influence of tartarized antimony and vessication in pneumonia.* Boston: Hillary Gray.

Mackay, C. (1852/1932). The magnetizers. In *Memoirs of extraordinary popular delusions and the madness of crowds* (pp. 304–345). New York: Farrar, Strauss and Giroux.

Massey, I. (1723). *A short and plain account of inoculation. With some remarks on the main argument made use of to recommend that practice, by Mr. Maitland and others. To which is added, a letter to the learned James Jurin, M. D. R. S. Secr. Col. Reg. Med. Lond. Soc. In answer to his letter to the learned Dr. Cotesworth, and his comparison between the mortality of natural and inoculated small pox* (2nd ed.). London: W. Meadows.

McCandless, P. (1992). Mesmerism and phrenology in antebellum Charleston: "Enough of the marvelous." *Journal of Southern History, 58*, 199–230.

McConkey, K. M., & Perry, C. (2002). Benjamin Franklin and mesmerism, revisited. *International Journal of Clinical and Experimental Hypnosis, 50*, 320–331.

McGrew, R. E., & McGrew, M. P. (1985). Mesmerism: animal magnetism. In *Encyclopedia of Medical History* (pp. 197–200). New York: McGraw-Hill Book Company.

Mermann, A. C. (1990). The doctor's critic: the unorthodox practitioner. *The Pharos* Winter, 9–13.

Mesmer, F. A. (1766/1980). Physical-Medical treatise on the influence of the planets. In G. Bloch (Ed.), *Mesmerism: A translation of the original scientific and medical writings of F.A. Mesmer* (pp. 1–22). Los Altos, CA: William Kaufmann, Inc.

Mesmer, F. A. (1775/1980). Letter from M. Mesmer, Doctor of Medicine at Vienna, to A.M. Unzer, Doctor of Medicine, on the medicinal usage of the magnet. In G. Bloch (1980) (Ed.), *Mesmerism: A translation of the original scientific and medical writings of F.A. Mesmer* (pp. 23–30). Los Altos, CA: William Kaufmann Inc.

Mesmer, F. A. (1779/1948). *Mesmerism by Doctor Mesmer (1779): Being the first translation of mesmer's historic "Mémoire sur la découverte du Magnétisme Animal" to appear in English.* London: Macdonald & Co. (Translated by Myers, V. R.).

Mesmer, F. A. (1779/1980). Dissertation on the discovery of animal magnetism. In G. Bloch (Ed.), *Mesmerism: A translation of the original scientific and medical writings of F. A. Mesmer* (pp. 41–78). Los Altos, CA: William Kaufmann, Inc.

Mesmer, F. A. (1781/2005). *Translation of the text of: Mesmer's proposal for a trial of the curative results of his treatment of patients by 'Animal magnetism' read to an assembly of the Faculté de Médecine de Paris by Deslon on behalf of Mesmer on 18 September 1780, and the Faculté's response. From: Mesmer, F.A. 1781 Précis Historique de Faits Relatifs au Magnétism Animal Jusqu'en Avril 1781* (pp. 111–114). The James Lind Library www.jameslindlibrary.org/trial_records/ 17th_18thCentury/mesmer/mesmer_translation.html (Accessed August 2, 2006).

Montagu, M. W. (1717/1861/1970). To Mrs. S. C [Miss Sarah Chiswell]. In *The letters and works of Lady Mary Wortley Montagu. Edited by her great-grandson, Lord Wharncliffe.* New York: AMS Press; 1970 (Reprint of the 1861 edition).

Moses, L. E. (1984). The series of consecutive cases as a device for assessing outcomes of intervention. *New England Journal of Medicine, 311*, 705–710.

Nettleton, T. (1722). A letter from Dr. Nettleton, physician at Halfax in Yorkshire, to Dr. Whitaker, concerning the inoculation of the small pox. *Philosophical Transactions of the Royal Society of London, 32*, 35–38.

Nettleton, T. (1723). Part of a letter from Dr. Nettleton, physician at Halifax, to Dr. Jurin, R. S. Sec concerning the inoculation of the small pox, and the mortality of that distemper in the natural way. *Philosophical Transactions of the Royal Society of London, 32*, 209–212.

Parish, D. (1990). Mesmer and his critics. *New Jersey Medicine, 87*, 108–110.

Pattie, F. A. (1956). Mesmer's medical dissertation and its debt to Mead's *De Imperio Solis ac Lunae. Journal of the History of Medicine and Allied Sciences, 11*, 275–287.

Pattie, F. A. (1994). *Mesmer and animal magnetism: A chapter in the history of medicine.* Hamilton, NY: Edmonston Publishing.

Perry, C., & McConkey, K. M. (2002). The Franklin commission report, in light of past and present understandings of hypnosis. *International Journal of Clinical and Experimental Hypnosis, 50*, 387–396.

Roth, N. (1977). Mesmerism in America. *Medical Instrumentation, 11*, 118–119.

Rush, B. (1818). Observations on the duties of a physician, and the methods of improving medicine: Accommodated to the present state of society and manners in the United States. In B. Rush (Ed.), *Medical Inquiries and Observations* (Vol. 1, 5th ed., pp. 251–264). Philadelphia: M. Carey & Sons.

Schneck, J. M. (1959). The history of electrotherapy and its correlation with Mesmer's animal magnetism. *American Journal of Psychiatry, 116*, 463–464.

Schneck, J. M. (1978). Benjamin Rush and animal magnetism, 1789 and 1812. *International Journal of Clinical and Experimental Hypnosis, 26*, 9–14.

Shermer, M. (2002). Mesmerized by magnetism. *Scientific American, 287*(5), 41.

Smith, B. (1985). Gullible's travails: Tuberculosis and quackery 1890–1930. *Journal of Contemporary History, 20*, 733–756.

Spottiswoode, R. (Director) (1993). Mesmer: Charlatan, fraud . . . or genius. Image Entertainment.

Stone, E. (1764). An account of the success of the bark of the willow in the cure of agues. *Philosophical Transactions, 53*, 195–200.

Tomlinson, W. K., & Perret, J. J. (1974). Mesmerism in New Orleans, 1845–1861. *American Journal of Psychiatry, 131*, 1402–1404.

Walsh, J. J. (1923). Mesmer and his cures. In *Cures: The story of the cures that fail* (pp. 88–96). New York: D. Appleton and Company.

Waterson, D. (1909). Mesmer and Perkin's tractors. *International Clinics, 3*, 16–23.

Webster's Universal College Dictionary. (1997). New York: Gramercy Books.

Wester, W. C., II. (1976). The Phreno-magnetic society of Cincinnati – 1842. *American Journal of Clinical Hypnosis, 18*, 277–281.

Winter, A. (1998). *Mesmerized: Powers of mind in Victorian Britain*. Chicago: University of Chicago Press.

Withering, W. (1785). *An account of the foxglove and some of its medical uses: with practical remarks on dropsy and other diseases*. London: J and J Robinson.

Wolf, J. K. (1980). An aside on quackery. In *Practical clinical neurology* (pp. 104–105). Garden City: New York: Medical Examination Publishing Co.

Woodward, J. (1714). An account, or history of the procuring the small pox by incision of inoculation, as it has for some time been practised at Constantinople. *Philosophical Transactions of the Royal Society of London, 29*, 72–92.

22
Hysteria in the Eighteenth Century

Diana Faber

Introduction

In her study "Hysteria: The history of a disease" (1965) Ilza Veith referred to the eighteenth century as "controversial." It was indeed controversial in that there were competing and changing theories of the nature of hysteria. These depended upon the scientific discoveries and changing theories that influenced medical thought. However, the term "hysteria," while convenient for the historian, did not reflect the most common usage during the seventeenth and eighteenth centuries.

The term "hysteria" has as its etymology the Greek and Latin term for the womb. However, the Greek physician, Galen (c. AD130–c. 200) noted that "hysterical passion is just one name; varied and innumerable, however, are the forms which it encompasses" (cited in Veith, 1965, p. 39). Indeed, physicians and laymen alike in the seventeenth and eighteenth centuries described the extraordinary behaviors that they witnessed as fits, suffocation of the mother, spleen, vapors, hysteric distemper, nerves, etc. As late as 1788, William Rowley (1742–1806) gave the following title to his work "A Treatise on female, nervous, hysterical, hypochondriacal, bilious, convulsive Diseases . . ." Thus the term "hysteria" raises problems. As Hunter and McAlpine (1963, p. 288) point out, "The concept of 'Vapours' or 'Hysterick Fits' which was popular in the later seventeenth and eighteenth centuries illustrates one of the pitfalls of tracing the history of mental illness down the centuries by terminology." The historian's difficult task is to find the underlying meanings and theories that underpin the medical language and also to examine and describe some of the different medical practices and therapeutics.

The confusion of ideas, theory, and medical practice during the late-seventeenth and eighteenth centuries coincided with the end to witch hunts, which often involved women who displayed hysteric-like behavior or other signs of mental disturbance. These women had been thought to be possessed by the devil by their inquisitors, sometimes armed with the "Malleus Maleficarum" as a guide.

The changing outlook stemmed from individuals such as Edward Jorden (1578–1632), who became involved in the trial of a woman accused of having put a spell on a 14-year-old girl. In his defence of the "witch," Jorden argued that the child was suffering from a natural disorder of the mind and behavior. Still, the court found the woman guilty of witchcraft. In 1603, Jorden wrote a treatise to acquaint the medical profession above all, with those behaviors that should be considered clinical signs and not the work of the devil. Thus, new explanations and theories for certain aberrant behaviors began to take hold in the seventeenth century.

Confusion and Controversies

This new scientific interest in hysteric behaviors led to much confusion and controversy on account of competing theories. Whatever the controversies, there was usually similarity in the descriptions of the outward behaviors of the women who suffered from hysteric disorders. Whereas hysteric symptoms were generally perceived to involve women, given the general belief that their seat lay in the womb, there were also signs of dissent from that view.

Thomas Willis (1621–1675) dismissed the idea of a "wandering womb," for in cases of hysteric behavior post-mortem observations revealed the womb to be intact and in its place. Willis went further by suggesting that men too could be subject to hysteric fits since hysteric symptoms involved the brain. Willis even stated that, when an unusual symptom arose, one often gave it the epithet of "hysteric" to cover up medical ignorance.

Nathaniel Highmore (1613–1685) was a friend of William Harvey (1578–1657) and was undoubtedly influenced by his theory of the circulation of the blood. According to Highmore (1670), animal spirits or minute particles were heated within the heart, and together with the blood, were transmitted through the arteries to the brain. There they separated from the blood and could infiltrate many parts of the body, causing serious disorders, including hysteric fits. Interestingly, Willis refuted Highmore's explanation, claiming that the irregularities of the movement of the blood were of secondary importance, being caused by visceral spasms. Moreover, while rejecting Highmore's assertion, Willis also set out his theory of the brain and an explanation of the convulsive disorders of hysteria and hypochondria (1670).

Thomas Sydenham (1624–1689), the so-called "English Hippocrates" revitalized some aspects of Hippocratic medicine, including careful bedside observations and a greater trust in the healing power of nature. Sydenham also preferred facts rather than the physiological theories of the humoralists, the iatrochemists, and those who were now emphasizing mechanics and solids. When he was asked to write on the subject of hysterical diseases, he wrote his "Epistolary Dissertation" (1682). Here he claimed that next to fevers, hysteric pathologies were the most commonly found among chronic disorders. He also proposed hypochondriacal symptoms as the male equivalent of hysteric.

Sydenham further pointed out that hysteric disorders had a remarkable ability to imitate a multiplicity of other disorders, and this obliged the clinician to make very careful differential diagnoses. This was particularly important given the plurality of diseases, both acute and chronic. Careful descriptions of symptoms, the course of an illness, and its outcome were thus of prime importance.

Moreover, Sydenham was one of the first physicians to recommend that human diseases be ordered in the same way as botanists were ordering plant species. He provided a nosological system, but situated 'hysteria' in parallel to other maladies by virtue of its power to deceive. The rise of nosological studies in the eighteenth century is usually attributed to Sydenham.

Vapors

In addition to citing nervous disorders as the cause of hysteric fits, the theory of vapors grew in importance. It arose with the theories of atmospheric pressure and of expansion and condensation of steam in the early-seventeenth century. For their part, chemists were also beginning to isolate and manipulate different gases when physicians introduced various theories of vapors to explain hysteric disorders.

In 1689, Lange put forward his clinical theory in "Traité des Vapeurs." He was probably inspired by the work of Jan van Helmont (1577–1644), who published work on gases and ferments in 1648. Of the four types of ferment that Lange described, it was the seminal ferment that produced hysteric vapors. As women's ferments were fixed and heavy, he wrote, they did not always reach their destination, the sexual organs. However, if they did, sexual relations were needed to prevent the ferments from rising from the uterus via the nerves to the brain. Indeed, if such a rise did take place it could give rise to hysteric convulsions, delirium, and mania by affecting other organs, provoking spasms, vomiting, functional losses of the sense organs, and commotions of the body.

Lange even had a theory to explain why some individuals were more susceptible than others to this pathology. He pointed out that gentle, placid, and amiable people with no sexual problems were generally not at risk. This category included most men whose vapors were too volatile to reach the brain. As for women, those who were married and whose lives were daily filled with hard physical work were less prone to hysteric fits than those of the leisured class. Such thinking called for marriage, a more active life, fresh air, and exercise. Yet vapors now held a certain 'cachet' or mark of distinction, and this social factor discouraged some women from following this advice.

Uterine theories, such as Lange's, found favor with specialists in women's diseases and midwives; but perhaps for many physicians, theories of hysteric symptoms remained marginal to their main practice. Moreover, much of the picture of hysteria remained ambiguous through the seventeenth century.

In Search of "Hysteria"

A perusal of some of the titles of medical treatises in the eighteenth century reveals the lack of consensus among the authors. In many cases, two or more disorders or conditions were confused or described. For Purcell (1674–1730), author of a "Treatise of Vapours, or, Hysterick Fits" (1702), epilepsies and hysteric fits are presented as almost equivalent – "an Epilepsy is Vapours arriv'd to a more violent degree" (see Hunter and Macalpine, 1963, p. 288).

In 1725, Richard Blackmore (1653–1729) published his "Treatise of the Spleen and Vapours: or, Hypochondriacal and Hysterical Affections." His title suggests hypochondriacal and hysterical to be separate pathologies; however, Blackmore claimed that they were the same 'malady,' but admitted that convention obliged him to separate them.

George Cheyne (1671–1743), in contrast, neatly entitled his work "The English Malady" (1733), but warned that the "spleen, vapours, lowness of spirits, hysterical, or hypochondriacal disorders," as described in Chap. 8, represented such a large field that it would require a volume in itself.

William Rowley (1742–1806) subtitled his treatise, "Female, Nervous, Hysterical, Hypochondriacal, Bilious, Convulsive Diseases" (1788), presumably emphasizing these problems in women. Hysteric fits often became associated or even conflated with epilepsy.

John Andrée (?1699–1806), founder and physician of the London Infirmary authored the first, or one of the first books on epilepsy. But his title, "Cases of the Epilepsy, Hysteric Fits, and St. Vitus Dance" (1746), suggested an association of three disorders, in essence a disease entity.

Medical Practice

When it came to medical practices, Hermann Boerhaave (1668–1738) was extremely influential in the eighteenth century. As professor of medicine at the School of Medicine at Leiden, where he spent all of his professional life, he had a profound effect on his students and followers, inspiring and initiating them into his practice of bedside teaching. Before the seventeenth century, physicians often had little contact with their patients and there was no systematic medical teaching in the universities. That situation changed in Britain when some of Boerhaave's students became founding members of the faculty of medicine at Edinburgh.

Robert Whytt (1714–1766) was the first professor of medicine at the University of Edinburgh. In 1767, in his "Observations on the Nature, Causes and Cure of those Disorders which have been Called Nervous, Hypochondriac, or Hysteric . . .", Whytt claimed that the disorders deserving the description of "nervous" were found in those who possessed an unusually delicate or unnatural state of the nerves. This view implied that females were the most affected by this weakness. Delicacy of nerves as the source of hysteric behavior was frequently quoted, and "nervous" assumed almost the status of refinement in a social sense. According to Whytt, patients who also suffered from indigestion, flatulence in the stomach and bowels, a lump in the throat, flying pains in the head, a sense of cold in its back part, frequent sighing, palpitations, inquietude, fits of salivation, or pale urine, etc. could be deemed "hysteric."

John Purcell (1674–1730) thought that the origin of fits lay in the "Stomach and Guts," and his list of the patient's symptoms even outnumbered those of Whytt. He referred to what later became known as prodromes or prodromata, the initial signs of an impending fit as

a Heaveness upon their Breast; a Grumbliing n their Belly; they Belch up, and sometimes Vomit, Sower, Sharp, Insipid, or Bitter Humours: They have a Difficulty in breathing; and think they feel something that comes up into their Throat, which is ready to Choake them; they Struggle; Cry out . . . (cited in Hunter & Macalpine, 1963, p. 289)

William Cullen (1712–1790) held chairs of medicine and chemistry at both Glasgow and Edinburgh, and became well known for his clinical lectures and nosological system. He assigned hysteria to one of his four main categories, "neuroses." At this time, neuroses did not mean mental disorders: Rather, they usually referred to certain somatic ailments that somehow seemed to affect sensation and motor functions. Cullen claimed that the fit began in the alimentary canal, rose to the brain, and finally affected a large part of the nervous system. He identified the preliminary symptoms and the process of the fit as large quantities of urine, a globus, muscular contractions, laughing, crying, and sometimes delirium.

Benjamin Rush (1745–1813), the famous American physician and psychiatrist, was taught and influenced by Cullen, but was more inclined to emphasize emotional influences on nervous disorders. Part of his fame rested on his therapeutics for hysteria. He used emetics, purgatives, and harsh bloodletting. In addition he favored "ducking," threats, and the use of a twirling or spinning chair until the patient became unconscious; the aim of this therapy being to rearrange the brain to reach normality. Rush was not alone: eighteenth-century physicians were trying many things, some new and some old, for various theoretical reasons, to treat hysteria.

Therapeutics

Many physicians in the seventeenth and eighteenth centuries were in agreement with the Hippocratic belief that first and foremost they should do no harm. They also believed in the healing power of nature itself. This meant waiting for nature to take its course and intervening only when necessary. Sceptics of the efficacy of certain remedies found reassurance in these adages.

A physician, George Young (1691–1757) was the author of "A Treatise on OPIUM Founded upon Practical Observations," dated 1753. He wrote that hysteric patients expected the doctor to prescribe some remedy. In his view, this remedy could be something innocuous, and it would probably succeed, "for the passions will subside in time, either with or without medicine" (Young, 1753, pp. 113–114). He pointed to cases where women, "seized with an hysteric vomiting, or colic as often they ere (sic) under any disappointment, anger or vexation, tho' the minute before in perfect health of both body and mind; yet one slight affront has set them immediately vomiting, with great difficulty of breathing, the whole train of hysteric symptoms." Young then added that the only medicines that worked were "strictly speaking the useless ones, mere placebos."

In the eighteenth century the word "placebo" indicated a drug used to please the patient, typically when there was uncertainty about the outcome of other medical procedures or drugs. Both Cullen and James Gregory used regular drugs at low doses as placebos. In a clinical lecture at the Edinburgh Infirmary, Cullen explained: "I make it a rule even in employing placebos to give what would have a tendency to be of use to the patient" (see Risse, 1986, p. 201).

One of the harshest treatments was the use of blisters. The blister dressings were usually applied to the crown of the shaven head where they produced running sores. The physician's aim was to remove harmful substances from the body. Later, when humoral therapies became less popular, the rationale for the use of blisters changed, and physicians used them as stimulants.

Opium was widely used with hysterics, often to relieve the pain of the so called "clavus hystericus" – like the hammering of a nail into the forehead. Apart from its analgesic property, it was seen as both a sedative and a stimulant.

There was a legacy of drugs and other remedies that became standard treatments, particularly in hospitals. For example, it was important to cleanse the digestive system. It was assumed by many that whatever the disorder, the digestive organs had to be implicated, and in particular the stomach, which was linked by "sympathy" to other parts of the body. This explains why routine hospital treatment, even for the hysteric sufferer, was to clear the stomach. The symptoms of stomach rumblings, etc. reinforced the use of cathartics, emetics, and expectorants. Cathartics came in three forms of varying strength, from the mildest to the strongest, i.e., laxatives, purgatives, and drastics. The mildest of these was often given in the form of rhubarb, and the latter consisted of mercurous chloride, which became known as the "Samson of medicine."

Among the tonics recommended for increasing tone, there were astringents, using alum, iron, copper, and lead preparations. Another tonic, one introduced into Europe in the seventeenth century, was Peruvian bark. Now known to contain quinine, it became one of the most popular tonics in the late eighteenth century. Among antispasmodic drugs, ether was found to be useful in treating cases of epilepsy and hysteria.

As for the popular practice of bloodletting, it was based in part on its power to lessen tension and spasms by eighteenth-century physicians, and was often recommended for both epilepsy and hysteria.

Nevertheless, not all physicians practiced these "heroic" therapies. For example, Francis Fuller (1670–1706) rejected the use of "Internal Physic," as well as purging and bleeding. Instead he insisted on the power of exercise to restore the tension or tone that had been lost through a "relaxation of the Solids" – the result of a life style that he deemed to be effete and languid. His popular treatise "Medicina Gymnastica" (1705) reached a ninth edition in 1777. The importance of exercise to promote health has ancient roots and had been put forward by Boerhaave who also recommended horseback riding. As for Sydenham, it was the ideal stimulant.

Diets could also be changed. There was a general prejudice against fish, fruit, and sometimes vegetables. In cases of hysteria, especially where amenorrhea was judged to be a contributing cause, hospital practitioners prescribed meat and red wine as stimulating and fortifying.

The most popular of the new therapies to emerge in the middle of the eighteenth century was electricity, a powerful stimulant that many physicians believed could tighten flaccid nerves (see Locke & Finger, this volume). Clinical trials with safe levels of electricity were endorsed by Benjamin Franklin, and the new therapy was enthusiastically accepted by John Wesley, the Methodist cleric who published his popular "Desideratum, or Electricity Made Plain and Useful" in 1760 (see Donat, this volume).

The Edinburgh Infirmary accepted its first delivery of an electrical apparatus in 1750, after which more deliveries followed. The warmth and exhilaration that followed electric therapy were also judged to be useful in overcoming a general debility that was often blamed for amenorrhea and hysteria. The method of "drawing sparks" across specific areas of the body was applied to the pelvic area for 2 or 3 min. In addition, sparks could be drawn from the throat to remove the sensation of the globus. Sparks were also applied to affected parts.

Electricity: A Case Study

One of the most interesting cases of hysteria was that of C.B., a young woman who from the age of about 14 in 1742, suffered from hysteric fits and convulsions that could reach nearly 40 in 24 h (see Finger, 2006; also this volume). She had undergone the usual therapies of bleeding, blisters, and "nervous medicine." As a result, the intervals between her fits and convulsions lengthened to 1 or 2 months, but she was still distressed by cramps and general convulsions. Finally, 10 years later, and now desperate, she approached Benjamin Franklin in Philadelphia with the intent of submitting herself to the therapy of electricity and its "wonderful" power.

C.B. reported that she underwent four strong strokes morning and evening, and her symptoms gradually decreased. Subsequently Franklin supplied her with a globe and bottle to enable her to "electrify" herself every day for 3 months. The therapy had long-lasting effects and she lived to the age of 79, never having a relapse. Interestingly, neither Franklin nor Evans, the young Philadelphia physician who worked with him on this case, explained why they thought electricity worked so well with C.B.

Nosology

As noted, the principle of ordering diseases into medical categories was initiated by Sydenham, who modeled his work on what the botanists were doing. He believed in the multiplicity of diseases, each to be recorded and set out according to its symptoms and course. According to Sydenham two-thirds of diseases are acute. Nevertheless, the remaining chronic ailments are the most difficult to classify, particularly those that present many symptoms.

Sydenham influenced the work of Boissier de Sauvages (1706–1771), a botanist and physician in Montpellier who devoted all his working life to nosology. He placed Hysteria in Class 4, "Spasms," which also included other convulsive diseases, such as epilepsy and chorea. In contrast, hypochondriasis, often judged to be the male equivalent of hysteria, appeared under Order 1, described as "disturbances of peripheral, extra-cerebral origin."

Cullen followed the systematizing trend, and in 1777 and 1796 he published "First Lines in the Practice of Physic." In this work he placed hysteria thus:

Class 2 NEUROSES (injury of the sense and motion, without an idiopathic pyrexia or any local affection).

Order 1 COMATA (diminution of voluntary motion, with sleep or a deprivation of the senses).

Genus 42 APOPLEXIA (Almost all voluntary motion diminished, with sleep more or less profound; the motion of the heart and arteries remaining).

Group B (symptomatic) 5. Hysteria and 6. Epilepsy.

Genus 63 Hysteria (rumbling of the bowels, a sensation of the globe turning itself in the belly, ascending (sic) to the stomach; sleep, convulsions, a great quantity of limpid urine, and the mind involuntarily fickle and mutable).

Species 1 from retention of the menses; **2** from menorrhagia; **3** from menorra serosa; **4** from obstruction of the viscera; **5** from fault of the stomach; **6** from too great salacity.

Comments: Under class 2, Neuroses referred to injury of the sense and motion. The Order 1 and Genus 42 indicate the decline and almost loss of voluntary motion. Group B which includes hysteria and are both to be defined by their different symptoms (see Meninger, 1963, pp. 434–439).

Hysteric Fits

One of the features of the hysteric fit was the way in which several parts of the body can be involved, and this required explanation. According to the old Galenic notion, many parts of the body may be affected by "sympathy" or "consensus." Jorden named this "communitie and consent." For example, vapors from the womb could reach out and affect another organ, or less specifically there could be a sympathetic action between two organs. This somewhat vague concept, popular during the seventeenth and eighteenth centuries, was supported by Whytt who hypothesized nerve-mediated links between the uterus, stomach, and brain. Similar explanations could offer support to uterine theories.

Another phenomenon that seemed to defy explanation was the contagion of hysteric fits. At some level imitation was involved. James Gregory commented upon the commotion caused in his ward at the Edinburgh Infirmary by such contagion among hysterical patients, and he suggested that probably more patients in the hospital had contracted hysteria than had been cured of it. Churches and theatres also witnessed epidemics of hysteria. In 1787,

Dr. St. Clare reported an "Epidemic of Hysterics" among 21 girl workers at a Lancashire cotton mill. He was able to put an end to the chaos by shocks from his portable electrical machine. In Paris the "Convulsionnaires de St. Médard" became famous, as did the convulsive dancing or tarantism in Italy. Such episodes called for social interventions.

The Edinburgh Infirmary

Given the geography, climate, and hospital population of the Edinburgh Infirmary, it may not have been typical of other hospitals in Britain, Europe, or elsewhere during the later part of the eighteenth century. Its winter climate was particularly harsh, and it differed perhaps from some other hospitals in that many of its patients came from a working population and suffered hardship, poverty, and probably malnutrition. In spite of these things, the specific examination of its practice in relation to hysteria can throw light on the problems of dealing with these patients. In the General Register, Hypochondriasis and Hysteria or Hysteric Complaints were classified under "Mental." Between 1770 and 1800 admissions for the categories of neurological and mental diseases had a high proportion of women because of the frequency of hysteria and headaches.

The case history of one patient, Isabel Gray, admitted to hospital on November 4, 1785, is informative. She was 24 years of age and unmarried when assigned to the care of James Gregory. He diagnosed hysteria, although the General Register of Patients listed her as suffering from dyspepsia. Her case was set out as follows:

November 6, 1785. Is liable at the interval of a month and most commonly about the time of menstruation of which she commonly has little or no warning. She, however, before their commencement has a sense of oppression about the precordia, and some degree of nausea. She is suddenly seized with vertigo and falls down insensible, is convulsed in all limbs and foams at the mouth. After continuing in this state for some time, she becomes insensible and does not return to her senses for a quarter of an hour or longer. She frequently complains of an headache attended with a throbbing of the temples, and a sense of knocking at her forehead. Has the sensation of a ball rising to her throat with a sense of suffocation and what she calls the creeping over her body. These symptoms are most urgent just before and after the fits. She is also troubled with tension and acidity of the stomach, and ruminates her food; appetite tolerable.

Tongue clean, pulse about 70, feeble; belly costive, menses regular. These symptoms came on about 2 years ago, and she ascribes them to the loss of her parents. Has been in the hospital since 12 (October?). Since when she has been purged, vomited, and has been using chalybeats (iron preparations) and the infusion (of gentian root) without benefit. When threatened yesterday with a fit, she had a draught of ether and laudanum which put it off.

November 7. No fits today but frequent threatenings of them. Repeat iron pills and bitter infusion. If paroxysm should recur, promptly administer vitriolated ether draught. Common enema in the afternoon.

November 8. Had a strong fit yesterday preceded by globus which last still continues but has produced no more fits. Did not get the draught as intended. Repeat previous medications. Apply electricity as sparks. Repeat draught as needed.

November 9. Had the sparks but not the shocks about the throat. Globus not removed at the time but almost gone now, she says. When a fit was beginning yesterday afternoon, she got the draught and the fit did not go on. Has had (the) threat (of fits) three times a day, took the draught first time and none of them (fits) went on. Omit electricity and ether. In the future wash (patient) with cold water.

November 10. No fits today or threatenings. Has used all her medicine except the pills. Omit the pills. Repeat the remaining (medication).

November 11. No more fits. Globus very troublesome; complains of general pains in her limbs. Skin cool. Pulse and countenance natural. Omit cold bathing. Repeat the remaining (medications).

November 12. No fits in 5 days. Still some slight remains of globus commonly in the throat, sometimes in the abdomen. Continue.

November 13. No more fits but much globus in the throat, inflation, pain, and borborygmi in the stomach and hypochondrium. Repeat draught with ether and vitriol. Stomach pills (rhubarb).

November 14. By mistake did not get the draught with ether as intended. Rhubarb pills operated thrice; still complains of gripes and borborygmi. Globus continuing and some nausea but no more fits. Ether draught as prescribed. Should have pill with 1 grain of opium at bedtime.

November 15. Continues easier in every way. Short, immediate relief from the draught. Goes out relieved.

Dr Gregory's commentary to the students:

This was plainly a nervous hysteria complaint. She was attacked about the time her menses appear every month. Her symptoms gave me a suspicion that epilepsy made a part of her disease but she had plainly some hysteric symptoms (such) as globus, etc. This disease may truly be called the protean form for every year I meet with it in a new shape. For this reason I take a great number of these patients in (so) that you see it in its different shapes. (Risse, 1988)

Commentary

Gregory's primary aim in the treatment of this patient was to control the fits that the troublesome globus announced. The patient also suffered other visceral disturbances that he associated with hysteria, in this case digestive problems that resulted in gripes. The iron pills were intended to fortify and electric sparks to stimulate recovery. The hospital procedure whereby the patient was discharged depended upon the medical outcome of "cured" or "relieved." In this situation Dr. Gregory had doubts about the efficacy of the treatment and therefore opted for the "relieved" category. He also suspected that epilepsy was involved – no doubt in view of the patient's vertigo, loss of consciousness, and foaming of the mouth. He may have also considered epigastric sensations, and movements of mastication that sometimes heralded an epileptic fit.

Symptomatology

In Cullen's nosological system, the category "symptomatic" allowed him to list the pathological signs of hysteria. One of the obvious features of the hysteric fit was the way in which several parts of the body were involved – the digestive problems, the pain in the head, the sensation of suffocation and of the bolus rising in the throat, all seem to defy explanation in spite of the theory of "sympathy."

The case of Isabel Gray required Gregory to make a diagnostic decision. Having initially diagnosed hysteria, he concluded this was an accurate judgment, given the "shapes" in which hysteria could appear. He had identified epileptic symptoms, i.e., vertigo and foaming at the mouth. This case illustrates the possibility of the conflation of two disorders or of their appearance in tandem.

Hysteria and epilepsy, as we have seen, had often been associated and recognized as a source of diagnostic confusion. For example, Cheyne (1733, p. 253) asserted that "I think it is pretty evident, that it (epilepsy) differs very little, or not at all, or at most, in a few Circumstances only, from *Hyperchondriacal* and *Hysterick Fits*." Such diagnostic confusion persisted in some form or another into the mid-nineteenth century and beyond.

In 1846, Hector Landouzy in his "Traité Complet de l'Hystérie" drew up a table of criteria to provide a differential diagnosis between epilepsy and hysteria. Even earlier, John Ferriar (1761–1815), a physician at the Manchester Infirmary, was perplexed by the variety of symptoms that accompanied the hysteric fit, and he introduced the term "hysterical conversions" to describe them. He observed that

cases of hysterical conversions . . . are very common sources of error to young practitioners, and sometimes deceive even the most experienced. Whoever would present us with a good book on the *fallacy of symptoms*, which is greatly wanted, must be completely master of this unaccountable disease. (Cited in Hunter & Macalpine, 1963, p. 546)

Thus, toward the end of the century (1792–1798) Ferriar was expressing an urgent need to clarify the situation that confused practicing physicians. To many individuals, however, there seemed to be no obvious or easy method of eradicating confusion. Meanwhile, in France, Philippe Pinel, like Ferriar, was aware of the problems that physicians faced in their diagnostic decisions and in how they treated their hysteric patients.

Philippe Pinel (1745–1826)

Pinel's life mostly spanned the post-revolutionary period in France that brought with it the ideals of reform and progress that he embraced. It was after Pinel received his medical degree in 1773 and his move to Paris that his attention and interest became drawn to the diseases of the mind. He first worked in a private psychiatric institute; later was appointed to Bicêtre, a large mental hospital for men, and then was put in charge of La Salpêtrière, an even larger hospital for women.

Work in these two institutions provided him with wide experience in the mental problems of the patient, and it was here that he could put into practice his ideas for a humane treatment, optimistic that there could be cures for at least partially mental disorders. Hence, he talked with his patients and took great care in recording their accounts alongside with his own observations. He found cases of hysteria where perhaps they might have passed unnoticed.

In Pinel's view, one of the impediments to progress in the study of hysteria was the number of associated disorders and symptoms attached to it. He wanted to define hysteria in its uncluttered or "pure" state. He included hysteria in his first major work, "Nosographie Philosophique" (1813). He placed hysteria in the fourth class "Neuroses," where he included satyriasis in men and its female equaivalent, "furor uterinus" under the heading of "Genital Neuroses of Women." He observed that such neuroses generally made their first appearance at the time of puberty.

At first, disgust with her daily life, frequent tears without cause . . . shortly afterwards she lost the ability to speak, her face became discolored, there was a periodic tightening of her throat, feelings of strangulation, congestion of the salivary glands; and afterwards, abundant salivation, as though mercury had been used, inability to open her mouth, because of muscle spasm; tetanus-like rigidity of the rest of the body, pulse scarcely perceptible, respiration slow but regular, bowels constipated and the urine limpid. (Veith, 1965, p. 181)

Once her natural functions were restored there was calm for about a week, but afterwards the symptoms would reappear with the same violence.

Cullen had included "salacity" as a cause of hysteria in his nosological system. Likewise, Pinel proposed nymphomania or "furor uterinus" as a cause or variety of hysteria. He described this disorder in its three stages. In the first, the patient's mind becomes obsessed by lascivious thoughts and desires that she attempts to combat. In the second phase she becomes sexually provocative, and finally she is "on fire" and exhibits all the symptoms of "aliénation mentale" (Veith, 1965, p. 179) - she has become mentally deranged. This dramatic description suggests a changing view of hysteric women, many of whom were unmarried. Physicians, including Pinel, spoke as if in one voice in recommending sexual relations (within marriage) for women. At the same time abstinence and repression were the social norms imposed upon young unmarried women. Consequently, from the stereotype of the woman of frail nerves and of a more emotional disposition than men, there emerged also an image of women of excessive sexual impulses and behavior.

Pinel's belief in the curability of mental disorders and his approach to hysteria marked a break with earlier studies. In the past, a somatic view generally prevailed. Pinel did not deny the reality of physical symptoms, but he shifted the emphasis from the prevailing approach to the contribution of emotional factors that he believed underlay hysteric disorders. For example, while physicians generally blamed menstrual problems for hysteric symptoms, Pinel suggested that emotional problems could be the initial cause of menstrual irregularities. He recognized the reality of physical symptoms, but his approach contrasted, for example, with cases like that of Isabel Gray at the Edinburgh Infirmary.

Conclusions

The eighteenth century is marked by controversy, particularly at the theoretical level with regard to the etiology and development of hysteria. The theory of vapors was originally related to uterine theories, in which vapors rose from a diseased womb. In 1689, Lange elaborated on a theory of vapors, but later, the term "vapors" was frequently used to refer to general convulsive movements and became equated with hysteria. Both Willis and Whytt related hysteria to the brain and the nervous system, but uterine theories still survived throughout the eighteenth and into the nineteenth centuries. However, in his post mortem analyses, Pinel found no organic changes in the brain, and so freed the nervous system of etiological involvement in cases of hysteria. It is therefore difficult to find consensus in regard to hysteria in the period under review except for its often repeated symptoms, set out in the titles of medical treatises and in medical case studies.

Cullen's nosological statements confirm the salience of symptoms. Hysteria and epilepsy are classed as symptomatic within the genus 42; and within genus 63 hysteria is described by the many symptoms that appear almost word for word in the authors' titles. However, the lack of definitional clarity did not affect the medical practitioners whose concerns were focussed on the disorders associated with the hysteric fits including menstrual problems, intense headaches, feelings of suffocation and digestive problems. These called for the use of the many therapeutics that were available. Of these, electricity was deemed the most remarkable. Generally, the medical approach to hysteria remained somatic throughout the eighteenth century.

Pinel recognized the somatic role in hysteria, but emphasized the role of the emotions, which, over a century earlier Jorden had considered; Pinel also intended to focus on hysteria in its own right without the interference of associated pathologies. Given the importance that Pinel placed on the role of emotions in hysteria, his method of talking and listening to his patients introduced a new element of care. This was termed "traitement moral" – a concentration on the mental and emotional aspects of the afflicted. In human terms this therapy represented progress.

Thus, there still remained problems in the diagnosis of hysteria and its differentiation from associated pathologies. There was still a recognition of the protean nature of hysteria and its ability to deceive. Hysteria and epilepsy were often associated, given the "fit" that characterized them both. In the seventeenth and early eighteenth centuries many epileptic sufferers were cared for at home. It was, however, in the medical institutions that the two disorders were more likely to be taken the one for the other. The association between the two pathologies was part of a complicated and confusing story in the nineteenth century. (see Faber, 1997). The situation was not helped by the cohabitation of epileptics and hysterics, lodged in the same section at the Salpêtrière hospital. This was the situation found by Jean-Martin Charcot (1825–1893), the leading figure in the history of hysteria in the second half of the nineteenth century.

References

Andrée, J. (1746). *Cases of the epilepsy, hysteric fits, and St. Vitus dance, with the process of cure*. London: W. Meadows and J. Clarke.

Blackmore, R. (1725). *A treatise of the spleen and vapours: Or hypochondriacal and hysterical affections*. London: Pemberton.

Cheyne, G. (1733). *The English malady*. London/New York: Tavestock/Routledge.

Cullen, W. (1796). *First lines of the practice of physic*. Edinburgh: Bell/ Bradfute.

Faber, D. (1997). Jean-Martin Charcot and the Epilepsy/Hysteria relationship. *Journal of the History of the Neurosciences, 6*(3), 275–290.

Finger, S. (2006). *Doctor Franklin's medicine*. Philadelphia: University of Pennsylvania Press.

Fuller, F. (1705). *Medicina Gymnastica*. London: John Matthews.

Highmore, N. (1670). *De pasione hysterica et hypochondriaca, responsio epistolaris ad Doctorem Willis*. London.

Hunter, R. & Macalpine, I. (1963). *Three Hundred Years of Psychiatry* (pp. 1535–1860). Oxford: Oxford University Press.

Jorden, E. (1603). *A brief discourse of a disease called the suffocation of the mother*. London: John Windet.

Landouzy, H. (1846). *Traité Complet de l'Hystérie*. Paris: Baillière.

Lange (1689). *Traité des Vapeurs*. Paris: Denis Nion.

Meninger, K. (1963). *The vital balance*. New York: Viking Press.

Pinel, P. (1813). *Nosographie Philosophique* (5th ed.). Paris: J.A. Brosson.

Purcell, J. (1702). *A treatise of vapours, or, hysterick fits*. London: Newman and Cox.

Risse, G. (1986). *Hospital life in enlightenment Scotland*. Cambridge: Cambridge University Press.

Risse, G. (1988). Hysteria at the Edinburgh infirmary: The construction and treatment of a disease, 1770–1800. *Medical History, 3*, 1–22.

Rowley, W. (1788). *A treatise on female, nervous, hysterical . . . diseases*. London: Nourse.

St. Clare, W. (1787). Country news. *The Gentleman's Magazine, 57* (Part 1), 1.

Sydenham, T. (1682). *Epistolary Dissertation*. 1742 *Entire Works*. J. Swan (Trans.). London: Edward Cave.

Veith, I. (1965). *Hysteria: The history of a disease*. Chicago: University of Chicago Press.

Wesley, J. (1760). *The desideratum: Or, electricity made plain and useful*. London: W. Flannel.

Whytt, R. (1767). *Observations on the nature, causes, and cure of those dDisorders which have been commonly called nervous, hypochondriac, or hysteric*. Edinburgh: J. Balfour.

Willis, T. (1670). *Affectionum quae dicuntur hystericae et hypochondriacae*. London.

Young, G. (1753). *A treatise on opium, founded upon practical observations*. London: Millar.

Section F
Cultural Consequences

Introduction

Thomas Trotter, writing at the end of the long eighteenth century, remarks on the hugely increased significance of nervous disorders: "... at the beginning of the 19th century we do not hesitate to affirm that nervous disorders ... may be justly reckoned two-thirds of the whole with which civilised society is affected" and Robert Whytt, forty years earlier, wrote "All diseases may, in some sense, be called affections of the nervous system ...". But both Trotter and Whytt agreed that nervous disorders are to be found principally in those of a delicate and sensitive nature, a euphemism for the higher strata of society. It is only to be expected, therefore, that this increased interest in the nervous system and nervous physiology should work its way through into the general culture of the time, especially into literature and that form of literature, the novel, which is itself largely an eighteenth-century creation.

These cultural consequences are sampled in the three chapters which make up this final section. First, Timo Kaitaro looks at the way in which ideas about the relation of brain to mind affected the work of three eighteenth-century French thinkers: Denis Diderot (1713–1784), Julien Offray de La Mettrie (1709–1751) and Charles Bonnet (1720–1793). Both Diderot and La Mettrie considered the Cartesian view that body and mind are totally disparate realms to be unsustainable. Both believed that mind, sensibility, was an emergent property (nowadays one might say a 'system property') of matter as it became organised in living systems. Charles Bonnet, in contrast, was a strict mechanist so far as neurophysiology was concerned and believed that the relation between this mechanism and the mind, although known to God, must remain a mystery to mankind. Kaitaro shows how the Diderot/La Mettrie view leads to a different understanding of perception from the strict dualism of Bonnet. He argues that the latter view has prevailed into our own times in ideas about the localisation of function in the brain and in contemporary mind–brain identity theories.

The second chapter in this final section is by Marjorie Lorch who discusses the part played by neuroscience in the pithy and sometimes bitter writings of Jonathan Swift (1667–1745). Swift, at the end of his life, descended into what might nowadays be diagnosed as Alzheimer's or, perhaps, Pick's disease. It may be that he had a premonition, for throughout his life his writings dwell on psychopathology and madness. Lorch discusses this interest and shows how it is related to the neuroscience of the time and to his own observations of madhouses, especially Bedlam, of which he was, for a short time, a governor. She also relates his writings to those of other English writers of the period, including journalists publishing in the widely read *Spectator*, and points out that the very fact that so many publications concerned themselves with insanity suggests a widespread interest in the topic. Lorch concludes her chapter with a brief account of Swift's tragic final illness which he seems to have foreseen in his *Verses on the Death of Dr Swift Written by Himself* composed in 1731, fourteen years before he died.

Finally, George Rousseau discusses the eighteenth-century origin of the concept of temperament, a major theme in the work of one of the most acclaimed of twenty-first-century social scientists,

Jerome Kagan. Rousseau argues that it was in the eighteenth century that the old Galenical ideas of the four humours and of the soul as a subtle substance inhabiting the nervous system were finally displaced by a physiology of the nerves and brain. As the eighteenth century wore on, it became generally accepted that *temperament* was formed through a combination of nervous physiology and life experience. Rousseau supports his argument with extensive reference to the literature of eighteenth-century England: Smollett, Richardson and in particular Laurence Sterne. The notion of temperament, taken to an extreme, can be related to the idea that minds become gripped by a 'ruling passion' – an idea which Alexander Pope develops in *Essay on Man* and which appears in *Tristram Shandy* under the guise of 'hobby-horses'. Rousseau's chapter forms a fitting conclusion to this section (and indeed to the entire book) in showing, as it does, how the working out of ideas originating with Thomas Willis and his pupil (later colleague) John Locke, via the development of nervous physiology in the eighteenth century, affected the way literate humans understood themselves. In other words, what one of the characters in Richardson's *Clarissa meant* by exclaiming "I am me, myself, no one else."

The Editors

23

Technological Metaphors and the Anatomy of Representations in Eighteenth-Century French Materialism and Dualist Mechanism

Timo Kaitaro

Introduction

Metaphors and explanatory models referring to cultural artefacts, especially machines and symbolic representations, are commonly used in modern neuroscience. One often pictures the brain as a complex machine. Today the brain is usually compared to a computer that manipulates symbolic representations according to syntactic rules. The retention of information is often explained by "memory traces" stored in the brain. Such technological and semiotic metaphors have ancient roots, but they were especially popular during the eighteenth century after Descartes had pictured the human body as a machine. Thus many philosophers and scientists attempted to give explanations of psychological phenomena by referring to underlying neural mechanism. It was also during the Enlightenment that one started to emphasize that mental functions should be studied as material phenomena. Philosophers who wanted to examine mental phenomena in the context of materialist ontology criticized the traditional dualist ontology that emphasized the distinctness of the mental and the material. Surprisingly, these two aspects of modern neuroscience, the first related to reductionist explanatory models and the second to the ontology of scientific explanations, were, however, not so closely coupled in the eighteenth century as they are today. On the contrary, in the eighteenth century, these features usually appeared in opposing camps. Dualists like Charles Bonnet were keen in

propounding explanations of mental phenomena in terms of neural mechanism. In this they differed from the medically oriented materialist philosophers like La Mettrie and Diderot who – although they used mechanical metaphors – were not so much interested in suggesting mechanistic explanatory models as they were in criticizing dualist metaphysics and the mechanistic explanatory models attached to it. In this article, I shall examine how the dualists and the materialists differed in their respective ways of relying on mechanistic metaphors and models.

The Cartesian Models

In the Cartesian conception of man, one can distinguish two different, although closely intertwined and interdependent strands, which have greatly influenced later conceptions concerning the mind and its relation to the brain. Firstly, even if it could be questioned whether Descartes' theory of mind was representational in the sense of postulating representational entities *in the mind*, he certainly postulated some kind of representations *in the body*. Thus Descartes refers to "ideas" in their material aspect, as patterns of movement or as traces preserved in the brain (see Kaitaro, 1999). Whether Descartes thought that animals too could be said to have representations need not concern us here (see Gaukroger, 2000), but it is evident that in so far as human beings are

335

concerned, some kind of representations in the nervous system seem to be required by the dualistic nature of Descartes' theorization concerning mental functions. In order to explain the interaction between the mind that represents and the body that moves, one needs entities that can mediate between these two by having both material and representational properties.

According to Descartes, although the soul is joined to the body as a whole, it exercises its functions particularly at the "seat of the soul", which Descartes localized in the pineal gland (*Passions de l'âme*: XXX–XXXI). Descartes speaks of ideas as figures traced on the surface of this gland (Descartes, AT XI, pp. 177–178, FA I, pp. 451–452). These figures traced by animal spirits on the pineal gland are obviously local. On the other hand, Descartes theory of corporeal memory treated memory traces as distributed (Kaitaro, 1999; Sutton, 1998). In either case, the traces are material, not in the sense that they would be corporeal bodies, but in the sense in which material figures or patterns of movement are material phenomena (see Beyssade, 1991).

The other influential strand in Descartes' dualism is his attempt to describe the actions of animals as soulless automata, in terms of mechanics without any psychological vocabulary or without any Aristotelian sensitive animal souls. Descartes also applies mechanistic explanations to the non-voluntary aspects of human physiology. In so far as he had recourse to mechanical models and metaphors, his emphasis was not so much on the finalism involved in such models, but rather on showing that these processes did not involve the soul and that they could and should be explained strictly in terms of mechanics. Now, in Descartes' case, his mechanism and his postulation of representational entities in the brain are in fact closely interdependent theoretical constructions, which are furthermore both motivated by his dualism. The postulation of representational local traces in the brain is needed for his account of the mind–brain interaction that takes place in the pineal gland. And the mechanization of the body is a consequence of his dualism, which deprives living bodies of all other properties except those studied by mechanics and attributes these other properties to another substance, the soul. So it is logical that those who criticized Cartesian dualism in the eighteenth century tended to discard both of these postulations – the existence of representations in the brain and mechanistic physiology – and that those who accepted dualism tended to accept both.

From Mechanisms to Functions

So the development of materialist accounts of mental functions necessarily involved a criticism of Cartesian mechanism. For example, La Mettrie returns to Aristotelian notions of substantial forms in the first version of his materialism, presented in his *Histoire naturelle de l'âme* (1745). Even though La Mettrie later criticizes himself for having recourse to the "ancient and unintelligible doctrine of *substantial forms*" (de La Mettrie, 1747, 1960, p. 189), he never returned to Descartes' mechanical and passive conception of matter, but persisted in insisting in his later work *L'homme machine* (1747) that matter is far from being the passive substance that Descartes conceived it to be. La Mettrie writes that "organized matter is endowed with a moving principle (*principe moteur*) which distinguishes it from matter that is not [organized]", and that "each small fibre, or part of organized body is moved by a principle which it has in itself (*lui est propre*)" (de La Mettrie, 1747, 1960, pp. 181–182, 189). He insists on the fact that the action of this principle does not depend on the nerves, thus making it likewise independent of the soul exercising its functions in the brain. But it was not only organized matter that was active. La Mettrie also claims – pointing his words expressly against the Cartesians, Stahlians, Malebranchists, etc. – that "matter moves itself, not only when it is organized, as for example in an entire heart, but also when this organization is destroyed" (de La Mettrie, 1747, 1960, p. 188). Thus La Mettrie discards both Cartesian substances: the passively conceived matter, which moves only by mechanical impulses from other bodies, and the soul which is needed in order to make a living body self-moving. Although La Mettrie confesses that he cannot explain how inert and simple matter becomes active and composed of organs (de La Mettrie, 1747, 1960, p. 189), the key concept for him in explaining the appearance of mental functions is obviously organization, on which the difference between men and animals as well as the diversity of the faculties of different animals depends (de La Mettrie, 1747, 1960, pp. 189–190, 195).[1]

However, despite the fact that he discards the Cartesian framework, La Mettrie uses the metaphor of a machine, and boldly titles his materialist pamphlet as *L'homme machine*. In this work he writes that "the human body is a clock(work)" (*le corps humain est une horloge*), and concludes bluntly that "man is a machine" (*l'homme est une machine*) (de La Mettrie, 1747, 1960, pp. 190, 197). Neither of these claims was, however, at his time in any way a radical or an original thesis. The first was a traditional Cartesian comparison and the latter a common metaphor mentioned in such respectable dictionaries as the *Dictionnaire de l'Académie*, whose 1694 edition gives, as an example of the metaphorical uses of the term "machine", the sentence *Que l'homme est une machine admirable* (Man is an admirable machine).

Some scholars have taken statements by La Mettrie quoted above quite literally and concluded that he was a mechanist in the sense that he would have taken as his methodological ideal the methods of investigation of mathematical physics or that he proposed to "restate the general problem of the mind as a problem of physics" (Vartanian, 1960, p. 13). But some others like Thomson (1988) have observed that in fact La Mettrie was not a strict mechanist despite the fact that he used the metaphor of a machine in speaking of men. In fact, what shows that he was not a mechanist is precisely the fact that he relied on this *metaphor* (as we shall see real mechanists did not just present metaphors but mechanistic explanatory models taken literally). For La Mettrie the metaphoric term "machine" did not have the mechanistic connotation that it had for the Cartesians for whom the body was a piece of machinery and explainable in purely mechanistic terms (see, for example, Kaitaro, 2001a; Smith, 2002). One should thus not let the title of La Mettrie's work to mislead one to think that it was some kind of mechanistic materialism that the author was presenting.[2]

---

[1] In addition to organizations La Mettrie refers to the importance of language and culture in forming the mental capacities of humans (de La Mettrie, 1747, 1960, pp. 160–165).
[2] For the history of these attributions, see Bloch (1979) and Kaitaro (1987).

What was radical in the work was not its mechanism but rather its "vitalism", i.e. the attribution of vital properties to matter, and its emphasis in the self-moving nature of matter (Kaitaro, 2001a). In fact, the work has nothing to do with the "mechanistic materialism" sometimes attributed to it. It does not contain detailed analyses of the physiological mechanisms that give rise to thought, or of how organized bodies acquire the capacity to feel or to think. Instead the philosopher–physician frankly admits his ignorance and writes that it would be foolish to search for the mechanisms which give rise to thought in organized and sentient bodies (de La Mettrie, 1747, 1960, pp. 188–189). So, La Mettrie does not have recourse to mechanistic metaphors or models in order to construct explanatory hypotheses, as he had done in his earlier work, which contained quite a lot of traditional mechanistic explanations of mental phenomena in terms of the physiology of animal spirits. In *L'homme machine*, he puts the machine-metaphor to a different use: he uses it to point out the uselessness of covering our ignorance with futile hypotheses, one of them being closely attached to Cartesian mechanism: the existence of an immaterial soul; and he refers to a specific machine to illustrate his point. Referring to Locke's famous argument about "thinking matter" he writes:

The metaphysicians, who have suggested that matter could well have the faculty of thinking haven't dishonoured their reason. Why? Because they have the advantage (because here it is one) of expressing themselves badly. In fact, asking if matter can think, without considering it otherwise than in itself, is tantamount to asking whether matter can mark the hours. One can see in advance that we avoid this pitfall into which Mr. Locke unfortunately fell. (op. cit.: 149)

Here the mechanical metaphor is not used to point out that thought is reducible to mechanics. Instead the argument claims that just like the clock, which can have functional properties that are not reducible to the concepts of mechanics, the human body can perform functions that are mental, even though it is formed of parts none of which has any mental properties. This point is made even more explicitly in some clandestine philosophical manuscripts that circulated before La Mettrie's work was published as well as in some contemporary printed works (Kaitaro, 1997, pp. 93–95). These tended to emphasize that organized matter can acquire properties that are emergent in the sense that the parts of which it consists of do not have them. For example, the physician Antoine Le Camus observes in his *Médicine de l'esprit* (1753) that it would be absurd to claim, merely on the basis of its physical properties, that brass could not be capable of marking the hours. Analogously, one should not conclude that the brain could not acquire the capacity to think, just because the fibres of which it is formed are incapable of thought or of having ideas (Le Camus, 1753, Vol. I, pp. 174–175). It is to be noted that such arguments do not presuppose that thought could be reduced to matter in motion any more than the predicate "showing that it is four o'clock" could be translated into a physical description; neither is it necessary to think that any of the parts of the mechanism of the clock represents the idea of "four a clock". It is the whole clock that represents time, its parts do not; they just act according to the laws of mechanics. Likewise, the parts of the brain, fibres or traces left there during learning need not represent anything; it is the organism as a whole that does this.

Diderot's materialism continues the materialist tradition exemplified by La Mettrie, but in his case the rejection of physiological mechanism is even more explicit. His theory is based on the ideas of the so-called Montpellierian "vitalists" (see Kaitaro, 1997). This eighteenth-century form of vitalism was fully compatible with materialism and should thus not be confused with later spiritualist forms of vitalism which postulated metaphysical "vital forces".[3] The Montpellierian school represented a critical reaction to and a rejection of Cartesian iatromecanism as well as of Stahlian animism, which in contrast to Descartes had recourse to the soul in explaining all physiological and vital processes. The theories of the Montpellierians provided Diderot with a general model of the organism as a hierarchy of interrelated functions, which can be localized in different organs or in organic centres formed of several organs. Brain was considered to be one of these organic centres with its proper functions. Diderot did not attempt to apply this model of organic centres to the analysis and localization of brain functions themselves. The fact that Diderot, who admitted the importance of conjectures in science and philosophy, did not indulge in speculating on the physiological mechanisms of mental functions or on cerebral localizations, was probably due to the fact that he had seen the vanity of speculative physiology and considered such questions as empirical matters to be solved only on the basis of anatomo-pathological evidence, in the manner of the surgeon La Peyronie, who localized the soul in the corpus callosum on the basis of medical and pathological observations (Kaitaro, 1996). Diderot discusses the problem of the "seat of the soul" in his contribution to the article "Ame" of the *Encyclopédie*, which treats the anatomical and physiological questions related to the soul (*Encyclopédie*, Vol. I, pp. 340a–343b). Despite the credit given to La Peyronie's empirical studies, the passage concerning the seat of the soul ends in a sceptical note. Diderot did not consider the evidence sufficient to warrant definite conclusions. But, in addition, Diderot was, as a materialist, obviously critical of the existing models of localization, which usually attempted to localize the "seat of the soul" or to identify specific fibres with specific ideas. These two models were both, as we shall soon see, closely linked with the dualistic paradigm.

The Real Mechanists

But if La Mettrie uses the notion of a machine merely as a metaphor, there is a less metaphorical and more specific way of making a connection between mental phenomena and mechanics. And this model works just the other way that the comparison of La Mettrie does. If one could join representations to separate parts of the brain, for example specific fibres or bundles of fibres, one could use mechanism as an explanatory model. Psychological regularities could be explained mechanically by the

---

[3] For the Montpellierians vitalism consisted not so much in positing occult vital principles in order to explain vital phenomena as in the emphasis on the autonomy of life as a subject of study and in the criticism of mechanistic and reductionistic theories. Their theories could as well be characterized by such terms as "holistic" and "organicistic" (see Rey, 1987, Vol. I, pp. 1, 9–11, 28, 34–35 and Vol. III, pp. 155–156). Though they were not all materialists, their vitalism was not incompatible with materialism, as the use that Diderot made of their doctrines in founding his materialism shows (Kaitaro, 1997). Dr. Théophile de Bordeu, who appears as one of the protagonists in Diderot's *Le rêve de d'Alembert*, probably was a materialist (see Rey, op. cit.: Vol. I, p. 269).

physical interaction between the physiological correlates of ideas. For example, if two ideas tend to evoke one another in association, this would be due to the fact that the fibres that correspond to the former tend to activate the second mechanically. And likewise for reasoning: the fibres that represent the premises cause movements in the fibres that represent the conclusion. In fact, in this case what we have is no longer a loose metaphor but an explanatory mechanistic model.

I am not just imagining a possible theoretical position: such mechanism was defended in the eighteenth century. It is perhaps Charles Bonnet, who developed this model most consistently, although he was not the only one, or the first to support it (see Kaitaro, 1999). Bonnet and the others who presented similar theories were, however, not "mechanist materialists", for the simple reason that they were dualists. The two aspects of Cartesian physiology we mentioned – the postulation of representations in the brain and physiological mechanism – were both present in Bonnet's theory, but in contrast to Descartes, the Swiss naturalist seems not to worry about leaving the soul the freedom and autonomy with regard to body that Descartes tended to emphasize. Instead he suggests that actions usually attributed to the soul are due to a "hidden mechanism" (Bonnet, 1760, p. 506).

Bonnet postulated systematic and law-like connections between the physiological and the mental instituted by God (see Bonnet, 1760, p. 119), and, like Descartes, he was a strict mechanist in his physiology. Although Bonnet admitted that the union of the two substances is a mystery for us (Bonnet, 1760, pp. 6, 46 and 126; see also Savioz, 1948, p. 141), the connections instituted by God are not arbitrary in so far as His will is determined by the essences of things. Thus pain and pleasure are associated to certain movements of fibres of animals in a way that contributes to their survival (Bonnet, 1760, p. 119). The fact that the Creator has built the mechanisms of the brain in view of establishing these functional connections between the body and the mind allows for picturing the brain as a machine. Even the nervous fibres can be seen as small machines whose internal structure is preformed in view of their specific function (Bonnet, 1769, Vol. I, p. 18). And in so far as the function of these fibres is to represent different sensations or ideas, Bonnet postulates specific and

localizable correlates for representations in the brain. According to him there are specific fibres for each simple idea of which more complex ideas are formed, so that complex ideas correspond to the movements of the fibres corresponding to the simple ideas they consist of (Bonnet, 1760, pp. 202–205). By postulating such identities between mental representations and bodily events (movements of the fibres), he is able to present mechanistic explanations of mental activities like association, reasoning, etc. He makes detailed suggestions of neural mechanisms that could explain phenomena related to learning and memory (see Kaitaro, 1999, 2001a).

Memory is, according to Bonnet, a corporeal function. This can be seen from the fact that corporeal causes like maladies or accidents can cause a loss or a weakening of memory (Bonnet, 1760, p. 57). Memory depends on the dispositions of the fibres to repeat the movements caused by sensed objects. And the changes in these dispositions depend either on the modifications in form of the molecules of which the fibres consist of or in changes in their relative positions (Bonnet, 1760, pp. 58, 66).[4] Bonnet accounts for the association of ideas by referring to the fact that fibres that have been simultaneously affected acquire a tendency to affect one another. A simple recall of one sensation by another is thus based on the communication between the corresponding sensible fibres. Bonnet leaves the exact nature of this communication unspecified but observes that he cannot conceive any other way a body could affect another except by communicating its movement to the other either immediately or mediated by some medium (for example by means of a fluid) (Bonnet, 1760, pp. 601, 637).[5] He postulates that habitual associations are based on physiological changes in the points of

---

[4] Bonnet also mentions the conjecture that he himself had put forth in his earlier work (*Essai de psychologie*): that memory would be based on the conservation of mechanical movements in the brain. He observes, however, that such movements are too short-lived to act as the basis of memory. As evidence of this he refers to the fact how fast the after-image of the sun disappears (Bonnet, 1760, p. 55).

[5] Bonnet observes that the transmission of sensations could take place by vibrations or, more likely, by means of a subtle fluid (animal spirits) whose rapid movements are comparable to those of the "electric fluid" (*fluide électrique*) (op. cit. p. 31).

communication between appropriate fibres, that is, those that represent the associated elements (Bonnet, 1760, p. 643).

Likewise Bonnet tries to figure out neural mechanisms which would explain how memory is able to retain the order of a series of perceptions, for example, of a series of sensations, words or numbers which affect the sense organs constantly in the same order. Bonnet hypothesizes that remembering the order depends on the disposition of the fibres corresponding to the elements of the series to move one after the other in the correct relative order (Bonnet, 1760, p. 631). In order to explain how this takes place, Bonnet refers to the fact that attention augments the intensity of the movements imprinted on the fibres. And when attention is directed successively on a series of objects, it is more divided than when it is fixed on a particular object. Thus its effects tend to be distributed along the elements of the series: it affects relatively less the appropriate fibres and proportionately more the fibres corresponding to the other elements of the series. Since the communication of the movements between fibres corresponding to contiguous elements in the series is always stronger than those that are more distant, the movements are always repeated in the same order (Bonnet, 1760, pp. 641–649). By this mechanism of spreading excitation the fibres acquire a tendency to move one after the other in the same order, and consequently the corresponding ideas are brought into mind in the same order (Bonnet, 1760, p. 632). By the same simple mechanism, Bonnet attempts to explain the inhibiting effect of learned associations on newer ones: an element interposed between two elements already associated is more difficult to learn, since the new association has to compete with connections already established, so that in order to learn the new association one has to unlearn a previous one (Bonnet, 1760, pp. 632–633, 650). Bonnet thus claims that the order of perceptions in our mind depends essentially on the dispositions of the appropriate nervous fibres to affect one another (Bonnet, 1760, p. 633).

In so far as the only conceivable way fibres transmit movement to another fibre is through mechanical interaction, facts related to learning can, according to Bonnet, be explained on the basis of the postulated underlying neural mechanisms. Since Bonnet postulates a strict correlation between that which takes place in the mind and that which takes place in the brain, he can, in his explanatory hypotheses, replace mental events, i.e. ideas, by movements in the brain and reason as if these movements were the ideas themselves (Bonnet, 1760, p. 75). Thus psychology ends up being a science concerned with movements in the brain. So when Bonnet observes that the psychologist must study man like the physician studies nature (Bonnet, 1760, p. 160), this can be taken quite literally.

Interestingly, as we have seen, there is, however, one properly psychological variable intervening in the mechanisms of learning Bonnet describes, e.g. attention. But not only are the effects of this attention bodily and physical affections that concern the nervous fibres and of which a "purely mechanical explanation can be given" (Bonnet, 1760, pp. 136–139, 141–143), but also attention itself is determined by the relative pleasure or displeasure that objects give us (Bonnet, 1760, pp. 130–131, 138, 140, 144). And pleasure and pain are themselves caused by the "vibrations" of the fibres: the pleasurableness or painfulness of sensations depends on the degree of these vibrations and on the relative mobility of the molecules of which the fibres consist (Bonnet, 1760, pp. 118, 120–122). It seems thus that there are no psychological facts that could not be explained by neural mechanisms: the mental variables which are able to determine neural events are themselves determined by other neural events.

Understandably, Bonnet is at pains to prove that his system is compatible with free will, but with the aid of some subtle philosophical distinctions he argues that we have free will – as do in fact, according to him, oysters, horses and apes (he admits that their freedom of will is different from ours, but he claims it is nonetheless as real as ours, pp. 152–153). For the Swiss sensualist, in contrast to Condillac, whose concept of freedom he criticizes, freedom of will does not consist so much in being able to decide freely what one does or does not do, but in being free to do what it *pleases* one to do (Bonnet, 1760, p. 159) – and what pleases us depends, as we have seen, on neural causes. Bonnet also refers to the "inner sentiment" that convinces us that we are the immediate authors of our actions, but observes that even if this sentiment would be an illusion this would not change anything: if we attribute our actions to the hypothetical mechanism of the machine, we are still in control of our actions – in so far as this machine is ourselves (Bonnet, 1760, pp. 4, 23, 25).

That Bonnet's hypothetical models are not always entirely convincing should not keep us from seeing that the way in which they try to account for psychological phenomena by hypothetical neural mechanism that could give rise to them, is entirely modern and is in principle similar to those that have been proposed later by people like Pavlov, Hebb and countless other theorists who have tried to account for psychological phenomena by making hypotheses concerning the underlying neural mechanisms.

Mechanism and Dualism

That it were the dualists who were mechanists and not the materialists may seem odd at first sight, but in fact it is to be expected if one analyses the requirements of dualist theories on the one hand and the nature of technological metaphors on the other hand. Both mechanistic models and the localizable presence of representations in the brain can be shown to work better in the context of dualism. Let us start with mechanistic models. First of all, it is to be expected that technological metaphors work in physiology. And this is true to a certain extent independently of whether one works within materialist or dualist metaphysics. What are tools and machines, if not artificial organs, developed by an intelligent animal? Thus it is not surprising that the Greek word *organon*, "tool", is the etymological root of our word for organ. Tools and organs are both characterized by finality: they perform functions that are subservient to the *telos* of the organism or the user, and it is by understanding how they contribute to this end that we understand and explain them and their mechanisms. Thus we can understand organs through their functions, as if they too were mechanisms made in view of a purpose. However, problems arise when we get to the level of the organism as a whole and try to apply technological metaphors to it: when we try to conceive the whole body as a machine or a tool. One is then led to ask whose tool the body is. A dualist has no problems in answering this question: it is the instrument of the soul (see Bonnet, 1755, p. 1). But a materialist who wants to avoid the postulation of

anything beyond the body is thus motivated *not* to see the body as a machine.

Thus when Diderot uses metaphors which have recourse to the same physical phenomena as Bonnet, vibrations, his use of these never really quits the metaphorical level in order to build concrete explanatory models. In the *Rêve de d'Alembert* Diderot compares the fibres of our organs to the "vibrating sensible strings" of an instrument (Diderot, AT II, p. 113). Instead of referring to the mechanisms of the instrument itself, he presents the mutual resonances of the strings (based on the series of overtones) as an analogy for the way ideas evoke one another – which of course makes these associations (i.e. resonances) less dependent on contingent neural connections and more dependent on the content of ideas (i.e. series of overtones). And Diderot does not refer to any concrete cerebral or neural mechanism. But he makes d'Alembert remark on a problem in this analogy: it seems imply the existence of a musician, who is judging the consonances and dissonances. But Diderot replies that there is an essential difference between the *instrument philosophe* and the harpsichord: in the former case the instrument itself is sentient. And furthermore, it is not actually our instrument, in so far as we *are* the instrument provided with sensibility and memory. The *instrument philosophe* is at the same time the musician and the instrument (Diderot, AT II, p. 114).[6] The collapsing of the distinction between the musician and the instrument changes, however, the metaphor into a paradox: an instrument that is not used by anyone for a purpose or a task (which more or less is its definition) but which acts on its own.

Now we get to the second feature of dualistic theories: the existence of representations in the body. If the body is the organ or the tool of the soul,

---

[6] Bonnet (1760, p. 23) had also compared his hypothetical mechanical man to a harpsichord. The identifications of Diderot and Bonnet may at first glance look similar: Bonnet is claming that the machine is ourselves, and Diderot that we are the instrument. But of course, to say that a harpsichord is an instrument is not the same as to say that the instrument is a harpsichord. Bonnet is suggesting that a machine could be and act just like us, but Diderot is making a distinction between two kinds of machines, man-made instruments used by musicians and "instruments" in a metaphorical sense, instruments that unlike man-made machines are living and sensing.

this requires that the soul is able to interact with it. This is possible if there is a point of contact (the seat of the soul) and some law-like connections between ideas and physiological events, so that the perceptions of the soul correspond regularly to the stimulations of sense organs and so that ideas or representations of movement in the soul cause regularly corresponding movements in the muscles (see Kaitaro, 2001a). The representations postulated in the brain work as intermediate entities which have both mental, i.e. representational properties and material properties, like extension and locality, and which by this fact are able to mediate between the mental and the physical in a way planned by the author of the machine.

In his posthumous and unfinished *Éléments de physiologie*, Diderot discusses accounts of memory which postulate semiotic elements in the brain and turns the traditional Platonic metaphor of a memory trace (compared to an imprint on wax, *Theaetetus* 191c–195a) into a paradox exactly in the same way as he had done on the occasion of the metaphor of body as an instrument. He observes that the distinction between signs (as things which mean something for someone) and their interpreter collapses just like the distinction between the musician and the instrument had. He simply poses the question: if the brain is considered as a book, where is the reader, and answers: it is the book itself (Diderot, 1964, p. 243). Thus giving up the idea of a separate entity, i.e. the soul, using the body as an instrument seems to oblige the materialist to "deconstruct" metaphors based on cultural artefacts, technological or semiotic. Such artefacts cannot really be understood without reference to human bodies whose artificial extensions they are. Thus applying them metaphorically to the human body itself is bound to lead to paradoxes. Thus a materialist like Diderot had – having refused to get rid of these problems by postulating of an immaterial soul – real difficulties in taking such metaphors seriously, i.e., literally.

So, if there is no "ghost" in the machine, there is no need to postulate any representations in the brain since there is no one there to "read" these representations or traces anyway. And of course if there is no ghost, there is no need to postulate a machine that it could be in control of. Thus I find it completely natural that the eighteenth-century medical materialists like La Mettrie and Diderot

were not mechanists and had no need to postulate specific representational traces in the brain. Instead, I think it quite odd that this theory of representations in the brain surfaced again in the nineteenth-century localizationist doctrines, but this time in a materialist context (Kaitaro, 2001b). But I do not think that it is surprising that the nineteenth-century theorists, like Charcot, who had recourse to localizable images and representation stored in the cortex in explaining facts related to functional localization, ended up postulating an entity corresponding to the seat of the soul – the "centre of ideas", or "ideational centre" – for whose existence no clinical or pathological evidence existed but which was required by the anatomy of representations inherited from the dualistic tradition.

When we now look back we can easily see Bonnet, who built his theories mostly on the basis of a priori speculation, as a "precursor" in whose theory we can recognize some of the central elements of modern models of cerebral functions. If we replace the movements of his fibres with electric impulses and changes in their internal structure or in their points of communication with structural changes in nerve cells and in synaptic transmission, his explanations resemble those that we can read in twentieth-century textbooks. But if we look at the writings on brain function of a non-mechanist materialist like Diderot we can find no such thing.[7] Thus whatever one might think of dualist metaphysics, one cannot deny that it seems to be useful in producing mechanistic models of brain function. And it is certainly one of the ironies of history that Bonnet, who was not a materialist, won the day in so far as the influence of the models he supported on materialist theories is concerned: not only the nineteenth-century medical materialists, but also the twentieth-century Anglo-Saxon materialist philosophers, the so-called *identity theorists*, seem to follow closely in his footsteps (Kaitaro, 2004).

---

[7] However, in Diderot's case we might find analogies on a different level, not on the level of basic mechanisms, but rather on the level of notions related to the hierarchic integration of functions. Like Huglings-Jackson later, Diderot saw the organism as a hierarchy of functions, the levels of which range from organic molecules, via fibres, organs and organic centres to the organism as a whole (Kaitaro, 1997, pp. 130–133).

The reductionist materialism of these theorists bears closer resemblance to Bonnet's mechanism than to Diderot's non-reductionist materialism. This is perhaps just to say that our contemporary theories are perhaps more dualist than we realize and that, even if Descartes himself was not the strict Cartesian dualist he is often pictured as, we might all be more or less Cartesians.

Conclusions

So, although reductive materialist theories that postulate identities between mental states and neural events have turned out to be philosophically problematic (MacDonald, 1989), the forms of mechanistic explanation that Bonnet and his Cartesian contemporaries and predecessors relied on have been rather successful in the neurosciences. One might wonder what makes dualist explanatory schemes so successful even after one has discarded their metaphysical foundations or even if there were no type–type identities between the mental and the physical that they postulate.[8] The reason may be simple: by examining the body or the brain as a mechanism, the dualist is able to treat it as a cultural artefact, as a machine, which is characterized by a function (it is built for something) and then – after he has analysed its structure in terms of the functions of its parts – proceed by forgetting this teleological aspect when he examines it in terms of mechanical interactions between its parts. And in the same way the mechanistic dualist who identifies mental events with neural states is able to attribute semantic properties to neural events, but can then in proposing neural explanations ignore these meanings and treat these representations in the nervous systems as if they were merely physical things interacting with each other in the brain. Thanks to the postulated type–type identities, these representations can, unlike real representations (such as linguistic propositions or pictures), be identified by their physical properties or by their location independently of their representational qualities. Modern computational models have the same structure (but without necessarily implying any type–type identities): on the one hand, we have a description of the program (corresponding to the immaterial soul of the dualist) and, on the other hand, we have the machine whose workings can be described and explained in purely mechanistic terms.

Thus what dualism actually does is to delimit the sphere of neuroscientific explanations concerning the neural mechanism underlying mental functions. By permitting to separate these mechanisms from their mental and representative aspects for the purposes of mechanical explanation, they define an autonomous field of study independent of the interpretation of meanings. Questions related to intentions and meanings can thus be delegated to another faculty – that of humanities. In so far as we can neither answer the question, "What is the meaning or reference of the ambiguous word 'tu' (personal pronoun or past perfect of a verb) in the sonnet 'À la nue accablante tu …' by Mallarmé?", by studying its neural correlates in the brain of the reader (or the writer), nor know how the phonemic structure of words is processed in the brain by analysing the phonological structure of language, this kind of institutional dualisms is valid. But we should not, of course, expect the world to be devised according to our faculty divisions. In fact, Descartes himself observes that his metaphysical divisions are related and limited to certain explanatory contexts and that the actions of human persons are not completely understandable neither as mere matter in motion, nor as mental acts of immaterial souls, nor as a combination of these two – as one might expect a Cartesian dualist to think (Descartes, AT III, pp. 691–694). Though it is fashionable to consider that it is just modern neuroscientific explanations that undermine Cartesian substance dualism (see, for example Churchland, 2002, p. 44), one should also remember that in fact these explanations owe much to Cartesian dualism and to the mechanism that is part and parcel of it.

---

[8] Type–type identity is an identity between a type of mental event (for example, pain) and a type of physiological event (for example, the firing specific fibres in the brain). Identity theories of mind often postulate brain–mind identities of this variety and not merely so-called token–token identities. The latter kind of identities only imply that all mental states are physically realized without making the further claim that the tokens of the same type of mental state are necessarily always realized by the same type of physical events in the brain. Thus, for example, the functionalists accept the multiple realizability of mental states.

References

Beyssade, J.-M. (1991). Le sens commun dans la règle XII, corporel et incorporel. *Revue de Métaphysique et de Morale, 4*, 1–19.

Bloch, O. (1979). Sur l'image du matérialisme français du XVIIIe siècle dans l'historiographie philosophique de la première moitié du XIXe siècle: autour de Victor Cousin. In O. Bloch (Ed.), *Images au XIXe siècle du matérialisme du XVIIIe siècle* (pp. 37–54). Paris: Desclée.

Bonnet, Ch. (1755). *Essai de psychologie; ou considérations sur les opérations de l'âme, sur l'habitude et sur l'éducation. Auxquelles on a ajouté des principes philosophiques sur la cause première et sur son effet.* Londres.

Bonnet, Ch. (1760). *Essai analytique sur les facultés de l'âme.* Copenhagen: Frères Ch. & Ant. Philibert.

Bonnet, Ch. (1769). *Palingénesie philosophique* (2 Vols.). Genève: Chez Claude Philibert et Barthelemi Chirol.

Churchland, P. S. (2002). *Brain-wise. Studies in neurophilosophy.* Cambridge, MA/London: A Bradford Book/The MIT Press.

Descartes, R. (AT). In Ch. Adam & P. Tannery (Eds.), *Oeuvres complètes* (12 Vols.). Paris: Cerf, 1897–1913.

Descartes, R. (FA). In F. Alquié (Ed.), *Oeuvres philosophiques* (3 Vols.). Paris: Garnier, 1963, 1967, 1973.

Dictionnaire de l'Académie françoise, Le. Paris: Jean-Baptiste Coignard, 1694.

Diderot, D. (AT). In J. Assézat ja & M. Tourneux (Eds.), *Oeuvres completes* (20 Vols.). Paris: Garnier, 1875–1879.

Diderot, D. (1964). In J. Mayer (Ed.), *Élements de physiologie.* Paris: Librairie Marcel Didier.

Encyclopédie, ou dictionnaire raisonné des sciences, des arts et des métiers. In D. Diderot & J. d'Alembert (Eds.) (17 Vols.). Paris: Briasson, David, Le Breton & Durand, 1751–1765.

Gaukroger, S. (2000). Les âmes des animaux et l'homme machine: la question de la cognition. In B. Bourgeois & J. Havet (Eds.), *L'Esprit Cartésien: Actes du XXVIe Congrès de l'Association des Sociétés de Philosophie de Langue Française* (pp. 312–318). Paris: J. Vrin.

Kaitaro, T. (1987). The eighteenth-century French materialists and "mechanistic materialism". In J. Alavuotunki, A. Leikola, J. Manninen, & A.-L. Räisänen (Eds.), *Aufklärung und Französische Revolution II. Reports from the Department of History* (Vol. 3, pp. 66–83). Oulu: University of Oulu.

Kaitaro, T. (1996). La Peyronie and the experimental search for the seat of the soul: Neuropsychological methodology in the eighteenth century. *Cortex, 32*, 557–564.

Kaitaro, T. (1997). *Diderot's holism: Philosophical antireductionism and its medical background.* Frankfurt am Main: Peter Lang.

Kaitaro, T. (1999). Ideas in the brain: The localization of memory traces in the eighteenth century. *Journal of the History of Philosophy, 37*, 301–322.

Kaitaro, T. (2001a). "Man is an admirable machine" – A dangerous idea? *La Lettre de la Maison française d'Oxford, 14*, Michaelmas Term 2001, 105–120.

Kaitaro, T. (2001b). Biological and epistemological models of localization in the nineteenth century: From gall to charcot. *Journal of the History of Neurosciences, 10*, 262–276.

Kaitaro, T. (2004). Brain–mind identities in dualism and materialism – A historical perspective. *Studies in History and Philosophy of Biology and Biomedical Sciences, 35*, 627–645.

de La Mettrie, J. O. (1747/1960). *L'homme machine.* Edited with an introductory monograph and notes by A. Vartanian. Princeton, NJ: Princeton University Press.

Le Camus, A. (1753). *Médicine de l'esprit* (2 Vols.). Paris: Ganeau.

MacDonald, C. (1989). *Brain–mind identity theories.* London: Routledge.

Rey, R. (1987). *Naissance et développement du vitalisme en France de la deuxième moitié du XVIIIème siècle à la fin du Premier Empire.* Typescript dissertation (3 Vols.). Université de Paris I (Panthéon-Sorbonne), Département d'Histoire. (An abridged version of the thesis has been published in *Studies on Voltaire and the Eighteenth Century 381*, Voltaire Foundation, Oxford, 2000.)

Savioz, R. (1948). *La philosophie de Charles Bonnet de Genève.* Paris: J. Vrin.

Smith, C. U. M. (2002). Julien Offray de La Mettrie (1709–1951). *Journal of the History of Neurosciences, 9*, 110–124.

Sutton, J. (1998). *Philosophy and memory traces: Descartes to connectionism.* Cambridge: Cambridge University Press.

Thomson, A. (1988). L'homme machine, mythe ou métaphore? *Dix-huitième siècle, 20*, 368–376.

Vartanian, A. (1960). *Introductory monograph to La Mettrie's L'homme machine.* Princeton, NJ: Princeton University Press.

24
Explorations of the Brain, Mind and Medicine in the Writings of Jonathan Swift

Marjorie Perlman Lorch

Introduction

Jonathan Swift (1667–1745) was one of the most celebrated political satirists of his age. However, embedded in his writing are numerous astute observations on the mind and brain. Today, Swift is perhaps best remembered as the literary author of *Gulliver's Travels* (1726). However, to his contemporaries he was considered a leading commentator on the politics of England's relations with Ireland, and a significant spiritual head of the Church as the Dean of St. Patrick's Cathedral, Dublin for over 30 years (from 1713 to the time of his death in 1745). An underlying theme that runs throughout many of his political and satirical writings (e.g. the *Tale of a Tub, The Battle of the Books, The Legion Club*) was an interest in madness and mental states. This chapter considers the numerous original insights and reflections on neuroscientific topics in Swift's writings.

Eighteenth Century Context

For a source reflecting public interests at a particular historical period, it is useful to consider the popular publications of the day. This can be provided by considering the newspapers circulating in the London coffeehouses of the early eighteenth century. The most prominent of these was *The Spectator*, a daily English newspaper founded in 1711 by Joseph Addison and Richard Steele. The stated goal of *The Spectator* was

to enliven morality with wit, and to temper wit with morality ... to bring philosophy out of the closets and libraries, schools and colleges, to dwell in clubs and assemblies, at tea-tables and coffeehouses.

It was a key barometer to the public concerns of London and Great Britain throughout the century.

It has been estimated that one in every four Londoners read this paper. Postman (2000) suggests that *The Spectator* is significant in publishing history for it represents the first true outpouring of expository prose in the English language. Jonathan Swift was one of the magazine's founding writers and a major champion of the language.

Throughout the seventeenth century, major writers of natural philosophy still produced their works in Latin. The turn of the century saw a shift in England so that by the mid-eighteenth century their successors wrote in their native tongue. Postman (2000) suggests that the connection between the growth of experimental science – in medicine and physiology, as well as chemistry and physics – and the growth of expository prose is not merely coincidental. His thesis is that "the scientific revolution in the eighteenth century was, as much as anything else, a revolution in language" (Postman, 2000, 61). In this context, it is not surprising that one interpretation of the underlying themes in *Gulliver's Travels* is the pitfalls and peculiarities of language.

Swift on Language and Mind

In 1712, Swift published his *Proposal for correcting, improving and ascertaining the English Tongue* in the form of a letter to Harley, Earl of Oxford. In this tract, he called for the formation of a language academy to fix standards similar to the

Académie Française. Swift complained of the increasing use of neologisms, abbreviations, slang, affectation, phonetic spelling in English and urged that a way be found for "ascertaining and fixing our language for ever". Swift said he saw "no absolute necessity why any language should be perpetually changing. . . . many words that deserve to be utterly thrown out of our language, many more to be corrected, and perhaps not a few long since antiquated which ought to be restored on account of their energy and sound" (Swift, 1712/etext.library. adelaide.edu.au/s/swift/jonathan/s97p/, 3). In some ways, this was an odd position to take for a writer who was responsible for coining a large number of new words in the language. (His writing is cited as the first appearance of 175 words in the Oxford English Dictionary.)

Although there is much linguistic insight in Swift's proposals, he failed to appreciate that language change might also reflect new ways of thinking about things. His proposal is surprising and in many ways paradoxical, for it was Swift who was constantly exploring the relation between language and thought, rationality and chaos. One who opposed Swift's proposal to set up an academy to regulate the English Language was John Oldmixon. He pointed out that: "it will be in vain to pretend to ascertain Language, unless they had the Secret of setting Rules for Thinking, and could bring Thought to a Standard too" (Quoted in Howatt, 1984, p. 109; Oldmixon, 1712).

A more radical suggestion is made in *Gulliver's Travels:*

. . . a scheme for entirely abolishing all words whatsoever . . . a great advantage in point of health as well as brevity. For it is plain that every word we speak is in some degree a diminution of our lungs by corrosion, and consequently contributes to the shortening of our lives . . . since words are only names for things, it would be more convenient for all men to carry about them such things as were necessary to express the particular business they are to discourse on . . . And this invention would certainly have taken place . . . if the women . . . the vulgar and illiterate, had not threatened to raise a rebellion . . . such constant irreconcilable enemies to science are the common people. (Swift, 1726, 1992, p. 140)

Swift also describes a scheme for teaching mathematics which employed propositions "written on a thin Wafer . . . the Student was to swallow upon a fasting Stomach" (Swift, 1726, 1992, p. 141).

Swift on Mind and Madness

Swift's first significant written work was produced in response to a controversy involving his patron Sir William Temple and William Wotton on the relative value of ancient (i.e. Ancient Greek and Roman) and modern learning. In support of Temple's position regarding the superiority of the Ancients, 1697, Swift wrote his contribution to the controversy, entitled *The Battle of the Books* (written 1697; published 1704), and *A Tale of a Tub, written for the universal improvement of mankind* (written 1702; published 1710). The first considerations of the issues surrounding madness and rationality figure prominently in these early works. Swift satirically discusses the origin, use and importance of madness in a commonwealth. Swift puts forward the argument that madness is only unconventional or antisocial thought:

. . . if the Moderns mean by Madness, only a Disturbance or Transposition of the Brain, by Force of certain Vapours issuing up from the lower Faculties; Then has this Madness been the Parent of all those mighty Revolutions, that have happened in Empire, in Philosophy, and in Religion. For, the Brain, in its natural Position and State of Serenity, disposeth its Owner to pass his Life in the common Forms, without any Thought of subduing Multitudes to his own *Power*, his *Reasons* or his *Visions*; and the more he shapes his Understanding by the Pattern of Human Learning, the less he is inclined to form Parties after his particular Notions; because that instructs him in his private Infirmities, as well as in the stubborn Ignorance of the People. (Swift, 1710, 1985, p. 108)

Swift was equally interested in the issues of madness from both a moral and philosophical point of view. In the *Ode to Dr. William Sancroft*, written by Swift in 1692, he observes, "For swords are madmen's tongues, and tongues are madmen's swords". This can also be seen in "A Digression Concerning Madness" which is contained in *A Tale of a Tub* (1710). This essay is crucially concerned with understanding madness and epistemology.

Now I would gladly be informed how it is possible to account for such imaginations as these in particular men, without recourse to my phenomenon of vapours ascending from the lower faculties to overshadow the brain, and there distilling into conceptions, for which the narrowness of our mother-tongue has not yet assigned any other name beside that of madness or frenzy. (Swift, 1710, 1985, p. 105)

Another reference to Galenic vapours is made in the Preface to the *Battle of the Books:*

There is a Brain that will endure but one Scumming: Let the Owner gather it with Discretion, and manage his little Stock with Husbandry; but of all things, let him beware of bringing it under the Lash of his Betters; because, That will make it all bubble up into Impertinence, and he will find no new Supply: Wit, without knowledge, being a sort of Cream, which gathers in a Night to the Top, and by a skilful Hand, may be soon whipt into Froth; but once scumm'd away, what appears underneath will be fit for nothing, but to be thrown to the Hogs. (Swift, 1704, 1985, p. 140)

In an analysis of this work, Smith (1979) suggests that this work includes a parody of John Wilkins's *Essay Towards a Real Character* (1668) which is about Universal Language. Smith (1979) suggests that certain passages contain descriptions that accurately portray schizophrenic language.

...eternally talking, sputtering, gaping, bawling, in a Sound without Period or Article ...In much and deep Conversation with himself ...has forgot the common *Meaning* of Words, but an admirable Retainer of the *Sound* ...very sparing in his Words, but somewhat over-liberal of his Breath ...Besides, there is something Individual in human Minds, that easily kindles at the accidental Approach and Collision of certain Circumstances, which tho' of paltry and mean Appearance, do often flame out into the greatest Emergencies of Life. For great Turns are not always given by Strong Hands, but by the lucky Adoption, and at proper Seasons; and it is of no import, where the Fire was kindled, if the Vapour has once got up into the Brain, For the *upper region* of Man, is furnished like the *middle region* of the Air; The Materials are formed from Causes of the widest Difference, yet produce at last the same Substance and Effect.... (Swift, 1704, 1985, p. 112)

It also contains the interesting use of neologism: e.g. "his Epidemical Diseases being Fastidiosity, Amorphy and Oscitation".

A Tale of a Tub also contains passages that may be a parody of the writings of Paracelsus, i.e. Philip Theophrastus Bombastus of Hohenheim. (Bombast is defined in the OED as grandiose but empty language.) Kiernan (1971) suggests that the discussion of the role of vital air is a direct parody of Paracelsus:

Besides, there is something Individual in human Minds, that easily kindles at the accidental Approach and Collision of certain Circumstances, which tho' of paltry and mean Appearance, do often flame out into the greatest Emergencies of Life. For great Turns are not always

given by Strong Hands, but by the lucky Adoption, and at proper Seasons; and it is of no import, where the Fire was kindled, if the Vapour has once got up into the Brain, For the *upper region* of Man, is furnished like the *middle region* of the Air; The Materials are formed from Causes of the widest Difference, yet produce at last the same Substance and Effect.... (Swift, 1704, 1985, p. 102)

Furthermore, LaCasce (1970) suggests Swift is referring to Paracelsus when he describes experiments to create a universal medicine called "Zibeta Occidentalis" by turning human excrement into perfume in the *Tale of a Tub*:

...hear the words of the famous Troglodyte philosopher [Swift]: 'Tis certain', (said he) 'some grains of folly are of course annexed as part of the composition of human nature, only the choice is left to us, whether we please to wear them inlaid or embossed, and we need not go very far to seek how that is usually determined, when we remember it is with human faculties as with liquors, the lightest will ever be on top. (Swift, 1704, 1985, p. 116)

However, Swift was also interested in the real circumstance of mad men. When he was living in the city, he made several visits to Bethlehem Royal Hospital, known as "Bedlam", the London asylum for the insane. Bedlam (originally founded in 1246) was newly rebuilt in 1675 at Moorfields with room for 150 patients. In a letter to Stella, dated 13 December 1710, Swift describes a visit to Bedlam as a Sunday tourist outing with Lords and Ladies. He relates that he saw "the Tower and saw all the sights, lions, etc. then to Bedlam, then dined at the Chophouse behind the Exchange; ... and concluded the night at the Puppet-Shew ..."

In *Hints Towards An Essay On Conversation* (1710), Swift draws on his experience of what he witnessed on those visits to Bedlam:

I say nothing here of the itch of dispute and contradiction, telling of lies, or of those who are troubled with the disease called the wandering of the thoughts, that they are never present in mind at what passeth in discourse; for whoever labours under any of these possessions is as unfit for conversation as madmen in Bedlam. (Swift, 1710, 1985, p. 252)

His continuing support of the plight of the insane was signalled by his election as a governor there on 26 February 1714. However, he returned to Ireland to take up the position of Dean of St. Patrick's in Dublin at this time and was never actively involved in the London hospital. While in Ireland, Swift became an active governor at the Dublin Blue Coat School Infirmary, and was also on the Board of Trustees of Dr. Steeven's Hospital from 1721 (Le Fanu, 1978). (This hospital was founded in 1717, but did not open until 1733.)

Swift was also interested in madness from the point of social responsibility. Later in life, Swift willed his estate to the founding of St. Patrick's Hospital for lunatics and idiots. This was to be the first institution of its kind in Ireland. Swift initiated the purchase of the site and commissioned the architect. He was personally involved in the design of the new Dublin asylum which he suggested should be along the lines of Bedlam (Malcolm, 1989). He also wrote a biting satire of this process, which again draws the analogy between the madness of the morally (and politically) corrupt and those who were insane. In *A Serious and Useful Scheme to make an Hospital for Incurables, whether the Incurable Disease were Knavery, Folly, Lying, or Infidelity* (1733), Swift discussed the necessity of dealing with the number of "... fools, knaves, scolds, scribblers, infidels and liars, not to mention the incurably vain, proud, affected and ten thousand others beyond cure". He suggested that he himself be admitted as one of the "scribbling incurables". He was happy to feel that no person would be offended by his scheme, "because it is natural to apply ridiculous characters to all the world, except ourselves". (Aitken, 1907–1921/://www. bartleby.com/219/0419.html)

One of Swift's last major pieces of writing was the long poem *The Legion Club* (1736). In this work, the Irish Parliament is described like a madhouse full of "brain-sick brutes". Many of the details appear to have been taken from the first hand observations Swift had of Bedlam and other asylums:

... When I saw the keeper frown,
Tipping him with half-a-crown,
... Tie them, keeper, in a tether,
Let them starve and sink together;
Both are apt to be unruly,
Lash them daily, lash them duly;
Though 'tis hopeless to reclaim them,
Scorpion rods, perhaps, may tame them.

As discussed above, Swift's interest in insanity and madness was both from a moral and philosophical point of view. In the late seventeenth and early eighteenth century were many English writers who concerned themselves with the relation between madness and reason, its causes and nature – from Hobbes and Locke, to Swift's contemporaries Pope and Defoe, and latterly Johnson and Sterne.

The greatest similarities are perhaps between the writings of Swift and Defoe on the topic of madness. Defoe's *Essays on Projects* (1697) includes one on a hospital for fools to be funded by a charity lottery that touches on many of the same points. However, for Defoe, the crux of the matter is the distinction between madmen who lose their souls (also used to refer to "mind" in the Ancient literature) and fools who are born without them:

... since as the soul in man distinguishes him from a brute, so where the soul is dead (for so it is as to acting) no brute so much a beast as a man. But since never to have it, and to have lost it, are synonymous in the effect, I wonder how it came to pass that in the settlement of that hospital they made no provision for persons born without the use of their reason, such as we call fools, or, more properly, naturals. (Defoe, 1697 http://etext.library. adelaide.edu.au/d/defoe/daniel/d31es/part13.html)

Defoe also wrote an essay calling for an Academy to be established to regulate the English language (*On Academies*, 1697). This parallels the essay written by Swift in 1712 *A Proposal for Correcting, Improving and Ascertaining the English Tongue* discussed above. Although Defoe and Swift appear to concern themselves with identical issues, their motivations and intellectual curiosity about these subjects were not the same. Defoe was a professional writer, supporting himself on the proceeds of his works thus following the fashion trends of popular culture, while Swift wrote from religious, political, and most fundamentally, moral motives (Glendinning, 1998).

Similarly, Swift parodied the contemporary concerns regarding the origin of human language and the effect of culture on human nature. It is not surprising that the case of a "wild child" from Germany also attracted Swift's attention. A boy named Peter, reported to have been raised by animals, had been brought to England at Queen Caroline's request. The boy was committed to the care of Swift's close friend Dr. John Arbuthnot. There is a poem describing Swift's response to seeing the case (also documented in *Letter from Swift to Stella*) (1726): "The Most Wonderful Wonder that ever appeared to the Wonder of the British Nation. Being an

Account of the Travels of Mynheer Veteranus, through the Woods of Germany: And an account of his taking a most monstrous She Bear, who had nursed up the Wild Boy; &c.", which is signed "written by the Copper-Farthing Dean".

Apart from a general interest in normalcy and pathology in mental development and behaviour, Swift also speculated on the very nature of brain function. His descriptions of memory changes in ageing that appear in *Gulliver's Travels* are astutely accurate descriptions:

In talking they forget the common appellation of things, and the names of persons, even of those who are their nearest friends and relations. For the same reason they never can amuse themselves with reading, because their memory will not serve to carry them from the beginning of a sentence to the end ... They were the most mortifying sight I ever beheld. (Swift, 1726, 1992, p. 161)

Several authors (e.g. Boller & Forbes, 1998; Crichton, 1993; Lewis, 1993; Schafer, 2003) take this passage to be an early description of Alzheimer's disease.

Consideration of the processing functions of different parts of the brain is also discussed in Gulliver's Travels. In Chapter VI, there is "A further Account of the Academy. The Author proposes some Improvements which are honourably received".

But, however, I shall so far do justice to this part of the Academy, as to acknowledge that all of them were not so visionary. There was a most ingenious doctor, who seemed to be perfectly versed in the whole nature and system of government. This illustrious person had very usefully employed his studies, in finding out effectual remedies for all diseases and corruptions to which the several kinds of public administration are subject, by the vices or infirmities of those who govern, as well as by the licentiousness of those who are to obey. ... When Parties in a State are violent, he offered a wonderful Contrivance to reconcile them. The Method is this. You take an Hundred Leaders of each Party, you dispose of them into Couples of such whose Heads are nearest of a size; then let two nice operators saw off the *Occiput* of each Couple at the same time, in such a manner that the Brain may be equally divided. Let the *Occiputs* thus cut off be interchanged, applying each to the Head of his opposite Party-man. It seems indeed to be a Work that requireth some exactness, but the Professor assured us, that if it were dexterously performed, the Cure would be infallible. For he argued thus; that the two half Brains being left to debate the Matter between themselves within the space of one Scull, would soon come to a good Understanding, and produce that Moderation, as well as Regularity of Thinking, so much to be wished for in the Heads of those who imagine they come into the World only to watch and govern its Motion: And as to the difference of Brains in Quantity or Quality, among those who are Directors in Faction; the Doctor assured us from his own knowledge, that it was a perfect Trifle. (Swift, 1726, 1992, p. 143)

The identification of the occiput as the part of the brain to be operated on is consistent with the Medieval Cell Doctrine which located memory in the posterior cell (Whitaker, 2006, personal communication). This parody is generally considered to be about the Royal Society (the earliest meetings of this group date from 1645; Royal Charter granted in 1660) which had members including Thomas Willis, Robert Boyle, John Wilkins and Robert Hooke. It is possible that while in London Swift had the practice of reading the *Proceedings of the Royal Society* from which he learned of the latest research findings of the day. Smith (1979) suggests that Swift had read Thomas Willis's *Practice of Physick*, which included descriptions of the anatomy of the brain. However, there is no record of a copy of Willis in Swift's library or those he was known to use. In addition, the related work by Willis's student John Locke that also might be pointed to as a source of some of Swift's material on mind and brain function was also not in the catalogue of Swift's books (Passmann & Vieken, 2003).

An alternative source for this knowledge of brain anatomy was through one of his medical friends. Swift was friends with many doctors of medicine including John Arbothnut and John Freind (1675–1728) who were all members of The Society Club founded in 1711 in London (Le Fanu, 1978). Freind wrote *The History of Physick*, which he worked on while in the Tower on treason charges in 1722. In this, he discusses Thomas Willis' ideas on the nervous system.

Some of the ideas in Gulliver's Travels and Swift's other writings on mind and brain might be traced to Francis Bacon. His utopian tale *New Atlantis* (1659) describes the conquest of disease and prolongation of life, which is echoed in the description of the Houyhnhnms (Porter, 2001).

Jonathan Swift's library (Passmann & Vienken, 2003) contained books by Paracelsus (Bombast von Hohenheim) Avicenna and Galen, Newton and

Pascal, Spinoza, Hobbes and Wotton. However, there were no copies of Willis, Descartes, or Burton, and the only book written by Locke was *On Money*. In the library of Sir William Temple, the mentor of his youth, there were copies of Burton's *Anatomy of Melancholy* and John Wilkins' *Essay towards a real character and a philosophy of language*. Temple's library also contained Decartes, Gassendi, Harvey and Pliny, while the library of Swift's close friend Thomas Sheridan contained a copy of George Cheyne's *Essay on Sickness and Health* (1725), but note that Cheyne's *The English Malady* was not published until 1733. It is likely that Burton's book on *The Anatomy of Melancholy* was a particularly influential source for Swift's ideas about mental illness (Canavan, 1973). Swift's *Discourse Concerning the Mechanical Operation of the Spirit* is a parody of Descartes and other philosophical treatise while at the same time seriously considering questions of ontology (Fletcher, 1962). Finally, there is Swift's (1727) poem *On Dreams* that includes the insightful passage:

Those dreams, that on the silent night intrude,
And with false flitting shades our minds delude,
Jove never sends us downward from the skies;
Nor can they from infernal mansions rise;
But are all mere productions of the brain,
And fools consult interpreters in vain.

Swift's Own Mental State

Swift's own state of health figured heavily in his letters and poems throughout his life. He had many close friends who were eminent physicians in London and Dublin including John Radcliffe, John Arbuthnot, James Grattan and Richard Helsham. Gulliver's occupation was as a ship's surgeon and there are frequent comments about physicians' (lack of) ability to treat diseases in *Gulliver's Travels*.

Swift himself suffered from an intermittent but progressive and untreatable illness, which had as its major features deafness, vertigo and nausea. He documented many of his symptoms in his numerous letters to friends and family. The first letters that mention his symptoms were written in the winter of 1708–1709 when he was 23-years-old. Victorian physicians subsequently identified Swift's complaint as Ménière's disease that was described

first in 1861. For Swift's chronic illness, Arbuthnot prescribed Dust of Valerian and Peruvian Bark.

In later years, Swift wrote a surprising poem that he composed and revised over a long period. In the "Verses on the Death of Dr. Swift written by Himself" published in 1731 when he was 64-years-old, he includes the stanza, which details the loss of language and cognitive abilities with ageing:

For poetry he's past his prime
He takes an hour to find a rhyme...
See how the Dean begins to break
Poor gentleman, he droops apace,
You plainly find it in his face;
That old vertigo in his head
Will never leave him till he's dead;
Besides, his memory decays,
He recollects not what he says.

As discussed above, in the eighteenth century the concepts of madness and rationality were a major social concern of the day. Thus, it should not be surprising that Swift's own behaviour in his last years were described by his many biographers as reflecting madness, insanity and imbecility in old age. For example, Samuel Johnson (1781) wrote:

... his ideas, therefore, being neither renovated by discourse, nor increased by reading, wore gradually away, and left his mind vacant to the vexations of the hour ... till at last his anger was heightened into madness ... He now lost distinction. His madness was compounded of rage and fatuity. (Johnson, 1781)

Sir Walter Scott (1814) describes Swift's behaviour during his final years as marked by "violent and furious lunacy", "frantic fits of passion", and "situation of a helpless changeling". However, these descriptions can be seen as reflections of the writers' own concepts and beliefs about mental illness as well as their feelings towards their subject. It is essential to stress that while Samuel Johnson (1709–1784) was living at somewhat the same time as Swift, Scott (1771–1832) was born almost three decades after Swift's death. Both of their biographies are based on extracts of letters, and reminisces, rumours and hearsay handed down over decades. They had in fact never met Swift in life, and were openly declared his political and literary rivals (*The Times*, 1883).

Indeed, early on in his literary career, Swift writes in *Thoughts on Various Subjects* what will be an accurate presentiment of his own treatment:

When a true genius appears, you can know him by this sign:

> that all the dunces are in confederacy against him.
> (Swift, http://etext.library.adelaide.edu.au/s/swift/jonathan/s97th/)

Dr. William Wilde (1849) was at pains to dispel the impression created by Swift's biographers that he was insane:

> ... neither in his expression, nor the tone of his writing, nor from an examination of any of his acts, have we been able to discover a single symptom of insanity, nor aught but the effects of physical disease, and the natural wearing and decay of a mind such as Swift's ...

(For a more detailed discussion of the 250 years of retrospective diagnosis for Swift's final illness, see Lorch, 2006.)

Recently, a series of letters on dementia as described in *Gulliver's Travels* appeared in *Lancet*. There is argument about whether Swift is describing Alzheimer's or Pick's disease. Lewis (1993) suggested that Swift's description of the Struldbrugges in *Gulliver's Travels* (quoted above) as archetypal of Alzheimer's disease based on Swift's own uncle Godwin who is said to have suffered memory loss with ageing. Lewis goes on to assert that Swift was being prescient about his own final mental state:

> Anyone acquainted with Alzheimer's disease will recognise a familiar ring in words written well over two centuries ago ... Since it seems unlikely that such a picture could have been drawn other than from reality, it is probable that Swift was describing in 1726 what we now know as Alzheimer's disease.... (Lewis, 1993)

Boller and Forbes (1998) also review references to the passage describing the Struldbrugs. Of this description they note that, the diagnosis inferred by medical historians has varied according to the times and neurological fashion. For their part, Boller and Forbes offer the diagnosis of Primary Progressive Aphasia.

Conclusion

As the excerpts from Swift's writings quoted above illustrated, throughout his life he was fascinated by how people spoke, thought, behaved and reasoned. Swift was in fact the focus of both lay and professional interest in his mental state and linguistic abilities throughout his life and, for centuries after his death, speculation continues.

He died in 1745 at the age of 78. He bequeathed the bulk of his fortune (£12,000) to build and endow Ireland's first lunatic asylum.

This review of Swift's writings has revealed many references to neuroscientific ideas. This reflects both his own personal concerns, and those of the general reading public of the day. The literary evidence suggests that consideration of mind and brain and sanity and madness were popular issues. The references to many texts from the Ancient and Modern natural philosophers in Swift's writings suggest that these were common currency, and as such, were readily recognisable and potent targets of satire.

References

Aitken, G. A. (1907–1921). Swift. In A. W. Ward, A. R. Waller, W. P. Trent, J. Erskine, E. S. P. Sherman, & C. Van Doren (Eds.), *The Cambridge history of English and American literature*. New York: G.P. Putnam's Sons. Volume 11, From Steele, and Addison to Pope and Swift, Chapter 9. E-text Bartleby.com, 2000 (http://www.bartleby.com/cambridge/) accessed 10 Feb 2006.

Boller, F., & Forbes, M. M. (1998). History of dementia and dementia in history: An overview. *Journal of the Neurological Sciences, 158*, 125–133.

Canavan, T. L. (1973). Robert Burton, Jonathan Swift, and the tradition of anti-Puritan invective. *Journal of the History of Ideas, 34*, 227–242.

Crichton, P. (1993). Jonathan Swift and Alzheimer's disease. *Lancet, 342*, 874.

Defoe, D. (1697). *An essay upon projects*. http://etext.library.adelaide.edu.au/d/defoe/daniel/d31es/part13.html accessed 10 Feb 2006.

Fletcher, P. (1962). Samuel Beckett et Jonathan Swift: vers une étude compare. *Littératures X: Annales publies par la Faculté des Lettres de Toulouse, 11*, 81–117. (English Translation, 2002 by the author Paul Fletcher http://www.themodernword.com/beckett/paper_fletcher.html).

Glendinning, V. (1998). *Jonathan Swift*. London: Hutchinson.

Howatt, A. P. R. (1984). *A history of English language teaching*. Oxford: Oxford University Press.

Johnson, S. (1781/1891). Jonathan Swift. In *The lives of the poets* (Vol. 3, pp. 2–45). London: George Bell and Sons. http://www.jaffebros.com/lee/gulliver/biography/johnslife.html accessed 10 Feb 2006.

Jonathan Swift. (21 July 1883). *The Times*, 4.

Kiernan, C. (1971). Swift and science. *The Historical Journal, 14*, 709–722.

LaCasce, S. (1970). Swift on medical extremism. *Journal of the History of Ideas, 31*, 599–606.

Le Fanu, R. (1978). Swift's medical friends. In E. O'Brien (Ed.), *Essays in honour of J.D.H. Widdess* (pp. 43–56). Dublin: Cityview Press.

Lewis, J. M. (1993). Jonathan Swift and Alzheimer's disease. *Lancet, 342*, 504.

Lorch, M. (2006). Language and memory disorder in the case of Jonathan Swift: Considerations on retrospective diagnosis. *Brain, 129*, 3127–3137.

Malcolm, E. (1989). *Swift's Hospital: A history of St. Patrick's hospital Dublin (1748–1989)*. Dublin: Gill and Macmillan.

Passmann, D. F., & Vieken, H. J. (2003). *The library and reading of Jonathan Swift: A bio-bibliographical handbook. Part 1, Swift's library in four volumes.* Oxford: Peter Lange.

Porter, R. (2001). Medical futures. *Interdisciplinary Science Reviews, 26*, 35–42.

Postman, N. (2000). *Building a bridge to the eighteenth century.* New York: Knopf.

Schafer, D. (2003). Gulliver meets Descartes: early modern concepts of age-related memory loss. *Journal of the History of Neurosciences, 12*, 1–11.

Scott, W. (1814/1971). Memoirs of Jonathan Swift, D. D. In D. Donoghue (Ed.), *Jonathan Swift: A critical anthology* (pp. 98–100). Baltimore: Penguin.

Smith, F. N. (1979). *Language and reality in Swift's a tale of a tub.* Columbus: Ohio State University Press.

Swift. J. (1692/1958). Ode to Dr. William Sancroft. In H. Williams (Ed.), *The poems of Jonathan Swift.* Oxford: Clarendon.

Swift, J. (1704/1985). Battle of the books. In K. Williams (Ed.), *A tale of a tub and other satires.* London: Dent.

Swift, J. (1710/1909–1914). Hints towards an essay on conversation in English essays: Sidney to Macaulay. *The Harvard classics.* New York: P.F. Collier & Son. http://www.bartleby.com/27/8.html accessed 10 Feb 2006.

Swift, J. (1710/1985). Tale of a tub. In K. Williams (Ed.), *A tale of a tub and other satires.* London: Dent.

Swift, J. (1711). *Thoughts on various subjects.* http://e.text.library.adelaide.edu.au/s/swift/jonathan/s97th/ accessed 10 Feb 2006.

Swift, J. (1712). *A proposal for correcting, improving and ascertaining the English tongue.* London: Benjamin Tooke. http://etext.library.adelaide.edu.au/s/swift/jonathan s97th/ accessed 10 Feb 2006.

Swift, J. (1726/1948). Letter to Stella. In H. Williams (Ed.), *Journal to Stella.* Oxford: Claredon.

Swift, J. (1726/1992). *Gulliver's travels.* Ware: Wordsworth Editions.

Swift, J. (1727/1958). On dreams. In H. Williams (Ed.), *The poems of Jonathan Swift.* Oxford: Clarendon.

Swift, J. (1731/1958). Verses on the death of Dr. Swift written by Himself. In H. Williams (Ed.) *The poems of Jonathan Swift.* Oxford: Clarendon.

Swift, J. (1733/1991). A serious and useful scheme to make an hospital for incurables, whether the incurable disease were knavery, folly, lying, or infidelity. In J. McMinn (Ed.), *Swift's Irish pamphlets: An introductory selection.* Savage, Maryland: Ulster editions and monographs, Barnes & Noble.

Wilde, W. R. (1849). *The closing years of Dean Swift's life; with remarks on Stella, and of some of his writings hitherto unnoticed.* Dublin: Hodges and Smith.

25

Temperament and the Long Shadow of Nerves in the Eighteenth Century

George Rousseau

"Temperament, [is] a moderate and proportionable mixture of any thing, but more peculiarly of the four humours of the body" (Phillips, 1658: sub "Temperament")

"His whole house [of Shandy] might take their turn from the humours and dispositions which were then uppermost" (Sterne, 1940, p. I. 1)

"To descend to a very low instance, and that only as to *personality*; hast thou any doubt, that thy strong-muscled bony face was as much admired by thy mother, as if it had been the face of a Lovelace?" (Richardson, 1748, VI. i. 5)

One of our most acclaimed social scientists, Harvard psychology professor Jerome Kagan, has spent much of his professional career studying temperament as the key to human individuality. The titles of his recent books speak for themselves: *Unstable ideas: Temperament, cognition, and self* (1989); *Galen's prophecy: Temperament in human nature* (1994); *Three seductive ideas* (1998), the salient one being *The Long Shadow of*

Temperament (2004); and dozens of articles featuring temperament as pivotal. Awarded this place of privilege, temperament is the superlative operative word in Kagan's conceptual framework. His theory, in brief, is that every human inherits a physiology determining the emotional temperament and shapes the larger psychological profile when combined with experience. The aim of this essay is to understand why a major psychologist of Kagan's international stature should have chosen temperament; and secondly to historicize Kagan's enterprise and inquire what, if anything, the eighteenth century contributed to temperament's historical map.[1]

This goal is worthwhile for several reasons: it explains the historical foundations on which one of the most prominent historical psychologists of our generation has worked and how he selected "temperament" above all other concepts; it serves to highlight Kagan's sites of originality and account for those that are derivative; it unpacks a concept – temperament – that recently has fallen into disuse and dropped out of the historical vocabulary despite Kagan; and – most crucial for this book – it focuses squarely on the eighteenth-century contributions to

---

[1] Jerome Kagan is Professor of Psychology Emeritus at Harvard University. He was Director of the Mind/Brain/Behavior Interfaculty Initiative at Harvard and is the author of numerous books. Even in his most recent book, *An Argument for Mind*, published in May 2006, he sustains his quest to understand the origins and development of temperament. He has spent a half century deciphering the behavior of infants and children as part of the search for the roots of temperament. His research has cast light on the influence of inborn temperament and on the limits of that influence: timid infants and toddlers do not all grow into shy teenagers. Kagan has also helped

explain milestones in child development – such as the onset of separation anxiety in babies – as due to changes in brain growth. Throughout this work he has retained esteem for the significance of brain and mind while researching the influence of environment and learning. But he also contends that the language of neuroscience can never replace the vocabulary of psychology; that despite important contemporary research on neurons, circuits, and hormones, we will always have to talk about morality, meaning, and love as psychological processes. Perhaps no other scholar of our time has placed such emphasis on the long shadow of temperament.

temperament. It also demonstrates why a book about eighteenth-century neuroscience, whose three key concepts are brain, mind, and medicine, cannot afford to omit temperament, or if it does, does so at its own peril. These goals may be accomplished in addition to glossing the larger perennial riddle about who we are and how we got to be that way.

Historicizing Kagan's Temperament

Temperament is an English word arising from the same class of words as temperance, temperature, temptation, temperamental, and other lexical derivatives indicating balance and proportion through mixture. This element of blending, based on the Latin word *temperare*, to mix, was its chief attribute throughout the early modern period. As an obscure translator made evident, "A Temperament is a proportion of the four chief Elementary Qualities proper for the true exercise of the Natural Functions" (La Framboisière, 1684, p. I. 18). In our sense temperament – this physiological mixture – enabled the life functions to work. By 1700 the word had clearly attached itself to mind. As poet-critic John Dryden claimed in the dedication to one of his great tragic plays, *Aureng-Zebe* (1676), "Our Minds are perpetually wrought on by the Temperaments of our Bodies" (Dryden, 1972, p. I. i. 22). By then the word had also embodied large clusters of concepts whose lexical origins derived from words about an assumed self and its properties: physical and biological (*temperature*), ethical (*temperance*), religious (*temptation*), and emotional (*temperamental*). Over the last three centuries it has also acquired other lexical compounds – *well-tempered, ill-tempered, short-tempered*, etc. – aiming to bestow nuance to the original *temperament* of an individual. If, for example, we claim that someone is *ill-tempered* we implicate the whole cluster of these categories of selfhood: physical and biological (*temperature*), ethical (*temperance*), religious (*temptation*), and emotional (*temperamental*). Much personification also occurred: for example, when composers of the early eighteenth century wrote, as Bach did, the "Well-Tempered Klavier" they intended not merely to innovate on the musical temperaments ("temperaments" then also denoting scales and what we would call musical keys) but

also on the emotional aspects, i. e., the idea of a clavier, or early keyboard instrument, capable of playing in all 24 keys of the scale, each emanating its own particular mood or emotion. So temperament was a far-reaching concept by 1700 with tentacles extending in many directions.

Kagan's use, however, remains to be understood, unpacked and historicized. He himself defines it on many occasions:

> The concept of temperament, as we use it, implies an inherited physiology that is preferentially linked to an envelope of emotions and behaviors (though the nature of that link is still poorly understood) . . . We believe, on the basis of the research described in this book, that inherited temperamental biases, combined with life experiences, create this variation in behavioral reaction to unfamiliar events. (Kagan, 2004, p. 35)

The idea would not have surprised thinkers in the long eighteenth century. All depends, of course, on Kagan's "inherited physiology" in relation to the crucial "life experiences" to which he refers in his books. The absence of soul or psyche is noteworthy: his system has no place for these concepts – soul and psyche – nor do they appear in his vocabulary. Mind fares better but nevertheless remains shadowy: a silhouette in the hinterland (see the prefaces to, and indexes in, Kagan 1989, 1994, 1998, 2004, 2005). Put otherwise, two components, and two components only, combine to constitute the individual's uniqueness: (1) inherited physiology and (2) life experiences.[2]

Kagan's map of selfhood is more scientific than any generated by the eighteenth century for its versions of the "inherited" component, as we shall see, of Kagan's "inherited physiology." Eighteenth-century philosophers were eager to explain how the infant detaches from its mother and gets to know itself, as in the manuals for mothers on child-rearing, and how the elusive, self-conscious "I" gets in touch with itself, as so many early Romantic poets from Gray and Collins to Wordsworth and Coleridge demonstrated (Cox, 1980). But they often omitted

---

[2] Kagan's *physiology* does not equal *anatomy*; it *includes* anatomical components, as it would in the research of any leading contemporary psychologist, but is not identical with a person's anatomical constitution; therefore any popular, reductionistic version of his theoretical frame such as "you are what you look like," or "you are your body," misrepresents it and must be resisted.

physiology from their concerns. Kagan, however, is fastidious to explicate that (1 above) is merely *predisposing* rather than determining. Hence "a child born with exactly the same temperament that favors bold, sociable behavior may become the head of a corporation, a politician, a trial lawyer, or a test pilot if raised by nurturing parents who socialize perseverance, control of aggression, and academic achievement" (Kagan, 2004, p. 5). This determining potential, or predisposing inclination combines with life experiences to turn the child into the adult he becomes. As Kagan writes in a most Lockean mood, qualified by equally Lockean caveats:

We wish to be clear from the start that an infant's temperament is only an initial potentiality for developing a coherent set of psychological characteristics. No temperamental bias determines a particular cluster of adult traits. Life experiences, acting in potent and unpredictable ways, select one profile from the envelope of possibilities. . . . A child born with exactly the same temperament is at some risk for a criminal career if his socialization fails to create appropriate restraints on behaviors that violate community norms (Kagan, 2004, p. 5).

Kagan's stance resembles the nature-nurture model of human identity but differs as a consequence of laying more weight than either camp does on the individual's neurophysiology. Furthermore, human behavior is understood linguistically as a category containing the whole of human identity (i. e., the person is what he does rather than the more Cartesian what he thinks: *cogito ergo sum*); indeed little is to be discovered about a person's *quintessence* beyond their behaviors. Secondly, Kagan makes perfectly clear that soul and psyche must be removed from his system, as are all other quasi-vitalistic characteristics (i. e., Bergson's *élan vitale* or life force). Finally, Kagan's model has been historically generated (even if insufficiently so for professional historians of science), to the degree that it permits the dominant ideologies of cultures to play a part in the shaping of individuals. As he writes:

Opinions regarding the relative contributions of biology and experience to behavior have cycled over the years, depending on other ideologies that happen to be ascendant in a given society. Eighteenth-century England, on the cusp of great economic and military power and eager to announce its intellectual separateness from Catholic France and Spain, celebrated the power of the human

mind to accomplish whatever it desired without metaphysical restrictions. John Locke satisfied that desire by declaring, without evidence, that each begins life intellectually innocent, free of both truth and doubt, so that postnatal experiences can create a Chaucer, Shakespeare, or Newton. (Kagan, 2004, p. 35)

Kagan's focus on Locke and the eighteenth century is intentional. It was the epoch when mind firmly incorporated itself into the trinity of selfhood and became the ascendant component of the body–mind–soul aggregate, leaving little room, it seems, for temperament. As indicated, the word temperament itself began to fade, as any lexical analysis shows. Despite its return in the next century, the nineteenth, it remained dormant until the late-twentieth century, in America as well as Europe, for reasons Kagan explains: "An important reason why America's first psychologists were not very interested in a child's temperament, even though Jung and Freud awarded it influence, was America's celebration with pragmatic value" (Kagan, 2004, p. 36). It may be so, but continues to beg the question raised earlier about the nature of temperament itself: what exactly is temperament in relation to the fundamental self, or selfhood, which casts each of us as individual or unique? If temperament is merely another term for soul, psyche, or unique self, then Kagan has not advanced the argument very far. But if, as he maintains, temperament is a *neuroscientific* concept indicating repetition and verifiability, a predisposition capable of being studied in laboratories in experiments, then the point is momentous:

Most scientists define temperament as a biological bias for particular feelings and actions that first appear during infancy or early childhood and are sculpted by environments into a large, but still limited, number of personality traits. The emotional components of a temperament, which many regard as seminal, refer to three qualities: variation in susceptibility to select emotional states, variation in the intensity of those states, and variation in the ability to regulate them. (Kagan, 2004, p. 40)

These constructions are based on modern nuances extending beyond the refinements of the eighteenth or nineteenth centuries. As such, it remains to be seen how such concepts evolved, and what the eighteenth century contributed to this debate about the "biological bias for particular feelings and actions that first appear during infancy or early childhood" (Kagan, 2004, p. 40).

What it Means to be a "Self" in the Eighteenth Century

"I am me, myself, no one else," a minor character in an eighteenth-century novel exclaims (Richardson, 1748, p. 768). It is not evident precisely what he means, or – if clear – how he would explain to himself how he came to be this unique self. Selves in every historical epoch of Western civilization have declared their difference from other selves, but each era produced explanations for the difference as unique as the selves (Morton, 1995). The Ancient Greek explanation depended on *psyche* (soul) in relation to the gods: their mood at any given moment. In the Medieval Christian world this dependency on godlike disposition was sustained, as when Beowulf proclaims in the Anglo-Saxon epic poem that "wurd gath ha sweal schall" (fate marches as it wills); but it was now overlaid by the sense that the individual soul had more responsibility, more free will, as Augustine and Aquinas argued, than previously for the influencing of godly attitude.

Medieval Catholic souls were transformed in the Renaissance as more anatomical than they had previously been, probably under the weight of widely disseminated drawings of Italian anatomists: Vesalius, Fabricius, Leonardo da Vinci, and others. But even when transformed in the mid-seventeenth century by Cartesian dualistic physiology into minds and bodies, these selves, as souls, remained nonphysiological. As a minor playwright of the period put it: "That the Soul is not a Temperament of Corporeal Humours is manifest" (Nicolas Ingelo, 1682, p. v. 203). More recently, if sweepingly, historian Roy Porter has claimed that it is useless to chart the early modern world's theories of body without equal attention paid to soul (Porter, 2004, p. 472).

Turn the chronological dials to the eighteenth-century Enlightenment and selves are more anatomically self-aware, as we shall see, than they had previously been. Their gradual awakening did not owe everything to education (crucial as education was) but to growing popular awareness about body parts. These historical "selves," from the ancient Greeks to the eighteenth century, if hard pressed to explain themselves would have accounted for their uniqueness – "I am me and no one else" – in terms dictated by their anatomical

chemistry. Such had been the weight of Galenic humoral theory for centuries: an overpowering model of explanation that had survived for eighteen hundred years. Indeed, it is impossible to understand the contexts of Kagan's "long shadow of temperament" without this glance into the distant Galenic past. No contextualization starting in the nineteenth century, or even the eighteenth, can account for his postulation, based on laboratory evidence, that the human being is an amalgam of "inherited physiology" and "life experiences": especially an inherited neurophysiology coupled to experience and shaped by free will apart from cultural, political, and socioeconomic conditions.

Galenic doctrine was based on four essential humors defining the individual temperament: hot, cold, wet, and dry. These were gendered as masculine (hot and dry) and feminine (cold and wet), and came in every combination, but they were not inherited. Indeed it is not at all clear why one person had one blend and someone else another. Parents may have been composed of a particular blend and their offspring of another (Temkin, 1973). Essential life functions, as well as momentous events and crises, were also explained in terms of humoral arrangements. During reproduction women warmed – literally their body temperature rose – to male heat in order to enter into the sexual intercourse leading to conception. Both birth (warmth) and death (the cold humors becoming further chilled) were humoral events, as were the more abstract individual character and personality, as we shall see. Once the brain was firmly implicated in the processes of mind during the late Renaissance under the influential neurophysiology of Descartes and Thomas Willis,[3] thought was also explained humorally and human identity chemically as the complex blend of these humors. It is no news that the humoral theory declined in the aftermath of Harvey's discovery of the circulation of the blood, wherein the blood and its vital fluids subsumed many of the functions of the humors. More consequential was the transition from these vital fluids, especially the animal spirits, to nerves as hollow tubes, and then to vibrations and electrical

---

[3] Galen and his followers had acknowledged the role brain plays in thought but had no model to account for its mechanisms, as did Descartes and Willis.

fluids. By approximately 1800 the humoral theory was effectively dead as an explanatory model of the fundamental individual temperament. Something replaced it – this *something*, and its operations, lies near the heart of this essay.

Before moving to the eighteenth-century "something else," however, I wish to pause on early modernist historian Michael Schoenfeldt's theory about "inwardness" in the Renaissance (Schoenfeldt, 1999). Without it one devalues the eighteenth-century debate and diminishes its contribution to Kagan's focus. For Schoenfeldt, "inwardness" – the whole inner world of a human being, his interior universe – was a secure key to a human being's temperament (Schoenfeldt, pp. 20–22). Inwardness did not equal the full extent of his melancholy, although Robert Burton of *Anatomy of Melancholy* (1621) fame might have protested, nor the composite picture of an individual's health based solely on his physiology, but indicated the person's ability to "manipulate" his anatomical-physiological blend. Without this ability to fashion the humors, no person could enjoy physical and mental health, the prerequisites for sound temperament (Schiesari, 1992; Paster, 2004). Hence Schoenfeldt encourages a rethinking of the crucial role of physiology in *all* Renaissance systems aiming to explain how persons became the individuals they were:

Galenic medicine led individuals to a kind of radical introspection, an introspection whose focus was physiological as well as psychological. The Galenic body achieves health not by shutting itself off from the world around it but by carefully monitoring and manipulating the inevitable and literal influences of the outside world, primarily through therapies of ingestion and excretion. (Schoenfeldt, 1999, p. 22)

Such introspection defined temperaments as selves and also produced a type of "nervous tension," even though the nerves were not as yet prominent (Paster, 1997, pp. 107–109); physiology – the aggregate of the unique rituals of every person's digestion and excretion – was fundamental in its formation. As Schoenfeldt comments: "The Galenic regime demands that one must pay careful attention to the self, in order to know its whims, its desires, and its weaknesses, in order to determine which foods are most appropriate to one's constitution" (Schoenfeldt, 1999, p. 20). Diet is elevated to the role of medicine, and intrinsically bound to selfhood and temperament. In Schoenfeldt's words,

"although frequently used to denote superficial characterological types then, the humoral vocabulary was capable of expressing the uniqueness of each individual, and of giving each individual the tools for introspection, as well as a reason for doing so" (Schoenfeldt, pp. 21–22).

What happened to these explanations during the demise of Galenism early in the eighteenth century, and what to the soul that had been so intrinsic to the processes of fashioning these humors? That is, not merely what, if anything, replaced the humors, but what replaced them sufficiently well to sustain concepts of individuality entrenched over centuries? If indeed the Galenic humors had represented the most basic level of explanation of the self's uniqueness (the "I am me alone"), as Temkin claimed (Temkin, 1973), a virtual paradigm shift was necessary to replace these humoral blends with an equivalent all-embracing model (Paster, 2004).

The "something else" I have been referring to was "nervous physiology", and I intentionally use the somewhat obsolete phrase to retain the historical flavor of the older pre nineteenth-century meaning. Our word is, of course, *neurophysiology*. By it we mean the whole branch of science dealing with the physiology of the nervous system. But the word neurophysiology first came into use as Darwin's *Origin of Species* was being displayed for the first time in London book shops in 1859, having then been used as a new-fangled adjective to describe Dr. Marshall Hall's "further researches and later views in Neuro-physiology" (Proceedings of the Royal Society, 1859, p. 55). Hall, similarly to many of our figures educated in William Cullen's Edinburgh, was one of the great "nerve doctors" of the early nineteenth century and a prolific author of scientific works. He had devoted much of his professional career to explaining nervous diseases and discovered the spinal reflex action. Unlike Thomas Willis, who coined the word "neurologie" as we shall see, Hall himself never used the word neurophysiology, having been satisfied with the older "nervous physiology," a phrase that continued to be in wide use until the 1850s. Indeed it was not until the 1870s and 1880s that "neuro-physiology" – always with the hyphen between its still newly amalgamated two components – began to appear with any frequency in the scientific literature, as it did in 1872, for example, in the second edition of Herbert Spencer's popular *Principles of*

Psychology. It is therefore anachronistic, at the very least, to impose this coinage on historical discussions of temperament before 1800. Provided we are aware when using it that we mean an early version of the modern neurophysiology, the older "nervous physiology" serves the purpose more accurately.

In just this sense "nervous physiology" replaced Schoenfeldt's Elizabethan inwardness and did so because it was so attractive to its time: the socially conscious eighteenth century with its new social-class divides and secular cults of commerce and consumerism (Brewer & Porter, 1993). Historians of science have by now meticulously traced its rise in the mechanization of the world championed by Descartes and Harvey and brain revolutionary Thomas Willis. As Roy Porter recently summarized this intellectual revolution:

Harvey's discovery of the circulation of the blood, Willis recognized, called for a rethinking of physiology at large. Building on painstaking anatomical investigations, especially of the nerves, vivisection experiments and clinical experience, his *Cerebre Anatomi* (Anatomy of the Brain, p. 1664) put "neurology" on the map; his *De Anima Brutorum* (On the Soul of Brutes, p. 1672) explained the workings of living creatures, while a further study, the *Pathologiae Cerebri et Nervosi Generis Specimen* (Pathology of the Brain, and Specimen of the Nature of the Nerves, p. 1667), proposed the neurological origins of epilepsy and other convulsive disorders. (Porter, 2004, pp. 55–56).

Temperament could not exist apart from this physiological "workings of living creatures," whether or not inherited, it was a phenomenon *inside* body and mind (now firmly lodged in the brain, as Willis demonstrated) and intrinsic to them. Only amidst the nonscientific illiterate, or those of religious persuasions claiming that physiology annihilated God, did soul remain the dominant factor in the new post-1700 theories of temperament. Soul was indeed a thorn: in a sense the genuine juggernaut at the centre of the debate about nervous physiology. But as time marched, increasing numbers of post-Willisian doctors and scientists overlooked its influence. They replaced it with "life experiences" that created variations in behavioral reaction to unfamiliar events (Rousseau, 2007).

Here Willis' new *nervous* physiology, whether or not inherited, was crucial. Given the role of the brain and its vessels – the nerves – in the workings of the body, no alternative existed. Locke, for his part, emphasized the role experience played in combining with the human's fundamental physiological makeup. Combine Harvey (circulating blood), Descartes (mind–body split), Willis (brain as the commander of the entire nervous apparatus), Locke (the crucial role of experience), and Newton (who cannot be omitted here because he mathematized these consequences in his new fluxions or calculus), and you have the blueprint of the "something else." Nervous physiology has carried the day in a way it never had among the chemical Galenists.

Less clear, however, was the role of nervous physiology in (what we call) the developing strands of psychology in the long eighteenth century: especially theories of identity, character, and personality feeding into temperament. Medically trained Locke was immersed in his Oxford tutor Willis' nervous physiology. It played a major part in Locke's generation of the theory of personal identity as the "continuation of experience," wherein Locke meant that every person considers who he fundamentally is by the ability to relate past to present experience through a process located as a steady-state self (Locke, 1977 [1690]; Yolton, 1984; Yolton & Locke, 1970). Locke did more to elevate experience (nurture) than any authority on human behavior. But temperament exceeds the domains of memory and experience if it is to account, as Kagan has demonstrated, for the *different ways* several people can interpret the same experience. As Kagan himself explains, temperament combines an "inherited physiology" that attaches with "life experiences" to create variations in behavioral reaction to unfamiliar events.

Eighteenth-century thinkers, in contrast, were ambivalent about the degree to which this physiology was inherited rather than acquired. Despite progress made in developing the double compound microscope then, physiologists had no basis for deciding between acquisition and inheritance. But there was almost monolithic agreement by mid-century, and certainly by 1800, that temperament was formed through *nervous* physiology coupled to life's experiences. Even contemporary novelists absorbed this perspective and reflected it in its fictions. Tobias Smollett's consumptive protagonist Matthew Bramble, an ailing country squire in search of health in Smollett's masterpiece, proclaims that no man should pretend to understand "the mysteries in physick [sic]" unless he has "taken regular courses in physiology" (Smollett and Knapp, 1966: letter of April 20). We read on to discover that Bramble means "nervous physiology" as he increasingly becomes convinced that all his ailments arise from "the irritable nerves of an invalid" (Smollett: letter of April 23). And Laurence Sterne's *Tristram Shandy* (1760–1767), the most physiologically self-aware novel of the long eighteenth century, contains whole sections on "nervous physiology" in its first two volumes as means to explaining how Tristram got to be the mess he is (Sterne, 1940, pp. 1–95).

We further glimpse the consequences of this developing nervous physiology by glancing at the eighteenth-century doctrine of the "ruling passion." It was predicated on a much older belief about the passions, or "affections of the mind" (the newer term after 1680). The Galenic view was that passions arose from humoral chemistry, refined by the later Stoic view that the gods themselves had guaranteed sufficient emotional diversity to ensure the world's work could be done. If everyone were melancholic (cold) or phlegmatic (dry), balance and equilibrium would not have existed; the four humors, instead, embodied the deity's master-plan. The later Stoics also claimed that the Galenists allowed self-will to play a greater role in the formation of the passions. After the seventeenth-century mechanists had done their work, a newer, physiologically based, doctrine developed that each person was guided by a "ruling passion" expressed in the form of an overpowering hobby or all-consuming obsession. Poet Alexander Pope pithily versified the theory in his monumental *Essay on Man* (1733–1734) in the terms of the old Galenic physiology:

So, cast and mingled with his very frame,
The Mind's disease, its ruling Passion came;
Each vital humour which should feel the whole,
Soon flows to this, in body and in soul.
(Pope, 1963, p. 520; Epistle II, lines 137–140)

Pope's "vital humour" is the physiological "spirit" lodged in the brain or heart, credited with nourishing body and soul and framing its "ruling passion." God-given (and thereby loosely tantamount to having been inherited), it is implanted at the moment of birth, inescapable and forever fixed in each human being. Pope analogizes it to a "disease" because its obsessive nature in each individual overruns his life. Laurence Sterne sustained this physio-psychological take on the "ruling passion" when he elevated it to a condition of "hobby-horses," as in Uncle Toby's obsession with fortifications and Walter Shandy's with names. Sterne presents the doctrine comically, but no one, certainly no physiologist of the era, was willing to explain precisely how it first enters the body other than in theological terms.

This was the quandary: most resorted to the age-old soul as some sort of God-given guide that ensured its entrance, but others were less certain. Pope himself, reflecting the dominant view of the early-eighteenth century, had begun his discussion of it by tracing its appearance in each person from "the moment of his birth" (Pope, 1963, p. 520; Epistle II, line 133) – this is as much as he would say. Nor would he counter Locke's view that any "ruling passion" had to be acquired through experience. If temperament included the mental faculties, so far as Locke was concerned it had to be acquired.

Locke's destruction of the notion of innate ideas also bore directly on these inherited characteristics of temperament by altering the fundamental sense of self. One can readily see what toll the *tabula rasa* took: if every baby were born with a blank mind, then experience alone – the composite of sensory impression, ideas associated from these impressions, higher abstract ideas such as freedom or nationalism, and, finally, the sense of self acquired from the temporal continuity of these processes through memory of them – could fill it up. But what role, if any, did physiology play among children ruled by blank slates waiting to be written on by experience (Kagan & Lamb, 1987)?

The circle of philosophers amongst whom Locke was a leading member recognized that the question led back to the age-old mind–body dualism; yet not even Cartesian physiology solved it. Descartes' pineal gland was largely ignored as a solution after 1700, in Britain partly on nationalistic grounds, rather than deployed as an organ on which to construct a new model of mind–body interaction. As the eighteenth century progressed the mind–body impasse widened, as did hard questions about the relation of inherited physiology to acquired ideas through "life's experience." Locke's whole thrust had been the demonstration that no innate ideas exist at birth, not even basic notions about God (God had to reveal himself also through *experience* rather than be taken for granted); for him all thought is acquired as experience (reading is also experience for Locke). Physiology predisposes creatures, Locke thought, to respond to sensory impressions and experience in differentiated ways – just along the lines Kagan suggests – but Locke is nowhere explicit about the precise mechanisms of this predisposition. Lesser thinkers were, especially among the Newtonian physiologists: for example, David Kinneir, a Scottish philosophical doctor – who aimed to demonstrate that the nerves predispose people to respond to experience in nuanced ways (Kinneir, 1737). Physiology among all these thinkers was philosophical to a degree common today only among neuroscientists; but not even similar physiological models could topple Locke's *tabula rasa*, which was highly influential despite any nervous physiology impinging upon it. No wonder that it filtered into the nineteenth-century psychologies inherited by Kagan.

The Long Shadow of Temperament in the Eighteenth Century: Further Historicizing Kagan

It would have been difficult to historicize Kagan's view of temperament if nervous physiology had not installed itself within two generations (1680–1740) as the dominant view. Later on, by c. 1770, and certainly continuing to 1800, it held sway contested only by radical dissenters and in fringe quarters. Moreover, a fundamental conceptual difference had existed between *humoral* and *nervous* physiology: both were physiologies, no matter how different their implied bodily chemistries, but there was a quantum leap from the former to the latter.

Briefly put, nervous physiology implicated the brain in ways unknown before Willis' revolution, complemented and augmented by his quondam student John Locke. It put forward a paradigm shift constructing a new model of human physio-anatomy, as has often been noted (Porter, 2001, 2004; Rousseau, 2004; Yolton, 1984, 1991), paving the way to the *new* view of "temperament." This was a sufficiently complex leap requiring amplification here before we can historicize the long shadow of temperament. It is insufficient, especially in a book such as this one, merely to

disentangle the brain's intrinsic anatomical structures as understood c. 1750. Brain remains the starting point – not the finishing – of any rigorous historicization of Kagan. The cultural, national, and political brain must be added to the anatomical one.

This giant step from humors to nerves also prompts us to take stock, for more than a substitution, or replacement, was involved. By approximately 1700 it was evident that the humoral fluids were on the way out. Harvey's circulation of the blood had called renewed attention to the bodily internal organs – the "solids" – because they were the anatomical repositories of those humors. As these "solids" claimed ever-more attention, digestion and passion (the word for our emotion) came to be seen as crucial components of the physiological operations of temperament, even in fiction, as in *Tristram Shandy*. Furthermore, as the brain claimed increased attention, not merely from Willis and his colleagues in the Royal Society but from others after the 1680s, it stole the show from the older chemical humors (Finger, 1994; Martensen, 2004). Willis' theory posited *one* brain – binary in its two anatomical lobes, localized in all its human functions from language to motion, fed by the body's nervous apparatus without which it expires[4] - and *two* souls: (1) corporeal or animal (*animus*) and (2) incorporeal (rational soul). The *animus* depended entirely on the material body: having been born with it, it also dies with it and has no afterlife whatever (Clarke, Dewhurst, & Aminoff, 1996). All its functions were corporeal and showed themselves materially, and in this sense it functions as a physiological captain lording over the body's regimen, especially digestion and the production of human passions. It performs its executive work in the mid-brain, while less loftily also commanding the anatomical organs (heart, midriff, bowels, limbs). Born of the equally material animal spirits at the moment of conception, during the act of sexual intercourse, its activity is thoroughly physiological; working through the nerves and blood through whose channels these animal spirits flowed.

The subtly refined immaterial fluid within the animal spirits – Willis' "*vital spirits*," which differed from the animal spirits – originated in the

---

[4] This because the animal spirits, containing the more ethereal "vital spirits,", flowed primarily through the nerves; hence remove nerves and brains dry up and die.

brain; brain alone could manufacture them from the moment of conception. This *animus,* or *anima,* was destined to infiltrate much eighteenth-century physiological thought and experiment, and was transformed as the base building-block of later theoretical systems: Haller's sensibility and irritability, Condorcet's sensory physiology, William Cullen's basic disease theory as nervous and many others. It remained so vibrant in the 1760s that even novelist Sterne targeted two-soul complexity for satire in a whole chapter of *Tristram Shandy* (II. 19):

As for that certain very thin, subtle, and very fragrant juice ... discovered in the cellulae of the occipital parts of the cerebellum, ... [as] the principal seat of the reasonable soul (for, you must know, in these latter and more enlightened ages, there are *two souls* in every man living, – the one according to the great Metheglingius,[5] being called the *Animus,* the other the *Anima*); ... my father could never subscribe to it by any means; the very idea of so noble, so refined, so immaterial, and so exalted a being as the *Anima,* or even the *Animus,* taking up her residence, and sitting dabbling, like a tad-pole, all day long, both summer and winter, in a puddle, – or in a liquid of any kind,[6] how thick or thin soever, he would say, shock'd his imagination; he would scarce give the doctrine a hearing.

Willis' second soul, incorporeal and rational, is unique to mankind: absent in lower creatures, immaterial, and immortal. It was the orthodox soul of Christian theology, implanted by God at the moment of conception and somehow attuned to the outside world. Nonscientific writers in Willis' generation went to lengths to demonstrate that this "second soul" was "not a Temperament of Corporeal Humours" (Ingelo, 1682, p. 203) but something else – so prevalent was their difference. But Willis also generated a complex counterpoint between the two: steeped in seventeenth-century

Anglican theology (incorporeal rational soul as the equivalent of 'spirit' in the Holy Trinity), post-Cartesian physiology, and immersed, linguistically, in analogy and metaphor, especially in the discursive temple he erected to discover every type of analogy for his newly privileged brain. But Willis has been poorly understood. One reason is that he cannot be adequately studied from one disciplinary perspective alone: his "brain science" (our term) and "neurologie" (Willis, Feindel, & Pordage, 1965), his word, is multidisciplinary in our current sense (Rousseau, 2004). Yet it was the brain as envisaged by Willis that, by allowing the science of two souls, ultimately enabled the development of Kagan's posthumoral temperament.

Willis also demonstrated that each soul conspires with the other to privilege the brain (Willis et al., 1965). That is, thought and imagination, intellection and fantasy, are man's unique gifts; to attain them each soul relies on the remarkable intricate human body. Even the rational soul required body despite its incorporeality. Persons – selves – are creatures who exceeded their two souls: composed of two inherently anatomical attributes – (1) a complex body given at birth (the composite of its physiology) and (2) life's experiences that would attach to this body until the moment of death – they required these two souls as preconditions to become selves. Here then, around 1700, is a version of Kagan in a complex early eighteenth-century key.

Willis never expatiated on the latter: life's experiences. This task was accomplished by his student and later colleague, Locke. Expound it Locke did, and so abundantly that his experiential psychology altered the course of the Enlightenment, paving the way to "temperamentalists" such as Kagan. Locke retained the two components – original physiology

---

[5] For the purposes of learned wit Sterne naughtily and ironically jumbles Willis' two souls (*anima* and rational) and attributes them to a person who never lived, Metheglingius; Sterne knew very well that Thomas Willis had been the originator of the two-soul theory and deeply respected his brain revolution.

[6] Willis claimed that the anatomic brain is composed primarily of salty liquid and sits surrounded by nervous fluid in the head, as if in a lake. Sterne's brain–soul chapter continues, page upon page, expunging any doubt that it was great anatomists of the prior generation – like Willis – he has been reading; thus he continues: "What,

therefore, seem'd the least liable to objections of any, was, that the chief Sensorium, or head-quarters [in his punning Sterne has created a neologism in head-quarters acknowledged by the *New Oxford English Dictionary*] of the soul, and to which place all intelligences were referred, and from whence all her mandates were issued, – was in, or near, the cerebellum, – or rather some-where about the *medulla oblagata,* wherein it was generally agreed by Dutch anatomists, that all the minute nerves from all the organs of the seven senses concentered, like streets and winding alleys, into a square" (Sterne, 1760, II. 19).

(including the attendant two souls) and life's experiences – constituting us as unique selves. Willis–Locke could have employed the old Renaissance word "temperament," or later "basic disposition," to amalgamate their two components. Steadfastly they did not. Wishing to distance themselves from any residual or vestigial Galenic humors, they eschewed these obsolete catch-alls, persuaded that their attentive readers would comprehend that the two souls, acting in tandem with the body, produced the totality of an individual's character, identity, and personality – the whole person.[7] The remarkable twist, which cannot be pursued here, is that nineteenth-century psychology restored "temperament," only to be dropped by the American pragmatists, as Kagan observes (Kagan, 2004, p. 36), and then reclaimed again by Kagan himself.

The Willisian Aftermath: Temperament and the Brain

No one knows to what degree Willis and his followers were read after 1700. What is clear is that his "neurologie" was immensely influential, especially when yoked to Locke's experiential psychology. As nervous physiology gained ground, Willis' bipartite soul narrowed into one: rational soul dropping out in favor of the corporeal *anima*, which appealed to increasingly secular and socially conscious people in the vanguard of contemporary life inquiring about their identities. This corporeal soul, or *anima*, was also Newtonized, its bodily operations upon occasion geometrically charted and explained through the new calculus. The use of the calculus by some iatrophysicists in

---

[7] Sixty years after Willis–Locke generated their leap, Sterne was still satirically commenting, exposing, and lambasting theories about nervous physiology, his "lucky or unlucky organization of the body" being the clue to human difference:

Now, as it was plain to my father [Walter Shandy], that all souls were by nature equal, – and that the great difference between the most acute and the most obtuse understanding, – was from no original sharpness or bluntness of thinking substance above or below another, – but arose merely from the lucky or unlucky organization of the body, in that part where the soul principally took up her residence, – he had made it the subject of his enquiry to find out the identical place (Sterne, 1760, p. II. 19)

constructing their theories cannot be treated here; suffice it to say it added to the allure of the corporeal *anima*. Diverse attempts were also made by some philosophical physicians (Bernard Mandeville, Mark Akenside, David Hartley, Cullen already mentioned) and scientists (Joseph Priestley and Erasmus Darwin, who was also a poet-physician) to resuscitate the incorporeal soul. But as time marched the attempt proved futile, except among dreamy poets, and most system-builders after 1780 began with the premise that the immortal part of selfhood did not lend itself to laboratory observation, measurement and experiment. In time Willis' rational soul went the way of all flesh – and certainly of the former humors: not entirely dead by 1800, when the young Wordsworth and Coleridge were compiling their literary manifestoes, but quickly expiring.

This corporeal soul had also caught on for reasons beside the above ones. It was concrete, visually captured, appealed to common sense, and was not lumbered by supernatural or magical baggage (Plate 1). Even the faraway Chinese of the early eighteenth century produced woodcuts like this one from an edition of 1726 (fourth year of the Yongzheng reign period of the Qing dynasty) showing a diagnosis chart of temperament based on body. The secret is lodged in the face, each of whose parts has significance. The left cheek corresponds to the liver and wood, the right cheek corresponds to the lungs and metal, the forehead to the heart and fire, the nose to the spleen and earth, and the chin to the kidney and water. There is no soul in this Chinese representation; had there been, one has the sense it might lodge – as it does in Walter Shandy's conception of his son's soul in *Tristram Shandy* – behind the nose. Likewise, in the developing Western corporeal soul nothing had to be taken on faith: gradually all its operations were becoming anatomically known, sometimes even seen under the microscope (more or less) by the naked eye. Slowly the organic world was dividing itself between creatures with, and without, nerves. When eighteenth-century microscopists such as Swiss physiologist Abraham Trembley studied the freshwater polyps, of which hydra is an example, they saw no nerves. But they recognized that any "nervous magic" existing within these primitive creatures must reside in their hidden physiologies rather than in supernatural impositions. The corporeal soul

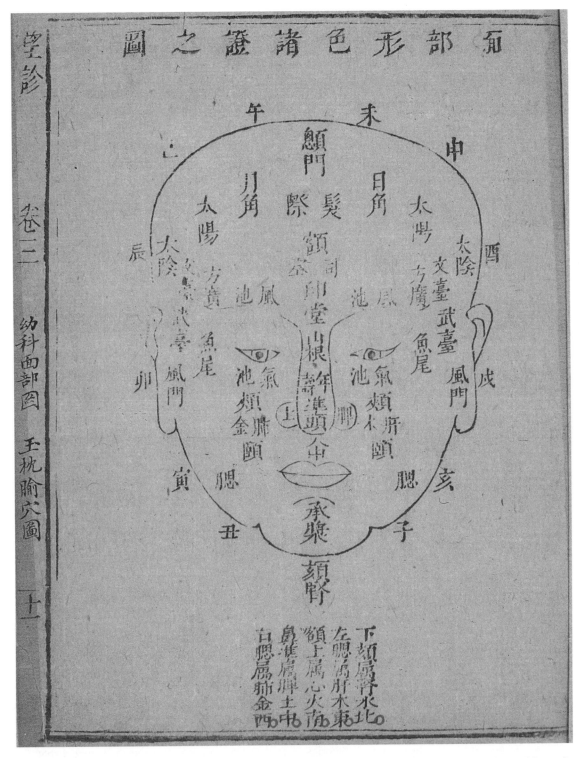

PLATE 1. "Chinese analysis of human temperament diagnosed by the nose." C. 1726. Wellcome Trust Centre Library for the History of Medicine, Historical Collection, Image Number L0038677. Sizhen juewei ("The Essentials of the Four Diagnostic Examinations"). This was one of four diagnostic examinations of temperament made among the eighteenth-century Chinese. Reprinted by kind permission of the Wellcome Trust Library

functioned, even according to church sermonists, because the brain, and its faithful attendant nerves, guided it. Without these nerves it was nothing.

By the 1750s nerves became the beneficiaries of the old humors to such degree that the ordinary person in the street could quip about them (Rousseau, 2004, pp. 29–38). As I have shown elsewhere with hard evidence (Rousseau, 2006), this development was not merely scientific or medical: it included nationhood (the growth of England and Scotland into world-class nation states), class structure (firm differentiation of upper and lower social classes to a degree unknown before), and economic aspects of the new pharmacology in a consumer society (quack nostrums on the rise). They converged to implement the practical side of the theoretical revolution so relevant to Kagan's much later enterprise.

One other factor played a part in this meteoric rise of the secular corporeal soul in the eighteenth century: the growing doctrines of sensibility and sympathy. Sensibility *de rigueur* required a nervous system: as the word itself denotes, sensibility, or sensitivity, is predicated on the *senses*, whether assessed in moral, ethical, or aesthetic categories, and the nerves in tandem with the brain must be ready to collaborate if its impressions are to amount to anything. Likewise for the more abstract empathy and sympathy. But by the 1740s Scottish philosophers and physicians – prominent figures such as George Cheyne, David Hume, and Adam Smith – were distinguishing between feeling and sympathy. If "feeling" held sway over the domains of human activity, "sympathy" was the coordination of its (i. e., "feeling's") components. To clarify further, sympathy was a "correspondence of parts": a yielding to, and bending from, emotional states.

A century earlier, English doctors such as Kenelm Digby, well known to Willis, had explained the precise mechanisms of these "correspondences": how one anatomical region bends to another, in health and illness (Digby, 1658). Diseased organs "correspond" *with* "healthy" ones geographically "far away," the "sympathetic correspondence" transforms the "body's regimen." Enlightenment Scottish philosophers and "system builders" imported these physiological theories, and by the time Cheyne and Hutcheson, Hume and Smith wrote from 1730 to 1770, "sympathy" was construed as intrinsic within man's physiological

body. This sympathy also feeds into the historicization of Kagan, and plays a fundamental role in defining Scottish Enlightenment. It is wrong to think this sympathy limited to physiological realms. These earlier "philosophical doctors" analogized from the "body of man" to the "civic body." What remained to be explained were the interstices: how one domain of thought, one "part of society," affected another. Even the "sympathy of nations," as Smith construed it (Smith, 1776), was a category of explanation for developing nation-states – so central was physiology to the dominant cultural discourse of the time.

Physiological "sympathy" rarely left "nervous sensibility" behind. It was only a small step from the "sympathy of parts" to a generalized "sensibility" that magnified the role of feelings within one person. The logic was straightforward: body parts "corresponded" through processes of anatomical sympathy; such "correspondence" proved that no part of man's body existed in isolation from the rest; yet similar "sympathetic correspondences" also existed between human beings. We are "sympathetically aroused" when we see a fellow creature in distress. Poverty, such as found among beggars and street urchins, "sympathetically" elicits feeling within us and drives us to "do good." No "corporeal soul" was needed for the empathic task: our nervous physiology did the necessary work. The more good we do, the more our experience confirms the social value of philanthropy.

By the 1770s physiological sensibility underlay these explanations for benevolence – "doing good" – deemed crucial to the fabric of a society steeped in as much poverty as wealth. And the anatomically savvy philosophers urged that "doing good" was not merely an act committed because the Samaritan had been told to do so in Sunday church. It was rather that his anatomical body was disposed to such good deeds through its nervous physiology. Even if originally ordained by the Maker, this fact of bodily construction is also germane to Kagan's "inherited physiology." Still, some folk were more disposed than others: this was the juncture at which the foci of civic sympathy and social sensibility were contested. Women and the upper classes were said to be more predisposed than men, especially rustics and workers. When asked why, the learned doctors' reply was that "their nerves were more exquisitely delicate" (Adair, 1787, 1790); their bodies more finely attuned to comply – consciously or unconsciously – with these "processes of correspondence and sympathy." The deduction was that individual "nerves" reflected their owner's race, class, gender, or age; a position not so far from Kagan's eventual all-embracing temperament (Cullen, 1773, 1786, 1797, 1800; Bell, 1802).

You could say that by c. 1780 the corporeal, nervous, palpating soul had extended its domain so far that it was now also accountable for social developments. Anatomy and neurophysiology permeated society's fabric, and the term *nervous* became the signifier of fashionability itself (Adair, 1787). However, it is an error to think that beneficent "sensibility," or "doing good," was limited to *persons*; it extended to countries, entire civilizations. The view, developed by Enlightenment anthropologists in Scotland and Germany (Kames, 1779; Ludwig, 1791), was that primitive societies had historically given way to more advanced ones in proportion to the numbers of "civilized members" within its population. By a process of logical deduction made from the argument of nervous physiology, the whole terraqueous globe was "sympathetically linked." How could these "sensible citizens" be distinguished from the rest? By the delicacy of their nerves, went the explanation. Willis and Cheyne had defined the basic frames for understanding; now the Scottish *philosophes* supplied the missing links. Kames focused on agricultural progress and legal advancement in his writing, but the bedrock of his belief assumed a progressively "nervous man" always lurking beneath the surface of his texts.

By the turn of the nineteenth century many of society's social ills were also being accounted for in physiological terms. The view was strong in the Scottish cities, especially in Edinburgh (Cullen & Thomson, 1827), hub of Enlightenment thought. For example, Dr. Thomas Trotter was one of the century's most distinguished naval physicians who reformed whole segments of the health of Nelson's navy before he was wounded in an accident and took early retirement. He was a "Cullenite" who remained faithful to his teacher's beliefs about the influence of the "nervous temperament" (Trotter, 1807). Trotter also emphasized nervous physiology and the role it played regulating mind and body in disease, especially in alcoholism, and he tried to apply his theories to gender and class (Trotter &

Porter, 1988). Like Cullen, he grasped the inherited components of nervous diseases, even when less impressed by his patients' social niche. He treated dozens of unstable male sailors and concluded that they were hysterical: suffering from nervous conditions in which it would be impossible to decide whether mind or body had been the cause – so intertwined were both. He also agreed that female nerves were inherently weaker than male, and that the nerves of the rich were more "exquisitely delicate" and therefore more easily capable of derangement, and sought to apply these views to the poor as well, particularly those who had dropped out and turned to the bottle (Trotter & Porter). On balance and assessing all these cases and diagnoses, his medical theory amounts to a "practical view of temperament" (Trotter, 1807, p. 113) determined by nervous physiology coupled to social ill: poor working conditions, squalor in accommodation, and the new urban sprawl. Trotter was visionary of the modern ills that would soon afflict society in the nineteenth century (Trotter & Porter).

Conclusions: The Politics of Temperament after 1800

None of these figures viewed the "nervous system" as a signifier for something else, certainly not as Kantian transcendental signifiers. Even when they decoded their theories of social impact, the nerves remained physical, and they remained so even to the German idealist philosophers after 1800. My historicizing of temperament has been largely Anglo-Saxon and it omits the Germans who made substantial contributions to its development after the 1780s (for example, Christian Wolff and Karl Philipp Moritz, who emphasized the psychological components of temperament). Yet even among the Germans the anatomical nerves were crucial for evolving theories of society and civilization, including aesthetic and ethical realms. Books about the "nervous bases of civilization," published in the English Regency, would not have seemed alien to any of these thinkers (Verity, 1837). One facet – the pharmacological – of this early anthropology of temperament remains to be understood as it pertains to the historicizing of Kagan's mindset.

By approximately 1800 the word temperament as the composite of character and personality was enshrined, if declining in the frequency of its use. Temperament, as Cheyne and his followers had demonstrated, developed as the result of multiple characteristics, including climate, geography, local community, and religion. In this belief Cheyne was an anthropologist of temperament no less than the later Scottish philosophers already discussed. The cause of so much British suicide, he reasoned, was in large part owing to the British climate: its weather of fogs and mists, endless darkness that oppressed the corporeal soul (Cheyne, 1733). Nevertheless, even Cheyne was quick to notice that "low spirits … and high spirits" could be managed by the application of pills and potions (Cheyne, pp. 5–8). He saw from the start that nervous physiology lent itself to pharmacological transformation more readily than the earlier humors: nervous physiology before chemical blend in the old Galenic sense. Time has borne him out, as has Kagan's emphasis on the brain.

Here, then, was an early model of the pharmacological management of temperament. A full demonstration of this eighteenth-century pharmacological enhancement would require another chapter, even a book, extending beyond Cheyne and demonstrating how the growth of quack and patent medicines catered to the new social stresses and strains of high living (Porter, 1992). Cheyne was perhaps not the very first to chart it – precedence here is beside the point – but he was certainly one of its early proponents. It is the view that a model of selfhood based largely on nervous physiology is necessary before pharmacological manipulation can kick in (Cheyne, 1733). It may seem odd to locate aspects of the historical antecedents of Kagan's overarching concern for temperament in Willis, Locke, Cheyne, and Scottish doctors and philosophers, but they stare the historian in the face.

To this approach it may be objected that Kagan's "temperament" cannot be historicized without full recourse to the historical development of its cognates in *character* and *personality*. It is a reasonable request. But the notion that these are post-1800 modern words is erroneous: they developed long before then, and gradually accreted themselves – like shells on the beach – to the very same psychological components about which Kagan has justifiably made much. From Medieval times character meant the basic mark of any object or thing, but by the 1500s it was used to denote

what was distinctive about someone's writing, and by 1700 to designate the specific mental or moral qualities of people. This moral sense was Alexander Pope's usage too when he opened a famous poem with the outrageous line "Most Women have no Characters at all" (Butt, 1963, p. 560). Even a generation earlier, in the time of Willis and Locke, character described a person's unique possession of specified qualities.

The same is true of personality. Roughly before 1600, personality was used to signify general attributes that make someone human as distinct from brute-like. But by 1660 it, too, had joined character to signify what precisely it is that makes person A, considered as the possessor of individual characteristics or qualities, different from person B. John Dryden claimed of the classes of personality that of the types who are "Morose, there are at least 9 or 10 different *Characters* and *humours* in the *Silent Woman*, all which persons have several concernments of their own" (Dryden, 1995, p. 68) – so here, already in the 1660s, the categories converge and are temperamentally complex.

The English Restoration was transformative in this sense just at the time when Willis was conducting his experiments on the brain. Ralph Cudworth, prominent among the Cambridge Platonists, speculated about personality in *The True Intellectual System of the Universe*: "Humane Souls," he wrote, "Minds, and Personalities, being unquestionably Substantial Things and Really Distinct from Matter" (Cudworth, 1678, p. i. V. 750). Personality was by then physical and included bioanatomical attributes, as when Richardson in *Clarissa Harlowe* used it to describe a unique face: "To descend to a very low instance, and that only as to personality; hast thou any doubt, that thy strong-muscled bony face was as much admired by thy mother, as if it had been the face of a Lovelace?" (Richardson, 1748, p. VI. i. 5).[8]

Continue down through the decades and hundreds of examples of psychological "personality" present themselves, as when sermonist Andrew Tucker claimed that "Personality is what makes a man to be himself, can never be divested … nor is

---

[8] Richardson liked the word and often uses it, as when Clarissa describes her motives for revenge: "Not invenom'd by personality, not intending to expose, or ridicule, or exasperate …" (Richardson, 1748, p. II. xxii. 137).

interchangeable with that of any other creature" (Tucker, 1777, p. 353). Our contemporary Professor Kagan cannot claim that he, or his colleagues, has invented these concepts. What has altered from their eighteenth-century usages – and it is a monumental development – is the superimposition of a theory of evolution coupled to rigorous laboratory experimentation about the combination of physiology and experience. In other words, if everything else is evolving, temperaments and personalities also are in highly predictive ways.

This recourse to early-modern etymology (linguistics) and pharmacology (transformation of bodies) demonstrates what has been at stake all along in historicizing Kagan. Crudely put, the combination may be called "the politics of temperament" for the way it suggests how the broad eighteenth-century currents of nervous physiology yielded in time to a nineteenth-century evolution of more psychologically based temperament implicated in contextualizing Kagan's overarching version. Both components, linguistic and pharmacological, are embedded in Kagan's contexts. The linguistic part without the pharmacological is incomplete. Once it is understood that by c. 1800, man's nervous physiology was taken for granted, there was no need, as there had been in the earlier generations of Willis and Cheyne, to establish the death of the humors. By mid-century the nerves held center-stage (Porter, 2001; Rousseau, 2004), and by 1800 even the layman – Everyman – took its truth for granted (Trotter, 1807). With nervous physiology on a firm footing, the newer concepts of character and personality as the basis of identity could claim the time and energy of psychologists.

In summary, identity, character, and personality were old words by 1800, each having been personified before Locke turned them into the stock and trade of *personal identity*. If each had been the special insignia, or defining mark, of a physical object (not necessarily human), by 1730 – when Locke's psychology of the self was being absorbed into general knowledge – they were routinely being applied to all major theories of selfhood. A person's outside and inside began to merge, as when Sterne's literary widow in *A Sentimental Journey* "wore a character of distress" (Sterne, 1768, p. 68), and when Bishop Joseph Butler, famous throughout the eighteenth century for his *Analogy of Religion, Natural and Revealed*, epitomized the point exquisitely when writing that "there is greater variety of parts in what we call a *character*, than there are features in a face" (Butler, 1874, p. II. 158). Soon human character would *equal* identity, and personal identity the hallmark of the stable, predictable character.

The remaining interstices from 1800 forward need historicizing too, especially during the nineteenth century when psychology grew into a powerful science. By the 1880s, when William James – like Kagan more recently – was espousing his ideas at Harvard, nervous temperament took a back seat among pragmatists who had more pressing concerns. Yet conceptually it is not so far from the Jamesian aftermath, in 1920 or 1940 (Sprott, 1925), to the mindset the young Kagan inhabited when he began his own explorations into temperament resulting in his influential books.

The point to make in conclusion is not merely the suggestion that language often antedates scientific concepts – words before ideas – or to argue, once again, that there was abundant psychology in the long eighteenth century despite its not yet being codified into a discipline (Rousseau, 1980). The larger matter is that Kagan's intuition is finely honed, perhaps as "prophetic" as Galen's (Kagan, 1994) and as nuanced as Freud's and Jung's were when they recognized they too must include temperament. Kagan may come at the end of a great wave in Western civilization about temperament's significance – "the long shadow of temperament from Galen," as he calls it – but not even he has exhausted its complexities. He has rather incorporated himself into the great tradition from Galen forward.

References

Adair, J. M. (1787). *A philosophical and medical sketch of the natural history of the human body and mind.* Bath: R. Cruttwell.

Adair, J. M. (1790). *Essays on fashionable diseases.* London: T. P. Bateman.

Bell, C. (1802). *The Anatomy of the Human Body,* London: T. N. Longman.

Brewer, J., & Porter, R. (1993). *Consumption and the world of goods.* London: Routledge.

Butler, J. (1874). *The works of the Right Reverend Joseph Butler, late lord bishop of Durham: To which is prefixed a preface, giving some account of the character and writings of the author.* Oxford: Clarendon Press.

Butt, J. (1963). *The Twickenham Abbreviated Edition of the Works of Alexander Pope*. New Haven: Yale University Press.

Cheyne, G. (1733). *The English Malady: Or, a treatise of nervous diseases of all kinds: As spleen, vapours, lowness of spirits, hypochondriacal, and hysterical distempers, &c. In three parts. . . . With the author's own case at large*. London: G. Strahan and J. Leake at Bath.

Clarke, E., Dewhurst, K., & Aminoff, M. J. (1996). *An illustrated history of brain function: Imaging the brain from antiquity to the present* (2nd ed.). San Francisco: Norman.

Cox, S. D. (1980). *"The stranger within thee": Concepts of the self in late-eighteenth-century literature*. Pittsburgh: University of Pittsburgh Press.

Cudworth, R. (1678). *The true intellectual system of the universe*: London: Richard Royston.

Cullen, W. (1773). *Lectures on the materia medica*. London: T. Lowndes.

Cullen, W. (1786). *First lines of the practice of physic*. Edinburgh: C. Elliot and T. Cadell.

Cullen, W. (1797). *Clinical lectures, delivered in the years 1765 and 1766, by William Cullen, M.D.* London: Lee and Hurst.

Cullen, W. (1800). *Nosology: Or, a systematic arrangement of diseases, by classes, orders, genera, and species*. Edinburgh: C. Stewart.

Cullen, W., & Thomson, J. (1827). *The Works of William Cullen: Containing his physiology, nosology, and first lines of the practice of physic*. Edinburgh: Blackwood.

Digby, K. (1658). *The power of sympathy*. London: R. Lowndes.

Dryden, J. (1995). *The Works of John Dryden*. Ware: Wordsworth Poetry Library.

Dryden, J., & Link, F. M. (1972). *Aureng-zebe*. London: Arnold.

Finger, S. (1994). *Origins of neuroscience: A history of explorations into brain function*. New York: Oxford University Press.

Ingelo, N. (1682). *Bentivolio and Urania: In six books*. London: Dorman Newman.

Kagan, J. (1989). *Unstable ideas: Temperament, cognition, and self*. Cambridge, MA: Harvard University Press.

Kagan, J. (1998). *Three seductive ideas*. Cambridge, MA: Harvard University Press.

Kagan, J. (2006). *An argument for mind*. Cambridge: Harvard University Press.

Kagan, J., & Lamb, S. (1987). *The emergence of morality in young children*. Chicago: University of Chicago Press.

Kagan, J., & Snidman, N. C. (2004). *The long shadow of temperament*. Cambridge, MA: Belknap Press of Harvard University Press.

Kagan, J., et al. (1994). *Galen's prophecy: Temperament in human nature*. Boulder, CO: Westview Press.

Kagan, J., et al. (2005). *A young mind in a growing brain*. Mahwah, NJ: Lawrence Erlbaum.

Kames, H. H. (1779). *Sketches of the history of man* (3rd ed.). Dublin: J. Williams.

Kinneir, D. (1737). *A new essay on the nerves; and the doctrine of the animal spirits rationally considered*. London: W. Innys and J. Leake at Bath.

La Framboisiére, N. A. (1684). *The art of physick made plain and easie*. London: Dorman Newman.

Locke, J., & Yolton, J. W. (1977). *An essay concerning human understanding*. London: Dent.

Ludwig, C. F. (1791). *Scriptores neurologici minores selecti sive opera minora ad anatomiam physiologiam et pathologiam nervorum spectantia*. Leipzig.

Martensen, R. L. (2004). *The brain takes shape: An early history*. Oxford: Oxford University Press.

Morton, J. (1995). *Persons, Bodies, Selves, Emotions*. Adelaide: University of Adelaide Press.

Paster, G. K. (1997). Nervous tension. In D. Hillman & C. Mazzio (Ed.), *The body in parts* (pp. 107–128). London: Routledge.

Paster, G. K. (2004). *Humoring the body: Emotions and the Shakespearean Stage*. Chicago, IL: University of Chicago Press.

Phillips, E. (1658). *The new world of English words: Or, a general dictionary, containing the interpretations of words . . .* London: E. Tyler.

Pope, A., & Butt, J. E. (1963). *The poems of Alexander Pope*. London: Methuen.

Porter, R. (1992). Addicted to modernity: Nervousness in the early consumer society. In J. Melling & J. Barry (Ed.), *Culture in history* (pp. 180–194). Exeter: University of Exeter Press.

Porter, R. (2001). Nervousness, eighteenth and nineteenth-century style: From luxury to labour. In M. Gijswijt-Hofstra, & Roy Porter (Eds.), *Cultures of neurasthenia* (pp. 31–50). London: Wellcome Series in the History of Medicine.

Porter, R. (2004). *Flesh in the age of reason*. New York: W.W. Norton & Co.

Proceedings of the Royal Society (1859). Catalogue of the Scientific Papers of Doctor Marshall Hall (Vol. 9, pp. 54–59). London: Published by the Royal Society.

Richardson, Samuel. (1748, rev. 1751), *Clarissa; or the history of a young lady* (7 Vols.). London: Published by the Editor of Pamela.

Rousseau, G. S. (1980). Psychology. In G. S. Rousseau & R. Porter (Ed.), *The ferment of knowledge: Studies in the historiography of eighteenth-century science* (pp. 143–210). Cambridge: Cambridge University Press.

Rousseau, G. S. (2004). *Nervous acts: Essays on literature, culture and sensibility*. Basingstoke: Palgrave Macmillan.

Rousseau, G. S. (2006). A sympathy of parts: Nervous science and Scottish society. In A. Patrizio & D. Kemp

(Ed.), *Anatomy acts: How we come to know ourselves* (pp. 17–30). Edinburgh: Birlinn.

Rousseau, G. S. (2007). Brainomania: Brain, mind and soul in the long eighteenth century in *British Journal for Eighteenth-Century Studies 30*, forthcoming.

Schiesari, J. (1992). *The gendering of melancholia: Feminism, psychoanalysis, and the symbolics of loss in Renaissance literature*. Ithaca NY: Cornell University Press.

Schoenfeldt, M. C. (1999). *Bodies and selves in early modern England: Physiology and inwardness in Spenser, Shakespeare, Herbert, and Milton*. Cambridge: Cambridge University Press.

Smith, A. (1776). *An inquiry into the nature and causes of the wealth of nations*. London: W. Strahan.

Smollett, Tobias and Knapp, Lewis M. (1966). *The Expedition of Humphry Clinker: Oxford English Novels*. London: Oxford University Press.

Sprott, W. J. H. (1925). *Physique and character: an investigation of the nature of constitution and of the theory of temperament* (2 Vols.). London: Kegan Paul.

Sterne, L. (1768). *A sentimental journey through France and Italy* (2nd ed.). London: Printed for T. Becket.

Sterne, L. (1940). *The life and opinions of Tristram Shandy, gentleman*. New York: Odyssey.

Temkin, O. (1973). *Galenism: Rise and decline of a medical philosophy*. Ithaca, NY: Cornell University Press.

Trotter, T. (1807). *A view of the nervous temperament: Being a practical enquiry into the increasing prevalence, prevention, and treatment of those diseases commonly called nervous, &c*. London: Longman.

Trotter, T., & Porter, R. (1988). *An essay, medical, philosophical, and chemical on drunkenness and its effects on the human body*. London: Routledge.

Tucker, A. (1777). *The light of nature pursued*. London: W. Oliver.

Verity, R. (1837). *Changes produced in the nervous system by civilization, considered according to the evidence of physiology and the philosophy of history*. London: n. p.

Willis, T., Feindel, W., & Pordage, S. (1965). *The anatomy of the brain and nerves* (Tercentenary ed.). Montreal: McGill University Press.

Yolton, J. W. (1984). *Thinking Matter: Materialism in eighteenth-century Britain*. Oxford: Blackwell.

Yolton, J. W. (1991). *Locke and French materialism*. Oxford: Clarendon Press.

Yolton, J. W., & Locke, J. (1970). *Locke and the compass of human understanding: A selective commentary on the essay*. Cambridge: Cambridge University Press.

Index